T0344985

Model-Based Processing

Model-Based Processing

An Applied Subspace Identification Approach

James V. Candy
Lawrence Livermore National Laboratory
University of California, Santa Barbara

Registered Office
John Wiley & Sons, Inc., 111 River Street, Hoboken, NJ 07030, USA

Editorial Office
111 River Street, Hoboken, NJ 07030, USA

For details of our global editorial offices, customer services, and more information about Wiley products visit us at www.wiley.com.

Wiley also publishes its books in a variety of electronic formats and by print-on-demand. Some content that appears in standard print versions of this book may not be available in other formats.

Library of Congress Cataloging-in-Publication Data applied for
ISBN: 9781119457763

Cover design by Wiley
Cover image: Courtesy of Patricia Candy

Set in 10/12pt WarnockPro by SPi Global, Chennai, India

Printed in the United States of America

V10008328_022019

Jesus replied, "I am the way, the truth and the life. No one comes to the Father except through me." (John: 14:6)

Contents

Preface

This text encompasses the basic idea of the model-based approach to signal processing by incorporating the often overlooked, but necessary, requirement of obtaining a model initially in order to perform the processing in the first place. Here we are focused on presenting the development of models for the design of model-based signal processors (MBSP) using subspace identification techniques to achieve a model-based identification (MBID) as well as incorporating validation and statistical analysis methods to evaluate their overall performance [1]. It presents a different approach that incorporates the solution to the system identification problem as the integral part of the model-based signal processor (Kalman filter) that can be applied to a large number of applications, but with little success unless a reliable model is available or can be adapted to a changing environment [2]. Here, using subspace approaches, it is possible to identify the model very rapidly and incorporate it into a variety of processing problems such as state estimation, tracking, detection, classification, controls and communications to mention a few [3, 4]. Models for the processor evolve in a variety of ways, either from first principles accompanied by estimating its inherent uncertain parameters as in parametrically adaptive schemes [5] or by extracting constrained model sets employing direct optimization methodologies [6], or by simply fitting a black-box structure to noisy data [7, 8]. Once the model is extracted from controlled experimental data, or a vast amount of measured data, or even synthesized from a highly complex truth model, the long-term processor can be developed for direct application [1]. Since many real-world applications seek a real-time solution, we concentrate primarily on the development of fast, reliable identification methods that enable such an implementation [9–11]. Model extraction/development must be followed by validation and testing to ensure that the model reliably represents the underlying phenomenology – a bad model can only lead to failure!

System identification [6] provides solutions to the problem of extracting a model from measured data sequences either time series, frequency data or simply an ordered set of indexed values. Models can be of many varieties ranging from simple polynomials to highly complex constructs evolving from nonlinear distributed systems. The extraction of a model from data is critical for a large number of applications evolving from the detection of submarines in a varying ocean, to tumor localization in breast tissue, to pinpointing the epicenter of a highly destructive earthquake, or to simply monitoring the condition of a motor as it drives a critical system component [1]. Each of these applications require an aspect of modeling and fundamental understanding (when possible) of the underlying phenomenology governing the process as well as the measurement instrumentation extracting the data along with the accompanying uncertainties. Some of these problems can be solved simply with a "black-box" representation that faithfully reproduces the data in some manner without the need to capture the underlying dynamics (e.g. common check book entries) or a "gray-box" model that has been extracted, but has parameters of great interest (e.g. unknown mass of a toxic material). However, when the true need exists to obtain an accurate representation of the underlying phenomenology like the structural dynamics of an aircraft wing or the untimely vibrations of a turbine in a nuclear power plant, then more sophisticated representations of the system and uncertainties are clearly required. In cases such as these, models that capture the dynamics must be developed and "fit" to the data in order to perform applications such as condition monitoring of the structure or failure detection/prediction of a rotating machine. Here models can evolve from lumped characterizations governed by sets of ordinary differential equations, linear or nonlinear, or distributed representations evolved from sets of partial differential equations. All of these representations have one thing in common, when the need to perform a critical task is at hand – they are represented by a mathematical model that captures their underlying phenomenology that must somehow be extracted from noisy measurements. This is the fundamental problem that we address in this text, but we must restrict our attention to a more manageable set of representations, since many monographs have addressed problem sets targeting specific applications [12, 13].

In fact, this concept of specialty solutions leads us to the generic state-space model of systems theory and controls. Here the basic idea is that all of the theoretical properties of a system are characterized by this fundamental set of models that enables the theory to be developed and then applied to any system that can be represented in the state-space. Many models naturally evolve in the state-space, since it is essentially the representation of a set of nth-order differential equations (ordinary or partial, linear or nonlinear, time (space) invariant or time (space) varying, scalar or multivariable) that are converted into a set of first-order equations, each of which is a state. For example, a simple mechanical system consisting of a single mass, spring, damper construct is characterized by a set of second-order, linear, time-invariant, differential equations that can simply be represented in state-space form by a set of two first-order equations, each one representing a state: one for displacement and one for velocity [12].

We employ the state-space representation throughout this text and provide sufficient background in Chapters 2 and 3.

System identification is broad in the sense that it does not limit the problem to various classes of models directly. For instance, for an unknown system, a model set is selected with some perception that it is capable of representing the underlying phenomenology adequately, then this set is identified directly from the data and validated for its accuracy. There is clearly a well-defined procedure that captures this approach to solve the identification problem [6–15]. In some cases, the class structure of the model may be known a priori, but the order or equivalently the number of independent equations to capture its evolution is not (e.g. number of oceanic modes). Here, techniques to perform order estimation precede the fitting of model parameters first, then are followed by the parameter estimation to extract the desired model [14]. In other cases, the order is known from prior information and parameter estimation follows directly (e.g. a designed mechanical structure). In any case, these constraints govern the approach to solving the identification problem and extracting the model for application. Many applications exist, where it is desired to monitor a process and track a variety of parameters as they evolve in time, (e.g. radiation detection), but in order to accomplish this on-line, the model-based processor must update the model parameters sequentially in order to accomplish its designated task. We develop these processors for both linear and nonlinear models in Chapters 4 and 5.

Although this proposed text is designed primarily as a graduate text, it will prove useful to practicing signal processing professionals and scientists, since a wide variety of case studies are included to demonstrate the applicability of the model-based subspace identification approach to real-world problems. The prerequisite for such a text is a melding of undergraduate work in linear algebra (especially matrix decomposition methods), random processes, linear systems, and some basic digital signal processing. It is somewhat unique in the sense that many texts cover some of its topics in piecemeal fashion. The underlying model-based approach of this text is the thread that is embedded throughout in the algorithms, examples, applications, and case studies. It is the model-based theme, together with the developed hierarchy of physics-based models, that contributes to its uniqueness coupled with the new robust, subspace model identification methods that even enable potential real-time methods to become a reality. This text has evolved from four previous texts, [1, 5] and has been broadened by a wealth of practical applications to real-world, model-based problems. The introduction of robust subspace methods for model-building that have been available in the literature for quite a while, but require more of a systems theoretical background to comprehend. We introduce this approach to identification by first developing model-based processors that are the prime users of models evolving to the parametrically adaptive processors that jointly estimate the signals along with the embedded model parameters [1, 5]. Next, we introduce the underlying theory evolving from systems theoretic realizations of state-space models along with unique representations (canonical forms) for multivariable structures [16, 17]. Subspace identification is introduced for these deterministic systems. With the theory and algorithms for these systems in hand, the algorithms are extended to the stochastic case, culminating with a combined solution for both model sets, that is, deterministic and stochastic.

In terms of the system identification area, this text provides the link between model development and practical applications in model-based signal processing filling this critical gap, since many identification texts dive into the details of the algorithms without completing the final signal processing application. Many use the model results to construct model-based control systems, but do not focus on the processing aspects. Again the gap is filled in the signal processing community, by essentially introducing the notions and practicalities of subspace identification techniques applied to a variety of basic signal processing applications. For example, spectral estimation, communications, and primarily physics-based problems, which this text will demonstrate in the final chapters. It is especially applicable for signal processors because they are currently faced with multichannel applications, which the state-space formulations in this text handle quite easily, thereby opening the door for novel processing approaches. The current texts are excellent, but highly theoretical, attempting to provide signal processors with the underlying theory for the subspace approach [9–12]. Unfortunately, the authors are not able to achieve this, in my opinion, because the learning curve is too steep and more suitable for control system specialists with a strong systems theoretical background. It is difficult for signal processors to easily comprehend, but by incorporating the model-based signal processing approach, which is becoming more and more known and utilized by the signal processing community as the connection will enable the readers to gently "bridge the gap" from statistical signal processing to subspace identification for subsequent processing especially in multichannel applications. This is especially true with readers familiar with our previous texts in model-based processing [1, 6]. It will also have an impact in the structural dynamics area due to our case studies and applications introducing structural/test engineers to the model-based identifiers/processors [16]. They already apply many of these identification techniques to their problem sets.

The approach we take is to introduce the concept of subspace identification by first discussing the ideas of signal estimation, identification to model-based signal processing (MBSP) leading to the concept of model-based identification (MBID) [1, 5]. Here the model set is defined, and a variety of techniques ranging from the black-box approach to well-defined structural models employing parameter estimation techniques are developed. After introducing these concepts in the first chapter, random signals and systems are briefly discussed leading to the concept of spectral estimation, which provides an underlying cornerstone of the original identification problem.

Next, state-space models are introduced in detail evolving from continuous-time, sampled-data to discrete-time systems leading to the stochastic innovations model linking the classical Wiener filter to the well-known Kalman filter [2]. With this in hand, multivariable (multiple-input/multiple-output) systems are developed simply as time series, to sophisticated canonical forms, leading to the matrix fraction transfer function descriptions. Chapter 3 is concluded with approximate nonlinear Gauss–Markov representations in state-space form.

Model-based processors are highlighted in the next two chapters 4 and 5 ranging from developments of the linear representations leading to the optimal Kalman filter [2]. Next the suite of nonlinear processors is developed initiated by the linearized processor leading to the special cases of the extended and unscented Kalman filters and culminating with the novel particle filter evolving from the Bayesian approach [5]. These techniques are extended to the joint signal/parameter estimation problem to create the parametric adaptive processors. Throughout these chapters, examples and case studies are introduced to solidify these fundamental ideas.

Next, we introduce the foundations for the heart of the text – subspace identification first constrained to deterministic systems. Here we develop the fundamental realization problem that provides the basis of subspace identification using Hankel matrices. Many of the underlying systems theoretical results introduced by Kalman [18] in the 1960s are captured by properties of the Hankel matrix. The problem is extended to the deterministic identification problem by incorporating input/output sequences [15]. Perhaps one of the most important contributions to realization theory is the concept of a balanced realization enabling the evolution of robust algorithms. All of these concepts are carefully developed in this chapter. Canonical realizations, that is, the identification of models in unique canonical forms is an important concept in identification [16–18]. Here, much of this effort has been ignored over the years primarily because the concept of a unique representation can lead to large errors when identifying the model. However, it is possible to show that they can also be considered a viable approach, since they transform the Hankel array to the so-called structural matrix enabling both the order and parameters to be identified simultaneously, leading to an invariant system description [17, 19]. Finally, we introduce the ideas of projection theory showing how orthogonal/oblique projections lead to popular deterministic identification techniques [9–11]. Chapter 6 is concluded with a detailed case study on the identification application for a mechanical/structural system.

Chapter 7 is the extension of the deterministic identification problem to the stochastic case. Here, in the realization context, covariance matrices replace impulse response matrices, while deterministic input/output sequences are replaced with noisy multichannel sequences – the real-world problem.

As in Chapter 6, we develop stochastic realization theory starting with the "indirect" realization approach [4] based on covariance matrices for infinite and finite sequences to develop the basis of stochastic realization theory. The main ideas evolve from the work of Akaike [20, 21] and the development of predictor spaces leading to the fundamental results from the systems theoretic viewpoint. The optimal solution to this problem proceeds from classical spectral factorization techniques leading to the steady-state Kalman filter and the fundamental innovations model that is an integral part of subspace realizations [1, 20–29]. Next, subspace methods are reintroduced for random vector spaces and provided as solutions to the stochastic realization problem followed by the so-called combined subspace technique extracting both deterministic and stochastic models simultaneously [9–13]. This chapter concludes with a case study discussing the design of a processor to detect modal anomalies in an unknown cylindrical structure.

The text concludes with a chapter describing sets of real-world applications of these techniques. The applications range from failure detection, to the threat detection of fission sources, to the identification of chirp signals for radar/sonar application, to the parametrically adaptive processor design for localization and tracking in the ocean environment, and to the design of an MBP chirp-based signals as well as a critical radiation system – the scintillator.

Appendices are included for critical review as well as problem sets and notes for the MATLAB software used in the signal processing/controls/identification areas at the end of each chapter.

References

1 Candy, J. (2006). *Model-Based Signal Processing*. Hoboken, NJ: Wiley/IEEE Press.

2 Kalman, R. (1960). A new approach to linear filtering and prediction problems. *Trans. ASME J. Basic Eng.* 82: 34–45.

3 Van Der Veen, A., Deprettere, E., and Swindlehurst, A. (1993). Subspace-based methods for the identification of linear time-invariant systems. *Proc. IEEE* 81 (9): 1277–1308.

4 Viberg, M. (1995). Subspace based signal analysis using singular value decomposition. *Automatica* 31 (12): 1835–1851.

5 Candy, J. (2016). *Bayesian Signal Processing: Classical, Modern and Particle Filtering Methods*, 2e. Hoboken, NJ: Wiley/IEEE Press.

6 Ljung, L. (1999). *System Identification: Theory for the User*, 2e. Englewood Cliffs, NJ: Prentice-Hall.

7 Ljung, L. and Soderstrom, T. (1983). *Theory and Practice of Recursive Identification*. Cambridge: MIT Press.

8 Soderstrom, T. and Stoica, P. (1989). *System Identification*. New York: Academic Press.

9 van Overschee, P. and De Moor, B. (1996). *Subspace Identification for Linear Systems: Theory, Implementation, Applications*. Boston, MA: Kluwer Academic Publishers.

10 Katayama, T. (2005). *Subspace Methods for System Identification*. London: Springer.

11 Verhaegen, M. and Verdult, V. (2007). *Filtering and System Identification: A Least-Squares Approach*. Cambridge: Cambridge University Press.

12 Juang, J. (1994). *Applied System Identification*. Upper Saddle River, NJ: Prentice-Hall PTR.

13 Aoki, M. (1990). *State Space Modeling of Time Series*, 2e. London: Springer.

14 Norton, J. (1986). *An Introduction to Identification*. New York: Academic Press.

15 Ho, B. and Kalman, R. (1966). Effective reconstruction of linear state variable models from input/output data. *Regelungstechnik* 14: 545–548.

16 Luenberger, D. (1967). Canonical forms for linear multivariable systems. *IEEE Trans. Autom. Control* AC-12: 290–293.

17 Candy, J., Warren, M., and Bullock, T. (1977). Realization of an invariant system description from Markov sequences. *IEEE Trans. Autom. Control* AC-23 (12): 93–96.

18 Chen, C. (1984). *Linear System Theory and Design*. New York: Holt, Rinehart & Winston.

19 Guidorzi, R. (1975). Canonical structures in the identification of multivariable systems. *Automatica* 11: 361–374.

20 Akaike, H. (1974). Stochastic theory of minimal realization. *IEEE Trans. Autom. Control* 19: 667–674.

21 Akaike, H. (1975). Markovian representation of stochastic processes by canonical variables. *SIAM J. Control* 13 (1): 162–173.

22 Faurre, P. (1976). Stochastic realization algorithms. In: *System Identification: Advances and Case Studies* (ed. R. Mehra and D. Lainiotis), 1–23. New York: Academic Press.

23 Larimore, W. (1990). Canonical variate analysis in identification, filtering and adaptive control. In: *Proceedings of the 29th Conference on Decision and Control*, Hawaii, USA, 596–604.

24 Tse, E. and Wiennert, H. (1975). Structure determination and parameter identification for Multivariable stochastic linear systems. *IEEE Trans. Autom. Control* 20: 603–613.

25 Glover, K. and Willems, J. (1974). Parameterizations of linear dynamical systems: canonical forms and identifiability. *IEEE Trans. Autom. Control* 19: 640–646.

26 Denham, M. (1974). Canonical forms for identification of multivariable linear systems. *IEEE Trans. Autom. Control* 19: 646–656.

27 Candy, J., Bullock, T., and Warren, M. (1979). Invariant system description of the stochastic realization. *Automatica* 15: 493–495.

28 Candy, J., Warren, M., and Bullock, T. (1978). Partial realization of invariant system descriptions. *Int. J. Control* 28 (1): 113–127.

29 Sullivan, E. (2015). *Model-Based Processing for Underwater Acoustic Arrays*. New York: Springer.

James V. Candy
Danville, CA

Acknowledgements

The support and encouragement of my wife, Patricia, is the major motivational element needed to undertake this endeavor. My family, extended family, and friends having endured the many regrets, but still offer encouragement in spite of all of my excuses. Of course, the constant support of my great colleagues and friends, especially Drs. S. Lehman, I. Lopez, E. Sullivan, and Mr. B. Beauchamp, who carefully reviewed the manuscript and suggested many improvements, cannot go without a hearty acknowledgment.

Glossary

ADC	analog-to-digital conversion
AIC	Akaike information criterion
AR	autoregressive (model)
ARMA	autoregressive moving average (model)
ARMAX	autoregressive moving average exogenous input (model)
ARX	autoregressive exogenous input (model)
AUC	area-under-curve (ROC curve)
BSP	Bayesian signal processing
BW	bandwidth
CD	central difference
CDF	cumulative distribution
CM	conditional mean
CRLB	Cramer–Rao lower bound
C-Sq	Chi-squared (distribution or test)
CT	continuous-time
CTD	concentration–temperature–density (measurement)
CVA	canonical variate analysis
EKF	extended Kalman filter
EM	expectation–maximization
FPE	final prediction error
GLRT	generalized likelihood ratio test
G-M	Gaussian mixture
GM	Gauss–Markov
G-S	Gaussian sum
HD	Hellinger distance
HPR	high probability region
IEKF	iterated–extended Kalman filter
i.i.d.	independent-identically distributed (samples)
KD	Kullback divergence
KL	Kullback–Leibler
KLD	Kullback–Leibler divergence

KSP	Kalman–Szego–Popov (equations)
LD	lower diagonal (matrix) decomposition
LE	Lyapunov equation
LKF	linear Kalman filter
LMS	least mean square
LS	least-squares
LTI	linear, time-invariant (system)
LZKF	linearized Kalman filter
MA	moving average (model)
MAICE	minimum Akaike information criterion
MAP	maximum a posteriori
MATLAB®	mathematical software package
MBID	model-based identification
MBP	model-based processor
MBSP	model-based signal processing
MC	Monte Carlo
MDL	minimum description length
MIMO	multiple-input/multiple-output (system)
MinE	minimum probability of error
ML	maximum likelihood
MOESP	multivariable output error state-space algorithm
MMSE	minimum mean-squared error
MSE	mean-squared error
MV	minimum variance
N4SID	numerical algorithm for subspace state-space system identification
NMSE	normalized mean-squared error
N-P	Neyman–Pearson (detector)
ODP	optimal decision (threshold) point
PDF	probability density function (continuous)
P-E	probability-of-error (detector)
PEM	prediction error method
PF	particle filter
PI-MOESP	past-input multivariable output error state-space algorithm
PMF	probability mass function (discrete)
PO-MOESP	past-output multivariable output error state-space algorithm
PSD	power spectral density
RC	resistor capacitor (circuit)
REBEL	recursive Bayesian estimation library
RLC	resistor–inductor–capacitor (circuit)
RLS	recursive least-squares
RMS	root mean-squared
RMSE	root minimum mean-squared error

ROC	receiver operating characteristic (curve)
RPE	recursive prediction error
RPEM	recursive prediction error method
SID	subspace identification
SIR	sequential importance sampling-resampling
SIS	sequential importance sampling
SMC	sequential Markov chain
SNR	signal-to-noise ratio
SPRT	sequential probability ratio test
SPT	sigma-point transformation
SSIS	sequential sampling importance sampling
SSP	state-space processor
SSQE	sum-squared error
SVD	singular-value (matrix) decomposition
UD	upper diagonal matrix decomposition
UKF	unscented Kalman filter
UT	unscented transform
WSSR	weighted sum-squared residual statistical test
W-test	whiteness test
Z	Z-transform
Z-M	zero-mean statistical test

1

Introduction

In this chapter, we introduce the idea of model-based identification, starting with the basic notions of signal processing and estimation. Once defined, we introduce the concepts of model-based signal processing, that lead to the development and application of subspace identification. Next, we show that the essential ingredient of the model-based processor is the "model" that must be available either through the underlying science (first principles) or through the core of this text – model-based identification.

1.1 Background

The development of processors capable of extracting information from noisy sensor measurement data is essential in a wide variety of applications, whether it be locating a hostile target using radar or sonar systems or locating a tumor in breast tissue or even locating a seismic source in the case of an earthquake. The nondestructive evaluation (NDE) of a wing or hull of a ship provides a challenging medium even in the simplest of arrangements requiring sophisticated processing especially if the medium is heterogeneous. Designing a controller for a smart car or a drone or for that matter a delicate robotic surgical instrument also depends on providing enhanced signals for feedback and error corrections. Robots replacing humans in assembly lines or providing assistance in mundane tasks must sense their surroundings to function in a such a noisy environment. Most "hi-tech" applications require the incorporation of "smart" processors capable of sensing their operational environment, enhancing noisy measurements and extracting critical information in order to perform a pre-assigned task such as detecting a hostile target and launching a weapon or detecting a tumor and extracting it. In order to design a processor with the required capability, it is necessary to utilize as much available a priori information as possible. The design may incorporate a variety of disciplines to achieve the desired results. For instance, the processor must be able to sense the operational environment, whether it be highly cluttered electromagnetic propagation

Model-Based Processing: An Applied Subspace Identification Approach, First Edition. James V. Candy.
© 2019 John Wiley & Sons, Inc. Published 2019 by John Wiley & Sons, Inc.

at an airport or a noisy ocean acoustic environment in a busy harbor. Array radiation measurements in the case of an active radar system targeting signals of great interest can detect incoming threats, while passive listening provided by an acoustic array aids in the detection of submarines or similarly tumors in the human body for ultrasonics as well. The ability of the processor to operate effectively in such harsh environments requires more and more sophistication, rather than just simple filtering techniques. It is here that we address not only the need, but also the a priori requirements for a design. For instance, the detection and localization of a quiet diesel submarine cannot be achieved without some representation of the noisy, varying ocean incorporated in the processing scheme. How does such information get embedded? This is the question for not only the signal processor, but also the ocean acoustician and sensor designer to ponder. The solution boils down to the melding of this information enabling the development of a processor capable of performing well. So we see that except in an exceptional case, the knowledge of the underlying phenomenology that governs just how a signal propagates in an uncertain medium or environment coupled with that of how a sensor can make a reasonable measurement to provide the desired information and a processor capable of extracting that information defines a "team" consisting of a phenomenologist, sensor designer and signal processor that can enable a solution to the problem at hand. In this text, we discuss such an approach that incorporates all of these capabilities. We start with the basic processor and then progress to a scheme capable of incorporating the underlying phenomenology, measurement systems, and uncertainties into the processor. In order to do so, we start with defining signal processing and signal estimation, followed by the fundamental model-based signal processor and then approaches to obtain the required model from experimental as well as application data sets.

1.2 Signal Estimation

Signal processing is based on one fundamental concept – extracting critical information from uncertain measurement data [1, 2]. Processing problems can lead to some complex and intricate paradigms to perform this extraction especially from noisy, sometimes inadequate measurements. Whether the data are created using a seismic geophone sensor from a monitoring network or an array of hydrophone transducers located on the hull of an ocean-going vessel, the basic processing problem remains the same – extract the useful information. Techniques in signal processing (e.g. filtering, Fourier transforms, time–frequency and wavelet transforms) are effective; however, as the underlying process generating the measurements becomes more complex, the resulting processor may require more and more information about the process phenomenology to extract the desired information. The challenge is to formulate a meaningful strategy that is aimed at performing the processing required, even in the face of these high uncertainties. This strategy can be as simple as a transformation of the measured data to another domain for analysis or as complex as embedding a full-scale propagation model into the processor [3]. For example, think of trying to extract a set of resonances (damped sinusoids) from accelerometer time series. It is nearly impossible to calculate zero-crossings from the time series, but it is a simple matter to transform the data to the spectral domain using a Fourier transform and then applying the property that sinusoids are impulse-like in Fourier space facilitating their extraction through peak detection. Finding a sinusoidal source propagating in the ocean is another matter that is quite complex due to the attenuation and dispersion characteristics of this harsh, variable environment. Here, a complex propagation model must be developed and applied to "unravel" the highly distorted data to reveal the source – a simple Fourier transform will no longer work. The aims of both approaches are the same – to extract the desired information and reject the extraneous and, therefore, develop a processing scheme to achieve this goal. The underlying signal processing philosophy is a "bottoms-up" perspective enabling the problem to dictate the solution, rather than vice versa.

More specifically, signal processing forms the basic nucleus of many applications. It is a specialty area that many researchers/practitioners apply in their daily technical regimen with great success such as the simplicity in Fourier analysis of resonance data or in the complexity of analyzing the time–frequency response of dolphin sounds. Applications abound with unique signal processing approaches offering solutions to the underlying problem. For instance, the localization of a target in the hostile underwater ocean acoustic environment not only challenges the phenomenologist but also taxes the core of signal processing basics, thereby requiring that more sophistication and a priori knowledge be incorporated into the processor. This particular application has led to many advances both in underwater signal processing and in the development of a wide variety of so-called model-based or physics-based processors. A prime example of this technology is the advent of the model-based, matched-field processor [3–5] that has led not only to a solution of the target localization problem, but also to many applications in other areas such as nondestructive evaluation and biomedical imaging. Therefore, the conclusion remains the same, signal processing is a necessary ingredient as a working tool that must be mastered by phenomenologists to extract the useful information from uncertain measurements. In fact, we define signal processing as a set of techniques "to extract the desired information and reject the extraneous from uncertain measurement data."

Signal processing relies on any prior knowledge of the phenomenology generating the underlying measurements. Characterizing this phenomenology and propagation physics along with the accompanying measurement instrumentation and noise are the preliminaries that all phenomenologists must tackle to solve such a processing problem. In many cases, this is much easier said than done. The first step is to determine what the desired information is and typically this is not the task of the signal processor, but that of the phenomenologist performing the study. In our case, we assume that the investigation is to extract information stemming from signals emanating from a source, whether it be an autonomous unmanned vehicle (AUV) on the highway or passively operating in the deep ocean, or a vibrating structure responding to ground motion. Applications can be very complex especially in the case of ultrasound propagating through complex media such as tissue in biomedical applications or through heterogeneous materials of critical parts in nondestructive evaluation (NDE) investigations or photons emanating from a radiation source [6]. In any case, the processing usually involves manipulating the measured data to extract the desired information, such as location and tracking of the AUV, failure detection for the structure, or tumor/flaw detection, and localization in both biomedical and NDE [3].

If a measured signal is free from extraneous variations and is repeatable from measurement to measurement, then it is defined as a *deterministic* signal (Chapter 2). However, if it varies extraneously and is no longer repeatable, then it is defined as *random* signal. This text is concerned with the development of processing techniques to extract pertinent information from random signals utilizing any a priori information available. We call these techniques signal estimation or signal enhancement, and we call a particular algorithm a *signal estimator* or just *estimator*. Symbolically, we use the "caret" (ˆ) notation to annotate an estimate (e.g. s → ŝ). Sometimes, estimators are called filters (e.g. Wiener filter) because they perform the same function as a deterministic (signal) filter except for the fact that the signals are random; that is, they remove unwanted disturbances. Noisy measurements are processed by the estimator to produce "filtered" data. To solidify these concepts, consider the following examples.

Example 1.1 A deterministic signal composed of two sinusoids: the information at 10 Hz and the disturbance at 20 Hz, with its corresponding Fourier spectrum shown in Figure 1.1a. From a priori information, it is known that the desired signal has no frequencies above 15 Hz; however, the raw spectrum reveals the disturbance at 20 Hz. Since the data are deterministic, a low-pass filter with a cutoff frequency of 12.5 Hz is designed to extract the desired information (10 Hz signal) and reject the extraneous (20 Hz disturbance). The filtered data are shown in Figure 1.1b, where we can see the filtered signal and the resulting spectrum.   □

Consider the output of the estimation filter designed to eliminate random noise from a transient signal; that is, the estimator is a function of the signal

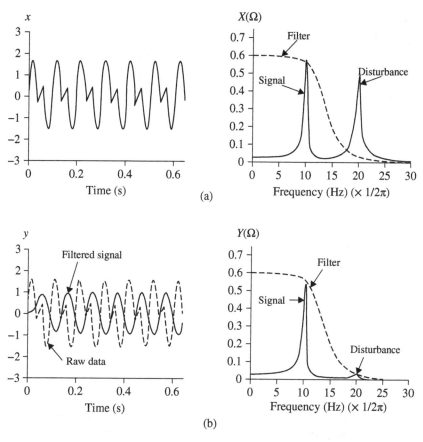

Figure 1.1 Processing of a *deterministic* signal. (a) Raw data and spectrum with signal at 10 Hz and disturbance at 20 Hz. (b) Processed data extracting the 10 Hz signal (desired) and rejecting the extraneous (20 Hz disturbance).

$x(t)$ and noise $n(t)$

$$\hat{y}(t) = a(x(t), n(t)).$$

Consider the following example to illustrate this processor.

Example 1.2 A random pulse-like signal contaminated by noise is shown in Figure 1.2a. Here, we see the measured data along with its Fourier spectrum. We design a signal estimator to extract the desired signal and remove the noise. The processed data are shown in Figure 1.2b, where we observe the results of the estimation *filter* and the corresponding enhanced spectrum. Here, we see how the filtered response has eliminated the random noise. We discuss the concepts of signal estimation using the modern parametric design methods in Chapter 4.   □

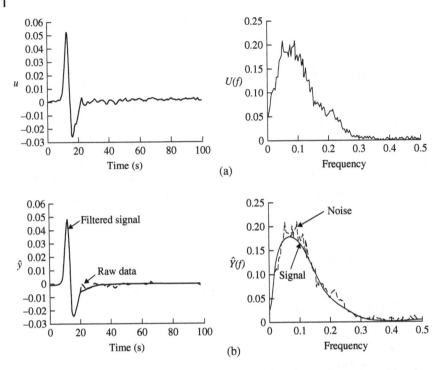

Figure 1.2 Processing of a *random* signal and noise. (a) Raw data and spectrum with noise. (b) Processed data extracting the signal (estimate) and rejecting the extraneous (noise).

If the estimation filter or more commonly the *estimator* employs a model of phenomenology or process under investigation, then it is considered a model-based processor. For example, suppose we use a so-called *Gauss–Markov* model (see Chapter 3) in our estimator design, that is, if we use

$$x(t) = Ax(t-1) + w(t-1)$$
$$y(t) = Cx(t) + v(t)$$

then the resulting signal estimator,

$$\hat{x}(t) = A\hat{x}(t-1) + K(t)e(t)$$

is called a model-based signal processor and in this case a *Kalman filter*. Model-based signal processing is discussed in Chapter 4. We shall see that random signals can be characterized by stochastic processes (Chapter 2), transformed to equivalent deterministic representations (covariance and power spectrum) and processed (model-based processors, Chapters 4 and 5) much the same as a deterministic signal.

Estimation can be thought of as a procedure made up of three primary parts:

- Criterion function
- Models
- Algorithm.

The criterion function can take many forms and can also be classified as deterministic or stochastic. Models represent a broad class of information, formalizing the a priori knowledge about the process generating the signal, measurement instrumentation, noise characterization, and underlying probabilistic structure. Finally, the algorithm or technique chosen to minimize (or maximize) the criterion can take many different forms depending on (i) the models, (ii) the criterion, and (iii) the choice of solution. For example, one may choose to solve the well-known least-squares problem recursively or with a numerical-optimization algorithm. Another important aspect of most estimation algorithms is that they provide a "measure of quality" of the estimator. Usually, what this means is that the estimator also predicts vital statistical information about how well it is performing.

Intuitively, we can think of the estimation procedure as follows:

- The specification of a criterion
- The selection of models from a priori knowledge
- The development and implementation of an algorithm.

Criterion functions are usually selected on the basis of information that is meaningful about the process or the ease with which an estimator can be developed. Criterion functions that are useful in estimation can be classified as deterministic and probabilistic. Some typical functions are as follows:

- *Deterministic:*
  - Squared error
  - Absolute error
  - Integral absolute error
  - Integral squared error
- *Probabilistic:*
  - Maximum likelihood
  - Maximum a posteriori (Bayesian)
  - Maximum entropy
  - Minimum (error) variance.

Models can also be deterministic as well as probabilistic; however, here we prefer to limit their basis to knowledge of the process phenomenology (physics) and the underlying probability density functions as well as the necessary statistics to describe the functions. Phenomenological models fall into the usual classes defined by the type of underlying mathematical equations and their structure, namely, linear or nonlinear, differential or difference, ordinary or partial, time invariant or varying. Usually, these models evolve to a stochastic model by the inclusion of uncertainty or noise processes.

Finally, the *estimation algorithm* can evolve from various influences. A preconceived notion of the structure of the estimator heavily influences the resulting algorithm. We may choose, based on computational considerations, to calculate an estimate recursively, rather than as a result of a batch process because we require an online, pseudo-real-time estimate. Also each algorithm must provide a measure of *estimation quality*, usually in terms of the expected estimation error. This measure provides a means for comparing estimators. Thus, the estimation procedure is a combination of these three major ingredients: criterion, models, and algorithm.

1.3 Model-Based Processing

Another view of the underlying processing problem is to decompose it into a set of steps that capture the strategic essence of the processing scheme. Inherently, we believe that the more "a priori" knowledge about the measurement and its underlying phenomenology we can incorporate into the processor, the better we can expect the processor to perform – as long as the information that is included is correct! One strategy called the *model-based approach* provides the essence of model-based signal processing (MBSP) [3]. Some believe that all signal processing schemes can be cast into this generic framework. Simply, the model-based approach is "incorporating mathematical models of both physical phenomenology and the measurement process (including noise) into the processor to extract the desired information." This approach provides a mechanism to incorporate knowledge of the underlying physics or dynamics in the form of mathematical process models along with measurement system models and accompanying noise as well as model uncertainties directly into the resulting processor. In this way, the model-based processor (MBP) enables the interpretation of results directly in terms of the problem physics. It is actually a modeler's tool enabling the incorporation of any a priori information about the problem to extract the desired information. This approach of selecting the appropriate model is depicted in the signal processing staircase of Figure 1.3. We note that as we progress up the "modeling" steps to increase the (SNR), the complexity of the model increases to achieve the desired results. This is our roadmap [3]. As depicted, the fidelity of the model incorporated into the processor determines the complexity of the model-based processor with the ultimate goal of increasing the inherent signal-to-noise ratio (SNR). These models can range from simple, implicit, nonphysical representations of the measurement data such as the Fourier or wavelet transforms to

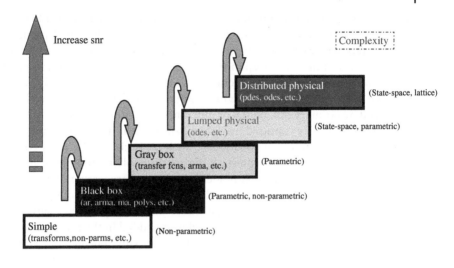

Figure 1.3 Model-based signal processing "model staircase": Step 1 – simple models. Step 2 – black-box models. Step 3 – gray-box models. Step 4 – lumped physical models. Step 5 – distributed physical models.

parametric black-box models used for data prediction, to lumped mathematical representations characterized by ordinary differential equations, to distributed representations characterized by partial differential equation models to capture the underlying physics of the process under investigation. The dominating factor of which model is the most appropriate is usually determined by how severe the measurements are contaminated with noise and the underlying uncertainties. In many cases, if the SNR of the measurements is high, then simple non-physical techniques can be used to extract the desired information; however, for low SNR measurements, more and more of the physics and instrumentation must be incorporated for the extraction. As we progress up the *modeling steps* to increase SNR, the model and algorithm complexity increases proportionally to achieve the desired results. Examining each of the steps individually leads us to realize that the lowest step using no explicit model (*simple*) essentially incorporates little a priori information and is used to analyze the information content (spectrum, time–frequency, etc.) of the raw measurement data to attempt to draw some rough conclusions about the nature of the signals under investigation. Progressing up to the next step, *black-box* models are basically used as data prediction mechanisms. They have a parametric form (e.g. polynomial, transfer function), but again there is little physical information that can be gleaned from their outputs. At the next step, the *gray-box* models evolve that can use the underlying black-box structures; however, now the parameters can be used extract limited physical information from the data. For instance, a black-box transfer function model fit to the data yields coefficient polynomials, which can be factored to extract resonance frequencies and damping coefficients characterizing the overall system response being measured. Progressing further up the steps, we finally reach the true model-based techniques that explicitly incorporate the process physics using a *lumped physical* model structure usually characterized by ordinary differential or difference equations. The top step leads us to processes that are captured by *distributed physical* model structures in the form of partial differential equations. This level is clearly the most complex, since much computer horsepower is devoted to solving the physical propagation problem. So we see that model-based signal processing offers the ability to operate directly in the physical space of the phenomenologist with the additional convenience of providing a one-to-one correspondence between the underlying phenomenology and the model embedded in the processor.

Various levels of models can be used for processing purposes. It is important to investigate what, if anything, can be gained by filtering the data. The amount of information available in the data is related to the precision (variance) of the particular measurement instrumentation used as well as to any signal-processing devices or algorithms employed to improve the estimates. As we utilize more and more information about the physical principles underlying the given data, we expect to improve our estimates (decrease estimation error) significantly.

A typical measurement Y_{meas} is depicted in Figure 1.4. Accordingly, a reasonable model of this simple measurement instrument might be given by

$$Y_{meas} = S_{true} + N_{noise} \qquad \text{(Measurement)}$$

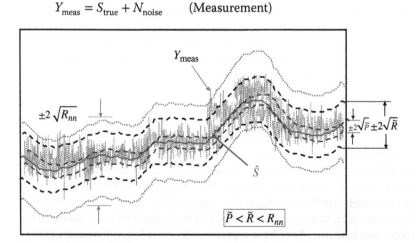

Figure 1.4 Model-based signal processing of a noisy measurement conceptually illustrating that the more "prior" information incorporated into the processor enables a reduction of error variance.

If we were to use Y_{meas} to estimate S_{true} (that is, $\hat{S}$), we have the noise lying within the $\pm 2\sqrt{\tilde{R}}$ confidence limits superimposed on the signal (see Figure 1.4). The best estimate of S_{true} we can hope for is only within the accuracy (bias) and precision (variance) of the instrument. If we include a model of the measurement instrument as well as its associated uncertainties, then we can improve the SNR of the noisy measurements. This technique represents the processing based on our instrument and noise (statistical) models given by

$$Y_{meas} = c(S_{true}) + N_{noise}, \quad N \sim \mathcal{N}(0, R_{nn}) \quad \text{(Measurement and Noise)}$$

where $c(S_{true})$ is the measurement system model and $\mathcal{N}(0, R_{nn})$ is the noise statistics captured by a *Gaussian* or *normal distribution* of zero-mean and variance specified by R_{nn}.

We can also specify the estimation error variance $\tilde{R}$ or equivalently the quality of this processor in estimating S. Finally, if we incorporate not only instrumentation knowledge but also knowledge of the physical process, then we expect to do even better; that is, we expect the signal estimation error defined by $(\tilde{S} := S_{true} - \hat{S})$ variance $\tilde{P}$ to be small (see Figure 1.4 for $\pm 2\sqrt{\tilde{R}}$ confidence limits or bounds) as we incorporate more and more knowledge into the processor, that is

$$\hat{S} = a(S_{true}) + W_{noise}, \quad W \sim \mathcal{N}(0, R_{ww})$$
$$\text{(Process Model and Noise)}$$
$$Y_{meas} = c(S_{true}) + N_{noise}, \quad N \sim \mathcal{N}(0, R_{nn})$$
$$\text{(Measurement Model and Noise)}$$

where $a(S_{true})$ is the process system model and $\mathcal{N}(0, R_{ww})$ is the process noise statistics captured by a zero-mean, Gaussian distribution with variance R_{ww}.

In fact, this is the case because it can be shown that

$$\text{Instrument variance} > \tilde{R} > \tilde{P}$$

This is the basic idea in model-based signal processing: "The more a priori information we can incorporate into the algorithm, the smaller the resulting error variance."

Consider the following example to illustrate these ideas.

Example 1.3 The voltage at the output of an RC-circuit is to be measured using a high impedance voltmeter shown in Figure 1.5. The measurement is contaminated with random instrumentation noise, which can be modeled as

$$e_{out} = K_e e + n$$

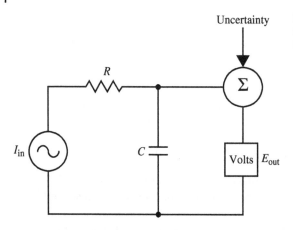

Uncertainty

Figure 1.5 Model-based signal processing "model" for an RC-circuit.

where

$$e_{\text{out}} = \text{measured voltage}$$
$$K_e = \text{instrument amplification factor}$$
$$e = \text{true voltage}$$
$$n = \text{zero-mean random noise of variance} \quad R_{nn}$$

This model corresponds to those described previously. A processor is to be designed to improve the precision of the instrument. Then, we have

- Measurement:

$$s \to e$$
$$c(\cdot) \to K_e$$
$$v \to n$$
$$R_{vv} \to R_{nn}$$

- and for the filter,

$$\hat{s} \to \hat{e}$$
$$\tilde{R}_{ss} \to \tilde{R}_{ee}$$

The precision of the instrument can be improved even further by including a model of the process (circuit). Writing the Kirchhoff node equations, we have

$$\dot{e} = \frac{1}{C}I_{\text{in}} - \frac{e}{RC} + q$$

where

R = resistance

C = capacitance

I_{in} = excitation current

q = zero-mean random noise of variance R_{qq}

The more sophisticated model-based processor employs both measurement and process models. Thus, we have

- Process:

$$\dot{s} \to \dot{e}$$
$$a(\cdot) \to -\frac{e}{RC} + \frac{1}{C}I_{in}$$
$$w \to q$$
$$R_{ww} \to R_{qq}$$

- Measurement:

$$s \to e$$
$$c(\cdot) \to K_e$$
$$v \to n$$
$$R_{vv} \to R_{nn}$$

- Therefore, the filter becomes

$$\hat{s} \to \hat{e}$$
$$\tilde{P}_{ss} \to \tilde{P}_{ee}$$

- such that

$$R_{vv} > \tilde{R}_{ee} > \tilde{P}_{ee}$$

This completes the example. □

Next, we illustrate the design of the model-based processor for this circuit.

Example 1.4 Suppose we are asked to design a simple RC-circuit as shown in Figure 1.5. The measurement is contaminated with random measurement noise as well as uncertainties in the values of the resistor and capacitor. Writing the Kirchhoff node equations, we have

$$I_{in} - \frac{e}{R} - C\dot{e} = 0$$

and the measurement given by the voltmeter is

$$e_{out} = K_e e.$$

Discretizing this equation using first differences with *sampling interval* ΔT and including white Gaussian noise sources for the parameter and measurement uncertainties, we develop the following discrete Gauss–Markov model:

$$e(t) = \underbrace{\left(1 - \frac{\Delta T}{RC}\right)}_{A} e(t-1) + \underbrace{\frac{\Delta T}{C}}_{B} I_{\text{in}}(t-1) + w(t-1)$$

$$e_{\text{out}}(t) = \underbrace{K_e}_{C} e(t) + v(t)$$

The resulting model-based processor (Kalman filter) is given by

$$\hat{e}(t|t) = \underbrace{\left(1 - \frac{\Delta T}{RC}\right)}_{A} \hat{e}(t-1|t-1) + K(t)e(t)$$

For $R = 3.3$ kΩ and $C = 1000$ μF, $\Delta T = 100$ ms, $K_e = 2.0$, $e_0 = 2.5$ V, $I_{\text{in}} = 100$ μA (negative step) with $w \sim \mathcal{N}(0, 0.0001)$ and $v \sim \mathcal{N}(0, 4.0)$ or

$$e(t) = 0.97e(t-1) + I_{\text{in}}(t-1) + w(t-1)$$
$$e_{\text{out}}(t) = 2e(t) + v(t)$$

The noisy measurement and corresponding model-based random signal estimate is shown in Figure 1.6. Here we see the noisy measurement data, the

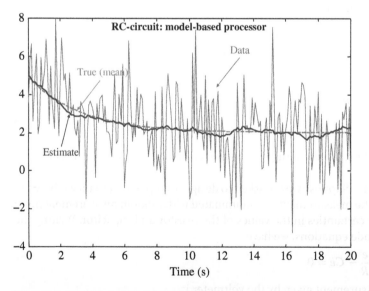

Figure 1.6 Model-based processor for RC-circuit: raw data (solid), true measurement (dashed), and model-based estimate (thick solid).

1.3 Model-Based Processing

noise-free (true) measurement (dashed line) along with the corresponding "filtered" or model-based estimate. Note how the estimate *tracks* the true measurement significantly increasing the SNR enhancing the signal. This completes the example. □

There are many different forms of model-based processors depending on the models used and the manner in which the estimates are calculated. For example, there are process model-based processors (Kalman filters [3]), statistical model-based processors (Box–Jenkins filters [7], Bayesian filters [6]), statistical model-based processors (covariance filters [2]), or even optimization-based processors (gradient filters [8]). In any case, many processors can be placed in a recursive form with various subtleties emerging in the calculation of the current estimate ($\hat{S}_{old}$). The standard technique employed is based on correcting or updating the current estimate as a new piece of measurement data becomes available. The estimates generally take the *recursive form*:

$$\hat{S}_{new} = \hat{S}_{old} + KE_{new}$$

where

$$E_{new} = Y - \hat{Y}_{old} = Y - C\hat{S}_{old}$$

Here we see that the new estimate is obtained by correcting the old estimate by a K-weighted amount. The error term E_{new} is the new information or innovation; that is, it is the difference between the actual measurement and the predicted measurement ($\hat{Y}_{old}$) based on the old estimate ($\hat{S}_{old}$). The computation of the weight K depends on the criterion used (e.g. mean-squared error, absolute error).

Consider the following example, which demonstrates how to recursively estimate the sample mean.

Example 1.5 The sample mean estimator can easily be put in the recursive form. The estimator is given by

$$\hat{S}(N) = \frac{1}{N} \sum_{t=1}^{N} y(t)$$

Extracting the Nth term from the sum, we obtain

$$\hat{S}(N) = \frac{1}{N} y(N) + \frac{1}{N} \sum_{t=1}^{N-1} y(t)$$

Identify $\hat{S}(N-1)$ from the last term, that is,

$$\hat{S}(N) = \frac{1}{N} y(N) + \left(\frac{N-1}{N} \right) \hat{S}(N-1)$$

The recursive form is given by

$$\underbrace{\hat{S}(N)}_{S_{new}} = \underbrace{\hat{S}(N-1)}_{S_{old}} + \underbrace{\frac{1}{N}}_{K} \underbrace{[y(N) - \hat{S}(N-1)]}_{E_{new}}$$ □

This completes the introductory discussion on model-based signal processing from a heuristic point of view. Next we discuss methods of *extracting* a model from measurement data in order to construct an MBP.

1.4   Model-Based Identification

Identification provides a solution to the problem of extracting a model from measured data sequences either time series, frequency data or simply an ordered set of indexed values. As illustrated in Figure 1.7, both inputs (when available) and measured outputs are used to perform the extraction in spite of any disturbances or parametric uncertainties. Since its fundamental evolution from control theory, identification or "system" identification is based on estimating a model of a *dynamic* system [7] from data. Models can be of many varieties ranging from simple polynomials to highly complex constructs evolving from nonlinear distributed systems. The extraction of a model from data is critical for a large number of applications evolving from the detection of submarines in a varying ocean, to tumor localization in breast tissue, to pinpointing the epicenter of a highly destructive earthquake or to simply monitoring the condition of a motor as it drives a critical system component [3].

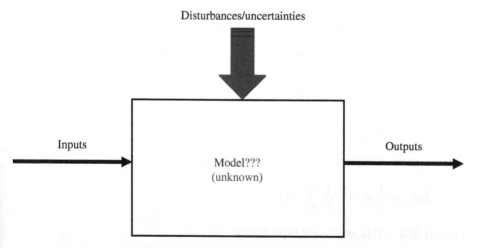

Figure 1.7 System identification problem: input (deterministic/random), disturbances (noise/parametric uncertainty), output (noisy measurement), model (class selected).

Each of these applications requires an aspect of modeling and fundamental understanding (when possible) of the underlying phenomenology governing the process as well as the measurement instrumentation extracting the data along with the accompanying uncertainties. Some of these problems can be solved with simply a "black-box" representation that faithfully reproduces the data in some manner without the need to capture the underlying dynamics (e.g. common check book entries) or to a "gray box" model that has been extracted, but has parameters of great interest (e.g. unknown mass of a toxic material). However, when the true need exists to obtain an accurate representation of the underlying phenomenology like the structural dynamics of an aircraft wing or the untimely vibrations of a turbine in a nuclear power plant, then more sophisticated representations of the system and uncertainties are clearly required. In cases such as these, models that capture the dynamics must be developed and "fit" to the data in order to perform applications such as condition monitoring of the structure or failure detection/prediction of a rotating machine. Here models can evolve from lumped characterizations governed by sets of ordinary differential equations, linear or nonlinear, or to distributed representations evolved from sets of partial differential equations. All of these representations have one thing in common, when the need to perform a critical task is at hand – they are represented by a mathematical model that captures their underlying phenomenology that must somehow be extracted from noisy measurements. This is the fundamental problem that we address in this text, but we must restrict our attention to a more manageable set of representations, since many monographs have addressed problem sets targeting specific applications [9, 10].

In fact, this concept of specialty solutions leads us to the generic *state-space* model (Chapter 3) of systems theory and controls. Here the basic idea is that all of the theoretical properties of a system are characterized by this fundamental model set that enables the theory to be developed and then applied to any system that can be represented in the state-space. Many models naturally evolve in the state-space, since it is essentially the representation of a set of nth order differential equations (ordinary or partial, linear or nonlinear, time (space) invariant or time (space) varying, scalar or multivariable) that are converted into a set of first-order equations, each one of which is a state. For example, a simple mechanical system consisting of a single mass, spring, damper construct is characterized by a set of second-order, linear, time-invariant, differential equations that can simply be represented in state-space form by a set of two first-order equations, each one representing a state: one for displacement and one for velocity [9]. We employ the state-space representation *throughout* this text and provide sufficient background in Chapter 3.

Here we are primarily focused on the development of models for the design and application of model-based signal processors (MBSP) using estimation/identification techniques to achieve a meaningful design. For model-based identification (MBID), we incorporate a *known structure*, rather than strictly a generic class of model (e.g. state-space, transfer function, polynomial); that is, we have possibly a physics-based structure and would like to "jointly" estimate the unknown or poorly known parameters embedded within that structure. The problem is similar to what the identification community might term a "gray-box" model [7] and processor to estimate parameters. Here we call it a parametrically adaptive processor in which the distinction can be made between three fundamental, but different, problems. The first problem is simply a parameter estimation, in which no particular model is embedded, such as in classic least-squares estimation [3, 11]. The second can be thought of as a simple signal estimation problem such as a model-based processor with known model as in the RC-circuit example. The third is the problem that we are most concerned and defines what we mean by model-based identification compared to subspace identification (discussed in the next section). We have defined this problem and the resulting processor in various contexts and applications as a *parametrically adaptive* processor [3, 6] in which both signal and parameters are estimated simultaneously. Therefore, we define the *model-based identification problem* as the joint estimation of unknown signal/parameters of a known structure from noisy measurement data. In essence, it is the development of a parametrically adaptive processor of known structure and capable of solving the joint signal/parametric estimation problem. This processor is known as a *model-based identifier* and is well known in the literature that was developed for a wealth of various applications [3, 6].

Identification is broad in the sense that it does not limit the problem to various classes of models directly. For instance, for an unknown system, a model set is selected with some perception that it is capable of representing the underlying phenomenology adequately, then this set is identified directly from the data and validated for its accuracy. There is clearly a well-defined procedure that captures this approach to solve the identification problem [7, 10]. In some cases, the class structure of the model may be known a priori, but the order or equivalently the number of independent equations to capture its evolution is not (e.g. number of oceanic modes). Here, techniques to perform order estimation precede the fitting of model parameters first and are followed by the parameter estimation to extract the desired model [12]. In other cases, the order is known from prior information and parameter estimation follows directly (e.g. a designed mechanical structure). In any case, these constraints govern the approach to solving the identification problem and extracting the model for application. Many applications exist, where it is desired to monitor a process and track a variety of parameters as they evolve in time (e.g. radiation detection), but in order to accomplish this on-line, the model-based processor or more appropriately the model-based identifier must update the model parameters sequentially in order to accomplish its designated task. We develop these processors for both linear and nonlinear models in Chapters 4 and 5.

In Chapter 5, we present a different approach that incorporates the solution into the identification problem as the integral part of the model-based signal processor (Kalman filter) that can be applied to a large number of applications, but with little success unless a reliable model is available or can be adapted to a changing environment [13]. Using various approaches, it is possible to identify the model very rapidly and incorporate it into a variety of processing problems such as state estimation, tracking, detection, classification, controls, and communications to mention a few [14, 15].

As an example, consider the development of a processor to extract the unknown parameters from the RC-circuit.

Example 1.6 Suppose we investigate the RC-circuit and would like to dynamically estimate the circuit parameters as well as its output voltage. The model in state-space form with $x := e$ and $y := e_{out}$ is given by (see Example 1.4)

$$e(t) = \theta_1 e(t-1) + \theta_2 I_{in}(t-1) + w(t-1)$$
$$e_{out}(t) = \theta_3 e(t) + v(t)$$

The processor can be implemented in a number of ways, one of which as a least-squares parameter estimator using a strictly parametric estimator based on

$$\theta(t) = \theta(t-1) + w(t-1)$$
$$e_{out}(t) = \theta(t) + v(t)$$

or as a MBID-processor (nonlinear) embedding the circuit model and jointly estimating both states and parameters based on the augmented model $x := [e \mid \theta_1 \mid \theta_2 \mid \theta_3]'$

$$e(t) = \theta_1(t-1)e(t-1) + \theta_2(t-1)I_{in}(t-1) + w_x(t-1)$$
$$\theta(t) = \theta(t-1) + w_\theta(t-1)$$
$$e_{out}(t) = \theta_3(t)e(t) + v(t)$$

The results are shown in Figure 1.8, where we see each of the parameter estimates as well as the filtered measurement. In Figure 1.8a, we see the parameter estimates for both the system ($\hat{\theta}_1 \approx -0.97$) and input ($\hat{\theta}_2 \approx -0.1$) model parameters, while in Figure 1.8b we observe both the output ($\hat{\theta}_3 \approx 2.0$) parameter and enhanced (filtered) output voltage measurement ($\hat{y} \approx \hat{e}_{out}$). This completes the design of the MBID. □

Next we consider the heart of the text – subspace identification.

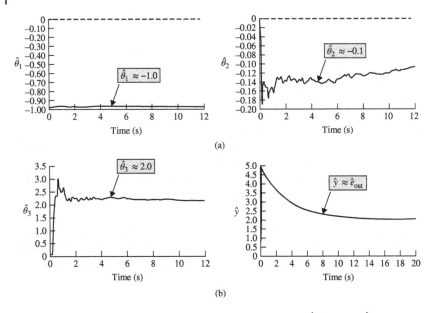

Figure 1.8 Model-based identification of RC-circuit parameters: $\hat{\theta}_1 \approx -0.97$, $\hat{\theta}_2 \approx -0.1$, $\hat{\theta}_3 \approx 2.0$ and $\hat{y} \approx \hat{e}_{out}$.

1.5 Subspace Identification

Models for the processor evolve in a variety of ways, either from first principles accompanied by estimating its inherent uncertain parameters as in parametrically adaptive schemes [3], or by extracting constrained model sets employing direct optimization methodologies [7], or by simply fitting a black-box structure to noisy data [8]. Once the model is extracted from controlled experimental data or data synthesized from a highly complex truth model, the long-term processor can be developed for direct application [6]. Since many real-world applications seek a real-time solution, we concentrate primarily on the development of fast, reliable identification methods that enable such an implementation [16–18]. Model extraction/development must be followed by validation and testing to ensure that the model reliably represents the underlying phenomenology – a bad model can only lead to failure!

The concept of subspace has a well-defined meaning in system identification. Simply stated, "a *subspace* is a space or subset of a parent space that inherits some or all of its characteristics." For instance, we can consider each layer of an onion a subset (subspace) of the onion itself, each inheriting its generic properties. In mathematics, we can consider the decomposition of a topological space into subsets, each inheriting properties of the previous region or subregion creating a hierarchy. For example, a topological space with all of its properties can be decomposed into subregions (layers): first into a topological-metric-space, next into a topological-normalized-metric-space, and finally into a topological-normalized-metric-inner product-space with each subregion (layer) inheriting some of the properties of its predecessor. In identification, subspace refers primarily to a vector (state) space in which its essential characteristics can be captured by a "smaller" space – the subspace. For example, an Nth order state-space dynamic system can be approximated by an Mth order ($M < N$) system capturing the majority of its properties. This is the concept that we will employ throughout this text.

Subspace identification (SID) is a technique to extract (identify) a black-box model in generic state-space form from uncertain input/output data using robust, numerically stable, linear algebraic methods. SID (simply) performs a black-box identification of a generic state-space model (A, B, C, D) from noisy measurements as illustrated in Figure 1.9. Unlike MBID discussed previously, it does *not* require any specific structure of the state-space model, only its order (number of states) that can be derived directly from the embedded linear algebraic decomposition. SID methods are constrained to *linear* state-space models.

More specifically, compared to the MBID processors that iteratively estimate parameters of the previous section [3, 7], we investigate subspace identification techniques [16–18]. These techniques enable the capability to extract a large number of parameters compared to the prediction error (P-E) methods that are quite time-consuming and not practical for large parametric problems coupled with the need for potential real-time applications. Subspace identification techniques are robust and reliable and capable of real-time application,

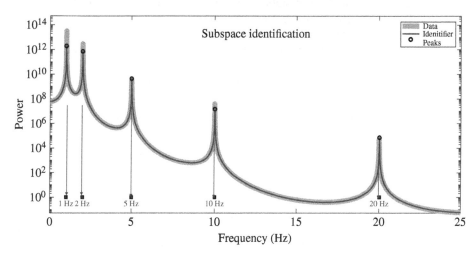

Figure 1.9 Identification of a sinusoidal model from noise-free data: spectral frequencies (identified) at 1 Hz, 2, 5, 10, and 20 Hz.

but they are not as accurate as the optimization-based P-E methods [8]. In fact, the SID methods are typically used as "starters" for P-E processors enabling a reasonable starting point for the numerical optimizers, thereby ensuring faster convergence. In this text, we concentrate on the SID approach, presented in Chapters 6 and 7.

Consider the following example of sinusoids in noise to illustrate this process.

Example 1.7 Suppose we are given an ideal measurement consisting of five sinusoids at frequencies ranging from 1 to 20 Hz and we would like to develop a model of this process, after performing an identification, that is, our underlying model is simply a sum of sinusoids of varying amplitudes such that

$$y(t) = \sum_{i=1}^{5} A_i \sin 2\pi \ f_i t \text{ with } f_i = \{1, 2, 5, 10, 20\} \text{ Hz} \tag{1.1}$$

We could simply perform a Fourier transform in this noise-free case and locate the peaks and magnitudes to extract the sinusoids or perform a parametric estimation using a nonlinear (sinusoid) optimization technique yielding a model-based identification. However, another approach is to perform a *black-box* identification choosing a generic model-class (e.g. state-space) that captures these dynamics in the form of a transfer function, estimate the underlying state-space model, and using it, calculate the corresponding impulse response. With this representation, the corresponding frequencies can be extracted from the eigenvalues (poles) of the identified model. The data were processed by an identifier using the proper order (20-states or 2 × No. sinusoids), and the results are shown in Figure 1.9, both in the Fourier (power) spectrum obtained from the simulated noise-free data along with the estimated (identified) frequencies (squares) annotated on the plot. There is perfect agreement. This example illustrates the concept of identification (black-box) and becomes more and more important as the SNR decreases and the spectral peaks are no longer available for detection and location. □

This completes the introduction to subspace identification. Next we discuss then notation that will be employed in this text.

1.6   Notation and Terminology

The notation used throughout this text is standard in the literature. Where necessary, vectors are represented by boldface, lowercase, **x**, and matrices by boldface, uppercase, **A**. The *adjoint* of the matrix **A** is annotated as adj(**A**) and its corresponding *determinant* by det(**A**). We define the notation $\underline{N}$ to be a shorthand way of writing $1, 2, \ldots, N$. It will be used in matrices, $A(\underline{N})$ to mean there are N-columns of A. We also use the equivalent MATLAB notation $(1:N)$. We denote the real part of a signal by Re and its imaginary part by Im. As mentioned previously, estimators are annotated by the caret, such as $\hat{x}$. We also define partial derivatives at the component level by $\frac{\partial}{\partial \theta_i}$, the N_θ- gradient vector by ∇_θ and higher order partials by ∇_θ^2.

The most difficult notational problem will be with the "time" indices. Since this text is predominantly discrete-time, we will use the usual time symbol t to mean a discrete-time index, i.e. $t \in \mathcal{I}$ for $\mathcal{I}$ the set of integers. However, and hopefully not too confusing, t will also be used for continuous-time, i.e. $t \in \mathcal{R}$ for $\mathcal{R}$ the set of real numbers denoting the continuum. When used as a continuous-time variable, $t \in \mathcal{R}$, it will be represented as a subscript to distinguish it, i.e. x_t. This approach of choosing $t \in \mathcal{I}$ primarily follows the system identification literature and for the ease of recognizing discrete-time variable in transform relations (e.g. discrete Fourier transform). The rule of thumb is therefore to "interpret t as a discrete-time index unless noted by a subscript as continuous in the text." With this in mind, we will define a variety of discrete estimator notations as $\hat{x}(t|t-1)$ to mean the estimate at time (discrete) t based upon all of the previous data up to $t-1$. We will define these symbols prior to their use with the text to assure no misunderstanding of its meaning.

With a slight abuse of notation, we will use the terminology *distribution* of X, $\Pr(X)$ in general, so as not to have to differentiate between density for continuous random variables or processes and mass for discrete variates. It will be obvious from the context which is meant. In some cases, we will be required to make the distinction between cumulative distribution function (CDF) and density (PDF) or mass (PMF) functions. Here we use the uppercase notation $P_X(x)$ for the CDF and lowercase $p_X(x)$ for the PDF or PMF.

Subsequently, we will also need to express a discrete PMF as a continuous PDF using impulse or delta functions as "samplers" much the same as in signal processing when we assume there exists an impulse sampler that leads to the well-known Nyquist sampling theorem [3]. Thus, corresponding to a discrete PMF we can define a continuous PDF through the concept of an *impulse sampler*, that is, given a discrete PMF defined by

$$p_X(x) \approx p(X = x_i) = \sum_i p_i \, \delta(x - x_i) \tag{1.2}$$

then we define the *equivalent continuous* PDF as $p_X(x)$. Moments follow from the usual definitions associated with a continuous PDF; for instance, consider the definition of the expectation or mean. Substituting the equivalent PDF and utilizing the sifting property of the impulse function gives

$$E\{x\} = \int_{-\infty}^{\infty} x \, p_X(x) \, dx = \int_{-\infty}^{\infty} x \left(\sum_i p_i \delta(x - x_i) \right) dx = \sum_i x_i p_i \tag{1.3}$$

which is precisely the mean of the discrete PMF(see Appendix A for more details).

Also, as mentioned, we will use the symbol $\sim$ to mean "distributed according to" as in $x \sim \mathcal{N}(m, v)$ defining the random variable x as Gaussian distributed with mean m and variance v. We may also use the extended notation: $\mathcal{N}(x : m, v)$ to include the random variable x as well. When *sampling*, we use the *non-conventional* right arrow "action" notation $\rightarrow$ to mean "draw a sample from" a particular distribution such as $x_i \rightarrow \Pr(x)$ – this again will be clear from the context. When *resampling*, that is, replacing samples with new ones, we use the "block" right arrow such as $x_j \leftrightarrow x_i$ meaning new sample x_j replaces current sample x_i. In a discrete (finite) probabilistic representation, we define a purely discrete variate as $x_k(t) := \Pr(x(t) = \mathcal{X}_k)$ meaning that x can only take on values (integers) k from a known set $\mathcal{X} = \{\mathcal{X}_1, \dots, \mathcal{X}_k, \dots, \mathcal{X}_N\}$ at time t. We also use the symbol $\square$ to mark the end of an example.

We define two projection operator projection operators: (i) orthogonal and (ii) oblique or parallel. The *orthogonal projection operator* "projects" $\bullet$ onto $\mathcal{Y}$ as $\mathcal{P}_{\bullet|\mathcal{Y}}$ or its complement $\mathcal{P}^{\perp}_{\bullet|\mathcal{Y}}$, while the *oblique projection operator* "projects" $\bullet$ onto $\mathcal{Y}$ "along" $\mathcal{Z}$ as $\mathcal{P}^{\|}_{\bullet|\mathcal{Y}\circ\mathcal{Z}}$. For instance, orthogonally projecting the vector $\mathbf{x}$, we have $\mathcal{P}_{\mathbf{x}|\mathcal{Y}}$ or $\mathcal{P}^{\perp}_{\mathbf{x}|\mathcal{Y}}$, while obliquely projecting $\mathbf{x}$ along $\mathcal{Z}$ is $\mathcal{P}_{\mathbf{x}|\mathcal{Y}\circ\mathcal{Z}}$ (see Appendix B). The matrix projection operator evolves from decompositions such as the LQ-decomposition (see Appendix C), where the operator is given in terms of its orthonormal bases as $\mathcal{P} = Q \times Q'$.

1.7 Summary

In this chapter, we introduced the concept of model-based signal processing (MBSP) based on the idea of incorporating more and more available a priori information into the processing scheme. Various signals were classified as deterministic or random. When the signal is random, the resulting processors are called estimation filters or estimators. A procedure was defined (estimation procedure) leading to a formal decomposition, which will be used throughout this text. After a couple of examples motivating the concept of model-based signal processing, a more formal discussion followed with a concrete RC-circuit example completing the chapter. Next we introduced the idea of model-based identification (MBID), which is equivalent to a variety of P-E methods, where a model developed from the underlying problem phenomenology is incorporated directly into the processor. After discussing the identification of the RC-circuit parameters, the subspace identification (SID) approach we introduced conceptually and illustrated through a sinusoidal identification example. Finally, the notation utilized throughout the text was discussed completing this introductory chapter.

MATLAB Notes

MATLAB is a command oriented vector–matrix package with a simple yet effective command language featuring a wide variety of embedded C language constructs making it ideal for signal processing and graphics. All of the algorithms we have applied to the examples and problems in this text are MATLAB-based in solution ranging from simple simulations to complex applications. We will develop these notes primarily as a summary to point out to the reader many of the existing commands that already perform the signal processing operations discussed in the presented chapter and throughout the text.

More specifically, MATLAB consists of a variety of "toolboxes" enabling the users to access a large number of numerically stable and accurate algorithms rapidly along with a corresponding suite of demonstration examples. For this text, the most important toolboxes are the signal processing (**signal**), system identification (**ident**), control system (**control**), and the statistics (**stats**), which can be used to solve the majority of problems introduced in this text. Throughout we will refer to specific commands that can be applied to a variety of problems and provide meaningful solutions.

References

1 Rossing, T., Dunn, F., Hartmann, W. et al. (2007). *Handbook of Acoustics*. New York: Springer.

2 Hayes, M. (1996). *Statistical Digital Signal Processing and Modeling*. New York: Wiley.

3 Candy, J. (2006). *Model-Based Signal Processing*. Hoboken, NJ: Wiley/IEEE Press.

4 Bucker, H. (1976). Use of calculated sound fields and matched-field detection to locate sound in shallow water. *J. Acoust. Soc. Am.* 59: 329–337.

5 Tolstoy, A. (1993). *Matched Field Processing for Ocean Acoustics*. Hackensack, NJ: World Scientific.

6 Candy, J. (2016). *Bayesian Signal Processing, Classical, Modern and Particle Filtering*, 2e. Hoboken, NJ: Wiley/IEEE Press.

7 Ljung, L. (1999, 2e). *System Identification: Theory for the User*. NJ: Prentice-Hall.

8 Ljung, L. and Soderstrom, T. (1983). *Theory and Practice of Recursive Identification*. Cambridge, MA: M.I.T. Press.

9 Juang, J. (1994). *Applied System Identification*. Upper Saddle River, NJ: Prentice-Hall PTR.

10 Aoki, M. (1990). *State Space Modeling of Time Series*, 2e. London: Springer.

11 Sorenson, H. (1986). *Parameter Estimation*. New York: IEEE Press.

12 Norton, J. (1986). *An Introduction to Identification*. New York: Academic Press.

13 Kalman, R. (1960). A new approach to linear filtering and prediction problems. *Trans. ASME J. Basic Eng.* 82: 34–45.

14 Van Der Veen, A., Deprettere, E.F., and Swindlehurst, A.L. (1993). Subspace-based methods for the identification of linear time-invariant systems. *Proc. IEEE* 81 (9): 1277–1308.

15 Viberg, M. (1995). Subspace based signal analysis using singular value decomposition. *Automatica* 31 (12): 1835–1851.

16 van Overschee, P. and De Moor, B. (1996). *Subspace Identification for Linear Systems: Theory, Implementation, Applications*. Boston, MA: Kluwer Academic Publishers.

17 Verhaegen, M. and Verdult, V. (2007). *Filtering and System Identification: A Least-Squares Approach*. Cambridge: Cambridge University Press.

18 Katayama, T. (2005). *Subspace Methods for System Identification*. London: Springer.

Problems

1.1 We are asked to estimate the displacement of large vehicles (semitrailers) when parked on the shoulder of a freeway and subjected to wind gusts created by passing vehicles. We measure the displacement of the vehicle by placing an accelerometer on the trailer. The accelerometer has inherent inaccuracies, which is modeled as

$$y = K_a x + n$$

with y, x, n the measured and actual displacement and white measurement noise of variance R_{nn} and K_a the instrument gain. The dynamics of the vehicle can be modeled by a simple mass–spring–damper. (a) Construct and identify the measurement model of this system. (b) Construct and identify the process model and model-based estimator for this problem.

1.2 Think of measuring the temperature of a liquid in a breaker heated by a burner. Suppose we use a thermometer immersed in the liquid and periodically observe the temperature and record it. (a) Construct a measurement model assuming that the thermometer is linearly related to the temperature, that is, $y(t) = k\Delta T(t)$. Also model the uncertainty of the visual measurement as a random sequence $v(t)$ with variance R_{vv}. (b) Suppose we model the heat transferred to the liquid from the burner as

$$Q(t) = C\,A\Delta T(t),$$

where C is the coefficient of thermal conductivity, A is the cross-sectional area, and $\triangle T(t)$ is the temperature gradient with assumed random uncertainty $w(t)$ and variance R_{ww}. Using this process model and the models developed above, identify the model-based processor representation.

1.3 We are given an RLC series circuit driven by a noisy voltage source $V_{in}(t)$ and we use a measurement instrument that linearly amplifies by K and measures the corresponding output voltage. We know that the input voltage is contaminated by and additive noise source, $w(t)$ with covariance, R_{ww} and the measured output voltage is similarly contaminated with noise source, $v(t)$ with R_{vv}.
(a) Determine the model for the measured output voltage, $V_{out}(t)$ (measurement model).
(b) Determine a model for the circuit (process model).
(c) Identify the general model-based processor structures. In each scheme, specify the models for the process, measurement, and noise.

1.4 A communications satellite is placed into orbit and must be maneuvered using thrusters to orientate its antennas. Restricting the problem to the single axis perpendicular to the page, the equations of motion are

$$J\frac{d^2\theta}{dt^2} = T_c + T_d$$

where J is the moment of inertia of the satellite about its center of mass, T_c is the thruster control torque, T_d is the disturbance torque, and θ is the angle of the satellite axis with respect to the inertial reference (no angular acceleration) A. Develop signal and noise models for this problem and identify each model-based processor component.

1.5 Consider a process described by a set of linear differential equations

$$\frac{d^2c}{dt^2} + \frac{dc}{dt} + c = Km.$$

The process is to be controlled by a proportional–integral–derivative (PID) control law governed by the equation

$$m = K_p\left(e + \frac{1}{T_i}\int e\,dt + T_d\frac{de}{dt}\right)$$

and the controller reference signal r is given by

$$r = e + c$$

Suppose the reference is subjected to a disturbance signal, and the measurement sensor, which is contaminated with additive noise, measures the "square" of the output. Develop the model-based signal and noise models for this problem.

1.6 The elevation of a tracking telescope is controlled by a DC motor. It has a moment of inertia J and damping B due to friction, and the equation of motion is given by

$$J\frac{d^2\theta}{dt^2} + B\frac{d\theta}{dt} = T_{\mathrm{m}} + T_{\mathrm{d}},$$

where T_{m} and T_{d} are the motor and disturbance torques and θ is the elevation angle. Assume a sensor transforms the telescope elevation into a proportional voltage that is contaminated with noise. Develop the signal and noise models for the telescope and identify all of the model-based processor components.

1.7 Suppose we have a two-measurement system given by

$$y = \begin{bmatrix} 3 \\ 4 \end{bmatrix} + v$$

where $R_{vv} = \mathrm{diag}[1, \ 0.1]$.
(a) What is the batch least-squares estimate ($W = I$) of the parameter x, if $y = [7 \ 21]'$?
(b) What is the batch weighted least-squares estimate of the parameter x with W selected for minimum variance estimation?

1.8 Calculate the batch and sequential least-squares estimate of the parameter vector x based on two measurements $y(1)$ and $y(2)$ where

$$y(1) = C(1)x + v(1) = \begin{bmatrix} 2 \\ 1 \end{bmatrix}$$

$$y(2) = c'x + v(2) = 4$$

$$C = \begin{bmatrix} 1 & 1 \\ 0 & 1 \end{bmatrix}, \quad c'(1) = [1 \ 2], \quad W = I$$

2

Random Signals and Systems

2.1 Introduction

We can apply a filter to a random signal, but since its output is still random, we must find a way to eliminate or reduce this randomness in order to employ the powerful techniques available from systems theory. We shall show that techniques from statistics combined with linear systems theory can be applied to extract the desired signal information and reject the disturbance or noise. In this case, the filter is called an estimation filter or simply an *estimator* that is required to extract the useful information (signal) from noisy or random measurements.

Techniques similar to linear deterministic systems theory hold when the random signal is transformed to its *covariance sequence* and its Fourier spectrum is transformed to its *power spectrum*. Once these transformations are accomplished, then the techniques of linear systems theory can be applied to obtain results similar to deterministic signal processing. In fact, we know that the covariance sequence and power spectrum are a discrete Fourier transform (DFT) pair, analogous to a deterministic signal and its corresponding spectrum, that is, we have that

$$\text{Fourier Transform}: \quad R_{xx}(k) \longleftrightarrow S_{xx}(\Omega)$$

and as in the deterministic case, we can analyze the spectral content of a random signal by investigating its power spectrum.

The power spectral density (PSD) function for a discrete random process is defined as:

$$S_{xx}(\Omega) = \lim_{N \to \infty} E\left\{ \frac{X(\Omega)X^*(\Omega)}{2N+1} \right\} \tag{2.1}$$

where $*$ is the complex conjugate. The expected value operation, $E\{.\}$, that can be thought of as "mitigating" the randomness. Similarly, the *covariance*[1] of the

---

1 It is also common to use the so-called *correlation* function, which is merely the mean-squared function and identical to the covariance function for a mean of zero.

Model-Based Processing: An Applied Subspace Identification Approach, First Edition. James V. Candy.
© 2019 John Wiley & Sons, Inc. Published 2019 by John Wiley & Sons, Inc.

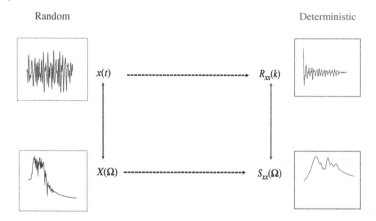

Figure 2.1 Random signal comparison: random signal and spectrum and the deterministic covariance and power spectrum.

process is defined by

$$R_{xx}(k) := E\{x(t)x(t+k)\} - E^2\{x(t)\} \tag{2.2}$$

In a sense, these relations replace the random signals, but play the same role as their deterministic counterparts. These concepts are illustrated in Figure 2.1 where we see the random signal and its random Fourier transform replaced by the deterministic covariance and its deterministic power spectrum.

Techniques of linear systems theory for random signals are valid, just as in the deterministic case where the covariance at the output of a system excited by a random signal $x(t)$ is given by the convolution relationship in the temporal or frequency domain as,

$$\text{Convolution}: \quad R_{yy}(k) = h(t) * h(-t) * R_{xx}(k)$$

$$\text{Multiplication}: \quad S_{yy}(\Omega) = H(\Omega)H^*(\Omega)S_{xx}(\Omega) = |H(\Omega)|^2 S_{xx}(\Omega) \tag{2.3}$$

Analogously, the filtering operation is performed by an estimation filter, $\hat{H}_f$, designed to shape the output PSD, similar to the deterministic filtering operation

$$\text{Filtering}: \quad S_{yy}(\Omega) = |\hat{H}_f(\Omega)|^2 S_{xx}(\Omega) \tag{2.4}$$

Consider the following example of analyzing a random signal using covariance and spectral relations.

Example 2.1 Suppose we have a measured signal given by

$$\underbrace{x(t)}_{\text{Measurement}} = \underbrace{s(t)}_{\text{Signal}} + \underbrace{n(t)}_{\text{Noise}}$$

and we would like to extract the signal and reject the noise, so we design an estimation filter,

$$\hat{y}(t) = \hat{h}_f(t) * x(t)$$

$$\hat{Y}(\Omega) = \hat{H}_f(\Omega)X(\Omega) \tag{2.5}$$

Our signal in this case will be the sinusoids at 10 and 20 Hz contaminated with additive random noise. In Figure 2.2a, we see the random signal and raw (random) discrete Fourier spectrum. Note that the noise severely obscures the sinusoidal signals, and there are many false peaks that could erroneously be selected as sinusoidal signals. Estimating the covariance of the random signal along with its corresponding power spectrum using statistical methods, we can now easily see the sinusoidal signals at the prescribed frequencies as

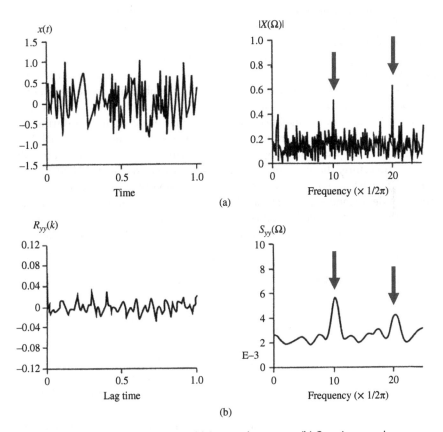

Figure 2.2 Random signal processing: (a) Data and spectrum. (b) Covariance and power spectrum.

depicted in Figure 2.2b. Therefore, analogous to deterministic signal processing techniques, the covariance and power spectra enable the powerful methods of linear systems theory to be applied to extract useful signal information for noisy data.   □

In this chapter, we are concerned with the development of analysis methods for random signals that will prove useful in applications to follow.

2.2   Discrete Random Signals

A random signal or equivalently a stochastic process finds its roots in probability theory. The idea of an underlying probability or sample space is essential to the treatment of random signals. Recall that a *probability/sample space*, Ξ, is a collection of samples or experimental outcomes, ξ, which are elements of that space. Certain subsets of Ξ or collections of outcomes are the *events*. For example, if we flip a fair coin, then the sample space consists of the set of outcomes, $\Xi = \{H, T\}$ and the events, $\{0, \{H\}, \{T\}\}$. When we attempt to specify how likely a particular event will occur during a trial or experiment, then we define the notion of *probability*. In terms of our probability space, we define a *probability function*, $\Pr(\cdot)$, as a function defined on a class of events, which assigns a number to the likelihood of a particular event occurring during an experiment. We must constrain the class of events on which the probability function is defined so that set operations performed produce sets that are still events. Therefore, the class of sets satisfying this constraint is called a *field (Borel)*. If we have a space, Ξ, a field, B, and a probability function, $\Pr(\cdot)$, we can precisely define and *experiment* as the triple, $\{\Xi, B, \Pr\}$. Stochastic processes can be thought of as a collection of indexed random variables. More formerly, a random signal or *stochastic process* is a two-dimensional function of t and ξ:

$$X(t, \xi) \qquad \xi \epsilon \Xi, \quad t \in T \tag{2.6}$$

where T is a set of index parameters (continuous or discrete) and Ξ is the sample space.

So we see that a *discrete random signal* can be precisely characterized by

$$x_i(t) := X(t, \xi_i) \tag{2.7}$$

A *realization* of a stochastic process and in fact each distinct value of t can be interpreted as a random variable. Thus, we can consider a stochastic process, simply a sequence of ordered random variables. A collection of realizations of a stochastic process is called an *ensemble*.

A common representation of a random process is a random walk given in the following example.

Example 2.2 (Random Walk) Suppose we toss a fair coin at each time instant, that is,

$$X(t, \xi_i), \qquad t \in T = \{0, 1, \dots, N - 1\}, \qquad \xi \epsilon \Xi = \{0, \{H\}, \{T\}\}$$

where the random variable is given by

$$X(t, \xi_i) = \begin{cases} K & \xi_1 = H \\ -K & \xi_2 = T \end{cases}$$

We define the corresponding probability mass function as

$$P_X(X(t, \xi_i) = \pm K) = 1/2$$

Here we see that for each $t \in T, X(t, \xi)$ is just a random variable. Now if we define the function,

$$y(N, \xi_i) = \sum_{t=0}^{K-1} X(t, \xi_i) \quad i = 1, 2, \dots$$

as the position at t for the ith realization, then for each t, y is a sum of discrete random variables, $X(., \xi_i)$, and for each i, $X(t, \xi_i)$ is the ith realization of the process (a time series). For a given trial, we may have $\xi_1 = \{HHTHTT\}$, and $\xi_2 = \{THHTTT\}$, then for these realizations we see the ensemble shown in Figure 2.3. Note that $y(5, \xi_1) = 0$, and $y(5, \xi_2) = -2K$, which can be predicted using the given random variables and realizations. This completes the example. □

In summary, we see that a random signal can be represented precisely as a discrete stochastic process that is a two-dimensional function of an index set (e.g. time) and samples of the probability space [1–3]. Since a random signal can be interpreted as a sequence of random variables, then we expect the basic statistics and properties to apply as well. In fact, this is the case, the resulting statistic merely becomes a function of the index parameter t (time) in our case. Basic statistics and properties hold for stochastic processes and are listed in Appendix A. Note that the expected value operation implies that for stochastic processes these basic statistics are calculated *across* the ensemble. For example, if we want to calculate the mean of a process, that is,

$$m_x(t) = E\{X(t, \xi_i) = x_i(t)\}$$

we simply take the values of $t = 0, 1, \dots$ and calculate the mean for each value of time across ($i = 1, 2, \dots$) the ensemble. So we see that dealing with stochastic processes is similar to dealing with random variables except that we must account for the (time) indices. Knowledge that a random signal is characterized by a stochastic process is not sufficient to make this information useful. We

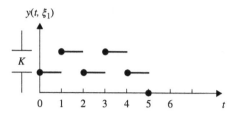

$y(t, \xi_1)$

Figure 2.3 Ensemble of random walk realizations.

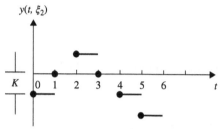

$y(t, \xi_2)$

$$y(5, \xi_1) = \sum_{t=0}^{4} x(t, \xi_1) = 0$$

$$y(5, \xi_2) = \sum_{t=0}^{4} x(t, \xi_2) = -2K$$

must consider properties of stochastic processes themselves that make them useful, but as before we must first develop the ideas from probability theory and then define them in terms of stochastic processes. Suppose we have more than one random variable, then we define the *joint* probability mass and distribution functions of an N-dimensional random variable as

$$p_X(x_1, \dots, x_N), \quad P_X(x_1, \dots, x_N)$$

All of the basic statistical definitions remain, except that we replace the single variate probability functions with the joint variate functions. Clearly, if we think of a stochastic process as a sequence of ordered random variables, then we are dealing with joint probability functions. Suppose we have two random variables, x_1 and x_2, and we know that the latter has already assumed a particular value, then we can define the *conditional* probability mass function of x_1 given that $X(\xi_2) = x_2$ has occurred by

$$p(x_1 \mid x_2) := p_X(X(\xi_1) \mid X(\xi_2) = x_2) \tag{2.8}$$

and it can be shown from basic probabilistic axioms (see Papoulis [1])that

$$p(x_1 \mid x_2) = \frac{p(x_1, x_2)}{p(x_2)} \tag{2.9}$$

Note also that this equation can also be written as

$$p(x_1, x_2) = p(x_2 \mid x_1)p(x_1) \tag{2.10}$$

Substituting this relation into above equation gives *Bayes' rule*, that is,

$$p(x_1 \mid x_2) = p(x_2 \mid x_1)\frac{p(x_1)}{p(x_2)} \tag{2.11}$$

If we use the definition of joint mass function, then we obtain the probabilistic *chain rule*:

$$p(x_1, \dots, x_N) = p(x_1 \mid x_2, \dots, x_N)p(x_2 \mid x_3, \dots x_N) \cdots p(x_{N-1} \mid x_N)p(x_N) \tag{2.12}$$

Along with these definitions follows the idea of conditional expectation, that is,

$$E\{x_i \mid x_j\} = \sum_i X_i p(x_i \mid x_j)$$

Note that the conditional expectation is also a *linear* operator that possesses many useful properties, that is,

$$E\{ax_i + b \mid x_j\} = aE\{x_i \mid x_j\} + b$$

With these definitions in mind, we can now examine some further properties of stochastic processes that will prove useful in characterizing random signals. A common class of useful stochastic processes is called Markov processes. Markov processes are defined in terms of a sequence of random variables $\{X(t)\}$, we say that a process is *Markov*, if and only if

$$p(X(t+1) \mid X(t), X(t-1), \dots, X(1)) = p(X(t+1) \mid X(t)) \tag{2.13}$$

that is, the future $X(t+1)$ depends only on the present $X(t)$ and not the past $\{X(t-1), \dots, X(1)\}$. Thus, if a process is Markov, then the chain rule simplifies considerably, that is,

$$p(x_1, \dots, x_N) = p(x_1 \mid x_2)p(x_2 \mid x_3) \cdots p(x_{N-1} \mid x_N)p(x_N)$$

Just as in systems theory, the concept of a time-invariant system is essential for analytical purposes, so is the equivalent concept of stationarity essential for the analysis of stochastic processes. We say that a stochastic process is *stationary*, if the joint mass function is time-invariant, that is,

$$p(X(1), \dots, X(N)) = p(X(1+k), \dots, X(N+k)) \quad \forall k \tag{2.14}$$

If this relation is only true for values of $t \leq M < N$, then it is said to be *stationary of order M*. Special cases arising are:

- For $M = 1$, $p(X(t)) = p(X(t+k)) = p(X) \,\forall k \Rightarrow m_x(t) = m_x$ a constant (*mean-stationary*).

- For $M = 2$, $p(X(t_1), X(t_2)) = p(X(t_1 + k), X(t_2 + k))$ or for $k = t_2 - t_1$, $p(X(t_2), X(t_2 + k))$ (*covariance stationary*) which implies

$$R_{xx}(t, t + k) = R_{xx}(k) = E\{x(t)x(t + k)\} - m_x^2$$

- A process that is *both* mean and covariance stationary is called *wide-sense stationary*.
- A process is said to be *ergodic* if its time average, E_T, is identical to its ensemble average, E, that is,

$$E\{y(t, \xi_i)\} \iff E_T\{y(t)\} = \lim_{N \to 0} \frac{1}{2N + 1} \sum_{t=-N}^{N} y(t)$$

- An ergodic process *must* be wide-sense stationary, but the *inverse* does *not* necessarily hold.

This means that *all* of the statistical information in the ensemble can be obtained from *one realization* of the process. For instance, in order to calculate the ensemble mean of this process, we must average across the ensemble for each instant of time to determine $m_x(t)$; however, if the process is ergodic, then we may select *any member* (realization) of the ensemble and calculate its time average, $E_T\{x_i(t)\}$, to obtain the required mean, that is,

$$m_x(t) = m_x$$

This discussion completes the introductory concepts of relating random signals to the probabilistic notion of stochastic processes. In the course of engineering applications, the ideas of probability spaces, etc. $x(t)$ is defined as a stochastic process, then it is assumed that the reader will know that [1–3]

$$x(t) \to X(t, \xi_i) \qquad t \in T, \quad \xi \epsilon \Xi$$

We now have all of the probabilistic ingredients to begin to characterize useful random signals. In the next section, we discuss the covariance and power spectrum "tools" and show how they can be used to specify the properties of random signals.

2.3 Spectral Representation of Random Signals

In many engineering problems before the design of processing algorithms can proceed, it is necessary to analyze the measured signal. As we have stated, analogous to the deterministic case, the covariance function transforms the random signal and the power spectrum (transforms) its Fourier transform for analysis as illustrated in Figure 2.1. In this section, we derive the relationship between the covariance function and power spectrum and show how they can be used to characterize fundamental random signals.[2]

First we begin by defining the power spectrum of a discrete random signal. Recall that the discrete-time Fourier transform (DtFT) pair is given by [4]

$$X(e^{j\Omega}) := \text{DtFT}[x(t)] = \sum_{t=-\infty}^{\infty} x(t)e^{-j\Omega t}$$

$$x(t) = \text{IDtFT}[X(e^{j\Omega})] = \frac{1}{2\pi}\int_{2\pi} X(e^{j\Omega})e^{j\Omega t}d\Omega \qquad (2.15)$$

If $x(t)$ is random, then $X(e^{j\Omega})$ is *also* random because

$$x(t,\xi_i) \Longleftrightarrow X(e^{j\Omega},\xi_i) \quad \forall i$$

both are simply realizations of a random signal over the ensemble generated by i. Also, and more important, $X(e^{j\Omega})$ for stationary processes almost never exists because any nonzero realization $x(t,\xi_i)$ is not absolutely summable in the ordinary sense (these integrals can be modified [see [2] for details]).

Now let us develop the relationship between the PSD and its corresponding covariance. We have

$$S_{xx}(e^{j\Omega}) = \lim_{N\to\infty} \frac{1}{2N+1}E\{X_N(e^{j\Omega})X_N^*(e^{j\Omega})\} \qquad (2.16)$$

Substituting for the DtFT, moving the expectation operation inside the summation and introducing a change of variable gives the relation

$$S_{xx}(e^{j\Omega}) = \sum_{k=-\infty}^{\infty} E_T\{R_{xx}(m+k,m)\}e^{-j\Omega k} \qquad (2.17)$$

If we further assume that the process is wide-sense stationary, then $R_{xx}(m+k,m) \to R_{xx}(k)$ is no longer a function of time, but lag k and, therefore,

$$S_{xx}(e^{j\Omega}) = \sum_{k=-\infty}^{\infty} R_{xx}(k)e^{-j\Omega k} \qquad (2.18)$$

evolves as the well-known *Wiener–Khintchine* relation with the corresponding covariance given by inverse discrete-time Fourier transform (IDtFT)

$$R_{xx}(k) = \frac{1}{2\pi}\int_{2\pi} S_{xx}(e^{j\Omega})e^{j\Omega k}\,d\Omega \qquad (2.19)$$

Recall that the DtFT is just the $\mathcal{Z}$-transform of $x(t)$ evaluated on the unit circle, that is,

$$S_{xx}(\Omega) = S_{xx}(z)|_{z=e^{j\Omega}}$$

---

2 We shall return to more mathematically precise notation for this development, that is, $X(\Omega) = X(e^{j\Omega})$.

and we obtain the equivalent pair in terms of the $\mathcal{Z}$-transform as

$$S_{xx}(z) = Z[R_{xx}(k)] = \sum_{k=-\infty}^{\infty} R_{xx}(k)z^{-k}$$

and

$$R_{xx}(k) = \frac{1}{2\pi j} \int S_{xx}(z)z^{k-1} \, dz \qquad (2.20)$$

Therefore, we can define random signals in terms of their covariances and spectral densities. For example, a purely random or *white noise* sequence, $e(t)$, is a sequence in which all the $e(t)$ are mutually independent, that is, knowing $e(t)$ in no way can be used to predict $e(t + 1)$. A white sequence is called completely random or unpredictable or memoryless (no correlation). Sequences of this type have historically been called white because of their analogy to white light, which possesses all frequencies (constant power spectrum), that is, the PSD of white noise is

$$S_{ee}(\Omega) = R_{ee} \quad \text{(constant)}$$

with the corresponding covariance is given by

$$R_{ee}(k) = R_{ee}\delta(k) \quad \text{(impulse)}$$

where R_{ee} is the variance of the noise.

In fact, the white noise characterization of *random* signals is analogous to the impulse representation of *deterministic* signals, that is,

$$R_{ee}\delta(k) \Longleftrightarrow A\delta(t)$$

and

$$S_{ee}(\Omega) = R_{ee} \Longleftrightarrow H(\Omega) = A$$

It is the random counterpart for random systems of the impulse excitation for the analysis of linear, time-invariant (LTI) systems.

Note also that random or white sequences are also Markov, since they are uncorrelated:

$$p(e(t) \mid e(t - 1), \dots, e(1)) = p(e(t))$$

If each of the $e(t)$ is also Gaussian-distributed, then as a check for *whiteness* of a sequence, we can perform a statistical hypothesis test (see [3]) using its underlying covariance. The 95% confidence interval or so-called *whiteness test* is given by

$$I = \left[R_{yy}(k) - \frac{1.96R_{yy}(0)}{\sqrt{N}}, R_{yy}(k) + \frac{1.96R_{yy}(0)}{\sqrt{N}} \right] \qquad (2.21)$$

Table 2.1 Properties of covariance and spectral functions.

	Covariance	Power spectrum
1.	*Average power* $R_{xx}(0) = E\{x^2(t)\}$	$R_{xx}(0) = \frac{1}{2\pi}\int_{2\pi} S_{xx}(z)z^{-1}\,dz$
2.	*Symmetry* $R_{xx}(k) = R_{xx}(-k)$ (even) $R_{xy}(k) = R_{yx}(-k)$	$S_{xx}(z) = S_{xx}(z^{-1})$ (even) $S_{xy}(z) = S_{yx}^*(z)$
3.	*Maximum* $R_{xx}(0) \geq \|R_{xx}(k)\|$ $\frac{1}{2}R_{xx}(0) + \frac{1}{2}R_{yy}(0) \geq \|R_{xy}(k)\|$ $R_{xx}(0)R_{yy}(0) \geq \|R_{xy}(k)\|^2$	$S_{xx}(e^{j\Omega}) \geq 0$
4.	*Real*	$S_{xx}(z) = E\{\|X(z)\|^2\}$ is real $S_{xy}(z) = E\{X(z)Y^*(z)\}$ is complex
5.	*Sum decomposition*	$S_{xx}(z) = S_{xx}^+(z) + S_{xx}^-(z) - R_{xx}(0)$

Some properties of the discrete covariance function for stationary processes along with accompanying properties of the PSD are given (without proof see [4] for details) in Table 2.1 for reference.

It is important to recognize that these properties are essential to analyze the information available in a discrete random signal. For instance, if we are trying to determine phase information about a particular measured signal, we immediately recognize that it is lost in both the autocovariance and corresponding output power spectrum.

With these properties in mind, we can now reexamine the expression for the PSD and decompose it further. From $\mathcal{Z}$-transform relations, we have

$$S(z) = \underbrace{\sum_{k=0}^{\infty} R(k)z^{-k}}_{S^+(z)} + \underbrace{\sum_{i=0}^{\infty} R(i)z^{i} - R(0)}_{S^-(z)}$$

which is the *sum decomposition* of the PSD given by

$$S(z) := S^+(z) + S^-(z) - R(0) \tag{2.22}$$

with the one-sided $\mathcal{Z}$-transforms, $S^+(z) = \mathcal{Z}_I\{R(k)\}$ and $S^-(z) = S^+(z^{-1})$. Note that $S^\pm(z)$ uses the $\pm$ to represent positive and negative time relative to inverse transforms. Also it can be shown that $S^+(z)$ has poles only *inside* the unit circle, while $S^-(z)$ has those only *outside* (see [5] or [6] for details).

Thus, the sum decomposition can be applied to calculate the power spectrum using one-sided transforms of the covariance function. The inverse process is more complicated, since the PSD must be decomposed into sums (usually by partial fractions): one having poles only inside the unit circle, the other poles only outside. The corresponding covariance is then determined using inverse $\mathcal{Z}$-transform methods.

In practice, the autocovariance and power spectrum find most application in the *analysis* of random signals yielding information about spectral content, periodicities, etc., while the cross-covariance and spectrum are used to estimate the properties of two distinct processes (e.g. input and output of a system) to follow.

So we see that the properties of covariance and spectra not only can be used to analyze the information available in random signal data, but also to estimate various signal characteristics. In the next section, we investigate the properties of linear systems excited by random inputs and use the properties of covariance and spectra to analyze the results.

2.4 Discrete Systems with Random Inputs

When random inputs are applied to linear systems, then covariance and power spectrum techniques must be applied transforming the signal and its Fourier spectrum in deterministic signal theory (see Figure 2.1). In this section, we develop the relationship between systems and random signals. From linear systems theory, we have the convolution or equivalent frequency relations

$$y(t) = h(t) * x(t) = \sum_{k=0}^{\infty} h(k)x(t - k) \tag{2.23}$$

or taking DtFTs, we obtain

$$Y(\Omega) = H(\Omega)X(\Omega) \tag{2.24}$$

If we assume that x is a random signal, then as we have seen in the previous section, we must resort to spectral representations of random processes. Exciting a causal linear system with a zero-mean random signal, we obtain the output power spectrum

$$S_{yy}(\Omega) = E\{Y(\Omega)Y^*(\Omega)\} = E\{(H(\Omega)X(\Omega))X^*(\Omega)H^*(\Omega)\} = |H(\Omega)|^2 S_{xx}(\Omega) \tag{2.25}$$

or equivalently

$$S_{yy}(\Omega) = E\{Y(\Omega)Y^*(\Omega)\} = E\{Y(\Omega)(X^*(\Omega)H^*(\Omega))\} = S_{yx}(\Omega)H^*(\Omega) \tag{2.26}$$

Table 2.2 Linear system with random inputs: covariance/spectrum relationships.

Covariance	Spectrum
$R_{yy}(k) = h(k) * h(-k) * R_{xx}(k)$	$S_{yy}(z) = H(z)H(z^{-1})S_{xx}(z)$
$R_{yy}(k) = h(k) * R_{xy}(k)$	$S_{yy}(z) = H(z)S_{xy}(z)$
$R_{yx}(k) = h(k) * R_{xx}(k)$	$S_{yx}(z) = H(z)S_{xx}(z)$
where	
$R_{yy}(k) = \frac{1}{2\pi j} \oint S_{yy}(z)z^{k-1}dz$	$S_{yy}(z) = \sum_{k=-\infty}^{\infty} R_{yy}(k)z^{-k}$
$R_{xy}(k) = \frac{1}{2\pi j} \oint S_{xy}(z)z^{k-1}dz$	$S_{xy}(z) = \sum_{k=-\infty}^{\infty} R_{xy}(k)z^{-k}$

Perhaps one of the most important properties that have led to a variety of impulse response identification techniques is

$$S_{yx}(\Omega) = E\{Y(\Omega)X^*(\Omega)\} = E\{H(\Omega)(X(\Omega)X^*(\Omega))\} = H(\Omega)S_{xx}(\Omega) \quad (2.27)$$

solving for H provides the *Wiener solution* in the frequency domain [7].

Similar results can be obtained for autocovariances and cross-covariances and corresponding spectra. We summarize these linear system relations in Table 2.2.

2.4.1 Spectral Theorems

We can now present an important result that is fundamental for modeling of random signals. Suppose we have an LTI system that is asymptotically stable with rational transfer function. If we excite this system with white noise, we get the so-called *spectral factorization theorem*, which states that

GIVEN a stable, rational, linear system excited by unit variance white noise, then there EXISTS a rational $H(\Omega)$ such that

$$S_{yy}(\Omega) = H(\Omega)H^*(\Omega) = |H(\Omega)|^2 \quad (2.28)$$

where the poles and zeros of $H(\Omega)$ lie within the unit circle.

If we can represent spectral densities in a properly factored form, then *all* stationary processes can be thought of as outputs of linear dynamical systems. Synthesizing random signals with given spectra then requires *only* the generation of white noise sequences. We summarize this discussion with the *representation theorem*[3]:

---

3 This is really a restricted variant of the famous Wold decomposition for stationary random signals (see [5] for details).

GIVEN a rational spectral density $S_{yy}(\Omega)$, then there EXISTS an asymptotically stable linear system when excited by white noise that produces a stationary output $y(t)$ with this spectrum.

The simulation of random signals with given statistics requires the construction of the power spectrum from the sum decomposition followed by a spectral factorization. This leads to the *spectrum simulation procedure*: (i) Calculate $S_{yy}(z)$ from the sum decomposition; (ii) perform the spectral factorization to obtain $H(z)$; (iii) generate a white noise sequence of variance R_{xx}; and (iv) excite the system $H(z)$ with the sequence [7].

The most difficult part of this procedure is to perform the spectral factorization. For simple systems, the factorization can be performed by equating coefficients of the known $S_{yy}(z)$, obtained in step 1, with the unknown coefficients of the spectral factor and solving the resulting nonlinear algebraic equations, that is,

$$S_{yy}(z) = \begin{cases} S_{yy}^+(z) + S_{yy}^-(z) - R_{yy}(0) = \dfrac{N_s(z,z^{-1})}{D_s(z,z^{-1})} \\[2mm] H(z)H(z^{-1})R_{xx} = \dfrac{N_H(z)}{D_H(z)}\dfrac{N_H(z^{-1})}{D_H(z^{-1})}R_{xx} \end{cases} \tag{2.29}$$

$$N_s(z,z^{-1}) = N_H(z)N_H(z^{-1}) \quad \text{and} \quad D_s(z,z^{-1}) = D_H(z)D_H(z^{-1})$$

For higher order systems, more efficient iterative techniques exist including multidimensional systems (see [8–10]). Next we consider a versatile model that can be applied to represent random signals.

2.4.2 ARMAX Modeling

We have just shown the relationship between a linear system and spectral shaping, that is, generating processes with specified statistics. To generate such a sequence, we choose to apply two models that are equivalent – the input/output (ARMAX) or state-space model (next chapter). Each model class has its own advantages: the input/output models are easy to use, while the state-space models are easily generalized.

The *input/output* or *transfer function* model is familiar to engineers and scientists because it is usually presented in the frequency domain with Laplace transforms. Similarly, in the discrete-time case, it is called the *pulse transfer function* model and is given by the $\mathcal{Z}$-transform

$$H(z) = \frac{B(z^{-1})}{A(z^{-1})} \tag{2.30}$$

where A and B are polynomials in z^{-1}.

$$A(z^{-1}) = 1 + a_1 z^{-1} + \cdots + a_{N_a} z^{-N_a} \tag{2.31}$$

$$B(z^{-1}) = b_0 + b_1 z^{-1} + \cdots + b_{N_b} z^{-N_b} \tag{2.32}$$

If we consider the equivalent time-domain representation, then we have a *difference equation* relating the output sequence $\{y(t)\}$ to the input sequence $\{x(t)\}$.

We define the *backward shift operator* q with the property that $q^{-k}y(t) :=$ $y(t-k)$.

$$A(q^{-1})y(t) = B(q^{-1})x(t) \tag{2.33}$$

or

$$y(t) + a_1 y(t-1) + \cdots + a_{N_a} y(t - N_a) = b_0 x(t) + \cdots b_{N_b} x(t - N_b) \tag{2.34}$$

When the system is excited by random inputs, the model is specified by the *autoregressive-moving average model with exogenous inputs* (ARMAX)[4]

$$\underbrace{A(q^{-1})y(t)}_{\text{AR}} = \underbrace{B(q^{-1})x(t)}_{\text{X}} + \underbrace{C(q^{-1})e(t)}_{\text{MA}} \tag{2.35}$$

where A, B, C, are polynomials, and $\{e(t)\}$ is a white noise source, and

$$C(q^{-1}) = 1 + c_1 q^{-1} + \cdots + c_{N_c} q^{-N_c}$$

The ARMAX model, usually abbreviated by ARMAX(N_a, N_b, N_c) represents the general form for popular time-series and digital filter models, that is,

- Pulse transfer function or Infinite Impulse Response (IIR) model: $C(\cdot) = 0$, or ARMAX($N_a, N_b, 0$), that is,

$$A(q^{-1})y(t) = B(q^{-1})x(t)$$

- Finite impulse response (FIR) or all-zero model: $A(\cdot) = 1, C(\cdot) = 0$, or ARMAX($1, N_b, 0$), that is,

$$y(t) = B(q^{-1})x(t)$$

- Autoregressive (AR) or all-pole model: $B(\cdot) = 0, C(\cdot) = 1$, or ARMAX ($N_a, 0, 1$), that is,

$$A(q^{-1})y(t) = e(t)$$

- Moving average (MA) or all-zero model: $A(\cdot) = 1, B(\cdot) = 0$, or ARMAX ($1, 0, N_c$), that is

$$y(t) = C(q^{-1})e(t)$$

---

4 The ARMAX model can also be interpreted in terms of the Wold decomposition of stationary times series, which states that a time series can be decomposed into a predictable or deterministic component ($x(t)$) and nondeterministic or random component ($e(t)$) [5].

- Autoregressive-moving average (ARMA) or pole–zero model: $B(\cdot) = 0$ or ARMAX($N_a, 0, N_c$), that is,

$$A(q^{-1})y(t) = C(q^{-1})e(t)$$

- Autoregressive model with exogeneous input (ARX) or pole–zero model: $C(\cdot) = 1$ or ARMAX($N_a, N_b, 1$), that is,

$$A(q^{-1})y(t) = B(q^{-1})x(t) + e(t)$$

ARMAX models can easily be used for signal processing purposes, since they are basically digital filters with known deterministic ($x(t)$) and random ($e(t)$) excitations. In fact, as we shall see subsequently, this model provides the basis for the parametric or modern approach to spectral estimation.

2.5 Spectral Estimation

Spectral estimation techniques have been developed and improved over the years with one major task in mind – the analysis of random data. Based on the previous discussion in this chapter, a majority of the initial effort was focused on applying the Wiener–Khintchine theorem and transform theory, while modern parametric techniques evolved primarily from the "speech" community [11]. In this section, we begin with the classical (nonparametric) methods that are viable when *long* data records are available and then evolve to the development of a select group of popular parametric techniques [4]. We make no attempt to provide detailed derivations of the algorithms that are available in other texts [4, 12, 13], but just follow a brief outline of the approach and present the final results. We conclude this chapter with a case study of noisy, bandpass filtered, sinusoidal signals.

2.5.1 Classical (Nonparametric) Spectral Estimation

With the initial application of Fourier analysis techniques to raw sun-spot data over 200 years ago, the seeds of spectral estimation were sown by Schuster [14]. Fourier analysis for random signals evolved rapidly after the discovery of the Wiener–Khintchine theorem relating the covariance and power spectrum. Finally with the evolution of the fast Fourier transform (see Cooley and Tukey [15]) and digital computers, all of the essential ingredients were present to establish the classical approach to nonparametric spectral estimation. Classical spectral estimators typically fall into two categories: direct and indirect. The direct methods operate directly on the raw data to transform it to the frequency domain and produce the estimate. Indirect methods first estimate the covariance sequence and then transform to the frequency domain – an application of the Wiener–Khintchine theorem. We develop two basic nonparametric spectral estimation techniques: the correlation method (indirect) and the periodogram method (direct).

2.5.1.1 Correlation Method (Blackman–Tukey)

The *correlation* *method* or sometimes called the Blackman–Tukey method is simply an implementation of the Wiener–Khintchine theorem: the covariance is obtained using a sample covariance estimator and then the PSD is estimated by calculating the DFT. The DFT transform pair is defined by

$$X(\Omega_m) := \mathrm{DFT}[x(t)] = \sum_{t=0}^{M-1} x(t)e^{-j\Omega_m t}$$

$$x(t) = \mathrm{IDFT}[X(\Omega_m)] = \frac{1}{M}\sum_{t=0}^{M-1} X(\Omega_m)e^{j\Omega_m t} \tag{2.36}$$

for $\Omega_m = \frac{2\pi}{M}m$ where it can be thought of as the DtFT with $\Omega \to \Omega_m$, that is, the DtFT sampled uniformly around the unit circle [4].

Therefore, we have that

$$\hat{S}_{xx}(\Omega_m) = \mathrm{DFT}[\hat{R}_{xx}(k)]$$
$$\hat{R}_{xx}(k) = \mathrm{IDFT}[\hat{S}_{xx}(\Omega_m)]$$

This technique tends to produce a noisy spectral estimate; however, a smoothed estimate can be obtained by multiplying R_{xx} by a *window function*, W usually called a *lag window*. The window primarily reduces spectral leakage and therefore improves the estimate. It is also interesting to note that a sample covariance estimator does not guarantee the positivity of the PSD (auto) when estimated directly from the Wiener–Khintchine theorem [4]. However, if the estimator is implemented directly in the Fourier domain, then it will preserve this property, since it is the *square* of the Fourier spectrum. We summarize the *correlation method* (Blackman–Tukey) of spectral estimation by[5]

- *Calculate* the DFT of $x(t)$, that is, $X(\Omega_m)$
- *Multiply* $X(\Omega_m)$ by its conjugate to obtain, $X(\Omega_m)X^*(\Omega_m)$
- *Estimate* the covariance from the IDFT, $\hat{R}_{xx}(k) = \mathrm{IDFT}[|X(\Omega_m)|^2]$
- *Multiply* the covariance by the lag window $W(k)$
- *Estimate* the PSD from the DFT of the windowed covariance, $\hat{S}_{xx}(\Omega_m) = \mathrm{DFT}[\hat{R}_{xx}(k)W(k)]$.

---

5 Note also if we replace X^* by Y^* we can estimate the cross-correlation $\hat{R}_{xy}(k)$ and the corresponding cross-spectrum $\hat{S}_{xy}(\Omega_m)$ using this method.

These correlation spectral estimates are statistically improved by using a lag or equivalently spectral window.[6] With practical window selection and long data records, the correlation method can be effectively utilized to estimate the PSD (see [4] for more details).

2.5.1.2 Average Periodogram Method (Welch)

Next we consider a more direct approach to estimate the PSD. We introduce the concept of a periodogram estimator with statistical properties equivalent to the correlation method, and then we show how to improve these estimates by statistical averaging and window smoothing leading to *Welch's method* of spectral estimation. The periodogram was devised by statisticians to detect periodicities in noisy data records [14]. The improved method of spectral estimation based on the so-called *periodogram* defined by

$$P_{xx}(\Omega_m) := \frac{1}{N}(X(\Omega_m) \ X^*(\Omega_m)) = \frac{1}{N}|X(\Omega_m)|^2$$

The samples of P_{xx} are uncorrelated, suggesting that one way of reducing the variance in P_{xx} is to *average* individual periodograms obtained by *sectioning* the original N point data record into K, L-point sections, that is,

$$\hat{S}_{xx}(\Omega_m) = \frac{1}{K} \sum_{i=1}^{K} \hat{P}_{xx}(\Omega_m, i) \tag{2.37}$$

where $\hat{P}_{xx}(\Omega_m, i)$ is the ith, L-point periodogram. If x is stationary, then it can be shown that this estimate is consistent, since the variance approaches zero as the number of sections become infinite [4]. For the periodogram estimator, we have

$$\text{var} \propto \frac{1}{K} \quad \text{and} \quad \text{bias} \propto \frac{K}{N}$$

So we see that for K large, the variance is inversely proportional to K, while the bias is directly proportional. Therefore, for a *fixed* record length N as the number of periodograms increases, variance decreases, but the bias increases. This is the basic trade-off between variance and resolution (bias), which can be used to determine a priori the required record length $N = LK$ for an acceptable variance. A *full* window, $W(t)$, can also be applied to obtain a smoothed spectral estimate.

Welch [16] introduced a modification of the original procedure. The data is sectioned into K records of length L; however, the window is applied directly

---

6 The window function is called a *lag window* in the time or lag domain $W(k)$ and a *spectral window* in the frequency domain $W(\Omega_m)$ with its maximum at the origin to match that property of the autocorrelation function; therefore, it is sometimes called a "half" window.

to the segmented records *before* periodogram computation. The modified periodograms are then

$$\hat{P}(\Omega_m, i) = \frac{1}{U}|DFT[x_i(t)W(t)]|^2 \quad i = 1, \ldots, K$$

where

$$U = \frac{1}{L}\sum_{t=0}^{L-1} W^2(t)$$

and

$$\hat{S}_{xx}(\Omega_m) = \frac{1}{K}\sum_{i=1}^{K} \hat{P}(\Omega_m, i)$$

We summarize the *average periodogram method* (Welch's procedure) by:

- *Section* the data, $\{x(t)\}$, $t = 1, \ldots, N$ into K sections each of length L, where $K = \frac{N}{L}$, that is,

$$x_i(t) = x(t + L(i-1)), \quad i = 1, \ldots, K, \quad t = 0, \ldots, L-1$$

- *Window* the data to obtain, $x_i(t) \times W_i(t)$
- *Estimate* K periodograms using the DFT as

$$\hat{P}(\Omega_m, i) = \frac{1}{U}|DFT[x_i(t)W(t)]|^2 \quad i = 1, \ldots, K$$

with $U = \frac{1}{L}\sum_{t=0}^{L-1} W^2(t)$
- *Estimate* the average spectrum using

$$\hat{S}_{xx}(\Omega_m) = \frac{1}{K}\sum_{i=1}^{K} \hat{P}(\Omega_m, i)$$

with $\text{var}\{\hat{S}_{xx}(\Omega_m)\} \propto \frac{1}{K}$ and $\text{bias}\{\hat{S}_{xx}(\Omega_m)\} \propto \frac{K}{N}$ adjusted for particular windows.

This completes the discussion of the classical or nonparametric methods of spectral estimation. Even though they are considered classical techniques with limited resolution capability, they still can provide us with reasonable information, if we have "long" data records with good signal levels. Next we consider the parametric methods of spectral estimation.

2.5.2 Modern (Parametric) Spectral Estimation

The parametric approach to spectral estimation is based on the underlying assumption that the measured data under investigation evolved from a process that can be represented by the selected model. The *parametric spectral estimation* approach is a three-step procedure, that is, (i) select the model set;

(ii) estimate the model parameters from the data or correlation lags; and (iii) obtain the spectral estimate by using the (estimated) model parameters. The major advantage of these parametric methods is that higher frequency resolution is achievable, since the data are not windowed [17]. It can be shown that these methods imply an infinite extension of the autocorrelation sequence [18]. We also note that this approach is indirect, since we must first find the appropriate model and then use it to estimate the PSD. Let us recall the basic spectral relations developed previously for linear systems with random inputs. The measurement of the output spectrum of a process is related to the input spectrum by the factorization relations:

$$S_{yy}(z) = H(z)H^*(z)S_{xx}(z) = |H(z)|^2 S_{xx}(z) \tag{2.38}$$

where

x is the input process

H is the linear system transfer function

y is the measurement.

Since we are taking the signal processing viewpoint, then we will assume that *only* the measured data, y, are available and not the excitation, x. In fact, if both x and y are available, then the problem is called the system identification problem (see Ljung and Soderstrom [19] for details) and the ARMAX model will be used. However, in this section, we restrict ourselves to purely signal representations and select the AR and ARMA model sets as representative for spectral estimation.

In summary, the modern method of *parametric spectral estimation* is given by:

- *Select* a representative model set (AR, ARMA).
- *Estimate* the model parameters from the data, that is given $\{y(t)\}$, find

$$\hat{\Theta} = \{\hat{A}(q^{-1}), \hat{C}(q^{-1})\} \quad \text{and} \quad \hat{R}_{ee}.$$

- *Estimate* the PSD using these parameters, that is,

$$S_{yy}(z) \approx \hat{S}_{yy}(z, \hat{\Theta}) = \hat{H}(z)\hat{H}^*(z)S_{ee}(z)$$

2.5.2.1 Autoregressive (All-Pole) Spectral Estimation

The AR or all-pole model is characterized by the difference equation

$$A(q^{-1})y(t) = \sqrt{R_{ee}}e(t) \tag{2.39}$$

where

y is the measured data

e is a zero-mean, white noise sequence with variance, R_{ee} and

A is an N_ath-order polynomial in backward shift operator q^{-1}.

Taking $\mathcal{Z}$-transforms, we have

$$H_{\text{AR}}(z) = \frac{Y(z)}{E(z)} = \frac{\sqrt{R_{ee}}}{A(z)} \tag{2.40}$$

where

$$A(z) = 1 + a_1 z^{-1} + \cdots + a_{N_a} z^{-N_a}$$

If we substitute H_{AR} for H, we obtain the AR power spectrum as

$$S_{yy}(z) = H_{\text{AR}}(z)H_{\text{AR}}^*(z)S_{ee}(z) = |H_{\text{AR}}(z)|^2 S_{ee}(z) \tag{2.41}$$

or

$$S_{yy}(z) = \left(\frac{1}{A(z)}\right)\left(\frac{1}{A(z^{-1})}\right)R_{ee}\Delta T = \frac{R_{ee}\Delta T}{|A(z)|^2}$$

which is the desired representation of the all-pole spectral density. Here recall that ΔT is the associated *sampling interval*. Note also that the spectral estimation procedure requires that we first estimate the AR model parameters, $(\{\hat{a}_i\}, \hat{R}_{ee})$ from the data and then form the estimate[7]

$$S_{\text{AR}}(\Omega) := \hat{S}_{yy}(\Omega) = \hat{S}_{yy}(z)|_{z=e^{j\Omega}} = \frac{\hat{R}_{ee}\Delta T}{|\hat{A}(e^{j\Omega})|^2} \tag{2.42}$$

The basic parameter estimation problem for the AR model, in the general (infinite covariance) case, is given as the minimum (error) variance solution to

$$\min_{\underline{a}} \quad J(t) = E\{e^2(t)\} \tag{2.43}$$

where the estimation error is defined by

$$e(t) := y(t) - \hat{y}(t)$$

and $\hat{y}$ is the minimum variance estimate obtained from the AR model

$$y(t) = -\sum_{i=1}^{N_a} a_i y(t-i) + e(t) \tag{2.44}$$

as

$$\hat{y}(t) = -\sum_{i=1}^{N_a} \hat{a}_i y(t-i) \tag{2.45}$$

Note that $\hat{y}$ is actually the one-step predicted estimate $\hat{y}(t|t-1)$ based on $[t-1]$ past data samples, hence, the popular name "linear (combination of data) predictor." The estimator can easily be derived by minimizing the

---

7 A popular method of "simulating" the spectrum at equally spaced frequencies is to zero-pad the coefficients and calculate the DFT of $A(e^{j\Omega})$, then multiply by its conjugate and divide by $\hat{R}_{ee}$.

so-called *prediction error* leading to the *orthogonality condition* and the well-known *normal or Yule–Walker* equations as [11, 13]

$$\sum_{i=1}^{N_a} a_i R_{yy}(i-j) = -R_{yy}(j) \quad \text{for} \quad j = 1, \ldots, N_a \tag{2.46}$$

Expanding these equations over j and solving, we obtain

$$\hat{\underline{a}}(N_a) = -\mathbf{R}_{yy}^{-1} \, \underline{R}_{yy}(N_a) \tag{2.47}$$

where $\mathbf{R}_{yy}$ is an $N_a \times N_a$ Toeplitz matrix, that is,

$$\begin{bmatrix} R_{yy}(0) & R_{yy}(1) & \cdots & R_{yy}(N_a-1) \\ \vdots & \vdots & \ddots & \vdots \\ R_{yy}(N_a-2) & R_{yy}(N_a-3) & \cdots & R_{yy}(1) \\ R_{yy}(N_a-1) & R_{yy}(N_a-2) & \cdots & R_{yy}(0) \end{bmatrix} \begin{bmatrix} a_1 \\ \vdots \\ a_{N_a-1} \\ a_{N_a} \end{bmatrix} = - \begin{bmatrix} R_{yy}(1) \\ \vdots \\ R_{yy}(N_a-1) \\ R_{yy}(N_a) \end{bmatrix}$$

we have

$$R_{ee} = J(t) = R_{yy}(0) + \sum_{j=1}^{N_a} a_j R_{yy}(j) \tag{2.48}$$

that gives the *prediction error variance*. Various methods of solution evolve based on the cost function selected and its associated limits [11].

We also note in passing that the main "knob" in AR spectral estimation is the selection of the *order* of the model. There are a number of criteria for model order selection [20], but there are still various opinions on which criterion is the most reasonable. If the order selected is too low, there will *not* be enough poles to adequately represent the underlying spectrum. Too high a choice of order will usually result in *spurious peaks* in the estimated spectrum. The most popular criteria for order selection were introduced by Akaike [21] and Rissanen [22]. The first criterion is the final prediction error (FPE) developed so that the selected order of the AR process minimizes the average error in the one-step prediction error. The FPE of order N_a is given by

$$\text{FPE}(N_a) = \left(\frac{N+1+N_a}{N-1-N_a} \right) \hat{R}_{ee} \tag{2.49}$$

where N is the number of data samples, N_a is the predictor order, and R_{ee} is the prediction error variance for the N_a order predictor. Note that as $N_a \to N$, then FPE increases reflecting the increase in the uncertainty in the estimate of R_{ee} or prediction error power. The *order selected* is the one for which FPE is *minimum*. Akaike suggested another criterion using a maximum likelihood approach to derive the so-called Akaike information criterion (AIC). The AIC determines the model order by minimizing an information theoretic function [7]. Assuming the process has Gaussian statistics, the AIC is

$$\text{AIC}(N_a) = -\ln \hat{R}_{ee} + 2 \left(\frac{N_a}{N} \right) \tag{2.50}$$

The first term is the prediction error variance, while the second consists of two parts, the $2/N$ is an additive constant accounting for the removal of the sample mean, while N_a is the penalty for the use of extra AR coefficients that do not result in a substantial reduction in prediction error power. The order selected is the one that minimizes Akaike information criterion estimate, that is, MAICE. As $N \to \infty$, AIC $\to$ FPE.

Another popular metric is the minimum description length (MDL) developed by Rissanen [22]. It is similar to the AIC, but theoretically has improved properties and is given by

$$\text{MDL}(N_a) = -\ln \hat{R}_{ee} + \left(\frac{N_a}{2} \ln N \right) \tag{2.51}$$

This completes the development of AR parameter estimation for spectral estimation. It should be noted that spectral estimation techniques utilizing an all-pole or AR model can be considered the class of *maximum entropy* MEM spectral estimators [18, 20].

2.5.2.2 Autoregressive Moving Average Spectral Estimation

The autoregressive moving average (ARMA) or *pole–zero* model is characterized by the difference equation

$$A(q^{-1})y(t) = C(q^{-1})e(t) \tag{2.52}$$

where

y is the measured data

e is a zero-mean, white noise sequence with variance, R_{ee}

A, C are N_ath and N_cth-order polynomials in the backward shift operator q^{-1}.

Taking $\mathcal{Z}$-transforms, we have

$$H_{\text{ARMA}}(z) = \frac{Y(z)}{E(z)} = \frac{C(z)}{A(z)} \tag{2.53}$$

where

$$C(z) = 1 + c_1 z^{-1} + \cdots + c_{N_c} z^{-N_c}$$
$$A(z) = 1 + a_1 z^{-1} + \cdots + a_{N_a} z^{-N_a}$$

Substituting for H above, we obtain

$$S_{yy}(z) = H_{\text{ARMA}}(z)H^*_{\text{ARMA}}(z)S_{ee}(z) = |H_{\text{ARMA}}(z)|^2 S_{ee}(z) \tag{2.54}$$

or

$$S_{yy}(z) = \left(\frac{C(z)}{A(z)} \right) \left(\frac{C(z^{-1})}{A(z^{-1})} \right) R_{ee} \Delta T = \left| \frac{C(z)}{A(z)} \right|^2 R_{ee} \Delta T \tag{2.55}$$

which is the desired representation of the rational (pole–zero) spectral density with the associated sampling interval ΔT. The parametric spectral estimation procedure requires that we obtain the ARMA parameter estimates, $(\{\hat{a}_i\}, \{\hat{c}_i\}, \hat{R}_{ee})$ from the data and then from the estimated power spectrum

$$S_{\text{ARMA}}(\Omega) := \hat{S}_{yy}(\Omega) = \left| \frac{\hat{C}(e^{j\Omega})}{\hat{A}(e^{j\Omega})} \right|^2 \hat{R}_{ee} \Delta T \tag{2.56}$$

The ARMA parameter estimation problem is nonlinear due to the presence of the MA polynomial. Many methods have been developed to estimate the parameters and most are based on various optimization schemes [19, 20, 23].

The "ideal" optimal solution to this problem for the ARMA model is based on minimizing the mean-squared error criterion and obtaining the orthogonality conditions as before

$$\min_{\Theta} \quad J = E\{e^2(t)\}$$

which leads to the following matrix equations:

$$\underbrace{\begin{bmatrix} R_{yy} & | & -R_{ey} \\ - - - & - - & - - - \\ -R_{ye} & | & R_{ee} \end{bmatrix}}_{R} \underbrace{\begin{bmatrix} \underline{a} \\ -- \\ \underline{c} \end{bmatrix}}_{\Theta} = \underbrace{\begin{bmatrix} -\underline{R}_{yy} \\ - - - \\ \underline{R}_{ye} \end{bmatrix}}_{R}$$

with solution

$$\hat{\Theta} = \mathbf{R}^{-1}\underline{R} \tag{2.57}$$

Of course, this relation only holds if we have the covariances; however, since we do not have $\{e(t)\}$ we cannot calculate the required covariances. A variety of methods have evolved to solve this problem, for example, the extended least-squares (ELS) methods (see [4] or [7] for details).

Note that when spectra have sharp lines as well as deep notches, then both poles (AR) and zeros (MA) are required to adequately describe the spectra owing to the importance of ARMA methods. Next we consider some entirely different approaches to the spectral estimation problem.

2.5.2.3 Minimum Variance Distortionless Response (MVDR) Spectral Estimation

The minimum variance distortionless response (MVDR) spectral estimation method relies on estimating the PSD by measuring the power from a set of narrowband filters as introduced in [24] for seismic arrays. It is similar to the classical correlation/periodogram techniques except that the narrowband filters of MVDR are different for each frequency, while those of the classical techniques are fixed. In this way, the narrowband filters of the MVDR approach can be thought of as adapting to the process for which PSD is desired. It is therefore considered a "data adaptive" method. The idea behind the MVDR spectral estimator is that for a given discrete frequency, Ω_m the signal to be extracted is characterized by a complex exponential (narrowband), that is,

$$x(t) = Ae^{j\Omega_m t} \tag{2.58}$$

and

$$X(\Omega) = A\delta(\Omega - \Omega_m)$$

The measured output signal is assumed contaminated with additive white noise

$$y(t) = x(t) + n(t) \tag{2.59}$$

An estimator is to be designed so that the output variance is minimized, and the frequency under consideration, Ω_m is passed undistorted. We assume that the estimator is of the form

$$\hat{y}(t) = h(t) * y(t) = \sum_{k=0}^{N} h(k)y(t-k) \tag{2.60}$$

Since the estimator is designed to pass the narrowband signal $x(t)$, we would like to pass $Ae^{j\Omega_m t}$ with unity gain or

$$H(\Omega)|_{\Omega=\Omega_m} = 1$$

In order to achieve this, we also require the estimate to be *unbiased*

$$E\{\hat{y}(t)\} = x(t) = Ae^{j\Omega_m t} = \sum_{k=0}^{N} h(k)E\{y(t-k)\}$$

but from this relation, we have that

$$Ae^{j\Omega_m t} = \sum_{k=0}^{N} h(k)E\{x(t-k) + n(t-k)\}$$

or

$$Ae^{j\Omega_m t} = \sum_{k=0}^{N} h(k)(Ae^{j\Omega_m(t-k)}) \tag{2.61}$$

Dividing, we obtain the desired result

$$1 = \sum_{k=0}^{N} h(k)e^{-jk\Omega_m} = H(\Omega)|_{\Omega=\Omega_m} \tag{2.62}$$

which implies that the estimator will pass the complex narrowband signal with unity gain when $\Omega = \Omega_m$. We can express this in vector notation by expanding the sum as

$$\begin{bmatrix} 1 & e^{-j\Omega_m} & \cdots & e^{-jN\Omega_m} \end{bmatrix} \underbrace{\begin{bmatrix} h(0) \\ h(1) \\ \vdots \\ h(N) \end{bmatrix}}_{\underline{h}} = \underline{V}'(\Omega_m)\underline{h} = 1 \tag{2.63}$$

The MVDR estimator is obtained by minimizing the variance

$$\min_{\underline{h}} J = \underline{h}' \mathbf{R}_{yy} \underline{h} \tag{2.64}$$

subject to the constraint that $\underline{V}'(\Omega_m)\underline{h} = 1$. This estimator can be derived using the Lagrange multiplier method of transforming the constrained problem to an unconstrained by augmenting the Lagrange multiplier λ into the cost function and minimizing [25]

$$\min_{\underline{h}} J = \underline{h}' \mathbf{R}_{yy} \underline{h} + \lambda(1 - \underline{V}'(\Omega_m)\underline{h}) \tag{2.65}$$

Performing the minimization and substituting for the constraint yields the filter weights as

$$\underline{h}_{\text{MVDR}} = \frac{\mathbf{R}_{yy}^{-1} \underline{V}(\Omega_m)}{\underline{V}'(\Omega_m)\mathbf{R}_{yy}^{-1} \underline{V}(\Omega_m)} \tag{2.66}$$

Substituting this expression into J gives the *MVDR spectral estimator* at Ω_m as

$$S_{\text{MVDR}}(\Omega_m) = \frac{\Delta T}{\underline{V}'(\Omega_m)\mathbf{R}_{yy}^{-1} \underline{V}(\Omega_m)} = \Delta T \underline{h}' \mathbf{R}_{yy} \underline{h} \tag{2.67}$$

which implies that the spectral estimate at a given frequency is obtained by scaling the output covariance appropriately at the desired frequency. In practice, MVDR exhibits more resolution than the classical correlation/periodogram estimators, but less than the AR (MEM) estimator. We also note that the peak of MVDR spectral estimate for a narrowband process is proportional to the power, while for the MEM it is proportional to the square of the power. In fact, it can be shown that the MVDR and MEM spectra are related by *Burg's formula*: [18]

$$\frac{1}{S_{\text{MVDR}}(\Omega)} = \frac{1}{N_a} \sum_{i=1}^{N_a} \frac{1}{S_{\text{AR}}(\Omega, i)} \tag{2.68}$$

where $S_{\text{AR}}(\Omega, i)$ is the MEM PSD corresponding to the ith-order AR model. Thus, we see that the MVDR spectral estimate *averages all of the lower order (less resolution) and higher order AR models* to obtain its estimate, thus explaining its lower resolution. In fact, a suggested method of implementing this technique is to use an AR spectral estimator and then perform this averaging.

This completes the discussion of the MVDR PSD estimation technique. In the next section, we discuss methods designed primarily to extract sinusoidal signals from noise.

2.5.2.4 Multiple Signal Classification (MUSIC) Spectral Estimation

In this section, we investigate a solution to the harmonics (sinusoids) in noise parameter estimation problem, which is prevalent in a variety of applications especially in spatial processing for direction-of-arrival and target localization problems [26, 27].

A *complex harmonic model* is defined by

$$s(t) = \sum_{i=1}^{N_s} A_i e^{j\Omega_i t + \phi_i} \tag{2.69}$$

where A_i is complex, $A_i = |A_i| e^{-j\theta}$; ϕ_i is random and uniformly distributed as $\phi \sim \mathcal{U}(-\pi, \pi)$ and Ω_i is the harmonic frequency. The set of N_s-complex amplitudes and harmonic frequencies, $[\{A_i\}, \{\Omega_i\}]$, $i = 1, \ldots, N_s$ is assumed deterministic, but unknown.

The corresponding *measurement model* is characterized by the complex harmonic signal in additive random noise

$$y(t) = s(t) + n(t) \tag{2.70}$$

with $n(t)$, a zero-mean random sequence with variance, σ_n^2. Therefore, its autocorrelation function is given by

$$R_{yy}(k) = R_{ss}(k) + R_{nn}(k) = \sum_{i=1}^{N_s} P_i e^{j\Omega_i k} + \sigma_n^2 \quad \text{for } P_i := |A_i|^2 \tag{2.71}$$

since the signal is uncorrelated with the noise. Expanding this relation to incorporate the N_s-harmonics, we obtain

$$R_{yy}(k) = \begin{cases} \sum_{i=1}^{N_s} P_i + \sigma_n^2 \delta(k) & k = 0 \\ \sum_{i=1}^{N_s} P_i e^{j\Omega_i k} & k \neq 0 \end{cases} \tag{2.72}$$

There are a number of approaches in which this problem can be attacked such as maximum likelihood parameter estimation, Prony's technique, linear prediction, but here we choose to use the *power method* [20, 26], that is, the cost function is based on maximizing the output power in the signal and the resulting function is related to its power spectrum. All peaks of the spectrum locate the associated harmonic frequencies of the model.

Now performing the expansion of Eq. (2.72) over k, we obtain

$$R_{yy}(N) := R_{ss} + R_{nn} = V(\Omega)PV^H(\Omega) + \sigma_n^2 I \tag{2.73}$$

where $V(\Omega) \in C^{N \times N_s}$ is the harmonic signal matrix with corresponding power matrix, $P \in C^{N_s \times N_s}$ and $P = \text{diag}[P_1 \cdots P_{N_s}]$. The signal matrix constructed from the harmonic model is

$$V(\Omega) = [\mathbf{v}_1(\Omega) \mid \mathbf{v}_2(\Omega) \mid \cdots \mid \mathbf{v}_{N_s}(\Omega)]$$
$$\text{with } \mathbf{v}_i(\Omega) := [1 \ e^{j\Omega_i} \cdots e^{j(N-1)\Omega_i}]' \tag{2.74}$$

Therefore, the output correlation matrix can be decomposed into the constituent eigenvalue (Λ)-eigenvector (E) system as

$$R_{yy}(N) = E \ \Lambda \ E^H \tag{2.75}$$

or

$$R_{yy}(N) = E \begin{bmatrix} \lambda_1 & & 0 \\ & \ddots & \\ 0 & & \lambda_N \end{bmatrix} E^H = V(\Omega) \begin{bmatrix} \lambda_1^s & & 0 \\ & \ddots & \\ & & \lambda_{N_s}^s & \\ 0 & & & 0 \end{bmatrix} V^H(\Omega) + \sigma_n^2 I$$
$$\tag{2.76}$$

which follows from the fact that there are N_s-harmonics. Therefore, we have that $\lambda_1 > \lambda_2 > \cdots > \lambda_N$ and

$$\lambda_i = \begin{cases} \lambda_i^s + \sigma_n^2 & i = 1, \dots, N_s \\ \sigma_n^2 & i = N_s + 1, \dots, N \end{cases} \tag{2.77}$$

The output correlation can also be written in partitioned form as

$$R_{yy}(N) = [E(N_s) \mid E(N - N_s)] \begin{bmatrix} \Lambda(N_s) & \mid & 0 \\ --- & \mid & --- \\ 0 & \mid & \Lambda(N - N_s) \end{bmatrix} \begin{bmatrix} E^H(N_s) \\ --- \\ E^H(N - N_s) \end{bmatrix}$$
$$\tag{2.78}$$

where both $E(N_s) \in C^{N \times N_s}$ and $E(N - N_s) \in C^{N \times (N - N_s)}$ are eigenvector matrices. Multiplying out the partitions gives the relation

$$R_{yy}(N) = E(N_s)\Lambda(N_s)E^H(N_s) + E(N - N_s)\Lambda(N - N_s)E^H(N - N_s) \tag{2.79}$$

or

$$R_{yy}(N) = [\mathbf{e}_1 \cdots \mathbf{e}_{N_s}]] \begin{bmatrix} \lambda_1^s + \sigma_n^2 & & 0 \\ & \ddots & \\ 0 & & \lambda_{N_s}^s + \sigma_n^2 \end{bmatrix} \begin{bmatrix} \mathbf{e}_1^H \\ \vdots \\ \mathbf{e}_{N_s}^H \end{bmatrix}$$
$$+ [\mathbf{e}_{N_s+1} \cdots \mathbf{e}_N] \begin{bmatrix} \sigma_n^2 & & 0 \\ & \ddots & \\ 0 & & \sigma_n^2 \end{bmatrix} \begin{bmatrix} \mathbf{e}_{N_s+1}^H \\ \vdots \\ \mathbf{e}_N^H \end{bmatrix} \tag{2.80}$$

which implies the decomposition

$$R_{yy}(N) = \sum_{i=1}^{N_s} (\lambda_i^s + \sigma_n^2) \mathbf{e}_i \mathbf{e}_i^H + \sum_{i=N_s+1}^{N} \sigma_n^2 \mathbf{e}_i \mathbf{e}_i^H \qquad (2.81)$$

From this *spectral decomposition*, it is possible to define various subspaces as

1) The *signal subspace* is defined as the N_s-dimensional subspace spanned by the *signal eigenvectors*, $\{\mathbf{e}_i\}$; $i = 1, \dots, N_s$, corresponding to the N_s-largest eigenvalues.
2) The *noise subspace* is defined as the $(N - N_s)$-dimensional subspace spanned by the *noise eigenvectors*, $\{\mathbf{e}_i\}$; $i = N_s + 1, \dots, N$, corresponding to the remaining $(N - N_s)$-eigenvalues.
3) The *signal* and *noise subspaces* are *orthogonal*.

Applying this decomposition to the harmonic estimation problem, we see that the *harmonic signal vectors*, $\mathbf{v}_i(\Omega)$ lie in the signal subspace spanned by the signal eigenvectors, $\{\mathbf{e}_i\}$; $i = 1, \dots, N_s$, and the corresponding *noise vectors* lie in the orthogonal noise subspace spanned by $\{\mathbf{e}_i\}$; $i = N_s + 1, \dots, N$. Thus, the harmonic signal vectors lie in the subspace spanned by the first N_s-eigenvectors of $R_{yy}(N)$ and are *orthogonal* to each noise eigenvector, that is,

$$\mathbf{v}_i^H(\Omega)\mathbf{e}_j = 0 \quad \text{for } i = 1, \dots, N_s; j = N_s + 1, \dots, N \qquad (2.82)$$

which also implies that

$$V^H(\Omega)\mathbf{e}_j = 0 \text{ for } j = N_s + 1, \dots, N \quad \text{and} \quad V(\Omega) = [\mathbf{v}_1(\Omega) \mid \cdots \mathbf{v}_{N_s}(\Omega)] \qquad (2.83)$$

As an illustration of the orthogonality property of the signal and noise subspaces, consider the case of two harmonic signal vectors ($N_s = 2$) and one noise vector ($N = 3$), then the signal vectors, $\{\mathbf{v}_1(\Omega), \mathbf{v}_2(\Omega)\}$, lie in the subspace spanned by the two signal eigenvectors, $\{\mathbf{e}_1, \mathbf{e}_2\}$. The noise eigenvector is orthogonal to this space as shown in Figure 2.4.

The harmonic frequency *power estimator* [26] follows directly from the orthogonality property of Eq. (2.82) as

$$P(\Omega) := \frac{1}{\sum_{j=N_s+1}^{N} |\mathbf{v}^H(\Omega)\mathbf{e}_j|^2} \quad \text{for } P(\Omega) \to \infty$$
$$\text{whenever } \Omega = \Omega_i, \quad i = 1, \dots, N_s \qquad (2.84)$$

where we have defined the *frequency vector*, $\mathbf{v}(\Omega)$, to represent the corresponding temporal Fourier frequencies.[8] So we see that if we calculate the power

---

8 We use this frequency vector convention throughout. Note also that it differs from the harmonic signal vector in that it spans over all Ω not just the one corresponding to the estimated harmonic signal.

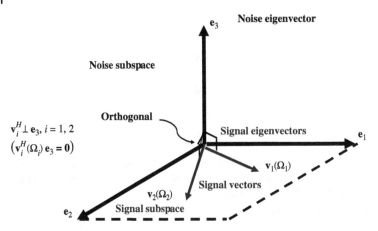

Figure 2.4 Signal and noise subspaces with two harmonic signal vectors and one noise eigenvector.

estimator over a set of frequencies, $\{\Omega\}$, then whenever this estimator passes through a harmonic frequency, $\Omega = \Omega_i$, a *peak* will occur in the power function. Simple peak detection then enables the estimation of the desired harmonic frequency. Another approach to estimate these frequencies is obtained by defining

$$E(\Omega) := \mathbf{v}^H(\Omega)\mathbf{e}(N) = [1 \ e^{-j\Omega} \ \cdots \ e^{-j(N-1)\Omega}]\begin{bmatrix} e(0) \\ e(1) \\ \cdots \\ e(N-1) \end{bmatrix} = \sum_{t=0}^{N-1} e(t)e^{-j\Omega t}$$

(2.85)

which is simply the DtFT of the eigen-filter, $\{e(t)\}$. Using the $\mathcal{Z}$-transform relations, we see that the harmonic frequencies correspond to the *roots* of

$$E(z) = \sum_{t=0}^{N-1} e(t)z^{-t} = \prod_{k=1}^{N_s}(1 - e^{j\Omega_k}z^{-1})$$

(2.86)

This approach to harmonic parameter estimation has evolved from the initial work of Pisarenko [26] and Schmidt [27] to the more modern approaches of subspace estimation [28–30]. We briefly mention the particular methods and refer the reader to the detailed derivations and performance properties of the referenced papers. However, it should be noted in passing that the derivations typically follow the eigen-decomposition approach we have outlined in this section.

The **MU**ltiple **SI**gnal **C**lassification (MUSIC) method was developed by Schmidt [27] as an extension of the original Pisarenko harmonic estimation technique. It evolved from the spatial (array) signal processing area. It follows directly from the decomposition of the measurement covariance matrix with the steps summarized as follows:

1) The eigen-decomposition of the autocorrelation matrix is performed, $R = E \Lambda E^H$.
2) The number of harmonics, N_s, are determined using the *order estimation* or *best rank approximation* methods [7] with the noise subspace dimensioned as $N - N_s$.
3) The *signal subspace* is defined by the set of largest eigenvalues and corresponding eigenvectors, $[\{\lambda_i\}, \{e_i\}]$, $i = 1, \ldots, N_s$.
4) The *noise subspace* is defined by the set of the remaining (smallest) eigenvalues and corresponding eigenvectors, $[\{\lambda_i\}, \{e_i\}]$, $i = N_s + 1, \ldots, N$.
5) The *power estimator* of Eq. (2.87) (to follow) is based on averaging over all of the *noise eigenvectors* to reduce the effects of spurious peaks, $|v_i^H(\Omega)e_j|^2 \rightarrow \sum_{j=N_s+1}^{N} |v^H(\Omega)e_j|^2$.
6) The power is found by solving the linear least-squares problem of Eq. (2.88).

Using the noise averaging, we have that the MUSIC power estimator is

$$P_{\text{MUSIC}}(\Omega) := \frac{1}{v^H(\Omega)[E(N - N_s)E^H(N - N_s)]v(\Omega)} = \frac{1}{\sum_{j=N_s+1}^{N} |v^H(\Omega)e_j|^2}$$

$$(2.87)$$

Note also that root-MUSIC avoids calculating the power estimator and just "roots" the resulting polynomial as in Eq. (2.73).

The power is found by solving Eq. (2.73) using least-squares for P as

$$\hat{P} = [V^H(\Omega)V(\Omega)]^{-1} V^H(\Omega)(R_{yy} - \sigma_N^2 I)V(\Omega)][V^H(\Omega)V(\Omega)]^{-H} \qquad (2.88)$$

where H is the Hermitian conjugate operation.

It should also be noted that the Eigenvector (EV) method in [31] is a weighted version of MUSIC, that is, the technique is identical except that instead of weighting by the identity as in MUSIC, the inverse of the noise eigenvalues is used. The result of this weighting is to reduce the effects of spurious peaks or roots even further.

2.6 Case Study: Spectral Estimation of Bandpass Sinusoids

Sinusoidal signals occur quite frequently in applications like radar and sonar primarily for target localization and tracking. In this section, we apply the spectral estimation techniques to noisy data generated by filtering a suite of sinusoidal signals (35, 50, 56, and 90 Hz) through a bandpass filter (30–70 Hz) in order to provide a test for each of the methods.

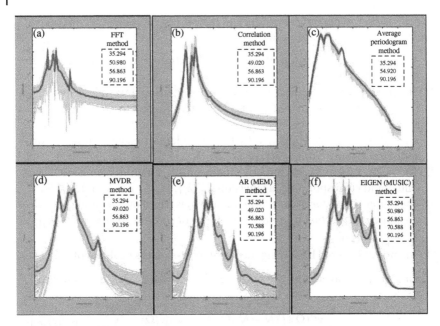

Figure 2.5 Bandpass sinusoidal problem (35,54,56,70,90 Hz) spectral estimation: (a) Discrete (fast Fourier) transform. (b) Correlation method (Blackman–Tukey). (c) Average Periodogram (Welch). (d) Minimum variance distortionless response (MVDR). (e) Autoregressive (maximum entropy). (f) Multiple signal classification (MUSIC).

Suppose we generate a 256-point test signal sampled at 0.001 sec by filtering sinusoids at 35, 50, 56, and 90 Hz contaminated with unit variance white noise (SNR = −20 dB)[9] through a bandpass Butterworth (third-order) filter with cutoff frequencies at 30 and 70 Hz. Here we generated a 100-member ensemble of these bandpass realizations and applied the classical and modern methods of spectral estimation with the results for each shown in Figure 2.5. Here the *average spectrum* was used to report the final results (box inset) using peak detection.

First, we merely perform a DFT using the fast Fourier transform with the results shown in Figure 2.5a. Even though it is noisy, the peaks are all present and can be discerned, if we believe them! Next, the correlation method is applied as shown in Figure 2.5b. Here we see a somewhat smoother spectrum with the sinusoids detected as well. The average periodogram method tends to oversmooth the data with only three of the four sinusoids detected as shown in Figure 2.5c. Next, we investigate the performance of the parametric methods. The MVDR results are shown in Figure 2.5d. Indicating both a smooth spectrum and good estimates of the sinusoids. Next the autoregressive method

---

9 Define $\text{SNR}_{out} := \frac{\text{Signal Energy}}{\text{Noise Variance}}$.

(MEM) was applied to the ensemble with the results shown in Figure 2.5e. The sinusoidal estimates are very prominent in the spectra and easily identified with similar results for the MUSIC methods in 2.5f. Again the sinusoids are clearly detected which both algorithms were designed to achieve; however, we note in both cases a false sinusoid was detected at 70 Hz – the cutoff frequency of the bandpass filter.

Summarizing we see that even though noisy, the classical methods are able to extract the sinusoids with the added advantage of some "smoothing" – a feature that requires long data records. The MVDR method lies between classical and parametric methods in performance. Here the sinusoids are detected along with the bandpass response. The modern methods have overestimated the number of sinusoids and provide false peaks at the bandpass filter cutoff frequency. This completes the case study.

2.7 Summary

In this chapter, we discussed the fundamental concepts underlying random signal analysis. We discussed the evolution of stochastic processes as models of phenomenological events ranging from coin-flipping to a linear dynamical system. We also discussed special processes (e.g., Gaussian, Markov) and properties of stochastic processes. Under the assumption of stationarity, we developed spectral representations and the concept of simulating a stationary process with given covariance by driving a linear system with white noise. We introduced a basic model, the autoregressive moving average model with exogeneous inputs (ARMAX). Using these ideas, we developed covariance estimators and employed them in the classical nonparametric spectral estimation schemes: the *correlation* and *periodogram* methods. We then developed the modern parametric methods of spectral estimation investigating the AR and ARMA techniques as well as the harmonic techniques of MVDR and MUSIC methods. Finally, we concluded with a case study involving the recovery of sinusoids in bandpass filtered noise.

Matlab Notes

MATLAB and its *Signal Processing Toolbox* (**signal**) can be used to design parametric spectral estimators. AR-models can be estimated from data using a variety of algorithms including the Levinson–Durbin recursion (**levinson**), the linear predictor forward filter (**lpc**), Burg forward–backward approach (**arburg**), the Covariance method (**arcov**), the modified Covariance method (**armcov**), the maximum entropy method (**pmem**), the Yule–Walker approach (**aryule**) and if the *System Identification Toolbox* (**ident**) is available the (**ar**) command has a suite of "approach" flags (e.g. *burg, yw, ls,* etc.). Note also these algorithms accommodate the lattice forms as well. ARMAX-models are available using the Prony technique (**Prony**) and the Steiglitz–McBride optimization algorithm (**stmcb**). Also in the *ID* toolbox ARMAX models can be estimated in batch mode as well as the ARX and general linear models using the Prediction Error Method (**PEM**), which can also accommodate multivariable models. A wide suite of models for model-based parameter estimation are also available in the *ID* toolbox. Recursive forms can also be applied for ARMAX (**RARMAX**), ARX (**rarx**), and the recursive PEM (**rpem**). Order estimation techniques are available (**AIC**) and (**FPE**), but other approaches discussed in this chapter are easily programmed in MATLAB (e.g. best rank approximation). The transfer function is estimated using the (**tfe**) command, which solves the Wiener filter problem in the frequency domain. A wealth of model-based spectral estimators is also available to solve for the harmonic temporal as well as spatial models. They are the Burg lattice algorithm (**pburg**), AR covariance, and modified covariance methods (**pcov**), (**pcovm**) as well as the Yule–Walker approach (**pyulear**). The spatial as well as temporal eigenvector methods form harmonic estimation are also available such as the MUSIC method (**pmusic**), its eigenvalue weighting variant (**peig**) as well as their root variants discussed in Chapter (**rootmusic**), (**rooteig**). Classical PSD methods are also available in the periodogram (**periodogram** and **pwelch**) and the multiple taper window method using prolate spheroidal functions (**pmtm**). In fact, a very nice interactive cross section of the PSD tools are available in the interactive (**sptool**), which is part of the *Signal Processing Toolbox*. As before, many of the other algorithms mentioned in this chapter can easily be implemented in MATLAB using its powerful vector–matrix language – ideal for signal processing applications.

References

1 Papoulis, A. (1965). *Probability, Random Variables and Stochastic Processes.* New York: McGraw-Hill.
2 Jazwinski, A. (1970). *Stochastic Processes and Filtering Theory.* New York: Academic Press.
3 Hogg, R. and Craig, A. (1970). *Introduction to Mathematical Statistics.* New York: MacMillan.
4 Candy, J. (1988). *Signal Processing: The Modern Approach.* New York: McGraw-Hill.
5 Astrom, K. (1970). *Introduction to Stochastic Control Theory.* New York: Academic Press.
6 Tretter, S. (1976). *Introduction to Discrete-Time Signal Processing.* New York: Wiley.

7 Candy, J. (2006). *Model-Based Signal Processing*. Hoboken, NJ: Wiley/IEEE Press.

8 Bauer, F. (1955). A direct iterative process for the Hurwitz decomposition of a polynomial. *Arch. Elect. Übertragung* 9: 285–290.

9 Kailath, T. and Rissanen, J. (1972). Partial realization of random systems. *Automatica* 8 (4): 389–396.

10 Rissanen, J. (1973). Algorithm for the triangular decomposition of block Hankel and Toeplitz matrices with application to factorizing positive matrix polynomials. *Math. Comput.* 17 (121): 147–154.

11 Makhoul, J. (1973). Spectral analysis of speech by linear prediction. *IEEE Trans. Audio Electroacoust.* AU-21 (3): 140–148.

12 Stoica, P. and Moses, R. (1997). *Introduction to Spectral Estimation*. Englewood Cliffs, NJ: Prentice-Hall.

13 Kay, S. (1988). *Modern Spectral Estimation: Theory and Applications*. Englewood Cliffs, NJ: Prentice-Hall.

14 Schuster, A. (1898). On the investigation of hidden periodicities with application to a supposed 26 day period of meteorological phenomena. *Terr. Magn.* 3 (1): 13–41.

15 Cooley, J. and Tukey, J. (1965). An algorithm for the machine calculation of complex Fourier series. *Math. Comput.* 19 (90): 297–301.

16 Welch, P. (1967). The use of fast Fourier transforms for the estimation of power spectra: a method based on time averaging over short modified periodograms. *IEEE Trans. Audio Electroacoust.* AU-15 (2): 70–73.

17 Childers, D. (ed.) (1978). *Modern Spectral Analysis*. New York: IEEE Press.

18 Burg, J. (1975). Maximum entropy spectral analysis. PhD dissertation. Stanford University.

19 Ljung, L. and Soderstrom, T. (1983). *Theory and Practice of Recursive Identification*. Boston, MA: MIT Press.

20 Kay, S. and Marple, L. (1981). Spectrum analysis – a modern perspective. *Proc. IEEE* 69 (11): 1380–1419.

21 Akaike, H. (1969). Fitting autoregressive models for prediction. *Ann. Inst. Statist. Math.* 21 (1): 243–247.

22 Rissanen, J. (1978). Modeling by shortest data description. *Automatica* 14 1978, 14 (5): 465–471.

23 Cadzow, J. (1980). High performance spectral estimator – a new ARMA method. *IEEE Trans. Acoust. Speech Signal Process.* ASSP-28 (5): 524–529.

24 Capon, J. (1969). High-resolution frequency-wavenumber spectrum analysis. *Proc. IEEE* 57 (8): 1408–1418.

25 Stewart, G. (1973). *Introduction to Matrix Computations*. New York: Academic Press.

26 Pisarenko, V. (1973). The retrieval of harmonics from a covariance function. *Geophys. J. R. Astrom. Soc.* 33 (5): 347–366.

27 Schmidt, R. (1986). Multiple emitter location and signal parameter estimation. *IEEE Antennas Propag. Mag.* 34 (3): 276–280.
28 van Overschee, P. and De Moor, B. (1996). *Subspace Identification for Linear Systems: Theory, Implementation, Applications.* Boston, MA: Kluwer Academic Publishers.
29 Katayama, T. (2005). *Subspace Methods for System Identification.* London: Springer.
30 Verhaegen, M. and Verdult, V. (2007). *Filtering and System Identification: A Least-Squares Approach.* Cambridge: Cambridge University Press.
31 Johnson, D. and Dudgeon, D. (1993). *Array Signal Processing: Concepts and Techniques.* Englewood Cliffs, NJ: Prentice-Hall.

Problems

2.1 Suppose the stochastic process $\{y(t)\}$ is generated by

$$y(t) = a \quad \exp(-t) + ct, \quad a, b \text{ random, then}$$

a) What is the mean of the process?
b) What is the corresponding covariance?
c) Is the process stationary, if $E\{a\} = E\{b\} = 0$ and $E\{ab\} = 0$.

2.2 Assume $y(t)$ is a zero-mean, ergodic process with covariance $R_{yy}(k)$, calculate the corresponding power spectra, $S_{yy}(z)$ if

a) $R_{yy}(k) = Ca^{|k|}$.
b) $R_{yy}(k) = C \cos(w|k|), \quad |k| < \frac{\pi}{2}$.
c) $R_{yy}(k) = C \exp(-a^{|k|})$.

2.3 Verify the covariance-spectral density pairs for the discrete process:
a) Bandlimited white noise
b) Triangular

2.4 Let the impulse response of a linear system with random input $u(t)$ be given by $h(t)$, then show that
a) $R_{yy}(k) = \sum_{m=0}^{\infty} \sum_{i=0}^{\infty} h(m)h(i)R_{uu}(k + i - m)$ and $S_{yy}(z) = H(z)H(z^{-1}) S_{uu}(z)$.
b) $R_{yy}(k) = \sum_{m=0}^{\infty} h(m)R_{yu}(k - m)$ and $S_{yy}(z) = H(z)S_{yu}(z)$.
c) $R_{uy}(k) = \sum_{m=0}^{\infty} h(m)R_{uu}(k - m)$ and $S_{uy}(z) = H(z)S_{uu}(z)$.

2.5 Derive the *sum decomposition* relation,

$$S_{yy}(z) = S_{yy}^+(z) + S_{yy}^-(z) - R_{yy}(0)$$

2.6 Develop a computer program to simulate the ARMA process

$$y(t) = -ay(t-1) + e(t)$$

where $a = 0.75$, $e \approx N(0, 0.1)$ for 100 data points.
a) Calculate the analytic covariance $R_{yy}(k)$.
b) Determine an expression to "recursively" calculate $R_{yy}(k)$.
c) Plot the simulated results and construct the $\pm 2\sqrt{R_{yy}(0)}$ bounds.
d) Do 95% of the samples fall within these bounds?

2.7 Develop a digital filter to simulate a sequence, $y(t)$, with covariance $R_{yy}(k) = 4e^{-3|k|}$.

2.8 Suppose we are given a linear system characterized by transfer function

$$H(z) = \frac{1 - 1/2z^{-1}}{1 - 1/3z^{-1}},$$

which is excited by discrete exponentially correlated noise

$$R_{xx}(k) = (1/2)^{|k|}$$

a) Determine the output PSD, $S_{yy}(z)$.
b) Determine the output covariance, $R_{yy}(k)$.
c) Determine the cross-spectrum, $S_{yx}(z)$.
d) Determine the cross-covariance, $R_{yx}(k)$.

2.9 Suppose we are given a causal LTI system characterized by its impulse response, $h(t)$. If this system is excited by zero-mean, unit variance white noise, then
a) Determine the output variance, $R_{yy}(0)$.
b) Determine the covariance, $R_{yy}(k)$ for $k > 0$.
c) Suppose the system transfer function is given by

$$H(z) = \frac{1 + b_0 z^{-1}}{1 + a_1 z^{-1} + a_2 z^{-2}}$$

d) Find a method to recursively calculate $h(t)$ and therefore $R_{yy}(0)$.

2.10 Given the covariance function

$$R_{yy}(k) = e^{-1/2|k|} \cos \pi |k|$$

Find the digital filter when driven by unit variance white noise produces a sequence $\{y(t)\}$ with these statistics.

2.11 Suppose we have a process characterized by difference equation

$$y(t) = x(t) + 1/2x(t-1) + 1/3x(t-2)$$

a) Determine a recursion for the output covariance, $R_{yy}(k)$.
b) If $x(t)$ is white with variance σ_{xx}^2, determine $R_{yy}(k)$.
c) Determine the output PSD, $S_{yy}(z)$.

2.12 Suppose we are given a linear system characterized by the difference equation

$$y(t) - 1/5y(t-1) = \frac{1}{\sqrt{3}}x(t)$$

and the system is excited by
 i) white Gaussian noise, $x \sim N(0, 3)$
 ii) exponentially correlated noise, $R_{ee}(k) = (1/2)^{|k|}$
In both cases find the following:
a) Output PSD, $S_{yy}(z)$
b) Output covariance, $R_{yy}(k)$
c) Cross-spectrum, $S_{ye}(k)$
d) Cross-covariance, $R_{ye}(k)$.

2.13 Suppose we have an MA process (two-point average)

$$y(t) = \frac{e(t) + e(t-1)}{2}$$

a) Develop an expression for $S_{yy}(z)$ when e is white with variance R_{ee}.
b) Let $z = \exp\{j\Omega\}$ and sketch the "response" of $S_{yy}(e^{j\Omega})$.
c) Calculate an expression for the covariance, $R_{yy}(k)$ in closed and recursive form.

2.14 Suppose we are given a zero-mean process with covariance

$$R_{yy}(k) = 10 \, \exp(-0.5|k|)$$

a) Determine a digital filter that when driven by white noise will yield a sequence with the above covariance.
b) Develop a computer program to generate $y(t)$ for 100 points.
c) Plot the results and determine of 95% of the samples fall within $\pm 2\sqrt{R_{yy}(0)}$.

2.15 Suppose a process is measured using an instrument with uncertainty $v \sim N(0; 4)$ such that

$$y(t) = 2x(t) + v(t)$$

a) Calculate the output covariance $R_{yy}(k)$.
b) Develop a computer program to simulate the output.

c) Plot the process $y(t)$ with the corresponding confidence limits $\pm 2\sqrt{R_{yy}(0)}$ for 100 data points, do 95% of the samples lie within the bounds?

2.16 Given the following ARMAX model

$$A(q^{-1})y(t) = B(q^{-1})u(t) + \frac{C(q^{-1})}{D(q^{-1})}e(t)$$

for q^{-1} the backward shift (delay) operator such that

$$A(q^{-1}) = 1 + 1.5q^{-1} + 0.7q^{-2}$$
$$B(q^{-1}) = 1 + 0.5q^{-1}$$
$$C(q^{-1}) = 1 + 0.7q^{-1}$$
$$D(q^{-1}) = 1 + 05q^{-1}$$

a) Find the pulse transfer representation of this process ($C = D = 0$). Convert it to the following equivalent *pole–zero* and *normal* state-space forms. Is the system controllable? Is it observable? Show your calculations.

b) Find the pole–zero or ARX representation of this process ($C = 1$, $D = 0$). Convert it to the equivalent state-space form.

c) Find the pole–zero or ARMAX representation of this process ($D = 0$). Convert it to the equivalent state-space form.

d) Find the all-zero or FIR representation of this process ($A = 1$, $C = D = 0$). Convert it to the equivalent state-space form.

e) Find the all-pole or IIR representation of this process ($B = 0$, $C = 0, D = 0$). Convert it to the equivalent state-space form.

f) Find the all-zero or MA representation of this process ($A = 1$, $B = 0, D = 0$). Convert it to the equivalent state-space form.

3

State-Space Models for Identification

3.1 Introduction

State-space models are easily generalized to multichannel, nonstationary, and nonlinear processes [1–23]. They are very popular for model-based signal processing primarily because most physical phenomena modeled by mathematical relations naturally occur in state-space form (see [15] for details). With this motivation in mind, let us proceed to investigate the state-space representation in a more general form to at least "touch" on its inherent richness. We start with continuous-time systems and then proceed to the sampled-data system that evolves from digitization followed by the purely discrete-time representation – the primary focus of this text.

3.2 Continuous-Time State-Space Models

We begin by formally defining the concept of state [1]. The *state* of a system at time t is the "minimum" set of variables (*state variables*) along with the *input* sufficient to uniquely specify the dynamic system behavior for all t over the interval $t \in [t_o, \infty)$. The *state vector* is the collection of state variables into a single vector. The idea of a *minimal set* of state variables is critical, and all techniques to define them must ensure that the smallest number of "independent" states have been defined in order to avoid possible violation of some important system theoretic properties [2, 3]. In short, one can think of states mathematically as simply converting an Nth-order differential equation into a set of N-first-order equations – each of which is a state variable. From a systems perspective, the states are the *internal* variables that may not be measured directly, but provide the critical information about system performance. For instance, measuring only the input/output voltages or currents of an integrated circuit board with a million embedded (internal) transistors – the internal voltages/currents would be the unmeasured states of this system.

Model-Based Processing: An Applied Subspace Identification Approach, First Edition. James V. Candy.

Let us consider a general *deterministic* formulation of a *nonlinear dynamic system* including the output (measurement) model in state-space form (continuous-time)[1]

$$\dot{x}_t = A(x_t, u_t) = a(x_t) + b(u_t)$$
$$y_t = C(x_t, u_t) = c(x_t) + d(u_t)$$

for x_t, y_t, and u_t the respective N_x-state, N_y-output, and N_u-input vectors with corresponding system (process), input, measurement (output), and feedthrough functions. The N_x-dimensional system and input functions are defined by $a(\cdot)$ and $b(\cdot)$, while the N_y-dimensional output and feedthrough functions are given by $c(\cdot)$ and $d(\cdot)$.

In order to specify the solution of the N_xth-order differential equations completely, we must specify the above-noted functions along with a set of N_x-initial conditions at time t_0 and the input for all $t \geq t_0$. Here N_x is the dimension of the "minimal" set of state variables.

If we constrain the state-space representation to be linear in the states, then we obtain the generic continuous-time, *linear, time-varying state-space* model given by

$$\dot{x}_t = A_t x_t + B_t u_t$$
$$y_t = C_t x_t + D_t u_t \tag{3.1}$$

where $x_t \in \mathcal{R}^{N_x \times 1}$, $u_t \in \mathcal{R}^{N_u \times 1}$, $y_t \in \mathcal{R}^{N_y \times 1}$ and the respective system, input, output, and feedthrough matrices are: $A \in \mathcal{R}^{N_x \times N_x}$, $B \in \mathcal{R}^{N_x \times N_u}$, $C \in \mathcal{R}^{N_y \times N_x}$, and $D \in \mathcal{R}^{N_y \times N_u}$.

An interesting property of the state-space representation is to realize that these models represent a complete generic form for almost any physical system. That is, if we have an *RLC* circuit or a mass-damper-spring (MCK) mechanical system, their dynamics are governed by the identical set of differential equations – only their coefficients differ. Of course, the physical meaning of the states is different, but this is the idea behind state-space – *many physical systems* can be captured by this *generic* form of differential equations, even though the systems are physically different.

Systems theory, which is essentially the study of dynamic systems, is based on the study of state-space models and is rich with theoretical results exposing the underlying properties of the system under investigation. This is one of the major reasons why state-space models are employed in signal processing, especially when the system is *multivariable*, having multiple inputs and multiple outputs (MIMO). Next we develop the relationship between the state-space representation and input/output relations – the transfer function.

---

1 We separate x_t and u_t for future models, but it is not really necessary.

For this development, we constrain the state-space representation above to be a linear time-invariant (LTI) *state-space* model given by

$$\dot{x}_t = A_c x_t + B_c u_t$$
$$y_t = C_c x_t + D_c u_t \tag{3.2}$$

where $A_t \to A_c$, $B_t \to B_c$, $C_t \to C_c$, and $D_t \to D_c$ their time invariant counterparts with the subscript, "c," annotating continuous-time matrices.

This LTI model corresponds the constant coefficient differential equation solutions, which can be solved using Laplace transforms. Taking the Laplace transform[2] of these equations and solving for $X(s)$, we have that

$$X(s) = (sI - A_c)^{-1} x_{t_0} + (sI - A_c)^{-1} B_c U(s) \tag{3.3}$$

where $I \in \mathcal{R}^{N_x \times N_x}$ is the identity matrix. The corresponding output is

$$Y(s) = [C_c(sI - A_c)^{-1} B_c + D_c]U(s) + C_c(sI - A_c)^{-1} x_{t_0} \tag{3.4}$$

From the definition of *transfer function* (zero initial conditions), we have the desired result

$$H(s)\big|_{t_0=0} := \frac{Y(s)}{U(s)} = C_c(sI - A_c)^{-1} B_c + D_c \tag{3.5}$$

Taking the inverse Laplace transform of this equation gives us the corresponding *impulse response matrix* of the LTI-system as [1]

$$H(t,\tau) = C_c e^{A_c(t-\tau)} B_c + D_c \delta(t) \quad \text{for} \quad t \geq \tau \tag{3.6}$$

So we see that the state-space representation enables us to express the input–output relations in terms of the internal variables or states. Note also that this is a multivariable representation compared to the usual single-input-single-output (SISO) (scalar) systems models that frequently appear in the signal processing literature.

Now that we have the multivariable transfer function representation of our LTI system, we can solve the state equations directly using inverse transforms to obtain the time-domain solutions. First, we simplify the notation by defining the Laplace transform of the state transition matrix or the so-called *resolvent matrix* of systems theory [1, 3] as

$$\Phi_c(s) := (sI - A_c)^{-1} \tag{3.7}$$

with

$$\Phi_c(t) = \mathcal{L}^{-1}[(sI - A_c)^{-1}] = e^{A_c t} \tag{3.8}$$

---

2 Recall the Laplace transform is defined as $X(s) := \mathcal{L}[x_t] = \int_{-\infty}^{\infty} x_t e^{-st} dt$ with inverse transform given by $x_t = \mathcal{L}^{-1}[X(s)] = \int_{-\infty}^{\infty} X(s) e^{st} ds$ with $s = \sigma + j\omega$, a complex variable.

Therefore, the *state transition matrix* is a critical component in the solution of the state equations of an LTI system given by

$$\Phi_c(t, t_0) = e^{A_c(t-t_0)} \quad t \geq t_0 \tag{3.9}$$

and we can rewrite the transfer function matrix as

$$H(s) = C_c\Phi_c(s)B_c + D_c \tag{3.10}$$

with the corresponding state-input transfer matrix given by

$$X(s) = \Phi_c(s)x_{t_0} + \Phi_c(s)B_cU(s) \tag{3.11}$$

Taking the inverse Laplace transformation gives the time-domain solution

$$x_t = \mathcal{L}^{-1}[X(s)] = \Phi_c(t, t_0)x_{t_0} + \Phi_c(t, t_0)B_c * u_t$$

or

$$x_t = \underbrace{\Phi_c(t, t_0)x_{t_0}}_{\text{zero-input}} + \underbrace{\int_{t_0}^{t} \Phi_c(t, \alpha)B_c u_\alpha d\alpha}_{\text{zero-state}} \tag{3.12}$$

with the corresponding output solution

$$y_t = C_c\Phi_c(t, t_0)x_{t_0} + \int_{t_0}^{t} C_c\Phi_c(t, \alpha)B_c u_\alpha d\alpha + D_c\delta(t) \tag{3.13}$$

Revisiting the continuous-time system of Eq. (3.12) and substituting the matrix exponential for the state transition matrix gives the LTI solution as

$$x_t = e^{A_c(t-t_0)}x_{t_0} + \int_{t_0}^{t} e^{A_c(t-\alpha)}B_c u_\alpha d\alpha \tag{3.14}$$

and the corresponding measurement system

$$y_t = C_c x_t + D_c u_t \tag{3.15}$$

In general, the continuous *state transition matrix* satisfies the following *properties*: [1, 2]

1) $\Phi_c(t, t_0)$ is uniquely defined for $t, t_0 \in [0, \infty)$ (Unique)
2) $\Phi_c(t, t) = I$ (Identity)
3) $\Phi_c(t)$ satisfies the matrix differential equation:

$$\dot{\Phi}_c(t, t_0) = A_t\Phi_c(t, t_0), \quad \Phi_c(t_0, t_0) = I, t \geq t_0 \tag{3.16}$$

4) $\Phi_c(t, t_0) = \Phi_c(t, \tau) \times \Phi_c(\tau, \alpha) \times \cdots \times \Phi_c(\beta, t_0)$ (Semi-Group)
5) $\Phi_c(t, \tau)^{-1} = \Phi_c(\tau, t)$ (Inverse)

Thus, the transition matrix plays a pivotal role in LTI systems theory for the analysis and prediction of the response of LTI and time-varying systems [2].

For instance, the *poles* of an LTI govern important properties such as stability and response time. The poles are the *roots* of the *characteristic* (polynomial) *equation* of A_c, which are found by solving for the roots of the determinant of the resolvent, that is,

$$|\Phi_c(s)| = |(sI - A_c)|_{s=p_i} = 0 \tag{3.17}$$

Stability is determined by assuring that all of the poles lie within the left half of the S-plane. Next we consider the sampled-data state-space representation.

3.3 Sampled-Data State-Space Models

Sampling a continuous-time system is commonplace with the advent of high-speed analog-to-digital converters (ADC) and modern computers. A sampled-data system lies somewhere between the continuous analog domain (physical system) and the purely discrete domain (stock market prices). Since we are strictly sampling a continuous-time process, we must ensure that all of its properties are preserved. The well-known *Nyquist sampling theorem* precisely expresses the required conditions (twice the highest frequency) to achieve "perfect" reconstruction of the process from its samples [15].

Thus, if we have a physical system governed by continuous-time dynamics and we "sample" it at given time instants, then a sampled-data model can be obtained directly from the solution of the continuous-time state-space model. That is, we know from Section 3.2 that

$$x_t = \Phi_c(t, t_0)x_{t_0} + \int_{t_0}^{t} \Phi_c(t, \alpha)B_c(\alpha)u_\alpha d\alpha$$

where $\Phi_c(\cdot, \cdot)$ is the continuous-time state transition matrix that satisfies the matrix differential equation of Eq. (3.16), that is,

$$\dot{\Phi}_c(t, t_0) = A_t\Phi_c(t, t_0), \quad \Phi_c(t_0, t_0) = I, \quad t \geq t_0$$

Sampling this system such that $t \to t_k$ over the interval $(t_k, t_{k-1}]$, then we have the corresponding *sampling interval* defined by $\Delta t_k := t_k - t_{k-1}$. Note this representation need *not* necessarily be equally-spaced – another important property of the state-space representation. Thus, the sampled solution becomes

$$x(t_k) = \Phi(t_k, t_{k-1})x(t_{k-1}) + \int_{t_{k-1}}^{t_k} \Phi(t_k, \alpha)B_c(\alpha)u_\alpha d\alpha \tag{3.18}$$

and therefore from the differential equation of Eq. (3.16), we have the solution

$$\Phi(t_k, t_{k-1}) = \int_{t_{k-1}}^{t_k} A(\alpha)\Phi(t_k, \alpha)d\alpha \quad \text{for} \quad \Phi(t_0, t_0) = I \tag{3.19}$$

where $\Phi(t_k, t_{k-1})$ is the *sampled-data* state transition matrix – the critical component in the solution of the state equations enabling us to calculate state evolution in time. Note that for an LTI sampled-data system, the state-transition matrix is $\Phi(t, t_k) = e^{A(t-t_k)}$ where A is the sampled-data system (process) matrix.

If we further assume that the input excitation is *piecewise constant* ($u_\alpha \rightarrow u(t_{k-1})$) over the interval $(t_k, t_{k-1}]$, then it can be removed from under the superposition integral in Eq. (3.18) to give

$$x(t_k) = \Phi(t_k, t_{k-1})x(t_{k-1}) + \left(\int_{t_{k-1}}^{t_k} \Phi(t_k, \alpha)B_c(\alpha)d\alpha \right) \times u(t_{k-1}) \qquad (3.20)$$

Under this assumption, we can define the sampled-data *input transmission* matrix as

$$B(t_{k-1}) := \int_{t_{k-1}}^{t_k} \Phi(t_k, \alpha)B_c(\alpha)d\alpha \qquad (3.21)$$

and therefore the *sampled-data state-space system* with equally or unequally sampled data is given by

$$x(t_k) = \Phi(t_k, t_{k-1})x(t_{k-1}) + B(t_{k-1})u(t_{k-1})$$
$$y(t_k) = C(t_k)x(t_k) + D(t_k)u(t_k) \qquad (3.22)$$

Computationally, sampled-data systems pose no particular problems when care is taken, especially since reasonable approximation and numerical integration methods exist [19]. This completes the discussion of sampled-data systems and approximations. Next we consider the discrete-time systems.

3.4 Discrete-Time State-Space Models

Discrete state-space models, the focus of subspace identification, evolve in two distinct ways: naturally from the problem or from sampling a continuous-time dynamical system. An example of a natural discrete system is the dynamics of balancing our own checkbook. Here the state is the evolving balance given the past balance and the amount of the previous check. There is "no information" between time samples, so this model represents a discrete-time system that evolves naturally from the underlying problem. On the other hand, if we have a physical system governed by continuous-time dynamics, then we "sample" it at given time instants. So we see that discrete-time dynamical systems can evolve from a wide variety of problems both naturally (checkbook) or physically (circuit). In this text, we are primarily interested in physical systems (physics-based models), so we will concentrate on sampled systems reducing them to a discrete-time state-space model.

We can use a *first-difference approximation* and apply it to the general LTI continuous-time state-space model of Eq. (3.2) to obtain a discrete-time system, that is,

$$\dot{x}_t \approx \frac{x(t) - x(t-1)}{\Delta T} \approx A_c x(t-1) + B_c u(t-1)$$
$$y_t \approx y(t) = C_c x(t) + D_c u(t)$$

Solving for $x(t)$, we obtain

$$x(t) = (I + A_c \Delta T)x(t-1) + B_c \Delta T u(t-1)$$
$$y(t) = C_c x(t) + D_c u(t) \tag{3.23}$$

Recognizing that the first-difference approximation is equivalent to a first-order *Taylor series* approximation of A_c gives the discrete system, input, output, and feedthrough matrices as

$$A \approx I + A_c \Delta T + O(\Delta T^2)$$
$$B \approx B_c \Delta T$$
$$C \approx C_c$$
$$D \approx D_c \tag{3.24}$$

The discrete, *linear, time-varying state-space representation* is given by the *system* or *process model* as

$$x(t) = A(t-1)x(t-1) + B(t-1)u(t-1) \tag{3.25}$$

and the corresponding discrete *output* or *measurement model* as

$$y(t) = C(t)x(t) + D(t)u(t) \tag{3.26}$$

where x, u, y are the respective N_x-state, N_u-input, N_y-output and A, B, C, D are the $(N_x \times N_x)$-system, $(N_x \times N_u)$-input, $(N_y \times N_x)$-output, and $(N_y \times N_u)$-feedthrough matrices.

The state-space representation for (LTI), discrete systems is characterized by constant system, input, output, and feedthrough matrices, that is,
$$A(t) \to A, \quad B(t) \to B \quad \text{and} \quad C(t) \to C, \quad D(t) \to D$$
and is given by the LTI system

$$x(t) = Ax(t-1) + Bu(t-1)$$
$$y(t) = Cx(t) + Du(t) \tag{3.27}$$

The discrete system representation replaces the Laplace transform with the Z-transform defined by the transform pair:

$$X(z) := \sum_{t=0}^{\infty} x(t)z^{-t}$$
$$x(t) = \int_{-\infty}^{\infty} X(z)z^{-1}dz \tag{3.28}$$

Time-invariant state-space discrete systems can also be represented in *input/output* or *transfer function* form using the $\mathcal{Z}$-transform to give

$$H(z) = C(zI - A)^{-1}B + D = C \left(\frac{\text{adj}(zI - A)}{\det(zI - A)} \right) B + D \tag{3.29}$$

where recall "adj" is the matrix adjoint (transpose of the cofactor matrix) and "det" is the matrix determinant.

We define the *characteristic equation* or *characteristic polynomial* of the N-dimensional system matrix A as

$$\alpha(z) := \det(zI - A) = z^N + \alpha_1 z^{N-1} + \alpha_2 z^{N-2} + \cdots + \alpha_{N-1} z + \alpha_N \tag{3.30}$$

with *roots* corresponding to the *poles* of the underlying system that are also obtained from the *eigenvalues* of A defined by

$$\lambda(A) := \text{eig}(A) = \prod_{n=1}^{N} (z - \lambda_n) \tag{3.31}$$

for λ_n is the nth system pole.

Taking inverse $\mathcal{Z}$-transforms of Eq. (3.29), we obtain the discrete impulse (or pulse) response matrix as

$$H_t = CA^{t-1}B + D\delta(t) \tag{3.32}$$

for δ the Kronecker delta function.

The solution to the state-difference equations can easily be derived by induction [3] and is given by

$$x(t) = \Phi(t, k)x(k) + \sum_{i=k+1}^{t} \Phi(t, i)B(i)u(i) \quad \text{for} \quad t > k \tag{3.33}$$

where $\Phi(t, k)$ is the *discrete-time state-transition* matrix. For time-varying systems, it can be shown (by induction) that the state-transition matrix satisfies

$$\Phi(t, k) = A(t - 1) \cdot A(t - 2) \cdots A(k)$$

while for time-invariant systems the state-transition matrix is given by

$$\Phi(t, k) = A^{t-k} \quad \text{for} \quad t > k$$

The *discrete state-transition matrix* possesses properties analogous to its continuous-time counterpart, that is,

1) $\Phi(t, k)$ is uniquely defined   (Unique)
2) $\Phi(t, t) = I$ (Identity)
3) $\Phi(t, k)$ satisfies the matrix difference equation:

$$\Phi(t, k) = A(t - 1)\Phi(t - 1, k), \quad \Phi(k, k) = I, t \geq k + 1 \tag{3.34}$$

4) $\Phi(t, k) = \Phi(t, t - 1) \times \Phi(t - 1, t - 2) \times \cdots \times \Phi(k + 1, k)$ (Semi-Group)
5) $\Phi^{-1}(t, k) = \Phi(k, t)$ (Inverse)

3.4.1 Linear Discrete Time-Invariant Systems

In this section, we concentrate on the discrete LTI system that is an integral component of the subspace identification techniques to follow. Here we develop a compact vector–matrix form of the system input/output relations that will prove useful in subsequent developments.

For a discrete LTI system, we have that the state transition matrix becomes

$$\Phi(t,k) = A(t-1) \cdot A(t-2) \cdots A(k) \quad \overbrace{=}^{A(t)\to A} \quad A^{t-k} \quad \text{for} \quad t > k$$

The discrete LTI solution of Eq. (3.33) is therefore

$$x(t) = A^t x(0) + \sum_{k=0}^{t-1} A^{t-k} B u(k); \quad t = 0, 1, \ldots, K \tag{3.35}$$

with the measurement or output system given by

$$y(t) = CA^t x(0) + \sum_{k=0}^{t-1} CA^{t-k} B u(k) + D\delta(t) \tag{3.36}$$

Expanding this relation further over K-samples and collecting terms, we obtain

$$
\begin{bmatrix} y(0) \\ y(1) \\ \vdots \\ y(K-1) \end{bmatrix} = \underbrace{\begin{bmatrix} C \\ CA \\ \vdots \\ CA^{K-1} \end{bmatrix}}_{\mathcal{O}} x(0) + \underbrace{\begin{bmatrix} D & \cdots & 0 \\ CB & D & \cdots & 0 \\ \vdots & \vdots & \vdots & \vdots \\ CA^{K-1}B & \cdots & CB & D \end{bmatrix}}_{\mathcal{T}} \begin{bmatrix} u(0) \\ u(1) \\ \vdots \\ u(K-1) \end{bmatrix}
$$

$$\tag{3.37}$$

where $\mathcal{O} \in \mathcal{R}^{KN_y \times N_x}$ is the *observability matrix* and $\mathcal{T} \in \mathcal{R}^{KN_y \times KN_u}$ is a Toeplitz matrix [15].

Shifting these relations in time $(0 \to t)$ yields

$$
\begin{bmatrix} y(t) \\ y(t+1) \\ \vdots \\ y(t+K-1) \end{bmatrix} = \mathcal{O}_K x(t) + \mathcal{T}_K \begin{bmatrix} u(t) \\ u(t+1) \\ \vdots \\ u(t+K-1) \end{bmatrix} \tag{3.38}
$$

leading to the vector input/output relation

$$y_K(t) = \mathcal{O}_K x(t) + \mathcal{T}_K u_K(t) \tag{3.39}$$

Catenating these m-vectors $(K \to m)$ of Eq. (3.39) to create a batch-data (*block Hankel*) matrix over the K-samples, we can obtain the "data equation," that is, defining the block matrices as

$$
\mathcal{Y}_{t,m|K} := \begin{bmatrix} y(t) & y(t+1) & \cdots & y(t+K-1) \\ y(t+1) & y(t+2) & \cdots & y(t+K) \\ \vdots & \vdots & \vdots & \vdots \\ y(t+m-1) & y(t+m) & \cdots & y(t+K+m-2) \end{bmatrix}
$$

$$
= [\mathbf{y}_m(t) \cdots \mathbf{y}_m(t+K-1)] \tag{3.40}
$$

$$
\mathcal{U}_{t,m|K} := \begin{bmatrix} u(t) & u(t+1) & \cdots & u(t+K-1) \\ u(t+1) & u(t+2) & \cdots & u(t+K) \\ \vdots & \vdots & \vdots & \vdots \\ u(t+m-1) & u(t+m) & \cdots & u(t+K+m-2) \end{bmatrix}
$$

$$
= [\mathbf{u}_m(t) \cdots \mathbf{u}_m(t+K-1)] \tag{3.41}
$$

$$
\mathcal{X}_{t,m|K} := [\mathbf{x}_m(t)\, \mathbf{x}_m(t+1) \cdots \mathbf{x}_m(t+K-1)] = [\mathbf{x}(t)\, A\mathbf{x}(t) \cdots A^{K-1}\mathbf{x}(t)] \tag{3.42}
$$

and, therefore, we have the vector–matrix *data equation* that relates the system model to the data (input and output matrices)

$$
\mathcal{Y}_{t,m|K} = \mathcal{O}_K \mathcal{X}_{t,m|K} + \mathcal{T}_K \mathcal{U}_{t,m|K} \tag{3.43}
$$

This expression represents the fundamental relationship for the input–state–output of an LTI state-space system. Next we discuss some of the pertinent system theoretic results that will enable us to comprehend much of the sub-space realization algorithms to follow.

3.4.2 Discrete Systems Theory

In this section we investigate the discrete state-space model from a systems theoretic viewpoint. There are certain properties that a dynamic system must possess in order to assure a consistent representation of the dynamics under investigation. For instance, it can be shown [2] that a necessary requirement of a measurement system is that it is *observable*, that is, measurements of available variables or parameters of interest provide enough information to reconstruct the internal variables or states.

Mathematically, a system is said to be *completely observable*, if for any initial state, say $x(0)$, in the state-space, there exists a finite $t > 0$ such that knowledge of the input $u(t)$ and the output $y(t)$ is sufficient to specify $x(0)$ uniquely. Recall that the linear *state-space* representation of a discrete system is defined by the following set of equations:

State Model: $\quad x(t) = A(t-1)x(t-1) + B(t-1)u(t-1)$

with the corresponding measurement system or output defined by

Measurement Model: $\quad y(t) = C(t)x(t) + D(t)u(t)$

Using this representation, the simplest example of an observable system is one in which each state is measured directly, therefore, and the measurement matrix C is a $N_x \times N_x$ matrix. In order to reconstruct $x(t)$ from its measurements $y(t)$, then from the measurement system model, C must be invertible. In this case, the system is said to be completely *observable*; however, if C is not invertible, then the system is said to be *unobservable*.

The next level of complexity involves the solution to this same problem when C is a $N_y \times N_x$ matrix, then a pseudo-inverse must be performed instead [1, 2]. In the general case, the solution gets more involved because we are not just interested in reconstructing $x(t)$, but $x(t)$ over all finite values of t; therefore, we must include the state model, that is, the dynamics as well.

With this motivation in mind, we now formally define the concept of observability. The solution to the state representation is governed by the state-transition matrix, $\Phi(t, 0)$, where recall that the solution of the state equation is [3]

$$x(t) = \Phi(t, 0)x(0) + \sum_{k=0}^{t-1} \Phi(t, k)B(k)u(k)$$

Therefore, premultiplying by the measurement matrix, the output relations are

$$y(t) = C(t)\Phi(t, 0)x(0) + \sum_{k=0}^{t-1} C(t)\Phi(t, k)B(k)u(k) \tag{3.44}$$

or rearranging we define

$$\tilde{y}(t) := y(t) - \sum_{k=0}^{t-1} C(t)\Phi(t, k)B(k)u(k) = C(t)\Phi(t, 0)x(0) \tag{3.45}$$

The problem is to solve this resulting equation for the initial state; therefore, multiplying both sides by $\Phi'C'$, we can infer the solution from the relation

$$\Phi'(t, 0)C'(t)C(t)\Phi(t, 0)x(0) = \Phi'(t, 0)C(t)\tilde{y}(t)$$

Thus, the observability question now becomes under what conditions can this equation uniquely be solved for $x(0)$? Equivalently, we are asking if the null space of $C(t)\Phi(t, 0)$ is $0 \in \mathcal{R}^{N_x \times 1}$. It has been shown [2, 4] that the following $N_x \times N_x$ *observability Gramian* has the identical null space, that is,

$$\mathcal{O}(0, t) := \sum_{k=0}^{t-1} \Phi'(k, 0)C'(k)C(k)\Phi(k, 0) \tag{3.46}$$

which is equivalent to determining that $\mathcal{O}(0, t)$ is nonsingular or rank N_x.

Further assuming that the system is LTI then over a finite time interval for $t = 0, \ldots, K - 1$ for $K > N_x$ leads to the $KN_y \times N_x$ *observability matrix* [4] given by

$$
\mathcal{O}_K := \begin{bmatrix} C \\ -\!-\!- \\ \vdots \\ -\!-\!- \\ CA^{K-1} \end{bmatrix}
\tag{3.47}
$$

Therefore, a necessary and sufficient condition for a system to be completely observable is that the *rank* of $\mathcal{O}$ or $\rho(\mathcal{O}_K)$ must be N_x. Thus, for the LTI case, checking that all of the measurements contain the essential information to reconstruct the states reduces to checking the rank of the observability matrix. Although this is a useful mathematical concept, it is primarily used as a rule of thumb in the analysis of complicated measurement systems.

Analogous to the system theoretic property of observability is that of controllability, which is concerned with the effect of the input on the states of the dynamic system. A discrete system is said to be *completely controllable* if for any $x(t), x(0) \in \mathcal{R}^{N_x \times 1}$, there exists an input sequence, $\{u(t)\}$ such that the solution to the state equations with initial condition $x(0)$ is $x(t)$ for some finite $t > 0$. Following the same approach as for observability, we obtain that the *controllability Gramian* defined by

$$
C(0, t) := \sum_{k=0}^{t-1} \Phi(0, k) B(k) B'(k) \Phi'(0, k)
\tag{3.48}
$$

is nonsingular or $\rho[C(0, t)] = N_x$

Again for the LTI system, then over a finite time interval for $t = 0, \ldots, K - 1$ for $K > N_x$ the $N_x \times KN_u$ *controllability matrix* defined by

$$
C_K := \begin{bmatrix} B | AB | \cdots | A^{K-1} B \end{bmatrix}
\tag{3.49}
$$

must satisfy the rank condition, $\rho(C_K) = N_x$ to be completely controllable [4].

If we continue with the LTI system description, we know from $\mathcal{Z}$-transform theory that the discrete transfer function can be represented by an infinite power series [4], that is,

$$
H(z) = C(zI - A)^{-1}B = \sum_{t=1}^{\infty} H_t z^{-t} \quad \text{for} \quad H_t = CA^{t-1}B
\tag{3.50}
$$

where H_t is the $N_y \times N_u$ unit impulse response matrix with $H_0 = D$. Here $\{H_t\}_0^{\infty}$ is defined as the *Markov sequence* with the corresponding set of *Markov parameters* given by the embedded system $\Sigma_{ABCD} = \{A, B, C, D\}$.

If the MIMO transfer function matrix is available, then the impulse response (matrix) sequence $\{H_t\}$ can be determined simply by long division of each matrix component (i, j) transfer function, that is, $H_{ij}(z) = N_{ij}(z)/D_{ij}(z)$. Consider the following example to illustrate this calculation that will prove useful in the classical realization theory of Chapter 6.

Example 3.1 Suppose we have a known $N_y \times N_u$ transfer function matrix with $N_y = N_u = 2$ –this system is unstable, but it is employed to illustrate a simple impulse response calculation.

$$H(z) = \begin{bmatrix} \dfrac{1}{z+2} & \dfrac{z-1}{(z+1)(z+2)} \\[2ex] \dfrac{z}{(z+1)(z+2)} & \dfrac{1}{z+1} \end{bmatrix}$$

Performing the long division of each element term-by-term, $H_{ij}(z) = N_{ij}(z)/D_{ij}(z); i = 1,2; j = 1,2$, we obtain the impulse response sequence as

$$H(z) = \begin{bmatrix} 1 & 1 \\ 1 & 1 \end{bmatrix} z^{-1} + \begin{bmatrix} -2 & -4 \\ -3 & -1 \end{bmatrix} z^{-2} + \begin{bmatrix} 4 & 10 \\ 7 & 1 \end{bmatrix} z^{-3}$$

$$+ \begin{bmatrix} -8 & -22 \\ -15 & -1 \end{bmatrix} z^{-4} + \begin{bmatrix} 16 & 46 \\ 31 & 1 \end{bmatrix} z^{-5}$$

□

We know from the *Cayley–Hamilton theorem* of linear algebra that an N-dimensional matrix satisfies its own characteristic equation [4]

$$A^N + \alpha_1 A^{N-1} + \alpha_2 A^{N-2} + \cdots + \alpha_{N-1} A + \alpha_N I = 0 \tag{3.51}$$

Therefore, pre- and postmultiplying this relation by the measurement and input transmission matrices C and B, respectively, it follows that the Markov parameters satisfy the recursion for the N-degree $\alpha(z)$

$$H_N = -\sum_{t=1}^{N} \alpha_t H_{N-t} = -\sum_{t=1}^{N} \alpha_t (CA^{N-t-1}B) \tag{3.52}$$

This result will have critical *realizability* conditions subsequently in Chapter 6.

The problem of determining the *internal description* Σ_{ABCD} from the *external description* ($H(z)$ or $\{H_t\}$) of Eq. (3.32) is called the *realization problem*. Out of all possible realizations, (A, B, C, D) having the same Markov parameters, those of smallest dimension are defined as *minimal realizations*. Thus, the dimension of the minimal realization is identical to the degree of the characteristic polynomial (actually the minimal polynomial for multivariable systems) or equivalently the degree of the transfer function (number of system poles).

In order to develop these relations, we define the $(K \times N_y N_u) \times (K \times N_y N_u)$ *Hankel matrix* by

$$\mathcal{H}_{K,K} := \begin{bmatrix} H_1 & H_2 & \cdots & H_K \\ H_2 & H_3 & \cdots & H_{K+1} \\ \vdots & \vdots & \cdots & \vdots \\ H_K & H_{K+1} & \cdots & H_{2K-1} \end{bmatrix} \tag{3.53}$$

Suppose the dimension of the system is N_x, then the Hankel matrix could be constructed such that $K = N_x$ using $2N_x$ impulse response matrices. Knowledge of the order N_x indicates the minimum number of terms required to exactly "match" the Markov sequence and extract the Markov parameters. Also the $\rho(\mathcal{H}_{K,K}) = N_x$ is the dimension of the *minimal realization*. If we did not know the dimension of the system, then we would let (in theory) $K \to \infty$ and determine the rank of $\mathcal{H}_{\infty,\infty}$. Therefore, the minimal dimension of an "unknown" system is the rank of the Hankel matrix.

In order for a system to be *minimal*, it must be *completely controllable* and *completely observable* [4]. This can be seen from the fact that the Hankel matrix factors as

$$
\mathcal{H}_{K,K} = \begin{bmatrix} CB & \cdots & CA^{K-1}B \\ \vdots & & \vdots \\ CA^{K-1}B & \cdots & CA^{2K-2}B \end{bmatrix} = \begin{bmatrix} C \\ \vdots \\ CA^{K-1} \end{bmatrix} \begin{bmatrix} B| \cdots |A^{K-1}B \end{bmatrix}
$$

(3.54)

or simply

$$
\mathcal{H}_{K,K} = \mathcal{O}_K \times \mathcal{C}_K
$$

(3.55)

From this factorization, it follows that the $\rho(\mathcal{H}_{K,K}) = \min[\rho(\mathcal{O}_K), \rho(\mathcal{C}_K)] = N_x$. Therefore, we see that the properties of controllability and observability are carefully woven into that of minimality and testing the rank of the Hankel matrix yields the dimensionality of the underlying dynamic system. This fact will prove crucial when we "identify" a system, $\Sigma_{ABCD} = \{A, B, C, D\}$, from noisy measurement data. We shall discuss realization theory in more detail in Chapter 6 using these results.

3.4.3 Equivalent Linear Systems

Equivalent systems are based on the concept that there are an infinite number of state-space systems that possess "identical" input/output responses. In the state-space, this is termed a set of coordinates; that is, we can change the state vectors describing a system by a *similarity transformation* such that the system matrices $\Sigma_{ABCD} := \{A, B, C, D\}$ are transformed to a different set of coordinates by the transformation matrix T [1–5], that is,

$$
\Sigma_{ABCD} \xrightarrow{T} \Sigma_{\overline{ABCD}}
$$

where we have

$$
\{A, B, C, D\} \to \left\{\overline{A}, \overline{B}, \overline{C}, \overline{D}\right\} := \left\{TAT^{-1}, TB, CT^{-1}, D\right\}
$$

yields an "equivalent" system from an input/output perspective, that is, the *transfer functions* are identical

$$
H(z) = \overline{C}(zI - \overline{A})^{-1}\overline{B} + \overline{D} = CT^{-1} \times (zI - TAT^{-1})^{-1} \times TB + D
$$
$$
= C(zI - A)^{-1}B + D
$$

as well as the corresponding impulse response matrices

$$H_t = \overline{CA}^{t-1}\overline{B} + \overline{D}\delta(t) = CT^{-1} \times TA^{t-1}T^{-1} \times TB + D\delta(t)$$
$$= CA^{t-1}B + D\delta(t)$$

There does exist unique representations of state-space systems termed "canonical forms," which we will see in Section 3.7 for SISO systems as well as for MIMO systems discussed subsequently in Chapter 6. In fact, much of control theory is based on designing controllers in a modal coordinate system with a diagonal system or modal matrix A_M and then transforming the modal system back to the physical coordinates using the inverse transform $T \to T_M^{-1}$ [4].

3.4.4 Stable Linear Systems

Stability of a linear system can also be cast in conjunction with the properties of controllability and observability [4–10]. For a homogeneous, discrete system, it has been shown that asymptotic stability follows directly from the eigenvalues of the system matrix expressed as $\lambda(A)$. These eigenvalues must lie within the unit circle or equivalently $\lambda_i(A) < 1 \, \forall \, i$.

Besides determining eigenvalues of A or equivalently factoring the characteristic equation to extract the system poles, another way to evaluate this condition follows directly from *Lyapunov stability* theory [9, 10] that incorporates observability.

An observable system is stable if and only if the *Lyapunov equation*

$$P = A'PA + C'C \quad \text{(Lyapunov Equation)} \tag{3.56}$$

has a *unique positive definite* solution, $P > 0$. The proof is available in [9] or [10]. As we shall see, this is an important property that will appear in subspace identification theory when extracting a so-called *balanced realization* as well as a unique *stochastic realization* [10].

This completes the section on discrete systems theory. It should be noted that all of the properties discussed in this section exist for continuous-time systems as well (see [2] for details).

3.5 Gauss–Markov State-Space Models

In this section, we extend the state-space representation to incorporate random inputs or noise sources along with random initial conditions. The discrete-time Gauss–Markov model will be applied extensively throughout this text.

3.5.1 Discrete-Time Gauss–Markov Models

Here we investigate the case when random inputs are applied to a discrete state-space system with random initial conditions. If the excitation is a random signal, then the state is also random. Restricting the input to be deterministic $u(t-1)$ and the noise to be zero-mean, white, Gaussian $w(t-1)$, the *Gauss–Markov model* evolves as

$$x(t) = A(t-1)x(t-1) + B(t-1)u(t-1) + W(t-1)w(t-1) \qquad (3.57)$$

where $w \sim \mathcal{N}(0, R_{ww})$ and $x(0) \sim \mathcal{N}(\bar{x}(0), P(0))$.

The solution to the Gauss–Markov equations can easily be obtained by induction to give

$$x(t) = \Phi(t,k)x(k) + \sum_{i=k}^{t-1} \Phi(t, i+1)B(i)u(i) + \sum_{i=k}^{t-1} \Phi(t, i+1)W(i)w(i)$$

$$(3.58)$$

which is Markov depending only on the previous state. Since $x(t)$ is just a linear transformation of Gaussian processes, it is also Gaussian. Thus, we can represent a Gauss–Markov process easily employing the state-space models.

When the *measurement* model is also included, we have

$$y(t) = C(t)x(t) + D(t)u(t) + v(t) \qquad (3.59)$$

where $v \sim \mathcal{N}(0, R_{vv})$. The model is shown diagrammatically in Figure 3.1.

Since the Gauss–Markov model of Eq. (3.57) is characterized by a Gaussian distribution, it is completely specified (statistically) by its mean and variance. Therefore, if we take the expectation of Eqs. (3.57) and (3.59) respectively, we obtain the *state mean vector* m_x as

$$m_x(t) = A(t-1)m_x(t-1) + B(t-1)u(t-1) \qquad (3.60)$$

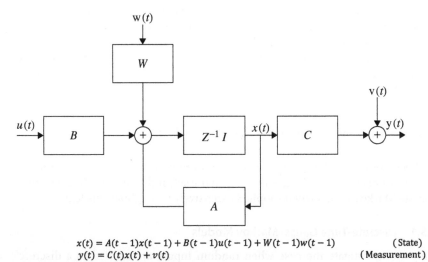

$$x(t) = A(t-1)x(t-1) + B(t-1)u(t-1) + W(t-1)w(t-1) \qquad \text{(State)}$$
$$y(t) = C(t)x(t) + v(t) \qquad \text{(Measurement)}$$

Figure 3.1 Gauss–Markov model of a discrete process.

and the *measurement mean vector* m_y as

$$m_y(t) = C(t)m_x(t) + D(t)u(t) \tag{3.61}$$

The *state variance*[3] $P(t) := \text{var}\{x(t)\}$ is given by the discrete Lyapunov equation:

$$P(t) = A(t-1)P(t-1)A'(t-1) + W(t-1)R_{ww}(t-1)W'(t-1) \tag{3.62}$$

and the measurement variance, $R_{yy}(t) := \text{var}\{y(t)\}$ is

$$R_{yy}(t) = C(t)P(t)C'(t) + R_{vv}(t) \tag{3.63}$$

Similarly, it can be shown that the *state covariance* propagates according to the following equations:

$$P(t, k) = \begin{cases} \Phi(t, k)P(k) & \text{for } t \geq k \\ P(t)\Phi'(t, k) & \text{for } t \leq k \end{cases} \tag{3.64}$$

We summarize the Gauss–Markov and corresponding statistical models in Table 3.1.

If we restrict the Gauss–Markov model to the stationary case, then

$$A(t) \to A, \quad B(t) \to B, \quad C(t) \to C, \quad W(t) \to W, \quad R_{ww}(t) \to R_{ww},$$
$$\text{and} \quad R_{vv}(t) \to R_{vv}$$

Table 3.1 Gauss–Markov representation.

State propagation

$\quad x(t) = A(t-1)x(t-1) + B(t-1)u(t-1) + W(t-1)w(t-1)$

State mean propagation

$\quad m_x(t) = A(t-1)m_x(t-1) + B(t-1)u(t-1)$

State variance/covariance propagation

$\quad P(t) = A(t-1)P(t-1)A'(t-1) + W(t-1)R_{ww}(t-1)W'(t-1)$

$\quad P(t, k) = \begin{cases} \Phi(t, k)P(k) & t \geq k \\ P(t)\Phi'(t, k) & t \leq k \end{cases}$

Measurement propagation

$\quad y(t) = C(t)x(t) + v(t)$

Measurement mean propagation

$\quad m_y(t) = C(t)m_x(t)$

Measurement variance/covariance propagation

$\quad R_{yy}(t) = C(t)P(t)C'(t) + R_{vv}(t)$

$\quad R_{yy}(t, k) = C(t)P(t)C'(t) + R_{vv}(t, k)$

---

3 We use the shorthand notation, $P(k) := P_{xx}(k, k) = \text{cov}\{x(k), x(k)\} = \text{var}\{x(k)\}$, throughout this text.

and the variance equations become

$$P(t) = AP(t-1)A' + WR_{ww}W'$$

with

$$R_{yy}(t) = CP(t)C' + R_{vv} \tag{3.65}$$

At steady state $(t \to \infty)$, we have

$$P(t) = P(t-1) = \cdots = P_{ss} := P$$

and therefore, the measurement covariance relations become

$$R_{yy}(0) = CPC' + R_{vv} \quad \text{for} \quad \text{lag} \quad k = 0 \tag{3.66}$$

By induction, it can be shown that

$$R_{yy}(k) = CA^{|k|}PC' \quad \text{for} \quad k \neq 0 \tag{3.67}$$

The measurement power spectrum is easily obtained by taking the $\mathcal{Z}$-transform of this equation to obtain

$$S_{yy}(z) = CS_{xx}(z)C' + S_{vv}(z) \tag{3.68}$$

where

$$S_{xx}(z) = T(z)S_{ww}(z)T'(z^{-1}) \quad \text{for} \quad T(z) = (zI - A)^{-1}W$$

with

$$S_{ww}(z) = R_{ww} \quad \text{and} \quad S_{vv}(z) = R_{vv}$$

Thus, using $H(z) = CT(z)$ the spectrum is given by

$$S_{yy}(z) = H(z)R_{ww}H'(z^{-1}) + R_{vv} \tag{3.69}$$

So we see that the Gauss–Markov state-space model enables us to have a more general representation of a multichannel stochastic signal. In fact, we are able to easily handle the multichannel and nonstationary statistical cases within this framework. Generalizations are also possible with the vector models, but those forms become quite complicated and require some knowledge of multivariable systems theory and canonical forms (see [1] for details). Before we leave this subject, let us consider a simple input/output example with Gauss–Markov models.

Example 3.2 Consider the following difference equation driven by random (white) noise:

$$y(t) = -ay(t-1) + e(t-1)$$

The corresponding state-space representation is obtained as

$$x(t) = -ax(t-1) + w(t-1) \quad \text{and} \quad y(t) = x(t)$$

Taking $\mathcal{Z}$-transforms (ignoring the randomness), we obtain the transfer function

$$H(z) = \frac{1}{1 - az^{-1}}$$

Using Eq. (3.62), the variance equation for the above model is

$$P(t) = a^2 P(t-1) + R_{ww}$$

Assume the process is stationary, then $P(t) = P$ for all t and solving for P it follows that

$$P = \frac{R_{ww}}{1 - a^2}$$

Therefore,

$$R_{yy}(k) = CA^{|k|}PC' = \frac{a^{|k|}R_{ww}}{1 - a^2} \quad \text{and} \quad R_{yy}(0) = CPC' + R_{vv} = \frac{R_{ww}}{1 - a^2}$$

for $C = 1$.

Choosing $R_{ww} = 1 - a^2$ gives $R_{yy}(k) = a^{|k|}$. Taking $\mathcal{Z}$-transforms the discrete power spectrum is given by

$$S_{yy}(z) = H(z)R_{ee}H'(z^{-1}) + R_{vv} = \frac{1}{1 - az^{-1}}R_{ww}\frac{1}{1 - az}$$

Therefore, we conclude that for stationary processes these models are equivalent.

Now if we assume a nonstationary process and let $a = -0.75$, $x(0) \sim \mathcal{N}(1, 2.3)$, $w \sim \mathcal{N}(0, 1)$, and $v \sim \mathcal{N}(0, 4)$, then the Gauss–Markov model is given by

$$x(t) = 0.75x(t-1) + w(t-1) \quad \text{and} \quad y(t) = x(t) + v(t)$$

The corresponding statistics are given by the mean relations

$$\begin{aligned} m_x(t) &= 0.75m_x(t-1) & m_x(0) &= 1 \\ m_y(t) &= m_x(t) & m_y(0) &= m_x(0) \end{aligned}$$

and the variance equations

$$P(t) = 0.5625P(t-1) + 1 \quad \text{and} \quad R_{yy}(t) = P(t) + 4$$

We apply the simulator available in MATLAB [15] to obtain a 100-sample realization of the process. The results are shown in Figure 3.2a through c. In Figure 3.2a,b we see the mean and simulated states with the corresponding confidence interval about the mean, that is,

$$[m_x(t) \pm 1.96\sqrt{P(t)}]$$

and

$$P = \frac{R_{ww}}{1 - a^2} = 2.286$$

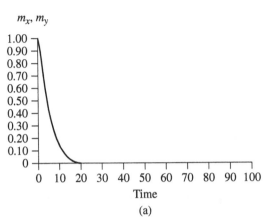

(a)

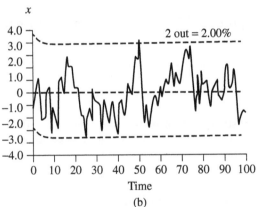

(b)

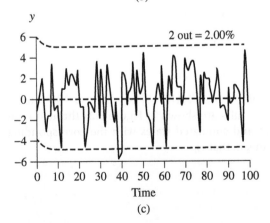

(c)

Figure 3.2 Gauss–Markov simulation of first-order process: (a) state/measurement means, (b) state with 95% confidence interval about its mean, and (c) Measurement with 95% confidence interval about its mean.

Using the above confidence interval, we expect 95% of the samples to lie within $(m_x \rightarrow 0)$ ±3.03 $(1.96 \times \sqrt{2.286})$ From the figure, we see that only 2 of the 100 samples exceed this bound, indicating a statistically acceptable simulation. We observe similar results for the simulated measurements. The steady-state variance is given by

$$R_{yy} = P + R_{vv} = 2.286 + 4 = 6.286$$

Therefore, we expect 95% of the measurement samples to lie within $(m_y \rightarrow 0)$ ±5.01$(1.96 \times \sqrt{6.286})$ at steady state. This completes the example. ◻

It should be noted that if the bounds are exceeded by more than 5% (< 95% lie within), then we must select a *new seed* and reexecute the simulation until the bound conditions are satisfied signifying a *valid* Gauss–Markov simulation.

3.6 Innovations Model

In this section, we briefly develop the innovations model that is related to the Gauss–Markov representation. The significance of this model will be developed throughout the text, but we take the opportunity now to show its relationship to the basic Gauss–Markov representation. We start by extending the original Gauss–Markov representation to the *correlated* process and measurement noise case and then show how the innovations model is a special case of this structure.

The *standard* Gauss–Markov model for *correlated process and measurement noise* is given by

$$x(t) = Ax(t-1) + Bu(t-1) + W(t-1)w^*(t-1)$$
$$y(t) = Cx(t) + v^*(t) \tag{3.70}$$

where $R^*(t,k) := R^*\delta(t-k)$ and

$$R^* := \begin{bmatrix} R_{w^*w^*} & | & R_{w^*v^*} \\ - & - & - \\ R_{v^*w^*} & | & R_{v^*v^*} \end{bmatrix} = \begin{bmatrix} WR_{ww}W' & | & WR_{wv} \\ - & - & - \\ R_{vw}W' & | & R_{vv} \end{bmatrix}$$

Here we observe that in the standard Gauss–Markov model, the $(N_x + N_v) \times (N_x + N_v)$ block covariance matrix, R^*, is full with cross-covariance matrices $R_{w^*v^*}$ on its off-diagonals. The usual standard model assumes that they are null (uncorrelated). To simulate a system with correlated $w(t)$ and $v(t)$ is more complicated using this form of the Gauss–Markov model because R^* must first be factored such that

$$R^* = \begin{bmatrix} R_1^* \\ R_2^* \end{bmatrix} \begin{bmatrix} R_1^{*\prime} & R_2^{*\prime} \end{bmatrix} \tag{3.71}$$

where R_i^* are matrix square roots [6, 7]. Once the factorization is performed, then the correlated noise is synthesized "coloring" the uncorrelated noise sources, $w(t)$ and $v(t)$ as

$$\begin{bmatrix} w^*(t) \\ v^*(t) \end{bmatrix} = \begin{bmatrix} R_1^{*\prime} w(t) \\ R_2^{*\prime} v(t) \end{bmatrix} \tag{3.72}$$

The innovations model is a constrained version of the correlated Gauss–Markov characterization. If we assume that $\{e(t)\}$ is a zero-mean, white, Gaussian sequence, that is, $e \sim \mathcal{N}(0, R_{ee})$, then the *innovations model* [11–15] evolves as

$$x(t) = A(t-1)x(t-1) + B(t-1)u(t-1) + K(t-1)e(t-1)$$
$$y(t) = C(t)x(t) + e(t) \tag{3.73}$$

where $e(t)$ is the N_y-dimensional innovations vector and $K(t-1)$ is the $(N_x \times N_y)$ weighting matrix with the innovations covariance specified by

$$R_{ee}^* := \mathrm{cov}\left(\begin{bmatrix} Ke(t) \\ e(t) \end{bmatrix} \right) = \begin{bmatrix} KR_{ee}K' & | & KR_{ee} \\ - & - & - & - & - \\ R_{ee}K' & | & R_{ee} \end{bmatrix} \delta(t-k)$$

It is important to note that the innovations model has implications in Wiener–Kalman filtering (spectral factorization) because R_{ee}^* can be represented in *factored* or *square-root* form ($R := \sqrt{R}\sqrt{R}'$) directly in terms of the weight and innovations covariance matrix as

$$R_{ee}^* := \begin{bmatrix} K\sqrt{R_{ee}} \\ \sqrt{R_{ee}} \end{bmatrix} \begin{bmatrix} \sqrt{R_{ee}}K' & \sqrt{R_{ee}} \end{bmatrix} \delta(t-k) \tag{3.74}$$

Comparing the innovations model to the Gauss–Markov model, we see that they are both equivalent to the case when w and v are correlated. Next we show the equivalence of the various model sets to this family of state-space representations.

3.7 State-Space Model Structures

In this section, we discuss special state-space structures usually called "canonical forms" in the literature, since they represent unique state constructs that are particularly useful. We will confine the models to SISO forms, while the more complicated multivariable structures will be developed in Chapter 6. Here we will first recall the autoregressive-moving average model with exogenous inputs ARMAX model of Chapter 2 and then its equivalent representation in the state-space form.

3.7.1 Time-Series Models

Time-series models are particularly useful representations used frequently by statisticians and signal processors to represent time sequences when no physics is available to employ directly. They form the class of black-box or gray-box models [15], which are useful in predicting data. These models have an input/output structure, but they can be transformed to an equivalent state-space representation. Each model set has its own advantages: the input/output models are easy to use, while the state-space models are easily generalized and usually evolve when physical phenomenology can be described in a model-based sense [15].

We have a *difference equation* relating the output sequence $\{y(t)\}$ to the input sequence $\{u(t)\}$ in terms of the backward shift operator as [4]

$$A(q^{-1})y(t) = B(q^{-1})u(t) \tag{3.75}$$

or

$$y(t) + a_1 y(t-1) + \cdots + a_{N_a} y(t - N_a) = b_0 u(t) + \cdots b_{N_b} u(t - N_b) \tag{3.76}$$

Recall from Section 2.4 that when the system is excited by random inputs, the models can be represented by an *ARMAX* and is abbreviated by ARMAX(N_a, N_b, N_c).

$$\underbrace{A(q^{-1})y(t)}_{AR} = \underbrace{B(q^{-1})u(t)}_{X} + \underbrace{C(q^{-1})e(t)}_{MA} \tag{3.77}$$

where A, B, C, are polynomials in the backward-shift operator q^{-1} such that $q^{-k}y(t) = y(t - k)$ and $\{e(t)\}$ is a white noise source with coloring filter

$$C(q^{-1}) = 1 + c_1 q^{-1} + \cdots + c_{N_c} q^{-N_c}$$

Since the ARMAX model is used to characterize a random signal, we are interested in its statistical properties (see [15] for details). We summarize these properties in Table 3.2.

This completes the section on ARMAX models.

3.7.2 State-Space and Time-Series Equivalence Models

In this section, we show the equivalence between the ARMAX and state-space models (for scalar processes). That is, we show how to obtain the state-space model given the ARMAX models by inspection. We choose particular coordinate systems in the state-space (canonical form) and obtain a relationship between entries of the state-space system to coefficients of the ARMAX model.

---

4 We change from the common signal processing convention of using $x(t)$ for the deterministic excitation to $u(t)$ and we include the b_0 coefficient for generality.

Table 3.2 ARMAX representation.

Output propagation
$$y(t) = (1 - A(q^{-1}))y(t) + B(q^{-1})u(t) + C(q^{-1})e(t)$$
Mean propagation
$$m_y(t) = (1 - A(q^{-1}))m_y(t) + B(q^{-1})u(t) + C(q^{-1})m_e(t)$$
Impulse propagation
$$h(t) = (1 - A(q^{-1}))h(t) + C(q^{-1})\delta(t)$$
Variance/covariance propagation
$$R_{yy}(k) = R_{ee} \sum_{i=0}^{\infty} h(i)h(i+k) \quad k \geq 0$$

---

y = the output or measurement sequence

u = the input sequence

e = the process (white) noise sequence with variance R_{ee}

h = the impulse response sequence

δ = the impulse input of amplitude $\sqrt{R_{ee}}$

m_y = the mean output or measurement sequence

m_e = the mean process noise sequence

R_{yy} = the stationary output covariance at lag k

A = the N_ath-order system characteristic (poles) polynomial

B = the N_bth-order input (zeros) polynomial

C = the N_cth-order noise (zeros) polynomial

An example is presented that shows how these models can be applied to realize a random signal. First, we consider the ARMAX to state-space transformation.

Recall from Eq. (3.77) that the general difference equation form of the ARMAX model is given by

$$y(t) = -\sum_{i=1}^{N_a} a_i y(t-i) + \sum_{i=0}^{N_b} b_i u(t-i) + \sum_{i=0}^{N_c} c_i e(t-i) \tag{3.78}$$

or equivalently in the frequency domain as

$$Y(z) = \left(\frac{b_o + b_1 z^{-1} + \cdots + b_{N_b} z^{-N_b}}{1 + a_1 z^{-1} + \cdots + a_{N_a} z^{-N_a}} \right) U(z)$$

$$+ \left(\frac{c_o + c_1 z^{-1} + \cdots + c_{N_c} z^{-N_c}}{1 + a_1 z^{-1} + \cdots + a_{N_a} z^{-N_a}} \right) E(z) \tag{3.79}$$

where $N_a \geq N_b$ and N_c and $\{e(t)\}$ is a zero-mean, white sequence with spectrum given by R_{ee}.

It is straightforward to show (see [5]) that the ARMAX model can be represented in *observer canonical form*:

$$x(t) = A_0 x(t-1) + B_0 u(t-1) + W_0 e(t-1)$$
$$y(t) = C_0' x(t) + b_0 u(t) + c_0 e(t) \tag{3.80}$$

where x, u, e, and y are the N_a-state vector, scalar input, noise, and output with

$$A_0 := \begin{bmatrix} 0 & | & -a_{N_a} \\ --- & | & \vdots \\ I_{N_a-1} & | & -a_1 \end{bmatrix} \quad B_0 := \begin{bmatrix} -a_{N_a} b_0 \\ \vdots \\ -a_{N_b+1} b_0 \\ --- \\ b_{N_b} - a_{N_b} b_0 \\ \vdots \\ b_1 - a_1 b_0 \end{bmatrix} \quad W_0 := \begin{bmatrix} -a_{N_a} c_0 \\ \vdots \\ -a_{N_c+1} c_0 \\ --- \\ c_{N_c} - a_{N_c} c_0 \\ \vdots \\ c_1 - a_1 c_0 \end{bmatrix}$$

$$C_0' := [0 \quad \cdots \quad 0 \quad 1]$$

Noting this structure we see that each of the matrix or vector elements $\{A_{i,N_a}, B_i, W_i, C_i\}$ $i = 1, \ldots, N_a$ can be determined from the relations

$$A_{i,N_a} = -a_i \quad i = 1, \ldots, N_a$$
$$B_i = b_i - a_i b_0$$
$$W_i = c_i - a_i c_0$$
$$C_i = \delta(N_a - i) \tag{3.81}$$

where

$$b_i = 0 \qquad \text{for} \quad i > N_b$$
$$c_i = 0 \qquad \text{for} \quad i > N_c$$
$$\delta(i-j) \quad \text{is the Kronecker delta}$$

Consider the following example to illustrate these relations.

Example 3.3 Let $N_a = 3$, $N_b = 2$, and $N_c = 1$; then the corresponding ARMAX model is

$$y(t) = -a_1 y(t-1) - a_2 y(t-2) - a_3 y(t-3) + b_0 u(t)$$
$$+ b_1 u(t-1) + b_2 u(t-2) + c_0 e(t) + c_1 e(t-1) \tag{3.82}$$

Using the observer canonical form of Eq. (3.80), we have

$$
x(t) = \left[\begin{array}{cc|c} 0 & 0 & -a_3 \\ \hline 1 & 0 & -a_2 \\ 0 & 1 & -a_1 \end{array} \right] x(t-1) + \left[\begin{array}{c} -a_3 b_0 \\ \hline b_2 - a_2 b_0 \\ b_1 - a_1 b_0 \end{array} \right] u(t-1)
$$

$$
+ \left[\begin{array}{c} -a_3 c_0 \\ -a_2 c_0 \\ \hline c_1 - a_1 c_0 \end{array} \right] e(t-1)
$$

$$
y(t) = [0 \quad 0 \quad 1] x(t) + b_0 u(t) + c_0 e(t)
$$

This completes the example. □

It is important to realize that if we assume that $\{e(t)\}$ is Gaussian, then the ARMAX model is equivalent to the innovations representation of Section 3.6, that is,

$$
x(t) = Ax(t-1) + Bu(t-1) + We(t-1)
$$
$$
y(t) = C'x(t) + b_0 u(t) + c_0 e(t) \tag{3.83}
$$

where, in this case, $K \to W$, $D \to b_0$, and $1 \to c_0$. Also, the corresponding covariance matrix becomes

$$
R_{ee}^* := \mathrm{cov}\left(\left[\begin{array}{c} We(t) \\ c_0 e(t) \end{array} \right] \right) = \left[\begin{array}{c|c} W R_{ee} W' & W R_{ee} c_0 \\ \hline c_0 R_{ee} W' & c_0 R_{ee} c_0 \end{array} \right] \delta(t-k)
$$

This completes the discussion on the equivalence of the general ARMAX to state-space. Next let us develop the state-space equivalent models for some of the special cases of the ARMAX model presented in Section 2.4.

We begin with the moving average (MA)

$$
y(t) = \sum_{i=1}^{N_c} c_i e(t-i) \quad \text{or}
$$

$$
Y(z) = C(z)E(z) = (1 + c_1 z^{-1} + \cdots + c_{N_c} z^{-N_c}) E(z)
$$

Define the state variable as

$$
x_i(t-1) := e(t-i-1), \quad i = 1, \dots, N_c \tag{3.84}
$$

and therefore,

$$x_i(t) = e(t - i) = x_{i-1}(t - 1), \quad i = 1, \ldots, N_c \tag{3.85}$$

Expanding this expression, we obtain

$$
\begin{aligned}
x_1(t) &= e(t - 1) \\
x_2(t) &= e(t - 2) = x_1(t - 1) \\
x_3(t) &= e(t - 3) = x_2(t - 1) \\
&\vdots \qquad \vdots \qquad \vdots \\
x_{Nc}(t) &= e(t - N_c) = x_{N_c-1}(t - 1)
\end{aligned}
\tag{3.86}
$$

or in vector–matrix form

$$
\begin{bmatrix} x_1(t) \\ x_2(t) \\ \vdots \\ x_{N_c}(t) \end{bmatrix}
=
\left[\begin{array}{cccc|c}
0 & \cdots & 0 & | & 0 \\
\hline
1 & \cdots & 0 & | & 0 \\
\vdots & \ddots & \vdots & | & \vdots \\
0 & \cdots & 1 & | & 0
\end{array}\right]
\begin{bmatrix} x_1(t-1) \\ x_2(t-1) \\ \vdots \\ x_{N_c}(t-1) \end{bmatrix}
+
\begin{bmatrix} 1 \\ 0 \\ \vdots \\ 0 \end{bmatrix} e(t-1)
$$

$$
y(t) = \begin{bmatrix} c_1 & c_2 & \cdots & c_{N_c} \end{bmatrix}
\begin{bmatrix} x_1(t-1) \\ x_2(t-1) \\ \vdots \\ x_{N_c}(t-1) \end{bmatrix}
+ c_0 e(t)
\tag{3.87}
$$

Thus, the general form for the MA state-space is given by

$$
\mathbf{x}(t) =
\left[\begin{array}{ccc|c}
0 & \cdots & 0 & | \; 0 \\
\hline
& & & | \; \vdots \\
& I_{N_c-1} & & | \; \vdots \\
& & & | \; 0
\end{array}\right]
\mathbf{x}(t-1) + \mathbf{b}e(t-1)
$$

$$
y(t) = \mathbf{c}'\mathbf{x}(t) + b_0 e(t)
\tag{3.88}
$$

with $N_x = N_c$, $\mathbf{b}, \mathbf{c} \in R^{N_x \times 1}$.

Next consider the autoregressive (AR) model (all-pole) given by

$$
\sum_{i=1}^{N_a} a_i y(t - i) = \sigma e(t) \quad \text{or} \quad Y(z) = \frac{\sigma E(z)}{1 + a_1 z^{-1} + \cdots + a_{N_a} z^{-N_a}}
\tag{3.89}
$$

Here the state vector is defined by $x_i(t - 1) = y(t - i - 1)$ and therefore, $x_i(t) = y(t - i) = x_{i+1}(t - 1)$; $i = 1, \ldots, N_a - 1$ with $x_{N_a}(t) = y(t)$. Expanding over i, we obtain the vector–matrix state-space model

$$
\begin{bmatrix} x_1(t) \\ x_2(t) \\ \vdots \\ x_{N_a}(t) \end{bmatrix} = \left[\begin{array}{c|cccc} 0 & 1 & 0 & \cdots & 0 \\ 0 & 0 & 1 & \cdots & 0 \\ \vdots & \vdots & \vdots & \ddots & \vdots \\ 0 & 0 & 0 & \cdots & 1 \\ \hline -a_{N_a} & -a_{N_a-1} & -a_{N_a-2} & \cdots & -a_1 \end{array} \right] \begin{bmatrix} x_1(t-1) \\ x_2(t-1) \\ \vdots \\ x_{N_a}(t-1) \end{bmatrix}
$$

$$
+ \begin{bmatrix} 0 \\ 0 \\ \vdots \\ \sigma \end{bmatrix} e(t-1)
$$

$$
y(t) = [0 \quad 0 \quad \cdots \quad 1] \begin{bmatrix} x_1(t-1) \\ x_2(t-1) \\ \vdots \\ x_{N_a}(t-1) \end{bmatrix} \tag{3.90}
$$

In general, we have the AR (*all-pole*) state-space model

$$
\mathbf{x}(t) = \left[\begin{array}{ccc|cccc} 0 & & & & & & \\ \vdots & & & & \mathbf{I}_{N_a-1} & & \\ 0 & & & & & & \\ \hline -a_{N_a} & -a_{N_a-1} & -a_{N_a-2} & \cdots & & & -a_1 \end{array} \right] \mathbf{x}(t-1) + \mathbf{b}e(t-1)
$$

$$
y(t) = \mathbf{c}'\mathbf{x}(t) \tag{3.91}
$$

with $N_x = N_a$, $\mathbf{b}, \mathbf{c} \in R^{N_x \times 1}$.

Another useful state-space representation is the *normal form* that evolves by performing a partial fraction expansion of a rational discrete transfer function model (ARMA) to obtain

$$
h(t) = \sum_{i=1}^{N_p} R_i(p_i)^{-t} \quad \text{or} \quad H(z^{-1}) = \frac{Y(z^{-1})}{E(z^{-1})} = \sum_{i=1}^{N_p} \frac{R_i}{1 - p_i z^{-1}} \tag{3.92}
$$

for $\{R_i, p_i\}; i = 1, \ldots, N_p$ the set of residues and poles of $H(z^{-1})$. Note that the normal form model is the decoupled or parallel system representation based on the following set of relations:

$$
y_i(t) - p_i y_i(t-1) = e(t), \quad i = 1, \ldots, N_p
$$

Defining the state variable as $x_i(t) := y_i(t)$, then equivalently

$$
x_i(t) - p_i x_i(t-1) = e(t), \quad i = 1, \ldots, N_p \tag{3.93}
$$

and therefore, the output is given by

$$y(t) = \sum_{i=1}^{N_p} R_i y_i(t) = \sum_{i=1}^{N_p} R_i x_i(t), \quad i = 1, \dots, N_p \tag{3.94}$$

Expanding these relations over i, we obtain

$$\begin{bmatrix} x_1(t) \\ x_2(t) \\ \vdots \\ x_{N_p}(t) \end{bmatrix} = \begin{bmatrix} p_1 & 0 & \cdots & 0 \\ 0 & p_2 & \cdots & 0 \\ \vdots & \vdots & \ddots & \vdots \\ 0 & 0 & \cdots & p_{N_p} \end{bmatrix} \begin{bmatrix} x_1(t-1) \\ x_2(t-1) \\ \vdots \\ x_{N_a}(t-1) \end{bmatrix} + \begin{bmatrix} 1 \\ 1 \\ \vdots \\ 1 \end{bmatrix} e(t-1)$$

$$y(t) = [R_1 \quad R_2 \quad \cdots \quad R_{N_p}] \begin{bmatrix} x_1(t-1) \\ x_2(t-1) \\ \vdots \\ x_{N_p}(t-1) \end{bmatrix} \tag{3.95}$$

Thus, the general decoupled form of the *normal state-space model* is given by

$$\mathbf{x}(t) = \begin{bmatrix} p_1 & 0 & \cdots & 0 \\ 0 & p_2 & \cdots & 0 \\ \vdots & \vdots & \ddots & \vdots \\ 0 & 0 & \cdots & p_{N_p} \end{bmatrix} \mathbf{x}(t-1) + \mathbf{b}e(t-1)$$

$$y(t) = \mathbf{c}'\mathbf{x}(t) \tag{3.96}$$

for $\mathbf{b} \in \mathcal{R}^{N_p \times 1}$ with $\mathbf{b} = \mathbf{1}$, an N_p-vector of unit elements. Here $\mathbf{c} \in \mathcal{R}^{1 \times N_p}$ and $\mathbf{c}' = [R_1 \quad R_2 \quad \cdots \quad R_{N_p}]$.

3.8 Nonlinear (Approximate) Gauss–Markov State-Space Models

Many processes in practice are nonlinear, rather than linear. Coupling the non-linearities with noisy data makes the signal processing problem a challenging one. In this section, we develop an approximate solution to the nonlinear modeling problem involving the linearization of the nonlinear process about a "known" reference trajectory. In this section, we limit our discussion to *discrete* nonlinear systems. Continuous solutions to this problem are developed in [6–15].

Suppose we model a *process* by a set of nonlinear stochastic vector difference equations in state-space form as

$$x(t) = a[x(t-1)] + b[u(t-1)] + w(t-1) \tag{3.97}$$

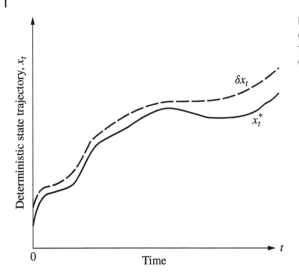

Figure 3.3 Linearization of a deterministic system using the reference trajectory defined by $(x^*(t), u^*(t))$.

with the corresponding *measurement* model

$$y(t) = c[x(t)] + d[u(t)] + v(t) \tag{3.98}$$

where $a[\cdot], b[\cdot], c[\cdot], d[\cdot]$ are nonlinear vector functions of $x, u,$ with $x, a, b, w \in R^{N_x \times 1}, y, c, d, v \in R^{N_y \times 1}$ and $w \sim \mathcal{N}(0, R_{ww}(t)), v \sim \mathcal{N}(0, R_{vv}(t))$.

Ignoring the additive noise sources, we "linearize" the process and measurement *models* about a known deterministic *reference trajectory* defined by $[x^*(t), u^*(t)]$ as illustrated in Figure 3.3,[5] that is,

$$x^*(t) = a[x^*(t-1)] + b[u^*(t-1)] \tag{3.99}$$

Deviations or perturbations from this trajectory are defined by

$$\delta x(t) := x(t) - x^*(t)$$
$$\delta u(t) := u(t) - u^*(t)$$

Substituting the previous equations into these expressions, we obtain the perturbation trajectory as

$$\delta x(t) = a[x(t-1)] - a[x^*(t-1)] + b[u(t-1)] - b[u^*(t-1)] + w(t-1) \tag{3.100}$$

---

5 In practice, the reference trajectory is obtained either by developing a mathematical model of the process or by simulating about some reasonable operating conditions to generate the trajectory using the state-space model.

The nonlinear vector functions $a[\cdot]$ and $b[\cdot]$ can be expanded into a first-order *Taylor series* about the reference trajectory $[x^*(t), u^*(t)]$ as[6]

$$a[x(t-1)] = a[x^*(t-1)] + \frac{da[x^*(t-1)]}{dx^*(t-1)}\delta x(t-1) + \text{HOT}$$

$$b[u(t-1)] = b[u^*(t-1)] + \frac{db[u^*(t-1)]}{du^*(t-1)}\delta u(t-1) + \text{HOT} \qquad (3.101)$$

We define the first-order *Jacobian matrices* as

$$A[x^*(t-1)] := \frac{da[x^*(t-1)]}{dx^*(t-1)} \quad \text{and} \quad B[u^*(t-1)] := \frac{db[u^*(t-1)]}{du^*(t-1)}$$
$$(3.102)$$

Incorporating the definitions of Eq. (3.102) and neglecting the higher order terms (HOT) in Eq. (3.101), the *linearized perturbation process model* in (3.100) can be expressed as

$$\delta x(t) = A[x^*(t-1)]\delta x(t-1) + B[u^*(t-1)]\delta u(t-1) + w(t-1) \quad (3.103)$$

Similarly, the measurement system can be linearized by using the reference measurement

$$y^*(t) = c[x^*(t)] + d[u^*(t)] \qquad (3.104)$$

and applying the Taylor series expansion to the nonlinear measurement model

$$c[x(t)] = c[x^*(t)] + \frac{d\ c[x^*(t)]}{dx^*(t)}\delta x(t) + \text{HOT}$$

$$d[u(t)] = d[u^*(t)] + \frac{d\ d[u^*(t)]}{du^*(t)}\delta u(t) + \text{HOT} \qquad (3.105)$$

The corresponding measurement perturbation model is defined by

$$\delta y(t) := y(t) - y^*(t) = c[x(t)] + d[u(t)] - c[x^*(t)] - d[u^*(t)] + v(t)$$
$$(3.106)$$

Substituting the first-order approximations for $c[x(t)]$ and $d[u(t)]$ leads to the linearized measurement perturbation as

$$\delta y(t) = C[x^*(t)]\delta x(t) + D[u^*(t)]\delta u(t) + v(t) \qquad (3.107)$$

where $C[x^*(t)]$ is defined as the measurement Jacobian.

Summarizing, we have linearized a deterministic nonlinear model using a first-order Taylor series expansion for the model functions, a, b, and c and then

---

6 We use the shorthand notation $\frac{d(\cdot)}{d\theta^*}$ to mean $\frac{d(\cdot)}{d\theta}|_{\theta=\theta^*}$.

developed a *linearized Gauss–Markov perturbation model* valid for small deviations from the reference trajectory given by

$$\delta x(t) = A[x^*(t-1)]\delta x(t-1) + B[u^*(t-1)]\delta u(t-1) + w(t-1)$$
$$\delta y(t) = C[x^*(t)]\delta x(t) + D[u^*(t)]\delta u(t) + v(t) \tag{3.108}$$

with A, B, C, and D the corresponding Jacobian matrices with w, v zero-mean, Gaussian.

We can also use linearization techniques to approximate the *statistics* of the process and measurements. If we use the first-order Taylor series expansion and expand about the *mean*, $m_x(t)$, rather than $x^*(t)$, then taking expected values

$$m_x(t) = E\{a[x(t-1)]\} + E\{b[u(t-1)]\} + E\{w(t-1)\} \tag{3.109}$$

gives

$$m_x(t) = a[m_x(t-1)] + b[u(t-1)] \tag{3.110}$$

which follows by linearizing $a[\cdot]$ about m_x and taking the expected value.

The variance equations $P(t) := \text{cov}(x(t))$ can also be developed in a similar manner (see [7] for details) to give

$$P(t) = A[m_x(t-1)]P(t-1)A'[m_x(t-1)] + R_{ww}(t-1) \tag{3.111}$$

Using the same approach, we arrive at the accompanying measurement statistics

$$m_y(t) = c[m_x(t)] + d[u(t)] \quad \text{and}$$
$$R_{yy}(t) = C[m_x(t)]P(t)C'[m_x(t)] + R_{vv}(t) \tag{3.112}$$

We summarize these results in the "approximate" Gauss–Markov model of Table 3.3.

Before we close, consider the following example to illustrate the approximation.

Example 3.4 Consider the discrete nonlinear process given by

$$x(t) = (1 - 0.05\Delta T)x(t-1) + 0.04\Delta Tx^2(t-1) + w(t-1)$$

with the corresponding measurement model

$$y(t) = x^2(t) + x^3(t) + v(t)$$

where $w(t) \sim \mathcal{N}(0, R_{ww})$, $v(t) \sim \mathcal{N}(0, R_{vv})$, and $x(0) \sim \mathcal{N}(\overline{x}(0), \overline{P}(0))$. Performing the differentiations, we obtain the following Jacobians:

$$A[x(t-1)] = 1 - 0.05\Delta T + 0.08\Delta Tx(t-1) \text{ and } C[x(t)] = 2x(t) + 3x^2(t)$$
□

Although the linearization approach discussed here seems somewhat extraneous relative to the previous sections, it becomes a crucial ingredient in the *classical approach* to (approximate) nonlinear estimation of the subsequent

Table 3.3 Approximate nonlinear Gauss–Markov model.

State propagation

$$x(t) = a[x(t-1)] + b[u(t-1)] + w(t-1)$$

State mean propagation

$$m_x(t) = a[m_x(t-1)] + b[u(t-1)]$$

State covariance propagation

$$P(t) = A[m_x(t-1)]P(t-1)A'[m_x(t-1)] + R_{ww}(t-1)$$

Measurement propagation

$$y(t) = c[x(t)] + v(t)$$

Measurement mean propagation

$$m_y(t) = c[m_x(t)]$$

Measurement covariance propagation

$$R_{yy}(t) = C[m_x(t)]P(t)C'[m_x(t)] + R_{vv}(t)$$

Initial conditions

$$x(0) \quad and \quad P(0)$$

Jacobians

$$A[x^*(t-1)] \equiv \left. \frac{da[x(t-1)]}{dx(t-1)} \right|_{x=x^*(t-1)} \qquad C[x^*(t)] \equiv \left. \frac{dc[x(t)]}{dx(t)} \right|_{x=x^*(t)}$$

---

chapters. We discuss the linear state-space approach (Kalman filter) to the estimation problem in Chapter 4 and then show how these linearization concepts can be used to solve the nonlinear estimation problem. There the popular "extended" Kalman filter processor relies heavily on these linearization techniques developed in this section for its development.

3.9 Summary

In this chapter, we have discussed the development of continuous-time, sampled-data, and discrete-time state-space models. The stochastic variants of the deterministic models were presented leading to the Gauss–Markov representations for both linear and (approximate) nonlinear systems. The discussion of both the deterministic and stochastic state-space models included a brief development of their second-order statistics. We also discussed the underlying discrete systems theory as well as a variety of time-series models (ARMAX, AR, MA, etc.) and showed that they can easily be represented in state-space form through the use of canonical forms (models). These models form the embedded structure incorporated into the model-based processors that will be discussed in subsequent chapters. We concluded the chapter with a brief development of a "linearized" nonlinear model leading to an approximate Gauss–Markov representation.

MATLAB Notes

MATLAB has many commands to convert to/from state-space models to other forms useful in signal processing. Many of them reside in the Signal Processing and Control Systems toolboxes. The matrix exponential is invoked by the **expm** command and is determined from Taylor/Padé approximants using the scaling and squaring approach. Also the commands **expmdemo1**, **expmdemo2**, and **expmdemo3** demonstrate the trade-offs of the Padé, Taylor, and eigenvector approaches to calculate the matrix exponential. The ordinary differential equation method is available using the wide variety of numerical integrators available (**ode***). Converting to/from transfer functions and state-space is accomplished using the **ss2tf** and **tf2ss** commands, respectively. ARMAX simulations are easily accomplished using the **filter** command with a variety of options converting from ARMAX-to/from transfer functions. The Identification Toolbox converts polynomial-based models to state-space and continuous parameters including Gauss–Markov to discrete parameters (**th2ss, thc2thd, thd2thc**).

References

1 Kailath, T. (1980). *Linear Systems*. Englewood Cliffs, NJ: Prentice-Hall.
2 Szidarovszky, F. and Bahill, A. (1980). *Linear Systems Theory*. Boca Raton, FL: CRC Press.
3 DeCarlo, R. (1989). *Linear Systems: A State Variable Approach*. Englewood Cliffs, NJ: Prentice-Hall.
4 Chen, C. (1984). *Introduction to Linear System Theory*. New York: Holt, Rinehart, and Winston.
5 Tretter, S. (1976). *Introduction to Discrete-Time Signal Processing*. New York: Wiley.
6 Jazwinski, A. (1970). *Stochastic Processes and Filtering Theory*. New York: Academic Press.
7 Sage, A. and Melsa, J. (1971). *Estimation Theory with Applications to Communications and Control*. New York: McGraw-Hill.
8 Maybeck, P. (1979). *Stochastic Models, Estimation and Control*, vol. 1. New York: Academic Press.
9 Anderson, B. and Moore, J. (2005). *Optimal Filtering*. Mineola, NY: Dover Publications.
10 Katayama, T. (2005). *Subspace Methods for System Identification*. London: Springer.
11 Goodwin, G. and Payne, R.L. (1976). *Dynamic System Identification*. New York: Academic Press.

12 Goodwin, G. and Sin, K. (1984). *Adaptive Filtering, Prediction and Control.* Englewood Cliffs, NJ: Prentice-Hall.

13 Mendel, J. (1995). *Lessons in Estimation Theory for Signal Processing, Communications, and Control.* Englewood Cliffs, NJ: Prentice-Hall.

14 Brown, R. and Hwang, P.C. (1997). *Introduction to Random Signals and Applied Kalman Filtering.* New York: Wiley.

15 Candy, J. (2006). *Model-Based Signal Processing.* Hoboken, NJ: Wiley/IEEE Press.

16 Robinson, E. and Silvia, M. (1979). *Digital Foundations of Time Series Analysis,* vol. 1. San Francisco, CA: Holden-Day.

17 Simon, D. (2006). *Optimal State Estimation Kalman, H_∞ and Nonlinear Approaches.* Hoboken, NJ: Wiley.

18 Grewal, M.S. and Andrews, A.P. (1993). *Kalman Filtering: Theory and Practice.* Englewood Cliffs, NJ: Prentice-Hall.

19 Moler, C. and Van Loan, C. (2003). Nineteen dubious ways to compute the exponential of a matrix, twenty-five years later. *SIAM Rev.* 45 (1): 3–49.

20 Golub, G. and Van Loan, C. (1989). *Matrix Computation.* Baltimore, MA: Johns Hopkins University Press.

21 Ho, B. and Kalman, R. (1966). Effective reconstruction of linear state variable models from input/output data. *Regelungstechnik* 14: 545–548.

22 Candy, J., Warren, M., and Bullock, T. (1977). Realization of an invariant system description from Markov sequences. *IEEE Trans. Autom. Control* 23 (7): 93–96.

23 Kung, S., Arun, K., and Bhaskar Rao, D. (1983). State-space and singular-value decomposition-based approximation methods for the harmonic retrieval problem. *J. Opt. Soc. Am.* 73 (12): 1799–1811.

Problems

3.1 Derive the following properties of conditional expectations:
a) $E_x\{X|Y\} = E\{X\}$ if X and Y are independent.
b) $E\{X\} = E_y\{E\{X|Y\}\}$.
c) $E_x\{g(Y)X\} = E_y\{g(Y)E\{X|Y\}\}$.
d) $E_{xy}\{g(Y)X\} = E_y\{g(Y)E\{X|Y\}\}$.
e) $E_x\{c|Y\} = c$.
f) $E_x\{g(Y)|Y\} = g(Y)$.
g) $E_{xy}\{cX + dY|Z\} = cE\{X|Z\} + dE\{Y|Z\}$.
(*Hint*: See Appendix A.1.)

3.2 Suppose x, y, z are Gaussian random variables with corresponding means m_x, m_y, m_z and variances R_{xx}, R_{yy}, R_{zz} show that:

a) If $y = ax + b$, a, b constants, then $y \sim N(am_x + b, a^2 R_{xx})$.
b) If x and y are uncorrelated, then they are independent.
c) If $x(i)$ are Gaussian with mean $m(i)$ and variance $R_{xx}(i)$, then for

$$y = \sum_i K_i x(i), \quad y \sim N\left(\sum_i K_i m(i), \sum_i K_i^2 R_{xx}(i) \right)$$

d) If x and y are jointly (conditionally) Gaussian, then

$$E\{x|y\} = m_x + R_{xy} R_{yy}^{-1}(y + m_y) \text{ and}$$
$$R_{x|y} = R_{xx} + R_{xy} R_{yy}^{-1} R_{yx}$$

e) The random variable $x = E\{x|y\}$ is orthogonal to y.
f) If y and z are independent, then

$$E\{x|y, z\} = E\{x|y\} + E\{x|z\} - m_x.$$

g) If y and z are not independent, show that

$$E\{x|y, z\} = E\{x|y, e\} = E\{x|y\} + E\{x|e\} - m_x$$

for $e = z - E\{x|y\}$.
(*Hint*: See Appendices A.1–A.3.)

3.3 Suppose we are given the factored power spectrum $S_{yy}(z) = H(z)H(z^{-1})$ with

$$H(z) = \frac{1 + \beta_1 z^{-1} + \beta_2 z^{-2}}{1 + \alpha_1 z^{-1} + \alpha_2 z^{-2}}$$

a) Develop the ARMAX model for the process.
b) Develop the corresponding Gauss–Markov model for *both* the standard and innovations representation of the process.

3.4 We are given the following Gauss–Markov model

$$x(t) = 1/3x(t - 1) + 1/2w(t - 1)$$
$$y(t) = 5x(t) + v(t)$$
$$w \sim N(0, 3) \quad v \sim N(0, 2)$$

a) Calculate the state power spectrum, $S_{xx}(z)$.
b) Calculate the measurement power spectrum, $S_{yy}(z)$.
c) Calculate the state covariance recursion, $P(t)$.
d) Calculate the steady-state covariance, $P(t) = \cdots = P = P_{ss}$.
e) Calculate the output covariance recursion, $R_{yy}(t)$.
f) Calculate the steady-state output covariance, R_{yy}.

3.5 Suppose we are given the Gauss–Markov process characterized by the state equations

$$x(t) = 0.97x(t-1) + u(t-1) + w(t-1)$$

for $u(t)$ a step of amplitude 0.03 and $w \sim N(0, 10^{-4})$ and $x(0) \sim N(2.5, 10^{-12})$.
a) Calculate the covariance of x, i.e. $P(t) = \text{Cov}(x(t))$.
b) Since the process is stationary, we know that

$$P(t+k) = P(t+k-1) = \cdots = P(0) = P$$

What is the steady-state covariance, P, of this process?
c) Develop a MATLAB program to simulate this process.
d) Plot the process $x(t)$ with the corresponding confidence limits $\pm 2\sqrt{P(t)}$ for 100 data points, do 95% of the samples lie within the bounds?

3.6 Suppose we are given the ARMAX model

$$y(t) = -0.5y(t-1) - 0.7y(t-2) + u(t) + 0.3u(t-1) + e(t)$$
$$+ 0.2e(t-1) + 0.4e(t-2)$$

a) What is the corresponding innovations model in state-space form for $e \sim N(0, 10)$?
b) Calculate the corresponding covariance matrix R_{ee}^*.

3.7 Given a continuous–discrete Gauss–Markov model

$$\dot{x}(t) = ax(t) + u(t) + w(t)$$
$$y(t_k) = \beta x(t_k) + v(t_k)$$

where $w(t)$ and $v(t_k)$ are zero-mean and white with respective covariances, R_{ww} and R_{vv}, along with a piecewise constant input, $u(t)$.
a) Develop the continuous–discrete mean and covariance propagation models for this system.
b) Suppose $w(t)$ is processed by a coloring filter that exponentially correlates it, $R_{ww}(\tau) = Ge^{-|\lambda|\tau}$. Develop the continuous–discrete Gauss–Markov model in this case.

3.8 Develop the continuous–discrete Gauss–Markov models for the following systems:
a) Wiener process: $\dot{z}(t) = w(t)$; $z(0) = 0$, w is zero-mean, white with R_{ww}.
b) Random bias: $\dot{z}(t) = 0$; $z(0) = z_o$ where $z_o \sim \mathcal{N}(0, R_{z_o z_o})$.

c) Random ramp: $\ddot{z}(t) = 0; \dot{z}(0) = z_1; z(0) = z_o$.
d) Random oscillation: $\ddot{z}(t) + \omega_o^2 z(t) = 0; \dot{z}(0) = z_1; z(0) = z_o$.
e) Random second order: $\ddot{z}(t) + 2\zeta\omega_n\dot{z} + \omega_n^2 z(t) = \omega_n^2 w(t); \quad \dot{z}(0) = z_1; z(0) = z_o$.

3.9 Develop the continuous–discrete Gauss–Markov model for correlated process noise, that is,

$$\dot{w}(t) = A_{cw}w(t) + B_{cw}u(t) + W_{cw}w^*(t) \quad \text{for} \quad w^* \sim \mathcal{N}(0, R_{w^*w^*})$$

3.10 Develop the approximate Gauss–Markov model for the following nonlinear state transition and measurement model are given by

$$x(t) = \frac{1}{2}x(t-1) + \frac{25x(t-1)}{1 + x^2(t-1)} + 8\cos(1.2(t-1)) + w(t-1)$$

$$y(t) = \frac{x^2(t)}{20} + v(t)$$

where $w \sim \mathcal{N}(0, R_{ww}(t))$ and $v \sim \mathcal{N}(0, R_{vv}(t))$. The initial state is Gaussian distributed with $\bar{x}(0) \sim \mathcal{N}(0, \bar{P}(0))$.

3.11 Consider the discrete nonlinear process given by

$$x(t) = (1 - 0.05\Delta T)x(t-1) + 0.04\Delta T x^2(t-1) + w(t-1)$$

with corresponding measurement model

$$y(t) = x^2(t) + x^3(t) + v(t)$$

where $w \sim \mathcal{N}(0, R_{ww}(t))$ and $v \sim \mathcal{N}(0, R_{vv}(t))$. The initial state is Gaussian distributed with $\bar{x}(0) \sim \mathcal{N}(0, \bar{P}(0))$.
Develop the approximate Gauss–Markov process model for this nonlinear system.

4

Model-Based Processors

4.1 Introduction

In this chapter, we introduce the fundamental concepts of model-based signal processing using state-space models [1, 2]. We first develop the paradigm using the linear, (time-varying) Gauss–Markov model of the previous chapter and then investigate the required conditional expectations leading to the well-known linear Kalman filter (LKF) processor [3]. Based on this fundamental theme, we progress to the idea of the linearization of a nonlinear, state-space model also discussed in the previous chapter, where we investigate a linearized processor – the linearized Kalman filter (LZKF). It is shown that the resulting processor can provide a solution (time-varying) to the nonlinear state estimation. We then develop the extended Kalman filter (EKF) as a special case of the LZKF, linearizing about the most available state estimate, rather than a reference trajectory. Next we take it one step further to briefly discuss the iterated–extended Kalman filter (IEKF) demonstrating improved performance.

We introduce an entirely different approach to Kalman filtering – the unscented Kalman filter (UKF) that evolves from a statistical linearization, rather than a dynamic model linearization of the previous suite of nonlinear approaches. The theory is based on the concept of "sigma-point" transformations that enable a much better matching of first- and second-order statistics of the assumed posterior distribution with even higher orders achievable.

Finally, we return to the basic Bayesian approach of estimating the posterior state distribution directly leading to the particle filter (PF). Here following the Bayesian paradigm, the processor, which incorporates the nonlinear state-space model, produces a nonparametric estimate of the desired posterior at each time-step. Computationally more intense, but it is capable of operating in both non-Gaussian and multimodal distribution environments. From this perspective, it is important to realize that the LKF is essentially an efficient recursive technique that estimates the conditional mean and covariance of the

Model-Based Processing: An Applied Subspace Identification Approach, First Edition. James V. Candy.
© 2019 John Wiley & Sons, Inc. Published 2019 by John Wiley & Sons, Inc.

posterior Gaussian distribution, but is incapable of dealing with multimodal, non-Gaussian problems effectively.

We summarize the results with a case study implementing a 2D-tracking filter.

4.2 Linear Model-Based Processor: Kalman Filter

We develop the model-based processors (MBP) for the dynamic estimation problem, that is, the estimation of state processes that vary with time. Here the state-space representation is employed as the basic model embedded in the algorithm. To start with, we present the Kalman filter algorithm in the predictor–corrector form. We use this form because it provides insight into the operation of this state-space (MBP) as well as the other recursive processors to follow.

The operation of the MBP algorithm can be viewed as a predictor–corrector algorithm as in standard numerical integration. Referring to the algorithm in Table 4.1 and Figure 4.1, we see its inherent timing in the algorithm. First, suppose we are currently at time t and have not received a measurement, $y(t)$ as yet. We have the previous filtered estimate $\hat{x}(t - 1|t - 1)$ and error covariance $\tilde{P}(t - 1|t - 1)$ and would like to obtain the best estimate of the state based on $[t - 1]$ data samples. We are in the "prediction phase" of the algorithm. We use the state-space model to predict the state estimate $\hat{x}(t|t - 1)$ and its

Table 4.1 Linear Kalman filter algorithm (predictor–corrector form).

Prediction

$\hat{x}(t|t - 1) = A(t - 1)\hat{x}(t - 1|t - 1) + B(t - 1)u(t - 1)$ (State Prediction)

$\hat{y}(t|t - 1) = C(t)\hat{x}(t|t - 1) + D(t)u(t)$ (Measurement Prediction)

$\tilde{P}(t|t - 1) = A(t - 1)\tilde{P}(t - 1|t - 1)A'(t - 1) + R_{ww}(t - 1)$ (Covariance Prediction)

Innovations

$e(t) = y(t) - \hat{y}(t|t - 1)$ (Innovations)

$R_{ee}(t) = C(t)\tilde{P}(t|t - 1)C'(t) + R_{vv}(t)$ (Innovations Covariance)

Gain

$K(t) = \tilde{P}(t|t - 1)C'(t)R_{ee}^{-1}(t)$ (Gain or Weight)

Correction

$\hat{x}(t|t) = \hat{x}(t|t - 1) + K(t)e(t)$ (State Correction)

$\tilde{P}(t|t) = [I - K(t)C(t)]\tilde{P}(t|t - 1)$ (Covariance Correction)

Initial conditions

$\hat{x}(0|0)$ $\tilde{P}(0|0)$

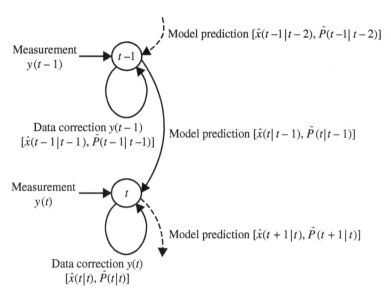

Figure 4.1 Predictor–corrector form of the Kalman filter: Timing diagram.

associated error covariance $\tilde{P}(t|t-1)$. Once the prediction based on the model is completed, we then calculate the innovations covariance $R_{ee}(t)$ and gain $K(t)$. As soon as the measurement at time t, that is, $y(t)$, becomes available, then we determine the innovations $e(t)$. Now we enter the "correction phase" of the algorithm. Here we correct or update the state based on the new information in the measurement – the innovations. The old, or predicted, state estimate $\hat{x}(t|t-1)$ is used to form the filtered, or corrected, state estimate $\hat{x}(t|t)$ and $\tilde{P}(t|t)$. Here we see that the error, or innovations, is the difference between the actual measurement and the predicted measurement $\hat{y}(t|t-1)$. The innovations is weighted by the gain matrix $K(t)$ to correct the old state estimate (predicted) $\hat{x}(t|t-1)$. The associated error covariance is corrected as well. The algorithm then awaits the next measurement at time $(t+1)$. Observe that in the absence of a measurement, the state-space model is used to perform the prediction, since it provides the best estimate of the state.

The covariance equations can be interpreted in terms of the various signal (state) and noise models (see Table 4.1). The first term of the predicted covariance $\tilde{P}(t|t-1)$ relates to the uncertainty in predicting the state using the model A. The second term indicates the increase in error variance due to the contribution of the process noise (R_{ww}) or model uncertainty. The corrected covariance equation indicates the predicted error covariance or uncertainty due to the prediction, decreased by the effect of the update (KC), thereby producing the corrected error covariance $\tilde{P}(t|t)$. The application of the LKF to the RC circuit was shown in Figure 1.8, where we see the noisy data, the "true" (mean) output voltage, and the Kalman filter predicted measurement voltage. As illustrated, the filter performs reasonably compared to the noisy data and tracks the true signal quite well. We derive the Kalman filter algorithm from the innovations viewpoint to follow, since it will provide the foundation for the nonlinear algorithms to follow.

4.2.1 Innovations Approach

In this section, we briefly develop the (linear) Kalman filter algorithm from the innovations perspective following the approach by Kailath [4–6]. First, recall from Chapter 2 that a model of a stochastic process can be characterized by the Gauss–Markov model using the state-space representation

$$x(t) = A(t-1)x(t-1) + B(t-1)u(t-1) + w(t-1) \qquad (4.1)$$

where w is assumed zero-mean and white with covariance R_{ww} and x and w are uncorrelated. The measurement model is given by

$$y(t) = C(t)x(t) + D(t)u(t) + v(t) \qquad (4.2)$$

where v is a zero-mean, white sequence with covariance R_{vv} and v is (assumed) uncorrelated with x and w.

The linear *state estimation problem* can be stated in terms of the preceding state-space model as

GIVEN a set of noisy measurements $\{y(i)\}$, for $i = 1, \dots, t$ characterized by the measurement model of Eq. (4.2), FIND the linear minimum (error) variance estimate of the state characterized by the state-space model of Eq. (4.1). That is, find the best estimate of $x(t)$ given the measurement data up to time t, $Y(t) := \{y(1), \dots, y(t)\}$.

First, we investigate the batch minimum variance estimator for this problem, and then we develop an alternative solution using the innovations sequence. The recursive solution follows almost immediately from the innovations. Next, we outline the resulting equations for the predicted state, gain, and innovations. The corrected, or filtered, state equation then follows. Details of the derivation can be found in [2].

Constraining the estimator to be linear, we see that for a batch of N-data, the minimum variance estimator is given by [1]

$$\hat{\underline{X}}_{MV} = K_{MV}\underline{Y} = R_{xy}R_{yy}^{-1}\underline{Y} \qquad (4.3)$$

---

1 Note that the complete form of the linear minimum variance estimator (see Appendix A) for nonzero mean is given by [2]

$$\hat{\underline{X}}_{MV} = \underline{m}_x + R_{xy}R_{yy}^{-1}(\underline{y} - \underline{m}_y)$$

where $\underline{\hat{X}}_{MV} \in R^{N_x N \times 1}$, R_{xy} and $K_{MV} \in R^{N_x N \times N_y N}$, $R_{yy} \in R^{N_y N \times N_y N}$, and $\underline{Y} \in R^{N_y N \times 1}$. Similarly, the linear estimator can be expressed in terms of the N data samples as

$$\underline{\hat{X}}_{MV}(N) = R_{xy}(N)R_{yy}^{-1}(N) = K_{MV}(N)\underline{Y}(N) \tag{4.4}$$

where $\underline{\hat{X}}_{MV}(N) = [\hat{x}'(1) \dots \hat{x}'(N)]'$, $\underline{Y}(N) = [y'(1) \dots y'(N)]'$, $\hat{x} \in R^{N_x \times 1}$, and $y \in R^{N_y \times 1}$.

Here we are investigating a "batch" solution to the state estimation problem, since all the N_y-vector data $\{y(1) \dots y(N)\}$ are processed in one batch. However, we require a recursive solution (see Chapter 1) to this problem of the form

$$\hat{X}_{new} = \hat{X}_{old} + KE_{new} \tag{4.5}$$

In order to achieve the recursive solution, it is necessary to transform the covariance matrix R_{yy} to be block diagonal. R_{yy} block diagonal implies that all the off-diagonal block matrices $R_{yy}(t, j) = 0$, for $i \neq j$, which in turn implies that the $\{y(t)\}$ must be uncorrelated or equivalently orthogonal. Therefore, we must construct a sequence of independent N_y-vectors, say $\{e(t)\}$, such that

$$E\{e(t)e'(k)\} = 0 \text{ for } t \neq k$$

The sequence $\{e(t)\}$ can be constructed using the orthogonality property of the minimum variance estimator such that

$$\underbrace{[y(t) - E\{y(t)|Y(t-1)\}]}_{\hat{y}(t|t-1)} \perp Y(t-1)$$

We define the *innovations* or new information [4] as

$$e(t) := y(t) - \hat{y}(t|t-1) \tag{4.6}$$

with the orthogonality property that

$$\text{cov}[y(T), e(t)] = 0 \quad \text{for } T \leq t - 1 \tag{4.7}$$

Since $\{e(t)\}$ is a time-uncorrelated N_y-vector sequence by construction, we have that the block diagonal innovations covariance matrix is $R_{ee}(N) = \text{diag}[R_{ee}(1) \cdots R_{ee}(N)]$ with each $R_{ee}(i) \in R^{N_y \times N_y}$.

The correlated measurement vector can be transformed to an uncorrelated innovations vector through a linear transformation, say L, given by

$$\underline{Y} = L\underline{e} \tag{4.8}$$

where $L \in R^{N_y N \times N_y N}$ is a nonsingular transformation matrix and $\underline{e} := [e'(1) \cdots e'(N)]'$. Multiplying Eq. (4.8) by its transpose and taking expected values, we obtain

$$R_{yy}(N) = LR_{ee}(N)L' \quad \text{and} \quad R_{yy}^{-1}(N) = (L')^{-1}R_{ee}^{-1}(N)L^{-1}$$

Similarly, we obtain

$$R_{xy}(N) = R_{xe}(N)L'$$

Substituting these results into Eq. (4.4) gives

$$\underline{\hat{X}}_{MV}(N) = R_{xy}(N)R_{yy}^{-1}(N)\underline{Y} = R_{xe}(N)R_{ee}^{-1}(N)\underline{e} \tag{4.9}$$

Since the $\{e(t)\}$ are time-uncorrelated by construction, $R_{ee}(N)$ is block diagonal. From the orthogonality properties of $e(t)$, $R_{xe}(N)$ is lower-block triangular, that is,

$$R_{xe}(N) = \begin{cases} R_{xe}(t, i) & t > i \\ R_{xe}(t, t) & t = i \\ 0 & t < i \end{cases}$$

where $e(i) \in R^{N_y \times 1}$, $R_{xe}(t, i) \in R^{N_x \times N_y}$, and $R_{ee}(i) := R_{ee}(i, i) \in R^{N_y \times N_y}$.

The recursive filtered solution now follows easily, if we realize that we want the best estimate of $x(t)$ given $Y(t)$; therefore, any block row can be written (for $N = t$) as

$$\hat{x}(t|t) := \underline{\hat{X}}_{MV}(t) = \sum_{i=1}^{t} R_{xe}(t, i)R_{ee}^{-1}(i)e(i)$$

If we extract the last (tth) term out of the sum (recall from Chapter 1), we obtain

$$\hat{X}_{new} = \hat{x}(t|t) = \underbrace{\sum_{i=1}^{t-1} R_{xe}(t, i)R_{ee}^{-1}(i)e(i)}_{\hat{X}_{old}} + \underbrace{R_{xe}(t, t)R_{ee}^{-1}(t)}_{K} \underbrace{e(t)}_{Error} \tag{4.10}$$

or

$$\hat{x}(t|t) = \hat{x}(t|t - 1) + K(t)e(t) \tag{4.11}$$

where $K(t) = R_{xe}(t)R_{ee}^{-1}(t)$ and $R_{xe}(t, t) = R_{xe}(t)$.

So we see that the recursive solution using the innovations sequence instead of the measurement sequence has reduced the computations to inverting a $N_y \times N_y$ matrix $R_{ee}(t)$ instead of a N_y $N \times N_y$ N matrix, $R_{yy}(N)$. Before we develop the expression for the filtered estimate of Eq. (4.11), let us investigate the innovations sequence more closely. Recall that the minimum variance estimate of $y(t)$ is just a linear transformation of the minimum variance estimate of $x(t)$, that is,

$$\hat{y}(t|t - 1) = C(t)\hat{x}(t|t - 1) + D(t)u(t) \tag{4.12}$$

Thus, the innovations can be decomposed using Eqs. (4.2) and (4.12) as

$$e(t) = y(t) - C(t)\hat{x}(t|t-1) - D(t)u(t) = C(t)[x(t) - \hat{x}(t|t-1)] + v(t)$$
$$= C(t)\tilde{x}(t|t-1) + v(t)$$

for $\tilde{x}(t|t-1) := x(t) - \hat{x}(t|t-1)$ – the predicted *state estimation error*. Consider the innovations covariance $R_{ee}(t)$ using this equation:

$$R_{ee}(t) = C(t)E\{\tilde{x}(t|t-1)\tilde{x}'(t|t-1)\}C'(t) + E\{v(t)\tilde{x}'(t|t-1)\}C'(t)$$

$$+ C(t)E\{\tilde{x}(t|t-1)v'(t)\} + E\{v(t)v'(t)\}$$

This expression yields the following, since v and $\tilde{x}$ are uncorrelated:

$$R_{ee}(t) = C(t)\tilde{P}(t|t-1)C'(t) + R_{vv}(t) \tag{4.13}$$

The cross-covariance R_{xe} is obtained as

$$R_{xe}(t) = E\{x(t)e'(t)\} = E\{x(t)[C(t)\tilde{x}(t|t-1) + v(t)]'\}$$
$$= E\{x(t)\tilde{x}'(t|t-1)\}C'(t)$$

Using the definition of the estimation error $\tilde{x}(t)$, substituting for $x(t)$ and from the orthogonality property of the estimation error for dynamic variables [3], that is,

$$E\{f(Y(T))\tilde{x}'(t|t-1)\} = 0 \quad \text{for } T \le t-1 \tag{4.14}$$

we obtain

$$R_{xe}(t) = \tilde{P}(t|t-1)C'(t) \tag{4.15}$$

Thus, we see that the weight or gain matrix is given by

$$K(t) = R_{xe}(t)R_{ee}^{-1}(t) = \tilde{P}(t|t-1)C'(t)R_{ee}^{-1}(t) \tag{4.16}$$

Before we can calculate the corrected state estimate, we require the predicted, or old estimate, that is,

$$\hat{X}_{\text{old}} := \hat{x}(t|t-1) = E\{x(t)|Y(t-1)\}$$

If we employ the state-space model of Eq. (4.1), then we have from the linearity properties of the conditional expectation

$$\hat{x}(t|t-1) = E\{A(t-1)x(t-1) + B(t-1)u(t-1) + w(t-1)|Y(t-1)\}$$

or

$$\hat{x}(t|t-1) = A(t-1)\hat{x}(t-1|t-1) + B(t-1)u(t-1) + \hat{w}(t-1|t-1)$$

However, from the orthogonality property (whiteness) of $w(\cdot)$, we have

$$\hat{w}(t|T) = E\{w(t)|Y(T)\} = 0 \quad \text{for } t \le T$$

which is not surprising, since the best estimate of zero-mean, white noise is zero (unpredictable). Thus, the prediction is given by

$$\hat{X}_{\text{old}} = \hat{x}(t|t-1) = A(t-1)\hat{x}(t-1|t-1) + B(t-1)u(t-1) \qquad (4.17)$$

To complete the algorithm, the expressions for the predicted and corrected error covariances must be determined. From the definition of predicted estimation error covariance, we see that $\tilde{x}(t|t-1)$ satisfies

$$\tilde{x}(t|t-1) = A(t-1)\tilde{x}(t-1|t-1) + w(t-1) \qquad (4.18)$$

Since w and $\tilde{x}$ are uncorrelated, the predicted error covariance $\tilde{P}(t|t-1) :=$ $\text{Cov}(\tilde{x}(t|t-1))$ can be determined from this relation [2] to give

$$\tilde{P}(t|t-1) = A(t-1)\tilde{P}(t-1|t-1)A'(t-1) + R_{ww}(t-1) \qquad (4.19)$$

The corrected error covariance $\tilde{P}(t|t)$ is calculated using the corrected *state estimation error* and the corresponding state estimate of Eq. (4.11) as

$$\tilde{x}(t|t) : = x(t) - \hat{x}(t|t) = x(t) - \hat{x}(t|t-1) - K(t)e(t)$$
$$= \tilde{x}(t|t-1) - K(t)e(t) \qquad (4.20)$$

Using this expression, we can calculate the required error covariance from the orthogonality property, $E\{\hat{x}(t|t-1)e'(t)\} = 0$, and Eq. (4.15) [2] yielding the final expression for the corrected error covariance as

$$\tilde{P}(t|t) = [I - K(t)C(t)]\tilde{P}(t|t-1) \qquad (4.21)$$

This completes the brief derivation of the LKF based on the innovations sequence (see [2] for more details). It is clear to see that the innovations sequence holds the "key" to unlocking the mystery of Kalman filter design. In Section 4.2.3, we investigate the statistical properties of the innovations sequence that will enable us to develop a simple procedure to "tune" the MBP.

4.2.2 Bayesian Approach

The Bayesian approach to Kalman filtering follows in this brief development leading to the *maximum a posteriori* (MAP) estimate of the state vector under the Gauss–Markov model assumptions (see Chapter 3). Detailed derivations can be found in [2] or [7].

We would like to obtain the MAP estimator for the state estimation problem where the underlying Gauss–Markov model with $w(t)$, $v(t)$, and $x(0)$ Gaussian distributed. We know that the corresponding state estimate will also be Gaussian, since it is a linear transformation of Gaussian variables.

It can be shown using Bayes' rule that the *a posteriori probability* [2, 7] can be expressed as

$$\text{Pr}\,(x(t)|Y(t)) = \frac{\text{Pr}\,(y(t)|x(t)) \times \text{Pr}\,(x(t)|Y(t-1))}{\text{Pr}\,(y(t)|Y(t-1))} \qquad (4.22)$$

Under the Gauss–Markov model assumptions, we know that each of the conditional distributions can be expressed in terms of the Gaussian distribution as follows:

- $\Pr(y(t)|x(t))$: $\quad\quad \mathcal{N}(C(t)x(t), R_{vv}(t))$
- $\Pr(x(t)|Y(t-1))$: $\quad \mathcal{N}(\hat{x}(t|t-1), \tilde{P}(t|t-1))$
- $\Pr(y(t)|Y(t-1))$: $\quad \mathcal{N}(\hat{y}(t|t-1), R_{ee}(t))$.

Substituting these probabilities into Eq. (4.22) and combining all constants into a single constant κ, we obtain

$$\Pr(x(t)|Y(t)) = \kappa \times \exp\left[-\frac{1}{2}(y(t) - C(t)x(t))'R_{vv}^{-1}(t)(y(t) - C(t)x(t))\right]$$

$$\times \exp\left[-\frac{1}{2}(x(t) - \hat{x}(t|t-1))'\tilde{P}^{-1}(t|t-1)(x(t) - \hat{x}(t|t-1))\right]$$

$$/ \exp\left[+\frac{1}{2}(y(t) - \hat{y}(t|t-1))'R_{ee}^{-1}(t)(y(t) - \hat{y}(t|t-1))\right]$$

Recognizing the measurement noise, state estimation error, innovations, and taking natural logarithms, we obtain the log a posteriori probability in terms of the Gauss–Markov model as

$$\ln \Pr(x(t)|Y(t)) = \ln \kappa - \frac{1}{2}v'(t)R_{vv}^{-1}(t)v(t) - \frac{1}{2}\tilde{x}'(t|t-1)$$

$$\times \tilde{P}^{-1}(t|t-1)\tilde{x}(t|t-1) + \frac{1}{2}e'(t)R_{ee}^{-1}(t)e(t) \quad\quad (4.23)$$

The MAP estimate is then obtained by differentiating this expression, setting it to zero and solving to obtain

$$\hat{X}_{\text{map}}(t) = [C'(t)R_{vv}^{-1}(t)C(t) + \tilde{P}^{-1}(t|t-1)]^{-1}$$

$$\times[\tilde{P}^{-1}(t|t-1)\hat{x}(t|t-1) + C'(t)R_{vv}^{-1}(t)y(t)] \quad\quad (4.24)$$

This relation can be simplified by using a form of the *matrix inversion lemma* [3], enabling us to eliminate the first bracketed term in Eq. (4.24) to give

$$\hat{X}_{\text{map}}(t) = \tilde{P}(t|t) \times [\tilde{P}^{-1}(t|t-1)\hat{x}(t|t-1) + C'(t)R_{vv}^{-1}(t)y(t)]$$

Solving for $\tilde{P}^{-1}(t|t-1)$ and substituting gives

$$\hat{X}_{\text{map}}(t) = \tilde{P}(t|t)$$

$$\times[(\tilde{P}^{-1}(t|t) - C'(t)R_{vv}^{-1}(t)C(t))\hat{x}(t|t-1) + C'(t)R_{vv}^{-1}(t)y(t)] \quad (4.25)$$

Multiplying out, regrouping terms, and factoring, this relation can be rewritten as

$$\hat{X}_{\text{map}}(t) = \hat{x}(t|t-1) + \underbrace{(\tilde{P}(t|t)C'(t)R_{vv}^{-1}(t))}_{K(t)}[y(t) - C(t)\hat{x}(t|t-1)] \quad\quad (4.26)$$

or finally

$$\hat{X}_{\text{map}}(t) = \hat{x}(t|t) = \hat{x}(t|t-1) + K(t)e(t) \qquad (4.27)$$

Further manipulations lead to equivalent expressions for the Kalman gain as

$$K(t) = \tilde{P}(t|t)C'(t)R_{vv}^{-1}(t) \equiv \tilde{P}(t|t-1)C'(t)R_{ee}^{-1}(t) \qquad (4.28)$$

which completes the Bayes' approach to the Kalman filter. A complete detailed derivation is available in [2] or [7] for the interested reader.

4.2.3 Innovations Sequence

In this subsection, we investigate the properties of the innovations sequence, which have been used to develop the Kalman filter [2]. It is interesting to note that since the innovations sequence depends directly on the measurement and is linearly related to it, then it spans the measurement space in which all of our data reside. In contrast, the states or internal variables are *not* measured directly and are usually *not* available; therefore, our designs are accomplished using the innovations sequence along with its statistical properties to assure optimality. We call this design procedure *minimal (error) variance design.*

Recall that the innovations or equivalently the one-step prediction error is given by

$$\mathbf{e}(t) = \mathbf{y}(t) - \hat{\mathbf{y}}(t|t-1) = \tilde{\mathbf{y}}(t|t-1) + \mathbf{v}(t) \qquad (4.29)$$

where we define $\tilde{\mathbf{y}}(t|t-1) := C(t)\tilde{\mathbf{x}}(t|t-1)$ for $\tilde{\mathbf{x}}(t|t-1) := \mathbf{x}(t) - \hat{\mathbf{x}}(t|t-1)$ – the state estimation error prediction. Using these expressions we can now analyze the statistical properties of the innovations sequence based on its orthogonality to the measured data. We will state each property first and refer to [2] for proof.

The innovations sequence is *zero-mean*:

$$E\{e(t)\} = E\{\tilde{y}(t|t-1) + v(t)\} = E\{\tilde{y}(t|t-1)\} + E\{v(t)\} = 0 \qquad (4.30)$$

The second term is null by definition and the first term is null because $\hat{x}(t|t-1)$ is an unbiased estimator.

The innovations sequence is *white* since

$$R_{ee}(t, t_\ell) = \begin{cases} C(t)\tilde{P}(t|t-1)C'(t) + R_{vv}(t) & t = t_\ell \\ 0 & t \neq t_\ell \end{cases}$$

or

$$\text{Cov}(e(t), e(t_\ell)) = R_{ee}(t, t_\ell)\delta(t - t_\ell)$$

The innovations sequence is determined recursively using the Kalman filter and acts as a "Gram–Schmidt orthogonal decomposer" or equivalently whitening filter.

The innovations sequence is also *uncorrelated with the deterministic input* $u(t-1)$ since

$$R_{eu}(t) = E\{e(t)u'(t-1)\} = E\{\tilde{y}(t|t-1) + v(t)\}u(t-1) = 0$$

Assuming that the measurement evolves from a Gauss–Markov process as well, then the innovations sequence is merely a linear transformation of Gaussian vectors and is therefore *Gaussian* with $e \sim \mathcal{N}(0, R_{ee}(t))$.

Finally, the innovations sequence is related to the measurement by an invertible linear transformation; therefore, it is an *equivalent sequence* under *linear transformations*, since either sequence can be constructed from knowledge of the second-order statistics of the other [2].

We summarize these properties of the *innovations sequence* as follows:

- Innovations sequence $\{e(t)\}$ is *zero-mean*.
- Innovations sequence $\{e(t)\}$ is *white*.
- Innovations sequence $\{e(t)\}$ is *uncorrelated* in time and with input $u(t-1)$.
- Innovations sequence $\{e(t)\}$ is *Gaussian* with statistics, $\mathcal{N}(0, R_{ee}(t))$, under the Gauss–Markov assumptions.
- Innovations $\{e(t)\}$ and measurement $\{y(t)\}$ sequences are *equivalent under linear invertible transformations*.

The innovations sequence spans the measurement or data space, but in the Kalman filter design problem, we are concerned with the state-space. Analogous to the innovations sequence in the output space is the predicted state estimation error $\tilde{x}(t|t-1)$ in the state-space. It is easy to show that from the orthogonality condition of the innovations sequence that the corresponding state estimation error is also orthogonal to $y(T)$, leading to the *state orthogonality condition* (see [2] for more details).

This completes the discussion of the orthogonality properties of the innovations sequence. Next we consider more practical aspects of processor design and how these properties can be exploited to produce a minimum variance processor.

4.2.4 Practical Linear Kalman Filter Design: Performance Analysis

In this section, we heuristically provide an intuitive feel for the operation of the Kalman filter using the state-space model and Gauss–Markov assumptions. These results coupled with the theoretical points developed in [2] lead to the proper adjustment or "tuning" of the processor. Tuning the processor is considered an art, but with proper statistical tests, the performance can readily be evaluated and adjusted. As mentioned previously, this approach is called the *minimum (error) variance* design. In contrast to standard filter design procedures in signal processing, the minimum variance design adjusts the statistical parameters (e.g. covariances) of the processor and examines the innovations sequence to determine if the LKF is properly tuned. Once tuned, then all of the statistics (conditional means and variances) are valid and may be used as reasonable estimates. Here we discuss how the parameters can be adjusted and what statistical tests can be performed to evaluate the filter performance.

Heuristically, the Kalman filter can be viewed simply by its update equation:

$$
\hat{X}_{\text{new}} = \overbrace{\hat{X}_{\text{old}}}^{\text{Prediction}} + \overbrace{K \times E_{\text{new}}}^{\text{Update}}
$$

where $\hat{X}_{\text{old}} \approx f(\text{model})$ and $E_{\text{new}} \approx f(\text{measurement})$.

Using this representation of the KF, we see that we can view the old or predicted estimate $\hat{X}_{\text{old}}$ as a function of the state-space model (A, B) and the prediction error or innovations E as a function primarily of the new measurement, as indicated in Table 4.1. Consider the new estimate under the following cases:

$$K \to \text{small} \qquad \hat{X}_{\text{new}} = \hat{X}_{\text{old}} = f(\text{model})$$

$$K \to \text{large} \qquad \hat{X}_{\text{new}} = K E_{\text{new}} = f(\text{measurement})$$

So we can see that the operation of the processor is pivoted about the values of the gain or weighting matrix K. For small K, the processor "believes" the *model*, while for large K, it believes the *measurement*.

Let us investigate the gain matrix and see if its variations are consistent with these heuristic notions. First, it was shown in Eq. (4.28) that the alternative form of the gain equation is given by

$$K(t) = \tilde{P}(t|t) C'(t) R_{vv}^{-1}(t)$$

Thus, the condition where K is *small* can occur in two cases: (i) $\tilde{P}$ is small (fixed R_{vv}), which is consistent because small $\tilde{P}$ implies that the model is adequate; and (ii) R_{vv} is large ($\tilde{P}$ fixed), which is also consistent because large R_{vv} implies that the measurement is noisy, so again believe the model.

For the condition where K is *large*, two cases can also occur: (i) K is large when $\tilde{P}$ is large (fixed R_{vv}), implying that the model is inadequate, so believe the measurement; and (ii) R_{vv} is small ($\tilde{P}$ fixed), implying the measurement is good (high signal-to-noise ratio [SNR]). So we see that our heuristic notions are based on specific theoretical relationships between the parameters in the KF algorithm of Table 4.2.

Summarizing, a Kalman filter is not functioning properly when the *gain* becomes small and the measurements still contain information necessary for the estimates. The filter is said to *diverge* under these conditions. In this case, it is necessary to detect how the filter is functioning and how to adjust it if necessary, but first we consider the tuned LKF.

Table 4.2 Heuristic notions for Kalman filter tuning.

Kalman filter heuristics

Condition	Gain	Parameter
Believe model	Small	$\tilde{P}$ small (model adequate)
		R_{vv} large (measurement noisy)
Believe measurement	Large	R_{vv} small (measurement good)
		$\tilde{P}$ large (model inadequate)

When the processor is "tuned," it provides an *optimal* or minimum (error) variance estimate of the state. The innovations sequence, which was instrumental in deriving the processor, also provides the starting point to check the KF operation. A *necessary and sufficient condition* for a Kalman filter to be optimal is that the innovations sequence is zero-mean and white (see Ref. [8] for the proof). These are the first properties that must be evaluated to ensure that the processor is operating properly. If we assume that the innovations sequence is ergodic and Gaussian, then we can use the sample mean as the test statistic to estimate, m_e, the population mean. The sample mean for the ith component of e_i is given by

$$\hat{m}_e(i) = \frac{1}{N} \sum_{t=1}^{N} e_i(t) \quad \text{for} \quad i = 1, \dots, N_y \tag{4.31}$$

where $\hat{m}_e(i) \sim \mathcal{N}(m_e, R_{ee}(i)/N)$ and N is the number of data samples. We perform a statistical hypothesis test to "decide" if the innovations mean is zero [2]. We test that the mean of the ith component of the innovations vector $e_i(t)$ is

$$H_0 : \quad m_e(i) = 0$$
$$H_1 : \quad m_e(i) \neq 0$$

As our test statistic, we use the sample mean. At the α-significance level, the probability of rejecting the null hypothesis H_0 is given by

$$\Pr\left(\left|\frac{\hat{m}_e(i) - m_e(i)}{\sqrt{R_{ee}(i)/N}}\right| > \frac{\tau_i - m_e(i)}{\sqrt{R_{ee}(i)/N}}\right) = \alpha \tag{4.32}$$

Therefore, the *zero-mean test* [2] on each component innovations e_i is given by

$$\hat{m}_e(t) \underset{\substack{< \\ \text{Accept} \quad H_0}}{\overset{\substack{> \\ \text{Reject} \quad H_0}}{}} \tau_i \tag{4.33}$$

Under the null hypothesis H_0, each $m_e(i)$ is zero. Therefore, at the 5% significance level ($\alpha = 0.05$), we have that the threshold is

$$\tau_i = 1.96\sqrt{\frac{\hat{R}_{ee}(i)}{N}} \tag{4.34}$$

where $\hat{R}_{ee}(i)$ is the *sample variance* (assuming ergodicity) estimated by

$$\hat{R}_{ee}(i) = \frac{1}{N}\sum_{t=1}^{N} e_i^2(t) \tag{4.35}$$

Under the same assumptions, we can perform a *whiteness test* [2], that is, check statistically that the innovations covariance corresponds to that of an uncorrelated (white) sequence. Again assuming ergodicity of the innovations sequence, we use the sample covariance function as our test statistic with the *i*th-component covariance given by

$$\hat{R}_{ee}(i,k) = \frac{1}{N}\sum_{t=k+1}^{N} (e_i(t) - \hat{m}_e(i))(e_i(t+k) - \hat{m}_e(i)) \tag{4.36}$$

We actually use the *normalized covariance* test statistic

$$\hat{\rho}_{ee}(i,k) = \frac{\hat{R}_{ee}(i,k)}{\hat{R}_{ee}(i)} \tag{4.37}$$

Asymptotically for large N, it can be shown that (see Refs. [2, 9])

$$\hat{\rho}_{ee}(i,k) \sim \mathcal{N}(0, 1/N)$$

Therefore, the 95% confidence interval estimate is

$$I_{\rho_{ee}} = \hat{\rho}_{ee}(i,k) \pm \frac{1.96}{\sqrt{N}} \quad \text{for } N > 30 \tag{4.38}$$

Hence, under the null hypothesis, 95% of the $\hat{\rho}_{ee}(i,k)$ values must lie within this confidence interval, that is, for each *component* innovations sequence to be considered statistically white. Similar tests can be constructed for the *cross-covariance* properties of the innovations [10] as well, that is,

$$\text{Cov}(e(t), e(k)) = 0 \text{ and } \text{Cov}(e(t), u(t-1)) = 0$$

The whiteness test of Eq. (4.38) is very useful for detecting model inaccuracies from *individual component* innovations. However, for complex systems with a large number of measurement channels, it becomes computationally burdensome to investigate each innovation component-wise. A statistic capturing *all* of the innovations information is the *weighted sum-squared residual* (WSSR) [9–11]. It aggregates all of the innovations *vector* information over some finite window of length N. It can be shown that the WSSR is related to a maximum-likelihood estimate of the normalized innovations variance [2, 9]. The WSSR test statistic is given by

$$\hat{\rho}(\ell) := \sum_{k=\ell-N+1}^{\ell} e'(k)R_{ee}^{-1}(k)e(k) \quad \text{for } \ell \geq N \qquad (4.39)$$

and is based on the hypothesis test

$H_0 : \quad \{e(t)\} \ is \ \text{white}$

$H_1 : \quad \{e(t)\} \ is \ not \ \text{white}$

with the WSSR test statistic

$$\hat{\rho}(\ell) \overset{\overset{\text{Reject} \quad H_1}{>}}{\underset{\underset{\text{Accept} \quad H_0}{<}}{}} \tau \qquad (4.40)$$

Under the null hypothesis, the WSSR is chi-squared distributed, $\rho(\ell) \sim \chi^2(N_y N)$. However, for $N_y N > 30$, $\rho(\ell)$ is approximately Gaussian $\mathcal{N}(N_y N, 2N_y N)$ (see [12] for more details). At the α-significance level, the probability of rejecting the null hypothesis is given by

$$\text{Pr}\left(\left|\frac{\rho(\ell) - N_y N}{\sqrt{2N_y N}}\right| > \left|\frac{\tau - N_y N}{\sqrt{2N_y N}}\right|\right) = \alpha \qquad (4.41)$$

For a level of significance of $\alpha = 0.05$, we have

$$\tau = N_y N + 1.96\sqrt{2N_y N} \qquad (4.42)$$

Thus, the WSSR can be considered a "whiteness test" of the innovations *vector* over a finite window of length N. Note that since $[\{e(t)\}, \{R_{ee}(t)\}]$ are obtained from the state-space MBP algorithm directly, they can be used for both *stationary* and *nonstationary* processes. In fact, in practice, for a large number of measurement components, the WSSR is used to "tune" the filter, and then the component innovations are individually analyzed to detect model mismatches. Also note that the adjustable parameter of the WSSR statistic is the window length N, which essentially controls the width of the window sliding through the innovations sequence.

Other sets of "reasonableness" tests can be performed using the covariances estimated by the LKF algorithm and sample variances estimated using Eq. (4.35). The LKF provides estimates of the respective processor covariances R_{ee} and $\tilde{P}$ from the relations given in Table 4.2. Using sample variance estimators, when the filter reaches steady state, (process is stationary), that is, $\tilde{P}$ is constant,

Table 4.3 Statistical tuning tests for Kalman filter.

Kalman filter design/validation

Data	Property	Statistic	Test	Assumptions
Innovation	$m_e = 0$	Sample mean	Zero mean	Ergodic, Gaussian
	$R_{ee}(t)$	Sample covariance	Whiteness	Ergodic, Gaussian
	$\rho(\ell)$	WSSR	Whiteness	Gaussian
	$R_{ee}(t, k)$	Sample cross-covariance	Cross-covariance	Ergodic, Gaussian
	$R_{eu}(t, k)$	Sample cross-covariance	Cross-covariance	Ergodic, Gaussian
Covariances	Innovation	Sample variance $(\hat{R}_{ee})$	$\hat{R}_{ee} \approx R_{ee}$	Ergodic
	Innovation	Variance $(\hat{R}_{ee})$	Confidence interval about $\{e(t)\}$	Ergodic
	Estimation error	Sample variance $(\hat{P})$	$\hat{P} \approx P$	Ergodic, X_{True} known
	Estimation error	Variance $(\hat{P})$	Confidence interval about $\{\tilde{x}(t\vert t)\}$	X_{True} known

the estimates can be compared to ensure that they are reasonable. Thus, we have

$$\hat{R}_{ee}(i) \approx R_{ee}(i) \text{ and } \hat{\tilde{P}}(i) \approx \tilde{P}(i) \tag{4.43}$$

Plotting the $\pm 1.96\sqrt{R_{e_i e_i}(t)}$ and $\pm 1.96\sqrt{\tilde{P}_i(t\vert t)}$ about the component innovations $\{e_i(t)\}$ and component state estimation errors $\{\tilde{x}_i(t\vert t)\}$, when the true state is known, provides an accurate estimate of the Kalman filter performance especially when simulation is used. If the covariance estimates of the processor are reasonable, then 95% of the sequence samples should lie within the constructed bounds. Violation of these bounds clearly indicates inadequacies in modeling the processor statistics. We summarize these results in Table 4.3 and examine the RC-circuit design problem in the following example to demonstrate the approach in more detail.

Example 4.1 Suppose we have the *RC* circuit as shown in Figure 1.5. We measure the voltage across the capacitor with a high-impedance voltmeter as shown. Since these measurements are noisy and the component values are imprecise ($\pm\Delta$), we require an improved estimate of the output voltage. We develop an LKF to solve this problem from first principles–a typical approach. Writing the Kirchhoff current equations at the node, we have

$$I_{in}(t) - \frac{e(t)}{R} - C\frac{de(t)}{dt} = 0$$

where e_o is the initial voltage and R is the resistance with C the capacitance. The measurement equation for a voltmeter of gain K_e is simply

$$e_{out}(t) = K_e e(t)$$

We choose to use the discrete formulation; therefore, approximating the derivatives with first differences and substituting into, we have

$$C\frac{e(t) - e(t-1)}{\Delta T} = -\frac{e(t-1)}{R} + I_{in}(t-1)$$

or

$$e(t) = \left(1 - \frac{\Delta T}{RC}\right)e(t-1) + \frac{\Delta T}{C}I_{in}(t-1)$$

where the measurement is given above. Suppose that for this circuit the parameters are as follows: $R = 3.3$ kΩ and $C = 1000$ μF, $\Delta T = 100$ ms, $e_o = 2.5$ V, $K_e = 2.0$, and the voltmeter is precise to within ± 4 V. Then transforming the physical circuit model into state-space form by defining $x = e$, $y = e_{out}$, and $u = I_{in}$, we obtain

$$x(t) = 0.97x(t-1) + 100u(t-1) + w(t-1)$$
$$y(t) = 2x(t) + v(t)$$

The process noise covariance is used to model the circuit parameter uncertainty with $R_{ww} = 0.0001$, since we assume standard deviations, ΔR, ΔC of 1%. Also, $R_{vv} = 4$, since two standard deviations are $\Delta V = 2\left(\frac{1}{2} 4V\right)$. We also assume initially that the state is $x(0) \sim \mathcal{N}(2.5, 10^{-12})$ and that the input current is a step function of $u(t) = 300$ μA.

With the data simulated using MATLAB (see Figure 1.6), we now consider the design of the Kalman filter. In the ideal problem, we are given the model set

$$\Sigma := \{A, B, C, D, R_{ww}, R_{vv}, x(0), P(0)\}$$

along with the known input $\{u(t)\}$ and the set of noisy measurements, $\{y(t)\}$, to construct the processor. The LKF for the RC-circuit problem is

$$\hat{x}(t|t-1) = 0.97\hat{x}(t-1|t-1) + 100u(t-1) \text{ (Predicted State)}$$
$$\tilde{P}(t|t-1) = 0.94\tilde{P}(t-1|t-1) + 0.0001 \text{ (Predicted Covariance)}$$
$$e(t) = y(t) - 2\hat{x}(t|t-1) \text{ (Innovation)}$$
$$R_{ee}(t) = 4\tilde{P}(t|t-1) + 4 \text{ (Innovations Covariance)}$$

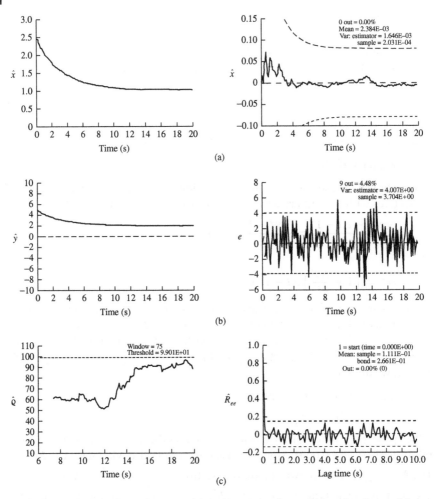

Figure 4.2 LKF design for *RC*-circuit problem. (a) Estimated state (voltage) and error. (b) Filtered voltage measurement and error (innovations). (c) WSSR and zero-mean/whiteness tests.

$$K(t) = 2\frac{\tilde{P}(t|t-1)}{4\tilde{P}(t|t-1)+4} \text{ (Gain)}$$

$$\hat{x}(t|t) = \hat{x}(t|t-1)) + K(t)e(t) \text{ (Updated State)}$$

$$\tilde{P}(t|t) = \frac{\tilde{P}(t|t-1)}{\tilde{P}(t|t-1)+1} \text{ (Updated Covariance)}$$

The estimator is also designed using MATLAB, and the results are shown in Figure 4.2. In Figure 4.2a, we see the estimated state (voltage) and estimation error as well as the corresponding confidence bounds. Note that the processor "optimally" estimates the voltage, since our models are exact. That is, it provides the minimum error variance estimate in the Gaussian case. Also since we have the true (mean) state, we can calculate the estimation error and use the corresponding error covariance to specify the bounds as shown. Note that the error is small and no samples exceed the bound as indicated by the overestimation of the variance compared with the sample variance (0.0017 > 0.0002). In Figure 4.2b, we see the filtered measurement ($\hat{y}(t|t-1)$) and the corresponding innovations sequence along with the confidence limits provided by the processor. Here only 4.5% of the samples exceed the bounds and the variance predicted by the filter is close to the sample variance estimate (4.0–3.7). The WSSR statistic, zero-mean, and whiteness tests are shown in Figure 4.2c. Here we see that using a window of 75 samples, the threshold is not exceeded, indicating a statistically white sequence. The innovations mean is small and well within the bound (0.11 < 0.27). The sequence is statistically white, since 0% of the normalized sample covariances exceed the bound completing this example. □

Next we discuss a special case of the KF – a steady-state Kalman filter that will prove a significant component of the subspace identification algorithms in subsequent chapters.

4.2.5 Steady-State Kalman Filter

In this section, we discuss a special case of the state-space LKF – the steady-state design. Here the data are assumed stationary, leading to a time-invariant state-space model, and under certain conditions a constant error covariance and corresponding gain. We first develop the processor and then show how it is precisely equivalent to the classical Wiener filter design. In filtering jargon, this processor is called the *steady-state Kalman filter* [13, 14].

We briefly develop the steady-state KF technique in contrast to the usual recursive time-step algorithms. By *steady state*, we mean that the processor has embedded time invariant, state-space model parameters, that is, the underlying model is defined by a set of parameters, $\Sigma = \{A, B, C, D, R_{ww}, R_{vv}, \hat{x}(0|0), \tilde{P}(0|0)\}$. We state without proof the fundamental theorem (see [3, 8, 14] for details).

If we have a stationary process implying the time-invariant system, Σ, and additionally the system is completely controllable and observable and stable (eigenvalues of A lie within the unit circle) with $\tilde{P}(0|0) > 0$, then the KF is *asymptotically stable*. What this means from a pragmatic point of view is that as time increases (to infinity in the limit), the initial error covariance is forgotten as more and more data are processed and that the computation of $\tilde{P}(t|t)$ is computationally stable. Furthermore, these conditions imply that

$$\lim_{t\to\infty} \tilde{P}(t|t) = \lim_{t\to\infty} \tilde{P}(t|t-1) \to \tilde{P}_{ss} \text{ (a constant)} \tag{4.44}$$

Therefore, this relation implies that we can define a *steady-state gain* associated with this covariance as

$$\lim_{t \to \infty} K(t) \to K_{ss} \text{ (a constant)} \tag{4.45}$$

Let us construct the corresponding *steady-state* KF using the correction equation of the state estimator and the steady-state gain with

$$e(t) = y(t) - C\hat{x}(t|t-1) - Du(t) \tag{4.46}$$

therefore, substituting we obtain

$$\hat{x}(t|t) = \hat{x}(t|t-1) + K_{ss}e(t) = (I - K_{ss}C)\hat{x}(t|t-1) - K_{ss}Du(t) + K_{ss}y(t) \tag{4.47}$$

but since

$$\hat{x}(t|t-1) = A\hat{x}(t-1|t-1) + Bu(t-1) \tag{4.48}$$

we have by substituting into Eq. (4.47) that

$$\hat{x}(t|t) = (I - K_{ss}C)A\hat{x}(t-1|t-1) + (I - K_{ss}C)Bu(t-1)$$
$$-K_{ss}Du(t) + K_{ss}y(t) \tag{4.49}$$

with the corresponding *steady-state gain* given by

$$K_{ss} = \tilde{P}_{ss}CR_{ee}^{-1} = \tilde{P}_{ss}C(C\tilde{P}_{ss}C' + R_{vv})^{-1} \tag{4.50}$$

Examining Eq. (4.49) more closely, we see that using the state-space model, Σ, and known input sequence, $\{u(t)\}$, we can process the data and extract the corresponding state estimates. The key to the steady-state KF is calculating K_{ss}, which, in turn, implies that we must calculate the corresponding steady-state error covariance. This calculation can be accomplished efficiently by combining the prediction and correction relations of Table 4.1. We have that

$$\tilde{P}(t|t-1) = A(I - \tilde{P}(t|t-1)C(C\tilde{P}(t|t-1)C' + R_{vv})^{-1})A' + R_{ww} \tag{4.51}$$

which in steady state becomes

$$\tilde{P}_{ss} = A(I - \tilde{P}_{ss}C(C\tilde{P}_{ss}C' + R_{vv})^{-1})A' + R_{ww} \tag{4.52}$$

There are a variety of efficient methods to calculate the steady-state error covariance and gain [8, 14]; however, a brute-force technique is simply to run the standard predictor–corrector algorithm implemented in UD-sequential form [15] until the $\tilde{P}_{ss}$ and therefore K_{ss} converge to constant matrix. Once they converge, it is necessary to only run the algorithm again to process the data using $\tilde{P}(0|0) = \tilde{P}_{ss}$, and the corresponding steady-state gain will be calculated directly. This is not the most efficient method to solve the problem, but it clearly does not require the development of a new algorithm.

Table 4.4 Steady-state Kalman filter algorithm.

Covariance

$$\tilde{P}_{ss} = A(I - \tilde{P}_{ss}C(C\tilde{P}_{ss}C' + R_{vv})^{-1})A' + R_{ww} \qquad \text{(steady-state covariance)}$$

Gain

$$K_{ss} = \tilde{P}_{ss}C(\tilde{C}P_{ss}C' + R_{vv})^{-1} \qquad \text{(steady-state gain)}$$

Correction

$$\hat{x}(t|t) = (I - K_{ss}C)A\hat{x}(t-1|t-1) + (I - K_{ss}C)Bu(t-1) - K_{ss}Du(t) + K_{ss}y(t)$$
$$\text{(state estimate)}$$

Initial Conditions

$$\hat{x}(0|0) \quad \tilde{P}(0|0)$$

We summarize the steady-state KF in Table 4.4. We note from the table that the steady-state covariance/gain calculations depend on the model parameters, Σ, *not* the data, implying that they can be *precalculated* prior to processing the actual data. In fact, the steady-state processor can be thought of as a simple (multi-input/multi-output) digital filter, which is clear if we abuse the notation slightly to write

$$\hat{x}(t+1) = (I - K_{ss}C)A\hat{x}(t) + (I - K_{ss}C)Bu(t) + K_{ss}y(t+1) \qquad (4.53)$$

For its inherent simplicity compared to the full time-varying processor, the steady-state KF is desirable and adequate in many applications, but realize that it will be *suboptimal* in some cases during the initial transients of the data. However, if we developed the model from first principles, then the underlying physics has been incorporated in the KF, yielding a big advantage over non-physics-based designs. This completes the discussion of the steady-state KF. Next let us reexamine our *RC*-circuit problem.

Example 4.2 We again use the tuned *RC*-circuit problem and replace the time-varying gain with their steady-state values obtained by running the algorithm of Table 4.1 (error covariance and gain) until they converge to constants. The steady-state values estimated are

$$\tilde{P}_{ss} = 1.646 \times 10^{-3} \quad \text{and} \quad K_{ss} = 8.23 \times 10^{-4}$$

Next the algorithm is executed with the initial error covariance set to its steady-state value ($\tilde{P}(0|0) = \tilde{P}_{ss} = 1.65 \times 10^{-3}$) giving the steady-state gain. The performance of the processor is not as good as the optimal during the transient stages of the processor. The errors are larger, for instance, the state estimation errors are larger. The zero-mean/whiteness test of the innovations indicates that both processors perform roughly the same, with the optimal performing slightly better with smaller mean (0.11 < 0.16) and whiteness (0% < 1%) of the samples exceeding the bound. Examining the innovations sequence itself, it is clear that there is a problem for the steady-state processor during the initial stage of the algorithm. So we see that there is a performance penalty for using a steady-state processor in lieu of the optimal time-varying KF; however, the computational advantages can be significant for large numbers of states. □

This completes the example. Next we investigate the steady-state KF and its relation to the Wiener filter.

4.2.6 Kalman Filter/Wiener Filter Equivalence

In this subsection, we show the relationship between the Wiener filter and its state-space counterpart, the Kalman filter. Detailed proofs of these relations are available for both the continuous and discrete cases [13]. Our approach is to state the Wiener solution and then show that the steady-state Kalman filter provides a solution with all the necessary properties. We use frequency-domain techniques to show the equivalence. We choose the frequency domain for historical reasons since the classical Wiener solution has more intuitive appeal. The time-domain approach will use the batch innovations solution discussed earlier.

The Wiener filter solution in the frequency domain can be solved by spectral factorization [8] since

$$H(z) = [S_{sy}(z)S_{yy}^{-1}(z^-)]_{cp}S_{yy}^{-1}(z^+) \tag{4.54}$$

where $H(z)$ has all its poles and zeros within the unit circle. The classical approach to Wiener filtering can be accomplished in the frequency domain by factoring the power spectral density (PSD) of the measurement sequence; that is,

$$S_{yy}(z) = H(z)H'(z^{-1}) \tag{4.55}$$

The factorization is unique, stable, and minimum phase (see [8] for proof).

Next we show that the *steady-state Kalman filter* or the *innovations model* (ignoring the deterministic input) given by

$$\hat{x}(t) = A\hat{x}(t-1) + Ke(t-1)$$
$$y(t) = C\hat{x}(t) + e(t) = \hat{y}(t) + e(t) \tag{4.56}$$

where e is the zero-mean, white innovations with covariance R_{ee}, is stable and minimum phase and, therefore, in fact, the *Wiener solution*. The "transfer function" of the innovations model is defined as

$$T(z) := \frac{Y(z)}{E(z)} = C(zI - A)^{-1}K \tag{4.57}$$

and the corresponding measurement PSD is

$$S_{yy}(z) = S_{\hat{y}\hat{y}}(z) + S_{\hat{y}e}(z) + S_{e\hat{y}}(z) + S_{ee}(z) \qquad (4.58)$$

Using the linear system relations of Chapter 2, we see that

$$S_{\hat{y}\hat{y}}(z) = CS_{\hat{x}\hat{x}}(z)C' = T(z)S_{ee}(z)T'(z^{-1}) \quad \text{with} \quad S_{ee}(z) = R_{ee}$$

$$S_{\hat{y}e}(z) = CS_{\hat{x}e}(z) = T(z)S_{ee}(z) \quad \text{and} \quad S_{e\hat{y}}(z) = S_{ee}(z)T'(z^{-1}) \qquad (4.59)$$

Thus, the *measurement* PSD is given in terms of the innovations model as

$$S_{yy}(z) = T(z)S_{ee}(z)T'(z^{-1}) + T(z)S_{ee}(z) + S_{ee}(z)T'(z^{-1}) + S_{ee}(z) \qquad (4.60)$$

Since $S_{ee}(z) = R_{ee}$ and $R_{ee} \geq 0$, then the following (Cholesky) factorization always exists as

$$R_{ee} = R_{ee}^{1/2}(R_{ee}')^{1/2} \qquad (4.61)$$

Substituting these factors into Eq. (4.60) gives

$$S_{yy}(z) = T(z)R_{ee}^{1/2}(R_{ee}')^{1/2}T'(z^{-1}) + T(z)R_{ee} + R_{ee}(z)T'(z^{-1}) + R_{ee} \qquad (4.62)$$

Combining like terms enables $S_{yy}(z)$ to be written in terms of its spectral factors as

$$S_{yy}(z) = \underbrace{(T(z)R_{ee}^{1/2} + R_{ee}^{1/2})}_{T_e(z)} \times \underbrace{((R_{ee}')^{1/2}T'(z^{-1}) + (R_{ee}')^{1/2})}_{T_e'(z^{-1})} \qquad (4.63)$$

or simply

$$S_{yy}(z) = T_e(z) \times T_e'(z^{-1}) \qquad (4.64)$$

which shows that the innovations model indeed admits a spectral factorization of the type desired. To show that $T_e(z)$ is the unique, stable, minimum-phase spectral factor, it is necessary to show that $|T_e(z)|$ has all its poles within the unit circle (stable). It has been shown (e.g. [4, 8, 16]) that $T_e(z)$ does satisfy these constraints.

This completes the discussion on the equivalence of the steady-state Kalman filter and the Wiener filter. Next we consider the development of nonlinear processors.

4.3 Nonlinear State-Space Model-Based Processors

In this section, we develop a suite of recursive, nonlinear state-space MBPs. We start by using the linearized perturbation model of Section 3.8 and the development of the linear MBP of Section 4.2 to motivate the development of the LZKF. This processor is important in its own right, if a solid reference trajectory is available (e.g. roadmap for a self-driving vehicle). From this processor, it is possible to easily understand the motivation for the "ad hoc" EKF that follows. Here we see that the reference trajectory is replaced by the most available state estimate. It is interesting to see how the extended filter evolves directly from this relationship. Next we discuss what has become a very popular and robust solution to this nonlinear state estimation problem – the so-called UKF. Here a statistical linearization replaces the state-space model linearization approach. Finally, we briefly develop the purely Bayesian approach to solve the nonlinear state estimation problem for both non-Gaussian and multimodal posterior distribution estimation problems followed by a 2D-tracking case study completing this chapter.

4.3.1 Nonlinear Model-Based Processor: Linearized Kalman Filter

In this section, we develop an approximate solution to the nonlinear processing problem involving the linearization of the nonlinear process model about a "known" reference trajectory followed by the development of a MBP based on the underlying linearized state-space model. Many processes in practice are nonlinear rather than linear. Coupling the nonlinearities with noisy data makes the signal processing problem a challenging one. Here we limit our discussion to *discrete* nonlinear systems. Continuous-time solutions to this problem are developed in [3–6, 8, 13–16].

Recall from Chapter 3 that our *process* is characterized by a set of nonlinear stochastic vector difference equations in state-space form as

$$x(t) = a[x(t-1)] + b[u(t-1)] + w(t-1) \qquad (4.65)$$

with the corresponding *measurement* model

$$y(t) = c[x(t)] + d[u(t)] + v(t) \qquad (4.66)$$

where $a[\cdot]$, $b[\cdot]$, $c[\cdot]$, $d[\cdot]$ are nonlinear vector functions of x, u, and y with $x, a, b, w \in R^{N_x \times 1}$, $y, c, d, v \in R^{N_y \times 1}$ and $w \sim \mathcal{N}(0, R_{ww}(t))$, $v \sim \mathcal{N}(0, R_{vv}(t))$.

In Section 3.8, we linearized a deterministic nonlinear model using a first-order Taylor series expansion for the functions, a, b, and c and developed a *linearized Gauss–Markov perturbation model* valid for small deviations given by

$$\delta x(t) = A[x^*(t-1)]\delta x(t-1) + B[u^*(t-1)]\delta u(t-1) + w(t-1)$$
$$\delta y(t) = C[x^*(t)]\delta x(t) + D[u^*(t)]\delta u(t) + v(t) \qquad (4.67)$$

with A, B, and C the corresponding Jacobian matrices and w, v zero-mean, Gaussian.

We used linearization techniques to approximate the statistics of Eqs. (4.65) and (4.66) and summarized these results in an "approximate" Gauss–Markov model of Table 3.3. Using this perturbation model, we will now incorporate it to construct a processor that embeds the $(A[\cdot], B[\cdot], C[\cdot], D[\cdot])$ Jacobian linearized about the reference trajectory $[x^*, u^*]$. Each of the Jacobians is deterministic and time-varying, since they are updated at each time-step. Replacing the (A, B) matrices and $\hat{x}(t|t-1)$ in Table 4.1, respectively, by the Jacobians and $\delta\hat{x}(t|t-1)$, we obtain the estimated state perturbation

$$\delta\hat{x}(t|t-1) = A[x^*(t-1)]\delta\hat{x}(t-1|t-1) + B[u^*(t-1)]\delta u(t-1) \quad (4.68)$$

For the Bayesian estimation problem, we are interested in the state estimate $\hat{x}(t|t-1)$ not in its deviation $\delta\hat{x}(t|t-1)$. From the definition of the perturbation defined in Section 4.8, we have

$$\hat{x}(t|t-1) = \delta\hat{x}(t|t-1) + x^*(t) \quad (4.69)$$

where the reference trajectory $x^*(t)$ was defined as

$$x^*(t) = a[x^*(t-1)] + b[u^*(t-1)] \quad (4.70)$$

Substituting this relation along with Eq. (4.68) into Eq. (4.69) gives

$$\hat{x}(t|t-1) = a[x^*(t-1)] + A[x^*(t-1)][\hat{x}(t-1|t-1) - x^*(t-1)]$$
$$+ b[u^*(t-1)] + B[u^*(t-1)][u(t-1) - u^*(t-1)] \quad (4.71)$$

The corresponding *perturbed* innovations can also be found directly

$$\delta e(t) = \delta y(t) - \delta\hat{y}(t|t-1) = (y(t) - y^*(t)) - (\hat{y}(t|t-1) - y^*(t))$$
$$= y(t) - \hat{y}(t|t-1) = e(t) \quad (4.72)$$

where reference measurement $y^*(t)$ is defined as

$$y^*(t) = c[x^*(t)] + d[u^*(t)] \quad (4.73)$$

Using the linear KF with deterministic Jacobian matrices results in

$$\delta\hat{y}(t|t-1) = C[x^*(t)]\delta\hat{x}(t|t-1) + D[u^*(t)]\delta u(t) \quad (4.74)$$

and using this relation and Eq. (4.73) for the reference measurement, we have

$$\hat{y}(t|t-1) = y^*(t) + C[x^*(t)]\delta\hat{x}(t|t-1) + D[u^*(t)]\delta u(t)$$
$$= c[x^*(t)] + C[x^*(t)]\delta\hat{x}(t|t-1) + d[u^*(t)]$$
$$+ D[u^*(t)]\delta u(t) \quad (4.75)$$

Therefore, it follows that the innovations is

$$e(t) = y(t) - c[x^*(t)] - C[x^*(t)][\hat{x}(t|t-1) - x^*(t)]$$
$$- d[u^*(t)] - D[u^*(t)][u(t) - u^*(t)] \quad (4.76)$$

The updated estimate is easily found by substituting Eq. (4.69) to obtain

$$\delta\hat{x}(t|t) = \delta\hat{x}(t|t-1) + K(t)e(t)$$
$$[\hat{x}(t|t) - x^*(t)] = [\hat{x}(t|t-1) - x^*(t)] + K(t)e(t) \quad (4.77)$$

Table 4.5 Linearized Kalman filter (LZKF) algorithm.

Prediction

$\hat{x}(t|t-1) = a[x^*(t-1)] + A[x^*(t-1)][\hat{x}(t-1|t-1) - x^*(t-1)] +$
$b[u^*(t-1)] + B[u^*(t-1)][u(t-1) - u^*(t-1)]$ (State Prediction)

$\tilde{P}(t|t-1) = A[x^*(t-1)]\tilde{P}(t-1|t-1)A'[x^*(t-1)] + R_{ww}(t-1)$

(Covariance Prediction)

Innovation

$e(t) = y(t) - c[x^*(t)] - C[x^*(t)][\hat{x}(t|t-1) - x^*(t)] - d[u^*(t)] - D[u^*(t)][u(t) - u^*(t)]$

(Innovations)

$R_{ee}(t) = C[x^*(t)]\tilde{P}(t|t-1)C'[x^*(t)] + R_{vv}(t)$ (Innovations Covariance)

Gain

$K(t) = \tilde{P}(t|t-1)C'[x^*(t)]R_{ee}^{-1}(t)$ (Gain or Weight)

Update

$\hat{x}(t|t) = \hat{x}(t|t-1) + K(t)e(t)$ (State Update)

$\tilde{P}(t|t) = [I - K(t)C[x^*(t)]]\tilde{P}(t|t-1)$ (Covariance Update)

Initial conditions

$$\hat{x}(0|0) \quad \tilde{P}(0|0)$$

Jacobians

$A[x^*(t-1)] \equiv \dfrac{da[x(t-1)]}{dx(t-1)}\Big|_{x=x^*(t-1)} \quad B[u^*(t-1)] \equiv \dfrac{db[u(t-1)]}{du(t-1)}\Big|_{u=u^*(t-1)}$

$C[x^*(t)] \equiv \dfrac{dc[x(t)]}{dx(t)}\Big|_{x=x^*(t)} \quad D[u^*(t)] \equiv \dfrac{d\ d[u(t)]}{du(t)}\Big|_{u=u^*(t)}$

---

which yields the identical update equation of Table 4.1. Since the state perturbation estimation error is identical to the state estimation error, the corresponding error covariance is given by $\delta\tilde{P}(t|\cdot) = \tilde{P}(t|\cdot)$ and, therefore,

$$\delta\tilde{x}(t|\cdot) = \delta x(t) - \delta\hat{x}(t|\cdot) = [x(t) - x^*(t)] - [\hat{x}(t|\cdot) - x^*(t)] = x(t) - \hat{x}(t|\cdot)$$

$$(4.78)$$

The gain is just a function of the measurement linearization, $C[x^*(t)]$ completing the algorithm. We summarize the discrete LZKF in Table 4.5.

Example 4.3 Consider the discrete nonlinear process approximating a nonlinear trajectory estimation problem [3] given by

$$x(t) = (1 - 0.05\triangle T)x(t-1) + 0.04\triangle Tx^2(t-1) + w(t-1)$$

with the corresponding measurement model

$$y(t) = x^2(t) + x^3(t) + v(t)$$

where $v(t) \sim \mathcal{N}(0, 0.09)$, $x(0) = 2.3$, $P(0) = 0.01$, $\Delta T = 0.01$ second, and $R_{ww} = 0$. The simulated measurements are shown in Figure 4.3c. The LZKF is designed from the following Jacobians:

$$A[x(t-1)] = 1 - 0.05\Delta T + 0.08\Delta T x(t-1) \text{ and } C[x(t)] = 2x(t) + 3x^2(t)$$

Observing the mean state, we develop a reference trajectory by fitting a line to the simulated state, which is given by

$$x^*(t) = 0.067\ t + 2.0 \quad 0 \le t \le 1.5 \quad \text{and} \quad u^*(t) = u(t) = 0.0 \quad \forall t$$

The LZKF algorithm is then given by

$$\hat{x}(t|t-1) = (1 - 0.05\Delta T)x^*(t-1)$$
$$+(1 - 0.05\Delta T + 0.08\Delta T x^*(t-1))[\hat{x}(t-1|t-1) - x^*(t-1)]$$
$$\tilde{P}(t|t-1) = [1 - 0.05\Delta T + 0.08\Delta T x^*(t-1)]^2 \tilde{P}(t-1|t-1)$$
$$e(t) = y(t) - (x^{*2}(t) - x^{*3}(t)) - (2x^*(t) + 3x^{*2}(t))[\hat{x}(t|t-1) - x^*(t)]$$
$$R_{ee}(t) = [2\hat{x}(t|t-1) + 3\hat{x}^2(t|t-1)]^2 \tilde{P}(t|t-1) + 0.09$$
$$K(t) = \frac{\tilde{P}(t|t-1)[2x^*(t) + 3x^{*2}(t)]}{R_{ee}(t)}$$
$$\hat{x}(t|t) = \hat{x}(t|t-1) + K(t)e(t)$$
$$\tilde{P}(t|t) = (1 - K(t)[2x^*(t) + 3x^{*2}(t)])\tilde{P}(t|t-1)$$
$$\hat{x}(0|0) = 2.3 \text{ and } \tilde{P}(0|0) = 0.01$$

An LZKF run is depicted in Figure 4.3. Here we see that the state estimate begins tracking the true state after the initial transient. The estimation error is good (0% lie outside confidence limits), indicating the filter is performing properly for this realization. The filtered measurement and innovations are shown in Figure 4.3b with their corresponding predicted confidence limits. Both estimates lie well within these bounds. The innovations are zero-mean ($3.9 \times 10^{-2} < 10.1 \times 10^{-2}$) and white (0% lie outside the limits) as shown in Figure 4.3c, indicating proper tuning. This completes the nonlinear filtering example. □

4.3.2 Nonlinear Model-Based Processor: Extended Kalman Filter

In this section, we develop the EKF. The EKF is ad hoc in nature, but has become one of the workhorses of (approximate) nonlinear filtering [3–6, 8, 13–17] more recently being replaced by the UKF [18]. It has found applicability in a wide variety of applications such as tracking [19], navigation [3, 13], chemical processing

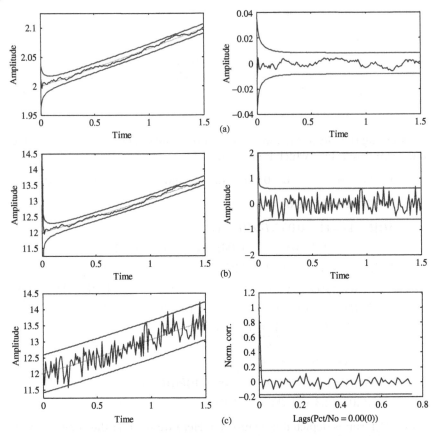

Figure 4.3 Nonlinear trajectory estimation: linearized Kalman filter (LZKF). (a) Estimated state (0% out) and error (0% out). (b) Filtered measurement (1% out) and error (innovations) (2.6% out). (c) Simulated measurement and zero-mean/whiteness test (3.9 × 10^{-2} < 10.1 × 10^{-2} and 0% out.

[20], ocean acoustics [21], seismology [22] (for further applications see [23]). The EKF evolves directly from the linearized processor of Section 4.3.1 in which the reference state, $x^*(t)$, used in the linearization process is replaced with the most recently available state estimate, $\hat{x}(t|t)$ – this is the step that makes the processor ad hoc. However, we must realize that the Jacobians used in the linearization process are deterministic (but time-varying) when a reference or perturbation trajectory is used. However, using the current state estimate is an approximation to the conditional mean, which is random, making these associated Jacobians and subsequent relations random. Therefore, although popularly ignored, most EKF designs should be based on *ensemble operations* to obtain reasonable estimates of the underlying statistics.

With this in mind, we develop the processor directly from the LZKF. If, instead of using the reference trajectory, we choose to linearize about each new state estimate as soon as it becomes available, then the EKF algorithm results. The reason for choosing to linearize about this estimate is that it represents the best information we have about the state and therefore most likely results in a better reference trajectory (state estimate). As a consequence, large initial estimation errors do not propagate; therefore, linearity assumptions are less likely to be violated. Thus, if we choose to use the current estimate $\hat{x}(t|\alpha)$, where α is $t - 1$ or t, to linearize about instead of the reference trajectory $x^*(t)$, then the EKF evolves. That is, let

$$x^*(t) \equiv \hat{x}(t|\alpha) \quad \text{for } t - 1 \leq \alpha \leq t \tag{4.79}$$

Then, for instance, when $\alpha = t - 1$, the predicted perturbation is

$$\delta\hat{x}(t|t - 1) = \hat{x}(t|t - 1) - x^*(t)|_{x^*=\hat{x}(t|t-1)} = 0 \tag{4.80}$$

Thus, it follows immediately that when $x^*(t) = \hat{x}(t|t)$, then $\delta\hat{x}(t|t) = 0$ as well.

Substituting the current estimate, either prediction or update into the LZKF algorithm, it is easy to see that each of the difference terms $[\hat{x} - x^*]$ is null, resulting in the EKF algorithm. That is, examining the *prediction* phase of the linearized algorithm, substituting the current available updated estimate, $\hat{x}(t - 1|t - 1)$, for the reference and using the fact that $(u^*(t) = u(t)\forall t)$, we have

$$\hat{x}(t|t - 1) = a[\hat{x}(t - 1|t - 1)] + A[\hat{x}(t - 1|t - 1)]$$
$$\times \underbrace{[\hat{x}(t - 1|t - 1) - \hat{x}(t - 1|t - 1)]}_{0}$$
$$+ b[u(t - 1)] + B[u(t - 1)]\underbrace{[u(t - 1) - u(t - 1)]}_{0}$$

yielding the prediction of the EKF

$$\hat{x}(t|t - 1) = a[\hat{x}(t - 1|t - 1)] + b[u(t - 1)] \tag{4.81}$$

Now with the predicted estimate available, substituting it for the reference in Eq. (4.76) gives the innovations sequence as

$$e(t) = y(t) - c[\hat{x}(t|t - 1)] - C[\hat{x}(t|t - 1)]\underbrace{[\hat{x}(t|t - 1) - \hat{x}(t|t - 1)]}_{0}$$
$$- d[u(t)] - D[u(t)]\underbrace{[u(t) - u(t)]}_{0}$$
$$= y(t) - c[\hat{x}(t|t - 1)] - d[u(t)] \tag{4.82}$$

where we have the new predicted or filtered measurement expression

$$\hat{y}(t|t - 1) \equiv c[\hat{x}(t|t - 1)] - d[u(t)] \tag{4.83}$$

The updated state estimate is easily obtained by substituting the predicted estimate for the reference $(\hat{x}(t|t-1) \to x^*(t))$ in Eq. (4.77)

$$\delta\hat{x}(t|t) = \delta\hat{x}(t|t-1) + K(t)e(t)$$

$$[\hat{x}(t|t) - \hat{x}(t|t-1)] = \underbrace{[\hat{x}(t|t-1) - \hat{x}(t|t-1)]}_{0} + K(t)e(t)$$

$$\hat{x}(t|t) = \hat{x}(t|t-1) + K(t)e(t) \tag{4.84}$$

The covariance and gain equations are identical to those in Table 4.5, but with the Jacobian matrices A, B, C, and D linearized about the predicted state estimate, $\hat{x}(t|t-1)$. Thus, we obtain the discrete EKF algorithm summarized in Table 4.6. Note that the covariance matrices, $\tilde{P}$, and the gain, K, are now functions of the current state estimate, which is the *approximate* conditional mean estimate and therefore is a single realization of a stochastic process. Thus, ensemble Monte Carlo (MC) techniques should be used to evaluate estimator performance. That is, for new initial conditions selected by a Gaussian random number generator (either $\hat{x}(0|0)$ or $\tilde{P}(0|0)$), the algorithm is executed generating a set of estimates, which should be averaged over the entire ensemble using this approach to get an "expected" state, etc. Note also in practice that this

Table 4.6 Extended Kalman filter (EKF) algorithm.

Prediction

$\hat{x}(t|t-1) = a[\hat{x}(t-1|t-1)] + b[u(t-1)]$ (State Prediction)

$\tilde{P}(t|t-1) = A[\hat{x}(t|t-1)]\tilde{P}(t-1|t-1)A'[\hat{x}(t|t-1)] + R_{ww}(t-1)$

(Covariance Prediction)

Innovation

$e(t) = y(t) - c[\hat{x}(t|t-1)] - d[u(t)]$ (Innovations)

$R_{ee}(t) = C[\hat{x}(t|t-1)]\tilde{P}(t|t-1)C'[\hat{x}(t|t-1)] + R_{vv}(t)$ (Innovations Covariance)

Gain

$K(t) = \tilde{P}(t|t-1)C'[\hat{x}(t|t-1)]R_{ee}^{-1}(t)$ (Gain or Weight)

Update

$\hat{x}(t|t) = \hat{x}(t|t-1) + K(t)e(t)$ (State Update)

$\tilde{P}(t|t) = [I - K(t)C[\hat{x}(t|t-1)]]\tilde{P}(t|t-1)$ (Covariance Update)

Initial conditions

$\hat{x}(0|0) \quad \tilde{P}(0|0)$

Jacobians

$A[x(t|t-1)] \equiv \frac{da[x(t-1)]}{d\hat{x}(t-1)}\big|_{x=\hat{x}(t|t-1)} \qquad C[\hat{x}(t|t-1)] \equiv \frac{dc[x(t)]}{dx(t)}\big|_{x=\hat{x}(t|t-1)}$

algorithm is usually implemented using sequential processing and UD (upper diagonal/square root) factorization techniques (see [15] for details).

Example 4.4   Consider the discrete nonlinear process and measurement system described in Example 4.3. The simulated measurement using MATLAB [2] is shown in Figure 4.4c. The EKF is designed from the following Jacobians:

$$A[x(t-1)] = 1 - 0.05\triangle T + 0.08\triangle Tx(t-1)$$

$$\text{and} \quad C[x(t)] = 2x(t) + 3x^2(t)$$

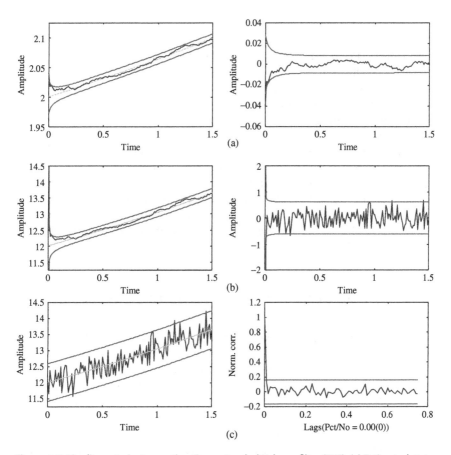

Figure 4.4 Nonlinear trajectory estimation: extended Kalman filter (EKF). (a) Estimated state (1% out) and error (1% out). (b) Filtered measurement (1.3% out) and error (innovations) (3.3% out). (c) Simulated measurement and zero-mean/whiteness test ($6.3 \times 10^{-2} <$ 11.8×10^{-2} and 0% out).

The EKF algorithm is then given by

$$\hat{x}(t|t-1) = (1 - 0.05\triangle T)\hat{x}(t-1|t-1) + 0.04\triangle T\hat{x}^2(t-1|t-1)$$
$$\tilde{P}(t|t-1) = [1 - 0.05\triangle T + 0.08\triangle Tx(t-1)]^2\tilde{P}(t-1|t-1)$$
$$e(t) = y(t) - \hat{x}^2(t|t-1) - \hat{x}^3(t|t-1)$$
$$R_{ee}(t) = [2\hat{x}(t|t-1) + 3\hat{x}^2(t|t-1)]^2\tilde{P}(t|t-1) + 0.09$$
$$K(t) = (\tilde{P}(t|t-1)[2\hat{x}(t|t-1) + 3\hat{x}^2(t|t-1)])R_{ee}^{-1}(t)$$
$$\hat{x}(t|t) = \hat{x}(t|t-1) + K(t)e(t)$$
$$\tilde{P}(t|t) = (1 - K(t)[2\hat{x}(t|t-1) + 3\hat{x}^2(t|t-1)])\tilde{P}(t|t-1)$$
$$\hat{x}(0|0) = 2.3 \text{ and } \tilde{P}(0|0) = 0.01$$

An EKF run is depicted in Figure 4.4. Here we see that the state estimate begins tracking the true state after the initial transient. The estimation error is reasonable ($\sim 1\%$ lie outside the limits), indicating the filter is performing properly for this realization. The filtered measurement and innovations are shown in Figure 4.4b and lie within the predicted limits. The innovations are zero-mean ($6.3 \times 10^{-2} < 11.8 \times 10^{-2}$) and white (0% lie outside the limits) as shown in Figure 4.4c, indicating proper tuning. Comparing the EKF to the linearized Kalman filter (LZKF) of Section 4.3.1 shows that it performs slightly worse in terms of predicted covariance limits for the estimated measurement and innovations. Most of this error is caused by the initial conditions of the processor.

Running an ensemble of 101 realizations of this processor yields similar results for the ensemble estimates: state estimation error increased to ($\sim 2\%$ outside limits), innovations zero-mean test increased slightly ($6.7 \times 10^{-2} < 12 \times 10^{-2}$), and whiteness was identical. This completes the example. □

4.3.3 Nonlinear Model-Based Processor: Iterated–Extended Kalman Filter

In this section, we discuss an extension of the EKF to the IEKF. This development from the Bayesian perspective coupled to numerical optimization is complex and can be found in [2]. Here we first heuristically motivate the technique and then apply it. This algorithm is based on performing "local" iterations (not global) at a point in time, t to improve the reference trajectory and therefore the underlying estimate in the presence of significant measurement nonlinearities [3]. A local iteration implies that the inherent recursive structure of the processor is retained providing updated estimates as the new measurements are made available.

To develop the iterated–extended processor, we start with the linearized processor update relation substituting the "linearized" innovations of Eq. (4.76) of the LZKF, that is,

$$\hat{x}(t|t) = \hat{x}(t|t-1) + K(t;x^*(t))$$
$$\times[y(t) - c[x^*(t)] - C[x^*(t)](\hat{x}(t|t-1) - x^*(t)) - d[u^*(t)]$$
$$-D[u^*(t)](u(t) - u^*(t))] \tag{4.85}$$

where we have explicitly shown the dependence of the gain on the reference trajectory, $x^*(t)$ through the measurement Jacobian. The EKF algorithm linearizes about the most currently available estimate, $x^*(t) = \hat{x}(t|t-1)$ and $u^*(t) = u(t)$ in this case. Theoretically, the updated estimate, $\hat{x}(t|t)$, is a better estimate and closer to the true trajectory. Suppose we continue and relinearize about $\hat{x}(t|t)$ when it becomes available and then recompute the corrected estimate and so on. That is, define the $(i+1)$th iterated estimate as $\hat{x}_{i+1}(t|t)$, then the corrected or updated *iterator* equation becomes

$$\hat{x}_{i+1}(t|t) = \hat{x}(t|t-1) + \overbrace{K(t;\hat{x}_i(t|t))}^{K_i(t)}[y(t) - c[\hat{x}_i(t|t)]$$
$$-C[\hat{x}_i(t|t)](\hat{x}(t|t-1) - \hat{x}_i(t|t))$$
$$\underbrace{-d[u(t)] - D[u(t)](u(t) - u(t))}_{0} \tag{4.86}$$

Now if we start with the 0th iterate as the predicted estimate, that is, $\hat{x}_0 \equiv \hat{x}(t|t-1)$, then the EKF results for $i = 0$. Clearly, the corrected estimate in this iteration is given by

$$\hat{x}_1(t|t) = \hat{x}(t|t-1) + K_0(t)\,[y(t) - c[\hat{x}(t|t)] - C[\hat{x}(t|t-1)]$$
$$\times \underbrace{(\hat{x}(t|t-1) - \hat{x}(t|t-1)) - d[u(t)]]}_{0} \tag{4.87}$$

where the last term in this expression is null leaving the usual innovations. Also note that the *gain* is reevaluated on each iteration as are the measurement function and Jacobian. The iterations continue until there is little difference in consecutive iterates. The *last* iterate is taken as the updated estimate. The complete (updated) iterative loop is given by

$$e_i(t) = y(t) - c[\hat{x}_i(t|t)] - d[u(t)]$$
$$R_{e_ie_i}(t) = C[\hat{x}_i(t|t)]\tilde{P}(t|t-1)C'[\hat{x}_i(t|t)] + R_{vv}(t)$$
$$K_i(t) = \tilde{P}(t|t-1)C'[\hat{x}_i(t|t)]R_{e_ie_i}^{-1}(t)$$
$$\hat{x}_{i+1}(t|t) = \hat{x}(t|t-1) + K_i(t)[e_i(t) - C[\hat{x}_i(t|t)](\hat{x}(t|t-1) - \hat{x}_i(t|t))]$$
$$\tilde{P}_i(t|t) = (I - K_i(t)C[\hat{x}_i(t|t)])\tilde{P}(t|t-1) \tag{4.88}$$

Table 4.7 Iterated–extended Kalman filter (IEKF) algorithm.

Prediction

$$\hat{x}(t|t-1) = a[\hat{x}(t-1|t-1)] + b[u(t-1)] \qquad \text{(State Prediction)}$$

$$\tilde{P}(t|t-1) = A[\hat{x}(t|t-1)]\tilde{P}(t-1|t-1)A'[\hat{x}(t|t-1)] + R_{ww}(t-1)$$

$$\text{(Covariance Prediction)}$$

$$\text{LOOP: } i = 1, \ldots, N_{\text{iterations}}$$

Innovation

$$e_i(t) = y(t) - c[\hat{x}_i(t|t)] - d[u(t)] \qquad \text{(Innovations)}$$

$$R_{e_i e_i}(t) = C[\hat{x}_i(t|t)]\tilde{P}(t|t-1)C'[\hat{x}_i(t|t)] + R_{vv}(t) \qquad \text{(Innovations Covariance)}$$

Gain

$$K_i(t) = \tilde{P}(t|t-1)C'[\hat{x}_i(t|t)]R_{e_i e_i}^{-1}(t) \qquad \text{(Gain or Weight)}$$

Update

$$\hat{x}_{i+1}(t|t) = \hat{x}(t|t-1) + K_i(t)[e_i(t) - C[\hat{x}_i(t|t)](\hat{x}(t|t-1) - \hat{x}_i(t|t))] \qquad \text{(State Update)}$$

$$\tilde{P}_i(t|t) = [I - K_i(t)C[\hat{x}_i(t|t)]]\tilde{P}(t|t-1) \qquad \text{(Covariance Update)}$$

Initial conditions

$$\hat{x}(0|0), \quad \tilde{P}(0|0), \quad \hat{x}_0(t|t) = \hat{x}(t|t-1)$$

Jacobians

$$A[\hat{x}(t|t-1)] \equiv \frac{da[x(t-1)]}{dx(t-1)}\Big|_{x=\hat{x}(t|t-1)} \qquad C[\hat{x}_i(t|t)] \equiv \frac{dc[x(t)]}{dx(t)}\Big|_{x=\hat{x}_{i(t|t)}}$$

Stopping rule

$$||\hat{x}_{i+1}(t|t) - \hat{x}_i(t|t)|| < \epsilon \text{ and } \hat{x}_i(t|t) \to \hat{x}(t|t)$$

A typical stopping rule is

$$||\hat{x}_{i+1}(t|t) - \hat{x}_i(t|t)|| < \epsilon \quad \text{and} \quad \hat{x}_i(t|t) \to \hat{x}(t|t) \qquad (4.89)$$

The IEKF algorithm is summarized in Table 4.7. It is useful in reducing the measurement function nonlinearity approximation errors, improving processor performance. It is designed for measurement nonlinearities and does not improve the previous reference trajectory, but it will improve the subsequent one.

Example 4.5 Again consider the discrete nonlinear trajectory process and measurement system described in Example 4.4. The simulated measurement using MATLAB is shown in Figure 4.5c. The IEKF is designed from the following Jacobians:

$$A[x(t-1)] = 1 - 0.05\triangle T + 0.08\triangle Tx(t-1) \quad \text{and}$$
$$\times C[x(t)] = 2x(t) + 3x^2(t)$$

The IEKF algorithm is then given by

$$\hat{x}(t|t-1) = (1 - 0.05\triangle T)\hat{x}(t-1|t-1) + 0.04\triangle T\hat{x}^2(t-1|t-1)$$
$$\tilde{P}(t|t-1) = [1 - 0.05\triangle T + 0.08\triangle Tx(t-1)]^2\tilde{P}(t-1|t-1)$$
$$e_i(t) = y(t) - \hat{x}_i^2(t|t) - \hat{x}_i^3(t|t)$$
$$R_{e_ie_i}(t) = [2\hat{x}_i(t|t) + 3\hat{x}_i^2(t|t)]^2\tilde{P}(t|t-1) + 0.09$$
$$K_i(t) = \tilde{P}(t|t-1)[2\hat{x}_i(t|t) + 3\hat{x}_i^2(t|t)]/R_{e_ie_i}(t)$$
$$\hat{x}_{i+1}(t|t) = \hat{x}(t|t-1) + K_i(t)[e_i(t) - [2\hat{x}_i(t|t) + 3\hat{x}_i^2(t|t)](\hat{x}(t|t-1) - \hat{x}_i(t|t))$$
$$\tilde{P}_i(t|t) = (1 - K_i(t)[2\hat{x}_i(t|t) + 3\hat{x}_i^2(t|t)])\tilde{P}(t|t-1)$$
$$\hat{x}(0|0) = 2.3 \text{ and } \tilde{P}(0|0) = 0.01$$

An IEKF run is depicted in Figure 4.5. Here we see that the state estimate ($\sim 0\%$ lie outside the limits) begins tracking the true state after the initial iterations (3). The estimation error is reasonable ($\sim 0\%$ out), indicating the filter is performing properly for this realization. The filtered measurement ($\sim 1\%$ out) and innovations ($\sim 2.6\%$ out) are shown in Figure 4.5b. The innovations are statistically zero-mean ($4 \times 10^{-2} < 10.7 \times 10^{-2}$) and white ($0\%$ lie outside the limits) as shown in Figure 4.5c, indicating proper tuning and matching the LZKF result almost exactly.

Executing an ensemble of 101-realizations of this processor yields similar results for the ensemble estimates: innovations zero-mean test decreased slightly ($3.7 \times 10^{-2} < 10.3 \times 10^{-2}$) and whiteness was identical. So the overall effect of the IEKF was to decrease the measurement nonlinearity effect especially in the initial transient of the algorithm. These results are almost identical to those of the LZKF. This completes the nonlinear filtering example. □

Next we consider a different approach to nonlinear estimation.

4.3.4 Nonlinear Model-Based Processor: Unscented Kalman Filter

In this section, we discuss an extension of the approximate nonlinear Kalman filter suite of processors that takes a distinctly different approach to the nonlinear Gaussian problem. Instead of attempting to improve on the linearized *model* approximation in the nonlinear EKF scheme discussed in Section 4.3.1 or increasing the order of the Taylor series approximations [16], a statistical *transformation* approach is developed. It is founded on the basic idea that "it is easier to approximate a probability distribution, than to approximate an arbitrary nonlinear transformation function" [18, 24, 25]. The nonlinear processors discussed so far are based on linearizing nonlinear (model) functions of the state and input to provide estimates of the underlying statistics (using Jacobians), while the transformation approach is based upon selecting a set

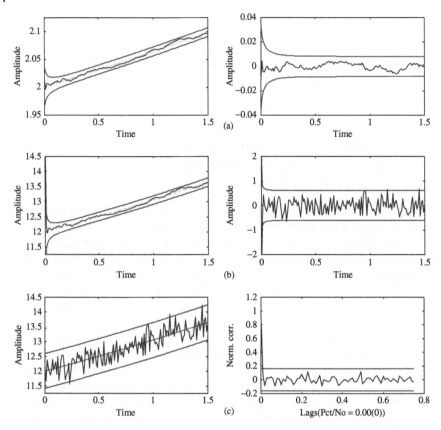

Figure 4.5 Nonlinear trajectory estimation: Iterated–extended Kalman filter (IEKF). (a) Estimated state (~ 0% out) and error (~ 0% out). (b) Filtered measurement (~ 1% out) and error (innovations) (~ 2.6% out). (c) Simulated measurement and zero-mean/whiteness test $(4 \times 10^{-2} < 10.7 \times 10^{-2}$ and 0% out).

of sample points that capture certain properties of the underlying probability distribution. This set of samples is then nonlinearly transformed or propagated to a new space. The statistics of the new samples are then calculated to provide the required estimates. Once this transformation is performed, the resulting processor, the UKF evolves. The UKF is a recursive processor that resolves some of the approximation issues [24] and deficiencies of the EKF of the previous sections. In fact, it has been called "unscented" because the "EKF stinks." We first develop the idea of nonlinearly transforming the probability distribution. Then apply it to our Gaussian problem, leading to the UKF algorithm. We apply the processor to the previous nonlinear state estimation problem and investigate its performance.

The *unscented transformation* (UT) is a technique for calculating the statistics of a random variable that has been transformed, establishing the foundation of the processor. A set of samples or *sigma-points* are chosen so that they capture the specific properties of the underlying distribution. Each of the σ-points is nonlinearly transformed to create a set of samples in the new space. The statistics of the transformed samples are then calculated to provide the desired estimates.

Consider propagating an N_x-dimensional random vector, $\mathbf{x}$, through an arbitrary nonlinear transformation $\mathbf{a}[\cdot]$ to generate a new random vector [24]

$$\mathbf{z} = \mathbf{a}[\mathbf{x}] \tag{4.90}$$

The set of σ-points, $\{\mathcal{X}_i\}$, consists of $N_x + 1$ vectors with appropriate weights, $\{\mathcal{W}_i\}$, given by $\Sigma = \{\mathcal{X}_i, W_i : i = 0, \dots, N_x\}$. The weights can be positive or negative, but *must satisfy* the *normalization constraint* that

$$\sum_{i=0}^{N_x} W_i = 1$$

The problem then becomes

GIVEN these σ-points and the nonlinear transformation $\mathbf{a}[\cdot]$, **FIND** the statistics of the transformed samples, $\mu_z \equiv E\{\mathbf{z}\}$ and $\mathbf{R}_{zz} = \text{Cov}(\mathbf{z})$

The *UT* approach is to

1) Nonlinearly *transform* each point to obtain the set of new σ-points: $\mathcal{Z}_i = \mathbf{a}[\mathbf{X}_i]$
2) Estimate the posterior *mean* by its weighted average: $\mu_z = \sum_{i=0}^{N_x} W_i \mathcal{Z}_i$
3) Estimate the posterior *covariance* by its weighted outer product:

$$\mathbf{R}_{zz} = \sum_{i=0}^{N_x} W_i (\mathcal{Z}_i - \mu_z)(\mathcal{Z}_i - \mu_z)'$$

The critical issues to decide are as follows: (i) N_x, the number of σ-points; (ii) W_i, the weights assigned to each σ-point; and (iii) *where* the σ-points are to be located. The σ-points should be selected to capture the "most important" properties of the random vector, $\mathbf{x}$. This parameterization captures the *mean* and *covariance* information and permits the direct propagation of this information through the arbitrary set of nonlinear functions. Here we accomplish this (approximately) by generating the discrete distribution having the same first and second (and potentially higher) moments where each point is directly transformed. The mean and covariance of the transformed ensemble can then be computed as *the estimate* of the nonlinear transformation of the original distribution (see [7] for more details).

The UKF is a recursive processor developed to eliminate some of the deficiencies (see [7] for more details) created by the failure of the linearization process to first-order (Taylor series) in solving the state estimation problem. Different from the EKF, the UKF does not approximate the nonlinear process and measurement models, it employs the true nonlinear models and approximates the underlying *Gaussian* distribution function of the state variable using the UT of the σ-points. In the UKF, the state is still represented as Gaussian, but it is specified using the minimal set of deterministically selected samples or σ-points. These points completely capture the true mean and covariance of the Gaussian distribution. When they are propagated through the true nonlinear process, the *posterior* mean and covariance are accurately captured to the second order for *any* nonlinearity with errors only introduced in the third-order and higher order moments as above.

Suppose we are given an N_x-dimensional *Gaussian distribution* having covariance, P_{xx}, then we can generate a set of σ-points having the same sample covariance from the columns (or rows) of the matrices $\pm\sqrt{(N_x + \kappa)\ P_{xx}}$. Here κ is a scaling factor. This set is zero-mean, but if the original distribution has mean μ_x, then simply adding μ_x to each of the σ-points yields a symmetric set of $2N_x$ samples having the desired mean and covariance. Since the set is symmetric, its odd central moments are null, so its first three moments are identical to those of the original Gaussian distribution. This is the *minimal* number of σ-points capable of capturing the essential statistical information. The basic UT technique for a multivariate Gaussian distribution [24] is

1) Compute the set of $2N_x$ σ-points from the rows or columns of $\pm\sqrt{(N_x + \kappa)\ P_{xx}}$. Compute $\mathcal{X}_i = \sigma + \mu_x$.

$$\mathcal{X}_o = \mu_x, \qquad\qquad W_o = \frac{\kappa}{(N_x + \kappa)}$$

$$\mathcal{X}_i = \mu_x + (\sqrt{(N_x + \kappa)\ P_{xx}})_i, \quad W_i = \frac{1 - W_o}{2(N_x + \kappa)}$$

$$\mathcal{X}_{i+N_x} = \mu_x - (\sqrt{(N_x + \kappa)\ P_{xx}})_i, \quad W_{i+N_x} = \frac{1 - W_o}{2(N_x + \kappa)}$$

where κ is a scalar, $(\sqrt{(N_x + \kappa)\ P_{xx}})_i$ is the ith row or column of the matrix square root of $(N_x + \kappa)P_{xx}$ and W_i is the weight associated with the ith σ-point.

2) Nonlinearly *transform* each point to obtain the set of σ-points: $\mathcal{Z}_i = \mathbf{a}[X_i]$.

3) Estimate the posterior *mean* of the new samples by its weighted average

$$\mu_z = \sum_{i=0}^{N_x} W_i \mathcal{Z}_i$$

4) Estimate the posterior *covariance* of the new samples by its weighted outer product

$$\mathbf{R}_{zz} = \sum_{i=0}^{N_x} W_i(\mathcal{Z}_i - \mu_z)(\mathcal{Z}_i - \mu_z)'$$

The discrete nonlinear process model is given by

$$x(t) = \mathbf{a}[x(t-1)] + \mathbf{b}[u(t-1)] + w(t-1) \tag{4.91}$$

with the corresponding measurement model

$$y(t) = \mathbf{c}[x(t)] + d[u(t)] + v(t) \tag{4.92}$$

for $w \sim \mathcal{N}(0, R_{ww})$ and $v \sim \mathcal{N}(0, R_{vv})$. The critical conditional Gaussian distribution for the *state variable* statistics is the prior [7]

$$x \sim \mathcal{N}(\hat{x}(t|t-1), \tilde{P}(t|t-1))$$

and with the eventual measurement statistics specified by

$$y \sim \mathcal{N}(\hat{y}(t|t-1), R_{\xi\xi}(t|t-1))$$

where $\hat{x}(t|t-1)$ and $\tilde{P}(t|t-1)$ are the respective corrected state and error covariance based on the data up to time $(t-1)$ and $\hat{y}(t|t-1)$, $R_{\xi\xi}(t|t-1)$ are the predicted measurement and residual covariance. The idea then is to use the "prior" statistics and perform the UT (under Gaussian assumptions) using *both* the process and nonlinear transformations (models) as specified above to obtain a set of transformed σ-points. Then, the selected σ-points for the Gaussian are transformed using the process and measurement models yielding the corresponding set of σ-points in the new space. The predicted means are weighted sums of the transformed σ-points and covariances are merely weighted sums of their mean-corrected, outer products.

To develop the UKF we must:

- PREDICT the next state and error covariance, $(\hat{x}(t|t-1), \tilde{P}(t|t-1))$, by UT transforming the prior, $[\hat{x}(t-1|t-1), \tilde{P}(t-1|t-1)]$, including the process noise using the σ-points $\mathcal{X}(t|t-1)$ and $\mathcal{X}(t-1|t-1)$.
- PREDICT the measurement and residual covariance, $[\hat{y}(t|t-1), R_{\xi\xi}(t|t-1)]$ by using the UT transformed σ-points $\mathcal{Y}(t|t-1)$ and performing the weighted sums.
- PREDICT the cross-covariance, $R_{\tilde{x}\xi}(t|t-1)$ in order to calculate the corresponding gain for the subsequent correction step.

We use these steps as our road map to develop the UKF. The σ-points for the UT transformation of the "prior" state information is specified with $\mu_x = \hat{x}(t-1|t-1)$ and $P_{xx} = \tilde{P}(t-1|t-1)$; therefore, we have the UKF algorithm given by

1) *Select* the $(2N_x + 1)$-points as:

$$\mathcal{X}_o = \mu_x = \hat{x}(t-1|t-1), \qquad\qquad W_o = \frac{\kappa}{(N_x + \kappa)}$$

$$\mathcal{X}_i = \mu_x + (\sqrt{(N_x + \kappa)\ P_{xx}})_i$$

$$= \hat{x}(t-1|t-1) + \left(\sqrt{(N_x+\kappa)\ (\tilde{P}(t-1|t-1)+R_{ww}(t-1))}\right)_i,$$

$$\times W_i = \frac{1}{2(N_x + \kappa)}$$

$$\mathcal{X}_{i+N_x} = \mu_x - (\sqrt{(N_x + \kappa)\ P_{xx}})_i,$$

$$= \hat{x}(t-1|t-1) - \left(\sqrt{(N_x+\kappa)\ (\tilde{P}(t-1|t-1)+R_{ww}(t-1))}\right)_i,$$

$$\times W_{i+N_x} = \frac{1}{2(N_x + \kappa)}$$

2) *Transform* (UT) process model:

$$\mathcal{X}_i(t|t-1) = \mathbf{a}[\mathcal{X}_i(t-1|t-1)] + \mathbf{b}[u(t-1)]$$

3) *Estimate* the posterior predicted (state) mean by:

$$\hat{x}(t|t-1) = \sum_{i=0}^{2N_x} W_i \mathcal{X}_i(t|t-1)$$

4) *Estimate* the posterior predicted (state) residual and error covariance by:

$$\tilde{\mathcal{X}}_i(t|t-1) = \mathcal{X}_i(t|t-1) - \hat{x}(t|t-1)$$

$$\tilde{P}(t|t-1) = \sum_{i=0}^{2N_x} W_i \tilde{\mathcal{X}}_i(t|t-1)\tilde{\mathcal{X}}'_i(t|t-1)$$

5) *Transform* (UT) measurement (model) with *augmented* σ-points to account for process noise as:

$$\hat{\mathcal{X}}_i(t|t-1) = [\mathcal{X}_i(t|t-1) \quad \mathcal{X}_i(t|t-1) + \kappa\sqrt{R_{ww}(t-1)}$$
$$\mathcal{X}_i(t|t-1) - \kappa\sqrt{R_{ww}(t-1)}]$$

$$\mathcal{Y}_i(t|t-1) = \mathbf{c}[\hat{\mathcal{X}}_i(t|t-1)] + d[u(t)]$$

6) *Estimate* the predicted measurement as:

$$\hat{y}(t|t-1) = \sum_{i=0}^{2N_x} W_i \mathcal{Y}_i(t|t-1)$$

7) *Estimate* the predicted residual and covariance as:

$$\xi_i(t|t-1) = \mathcal{Y}_i(t|t-1) - \hat{y}(t|t-1)$$

$$R_{\xi\xi}(t|t-1) = \sum_{i=0}^{2N_x} W_i \xi_i(t|t-1)\xi'_i(t|t-1) + R_{vv}(t)$$

8) *Estimate* the predicted cross-covariance as:

$$R_{\tilde{x}\xi}(t|t-1) = \sum_{i=0}^{2N_x} W_i \tilde{\mathcal{X}}_i(t|t-1)\tilde{\xi}_i'(t|t-1)$$

Clearly with these vectors and matrices available the corresponding gain and update equations follow immediately as

$$\mathcal{K}(t) = R_{\tilde{x}\xi}(t|t-1)R_{\xi\xi}^{-1}(t|t-1)$$
$$e(t) = y(t) - \hat{y}(t|t-1)$$
$$\hat{x}(t|t) = \hat{x}(t|t-1) + \mathcal{K}(t)e(t)$$
$$\tilde{P}(t|t) = \tilde{P}(t|t-1) - \mathcal{K}(t)R_{\xi\xi}(t|t-1)\mathcal{K}(t)$$

We note that there are *no* Jacobians calculated and the nonlinear models are employed directly to transform the σ-points to the new space. Also in the original problem definition, both process and noise sources (w, v) were assumed *additive*, but not necessary. For more details of the general process, see the following Refs. [26–28]. We summarize the UKF algorithm in Table 4.8.

Before we conclude this discussion, let us apply the UKF to the nonlinear trajectory estimation problem and compare its performance to the other nonlinear processors discussed previously.

Example 4.6 We revisit the nonlinear trajectory estimation problem of the previous examples with the dynamics specified by the discrete nonlinear process given by

$$x(t) = (1 - 0.05\Delta T)x(t-1) + 0.04x^2(t-1) + w(t-1)$$

and the corresponding measurement model

$$y(t) = x^2(t) + x^3(t) + v(t)$$

where recall that $v(t) \sim \mathcal{N}(0, 0.09)$, $x(0) = 2.0$, $P(0) = 0.01$, $\Delta T = 0.01$ seconds, and $R_{ww} = 0$. The simulated measurement is shown in Figure 4.6b. The UKF and EKF were applied to this problem. We used the square-root implementations of the EKF in MATLAB [7] and compared them to the square-root version of the UKF in recursive Bayesian estimation library (REBEL) [28]. The results are shown in Figure 4.6 where we see the corresponding trajectory (state) estimates in (a) and the "filtered" measurements in (b). From the figures, it appears that all of the estimates are quite reasonable with the UKF estimate (thick solid line) converging most rapidly to the true trajectory (dashed line). The EKF (thick dotted line) appears slightly biased. The measurements also indicate the similar performance. The zero-mean/whiteness tests confirm these observations. The EKF perform reasonably with zero-mean/whiteness values of $(1.04 \times 10^{-1} < 1.73 \times 10^{-1}/1\%$ out), and the IEKF performs similarly

Table 4.8 Discrete unscented Kalman filter (UKF) algorithm.

State: σ-points and weights

$\mathcal{X}_0 = \hat{x}(t-1|t-1),$ $\qquad W_0 = \frac{\kappa}{(N_x+\kappa)}$

$\mathcal{X}_i = \hat{x}(t-1|t-1) + (\sqrt{(N_x+\kappa)\ \tilde{P}(t-1|t-1)})_i,$ $\qquad W_i = \frac{1}{2(N_x+\kappa)}$

$\mathcal{X}_{i+N_x} = \hat{x}(t-1|t-1) - (\sqrt{(N_x+\kappa)\ \tilde{P}(t-1|t-1)})_i,$ $\qquad W_{i+N_x} = \frac{1}{2(N_x+\kappa)}$

State prediction

$\mathcal{X}_i(t|t-1) = \mathbf{a}[\mathcal{X}_i(t-1|t-1)] + \mathbf{b}[u(t-1)]$ (Nonlinear State Process)

$\hat{x}(t|t-1) = \sum_{i=0}^{2N_x} W_i \mathcal{X}_i(t|t-1)$ (State Regression)

State error prediction

$\tilde{\mathcal{X}}_i(t|t-1) = \mathcal{X}_i(t|t-1) - \hat{x}(t|t-1)$ (State Error)

$\tilde{P}(t|t-1) = \sum_{i=0}^{2N_x} W_i \tilde{\mathcal{X}}_i(t|t-1)\tilde{\mathcal{X}}_i'(t|t-1) + R_{ww}(t-1)$ (Error Covariance Prediction)

Measurement: σ-points and weights

$\hat{\mathcal{X}}_i(t|t-1) = \{\mathcal{X}_i(t|t-1),\ \mathcal{X}_i(t|t-1) + \kappa\sqrt{R_{ww}(t-1)},\ \mathcal{X}_i(t|t-1) - \kappa\sqrt{R_{ww}(t-1)}\}$

Measurement prediction

$\mathcal{Y}_i(t|t-1) = \mathbf{c}[\hat{\mathcal{X}}_i(t|t-1)] + d[u(t)]$ (Nonlinear Measurement)

$\hat{y}(t|t-1) = \sum_{i=0}^{2N_x} W_i \mathcal{Y}_i(t|t-1)$ (Measurement Regression)

Residual prediction

$\xi_i(t|t-1) = \mathcal{Y}_i(t|t-1) - \hat{y}(t|t-1)$ (Predicted Residual)

$R_{\xi\xi}(t|t-1) = \sum_{i=0}^{2N_x} W_i \xi_i(t|t-1)\xi_i'(t|t-1) + R_{vv}(t)$ (Residual Covariance Regression)

Gain

$R_{\tilde{x}\xi}(t|t-1) = \sum_{i=0}^{2N_x} W_i \tilde{\mathcal{X}}_i(t|t-1)\xi_i'(t|t-1)$ (Cross-covariance Regression)

$\mathcal{K}(t) = R_{\tilde{x}\xi}(t|t-1)R_{\xi\xi}^{-1}(t|t-1)$ (Gain)

State update

$e(t) = y(t) - \hat{y}(t|t-1)$ (Innovation)

$\hat{x}(t|t) = \hat{x}(t|t-1) + \mathcal{K}(t)e(t)$ (State Update)

$\tilde{P}(t|t) = \tilde{P}(t|t-1) - \mathcal{K}(t)R_{\xi\xi}(t|t-1)\mathcal{K}'(t)$ (Error Covariance Update)

Initial conditions

$\qquad \hat{x}(0|0) \quad \tilde{P}(0|0)$

with zero-mean/whiteness values of ($3.85 \times 10^{-2} < 1.73 \times 10^{-1}/0\%$ out), while the UKF is certainly comparable at ($5.63 \times 10^{-2} < 1.73 \times 10^{-1}/0\%$ out). This completes the example. □

4.3.5 Practical Nonlinear Model-Based Processor Design: Performance Analysis

The suite of nonlinear processor designs was primarily developed based on Gaussian assumptions. In the linear Kalman filter case, a necessary and

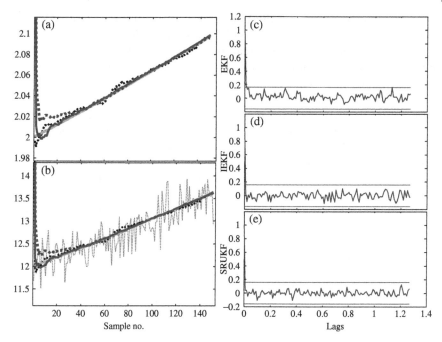

Figure 4.6 Nonlinear trajectory estimation: unscented Kalman filter (UKF). (a) Trajectory (state) estimates using the EKF (thick dotted), IEKF (thin dotted), and UKF (thick solid). (b) Filtered measurement estimates using the EKF (thick dotted), IEKF (thin dotted), and UKF (thick solid). (c) Zero-mean/whiteness tests for EKF: ($1.04 \times 10^{-1} < 1.73 \times 10^{-1}/1\%$ out). (d) Zero-mean/whiteness tests for IEKF: ($3.85 \times 10^{-2} < 1.73 \times 10^{-1}/0\%$ out). (e) Zero-mean/whiteness tests for UKF: ($5.63 \times 10^{-2} < 1.73 \times 10^{-1}/0\%$ out).

sufficient condition for optimality of the filter is that the corresponding innovations or residual sequence must be zero-mean and white (see Section 4.2 for details). In lieu of this constraint, a variety of statistical tests (whiteness, uncorrelated inputs, etc.) were developed evolving from this known property. When the linear Kalman filter was "extended" to the nonlinear case, the same tests can be performed based on approximate Gaussian assumptions. Clearly, when noise is additive Gaussian, these arguments can still be applied for improved design and performance evaluation.

In what might be termed "sanity testing," the classical nonlinear methods employ the basic statistical *philosophy* that "if all of the signal information has been removed (explained) from the measurement data, then the residuals (i.e. innovations or prediction error) should be uncorrelated"; that is, the model fits the data and all that remains is a white or uncorrelated residual (innovations) sequence. Therefore, the zero-mean, whiteness testing, that is, the confidence interval about the normalized correlations (innovations) and the (vector) WSSR tests of Section 4.2 are performed.

Perhaps one of the most important aspects of nonlinear processor design is to recall that processing results *only* in a *single realization* of its possible output. For instance, the conditional mean is just a statistic, but still a *random process*; therefore, if direct measurements are not available, it is best to generate an ensemble of realizations and extract the resulting ensemble statistics using a variety of methods (e.g. bootstrapping [29]). In any case, simple ensemble statistics can be employed to provide an average solution enabling us to generate "what we would expect to observe" when the processor is applied to the actual problem. This approach is viable even for the modern nonlinear unscented Kalman processors. Once tuned, the processor is then applied to the raw measurement data, and ensemble statistics are estimated to provide a thorough analysis of the processor performance.

Statistics can be calculated over the ensemble of processor runs. The *updated* state estimate is obtained directly from the ensemble, that is,

$$\tilde{\hat{x}}(t|t) \rightarrow \{\hat{x}_k(t|t)\}; \quad k = 1, \dots, K$$

and therefore, the updated state ensemble estimate is

$$\tilde{\hat{x}}(t|t) = \frac{1}{K} \sum_{k=1}^{K} \hat{x}_k(t|t) \tag{4.93}$$

The corresponding predicted state estimate can be obtained directly from the ensemble as well, that is,

$$\tilde{\hat{x}}(t|t-1) \rightarrow \{\hat{x}_k(t|t-1)\}; \quad k = 1, \cdots, K$$

and, therefore, the predicted state ensemble estimate is

$$\tilde{\hat{x}}(t|t-1) = \frac{1}{K} \sum_{k=1}^{K} \hat{x}_k(t|t-1) \tag{4.94}$$

The *predicted* measurement estimate is also obtained directly from the ensemble, that is,

$$\tilde{\hat{y}}(t|t-1) \rightarrow \{\hat{y}_k(t|t-1)\}, \quad k = 1, \dots, K$$

where recall

$$\hat{y}_k(t|t-1) = c[\hat{x}_k(t|t-1)] \tag{4.95}$$

Thus, the predicted measurement ensemble estimate is

$$\tilde{\hat{y}}(t|t-1) = \frac{1}{K} \sum_{k=1}^{K} \hat{y}_k(t|t-1) \tag{4.96}$$

Finally, the corresponding predicted innovations ensemble estimate is given by

$$\mathbf{e}_k(t) = \mathbf{y}(t) - \hat{\mathbf{y}}_k(t|t-1), \quad k = 1, \dots, K$$
$$\bar{\mathbf{e}}(t) \to \{\mathbf{e}_k(t)\}, \quad k = 1, \dots, K \tag{4.97}$$

and, therefore, the predicted innovations ensemble estimate is

$$\bar{\mathbf{e}}(t) = \frac{1}{K} \sum_{k=1}^{K} \mathbf{e}_k(t) \tag{4.98}$$

The ensemble residual or innovations sequence can be "sanity tested" for zero-mean, whiteness, and the mean-squared error (MSE) over the ensemble to ensure that the processor has been tuned properly. So we see that the ensemble statistics can be estimated and provide a better glimpse of the potential performance of the nonlinear processors for an actual problem. Note that the nonlinear processor examples have applied these ensemble estimates to provide the performance metrics discussed throughout.

4.3.6 Nonlinear Model-Based Processor: Particle Filter

In this section, we discuss particle-based processors using the state-space representation of signals and show how they evolve from the Bayesian perspective using their inherent Markovian structure along with importance sampling techniques as their basic construct. PFs offer an alternative to the Kalman MBPs discussed previously possessing the capability to not just characterize unimodal distributions but also to characterize multimodal distributions. That is, the Kalman estimation techniques suffer from two major shortcomings. First, they are based on *linearizations* of either the underlying dynamic nonlinear state-space model (LZKF, EKF, IEKF) or a linearization of a statistical transform (UKF), and, secondly, they are limited to *unimodal* probability distributions (e.g. Gaussian, Chi-squared). A PF is a completely different approach to nonlinear estimation that is capable of characterizing multimodal distributions and is not limited by any linearizations. It offers an alternative to approximate Kalman filtering for nonlinear problems [2, 30]. In fact, it might be easier to think of the PF as a histogram or kernel density-like estimator [7] in the sense that it generates an estimated empirical probability mass function (PMF) that approximates the desired *posterior distribution* such that inferences can be performed and statistics extracted directly. Here the idea is a change in thinking where we attempt to develop a nonparametric empirical estimation of the posterior distribution following a purely Bayesian approach using MC sampling theory as its enabling foundation [7]. As one might expect, the computational burden of the PF is much higher than that of other processors, since it must provide an estimate of the underlying state posterior distribution state-by-state at each time-step along with the fact that the number of random samples to characterize the distribution is equal to the number of particles (N_p) per time-step.

The PF is a *processor* that has *data* on input and produces an estimate of the corresponding *posterior distribution* on output. These *unequally spaced* random samples or particles are actually the "location" parameters along with their associated weights that gather or *coalesce* in the regions of *highest probabilities* (mean, mode, etc.) providing a nonparametric estimate of the empirical posterior distribution as illustrated in Figure 4.7. With these particles generated, the appropriate weights based on Bayes' rule are estimated from the data and lead to the desired result. Once the posterior is estimated, then inferences can be performed to extract a wealth of descriptive statistics characterizing the underlying process. In particle filtering, continuous distributions are approximated by "discrete" random measures composed of these weighted particles or point masses where the *particles* are actually random samples of the unknown or hidden states from the state-space representation and the

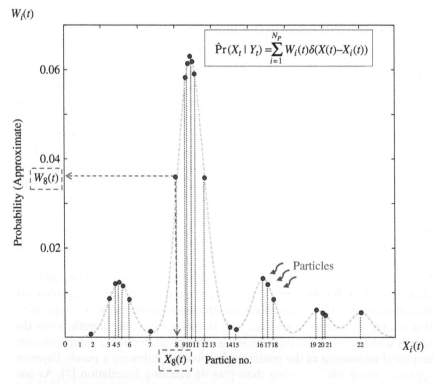

Figure 4.7 Particle filter representation of "empirical" posterior probability distribution in terms of weights (probabilities) and particles (samples).

weights are the approximate "probability masses" estimated recursively. Here in the figure, we see that associated with each particle, $x_i(t)$ is a corresponding weight or probability mass, $W_i(t)$ (filled circle). Knowledge of this *random measure*, $\{x_i(t), W_i(t)\}$, characterizes an estimate of the instantaneous (at each time t) filtering posterior distribution (dashed line). The estimated *empirical posterior distribution* is a weighted PMF given by

$$\hat{\Pr}(x(t)|Y_t) = \sum_{i=1}^{N_p} W_i(t)\delta(x(t) - x_i(t))$$

Thus, the PF does *not* involve linearizations around current estimates, but rather approximations of the desired distributions by these discrete random measures in contrast to the Kalman filter that sequentially estimates the conditional mean and covariance used to characterize the "known" (Gaussian) filtering posterior, $\Pr(x(t)|Y_t)$. PFs are a sequential MC methodology based on this "point mass" representation of probability distributions that only requires a state-space representation of the underlying process. This representation provides a set of particles that evolve at each time-step leading to an instantaneous approximation of the target posterior distribution of the state at time t given all of the data up to that time. The posterior distribution then evolves at each time-step as illustrated in Figure 4.8, creating an instantaneous approximation of t vs x_i vs $\hat{\Pr}(X_i(t)|Y_t)$ – generating an ensemble of PMFs (see [7] for more details). Statistics are then calculated *across* the ensemble at each time-step to provide estimates of the states. For example, the MAP estimate is simply determined by finding the particle $x_i(t)$ locating to the maximum posterior (weight) at each time-step across the ensemble as illustrated in the offset of Figure 4.8, that is,

$$\hat{X}_{\mathrm{MAP}}(t) = \arg\max_{x_i(t)} \hat{\Pr}(x(t)|Y_t) \tag{4.99}$$

The objective of the PF algorithm is to estimate these weights producing the posterior at each time-step. These weights are estimated based on the concept of importance sampling [7]. *Importance sampling* is a technique to compute statistics with respect to one distribution using random samples drawn from another. It is a method of simulating samples from a proposal or sampling (importance) distribution to approximate a targeted distribution (posterior) by appropriate weighting. For this choice, the weighting function is defined by the ratio

$$W(t) := \frac{\Pr[X_t|Y_t]}{\mathcal{I}(X_t|Y_t)} \tag{4.100}$$

where $\mathcal{I}(\cdot|\cdot)$ is the proposed sampling or *importance distribution* that leads to different PFs depending on its selection (e.g. prior $\rightarrow$ bootstrap).

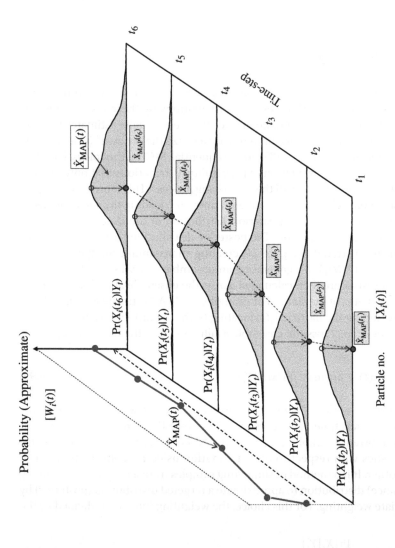

Figure 4.8 PF surface (X_i vs t vs $\hat{P}r(X_i(t)|Y_t)$) representation of posterior probability distribution in terms of time (index), particles (samples), and weights or probabilities (offset plot illustrates extracted MAP estimates vs t).

The batch *joint* posterior distribution follows directly from the *chain rule* under the Markovian assumptions on $x(t)$ and the conditional independence of $y(t)$ [31], that is,

$$\Pr(X_t|Y_t) = \overbrace{\prod_{i=0}^{t} \Pr(x(i)|x(i-1))}^{\text{State Transition}} \times \overbrace{\Pr(y(i)|x(i))}^{\text{Likelihood}} \tag{4.101}$$

for $\Pr(x(0)) := \Pr(x(0)|x(-1))$ where $X_t = \{x(0), x(1), \ldots, x(t)\}$ and $Y_t = \{y(0), y(1), \ldots, y(t)\}$.

In sequential Bayesian processing, the idea is to "sequentially" estimate the posterior at each time-step:

$$\cdots \Rightarrow \Pr(x(t-1)|Y_{t-1}) \Rightarrow \Pr(x(t)|Y_t) \Rightarrow \Pr(x(t+1)|Y_{t+1}) \Rightarrow \cdots$$

With this in mind, we start with a sequential form for the importance distribution using Bayes' rule to give

$$\mathcal{I}(X_t|Y_t) = \mathcal{I}(X_{t-1}|Y_{t-1}) \times \mathcal{I}(x(t)|X_{t-1}, Y_t)$$

The corresponding sequential importance sampling solution to the Bayesian estimation problem at time t can be developed first by applying Bayes' rule to give [7]

$$\Pr(X_t|Y_t) = \Pr(x(t), X_{t-1}|y(t), Y_{t-1}) = \Pr(x(t)|y(t), Y_{t-1}) \times \Pr(X_{t-1}|Y_{t-1})$$

substituting and grouping terms, we obtain

$$W(t) \propto \frac{\Pr(X_t|Y_t)}{\mathcal{I}(X_t|Y_t)} = \left(\frac{\Pr(X_{t-1}|Y_{t-1})}{\mathcal{I}(X_{t-1}|Y_{t-1})} \right) \times \frac{\Pr(x(t)|y(t), Y_{t-1})}{\mathcal{I}(x(t)|X_{t-1}, Y_t)}$$

where $\propto$ means "proportional to."

Therefore, invoking the chain rule (above) at time t and assuming the conditional independence of $y(t)$ and the decomposition of Eq. (4.101), we obtain the recursion

$$W(t) = W(t-1) \times \frac{\overbrace{\Pr(y(t)|x(t))}^{\text{Likelihood}} \times \overbrace{\Pr(x(t)|x(t-1))}^{\text{State Transition}}}{\mathcal{I}(x(t)|X_{t-1}, Y_t)} \tag{4.102}$$

The generic sequential importance sampling solution is obtained by drawing particles from the importance (sampling) distribution, that is, for the ith particle $x_i(t)$ at time t, we have

$$x_i(t) \sim \mathcal{I}(x(t)|X_{t-1}, Y_t)$$

$$W_i(t) = W_i(t-1) \times \frac{\Pr(y(t)|x_i(t)) \times \Pr(x_i(t)|x_i(t-1))}{\mathcal{I}(x_i(t)|X_{t-1}(i), Y_t)}$$

$$\mathcal{W}_i(t) = \frac{W_i(t)}{\sum_{i=1}^{N_p} W_i(t)} \tag{4.103}$$

The resulting posterior is then estimated by

$$\hat{\Pr}(x(t)|Y_t) = \sum_{i=1}^{N_p} \mathcal{W}_i(t) \times \delta(x(t) - x_i(t)) \tag{4.104}$$

Note that it can also be shown that the estimate converges to the true posterior as

$$\lim_{N_p \to \infty} \hat{\Pr}(x(t)|Y_t) \to \Pr(x(t)|Y_t) \tag{4.105}$$

implying that the MC error variance *decreases* as the number of particles *increase* [32].

There are a variety of PF algorithms available based on the choice of the importance distribution [33–36]. The most popular choice for the distribution is the state *transition prior*:

$$\mathcal{I}(x(t)|X_{t-1}, Y_t) \Rightarrow \Pr(x(t)|x(t-1)) \tag{4.106}$$

This prior is defined in terms of the state-space representation $A(x(t - 1), u(t - 1), w(t - 1))$, which is dependent on the known excitation $(u(\cdot))$ and process noise $(w(\cdot))$ statistics. For this implementation, we draw samples from the state *transition* distribution using the dynamic state-space model driven by the process uncertainty $\mathbf{w}_i(t)$ to generate the set of particles, $\{x_i(t)\}$ for each $i = 1, \ldots, N_p$. So we see that in order to generate particles, we execute the state-space simulation driving it with process noise $\{w_i(t)\}$ for the ith particle.

Substituting this choice into the expression for the weights gives

$$\begin{aligned} \mathcal{W}_i(t) &= \mathcal{W}_i(t-1) \times \frac{\Pr(y(t)|x_i(t)) \times \Pr(x(t)|x_i(t-1))}{\mathcal{I}(x(t)|X_{t-1}, Y_t))} \\ &= \mathcal{W}_i(t-1) \times \Pr(y(t)|x_i(t)) \end{aligned} \tag{4.107}$$

which is simply the *likelihood distribution*, since the priors cancel.

Note two properties for this choice of importance distribution. First, the weight does *not* use the most recent observation, $y(t)$ and second, it does not use the past particles $(x_i(t - 1))$, but *only* the likelihood. This choice is easily implemented and updated by simply evaluating the measurement likelihood, $\Pr(y(t)|x_i(t)); i = 1, \ldots, N_p$ for the sampled particle set.

This particular choice of importance distribution can lead to problems since the transition prior is not conditioned on the measurement data, especially the most recent. Failing to incorporate the latest available information from the most recent measurement to propose new values for the states leads to only a few particles having significant weights when their likelihood is calculated. The transition prior is a much broader distribution than the likelihood, indicating that only a few particles will be assigned a large weight. Thus, the algorithm will degenerate rapidly and lead to poor performance especially when data *outliers* occur or measurement noise is *small*. These conditions lead to a "mismatch" between the prior prediction and posterior distributions.

The basic "bootstrap" algorithm developed in [37] is one of the first practical implementations of this processor. It is the most heavily applied of all PF techniques due to its simplicity. We note that the importance weights are much simpler to evaluate with this approach that has been termed the *bootstrap* PF, the *condensation* PF, or the survival of the fittest algorithm [36–38]. Their practical implementation requires resampling of the particles to prevent degeneracy of the associated weights that increase in variance at each step, making it impossible to avoid the degradation. *Resampling* consists of processing the particles with their associated weights duplicating those of large weights (probabilities), while discarding those of small weights. In this way, only those particles of highest probabilities (importance) are retained, enabling a coalescence at the peaks of the resulting posterior distribution while mitigating the degradation. A measure based on the coefficient of variation is the *effective* particle sample size [32]:

$$N_{\text{eff}} = \frac{1}{\sum_{i=1}^{N_p} W_i^2(t)} \tag{4.108}$$

It is the underlying *decision statistic* such that when its value is less than a predetermined threshold resampling is performed (see [7] for more details).

The bootstrap technique is based on sequential sampling–importance–resampling (SIR) ideas and uses the *transition prior* as its underlying proposal distribution. The corresponding weight becomes quite simple and only depends on the likelihood; therefore, it is not even necessary to perform a sequential updating. Because the filter requires resampling to mitigate variance (weight) increases at each time-step [37], the new weights become

$$\mathcal{W}(t) \to \frac{1}{N_p}$$

revealing that there is *no need* to save the likelihood (weight) from the previous step! With this in mind, we summarize the simple *bootstrap* PF algorithm in Table 4.7. In order to achieve *convergence*, it is necessary to resample at every time-step. In practice, however, many applications make the decision to resample based on the effective sample-size metric of Eq. (4.108).

To construct the bootstrap PF, we assume the following: (i) $x_i(0) \sim \Pr(x(0))$ is *known*; (ii) $\Pr(x(t)|x(t-1))$, $\Pr(y(t)|x(t))$ are *known*; (iii) samples can be *generated* using the process noise input and the state-space model, $A(x(t-1), u(t-1), w(t-1))$; (iv) the *likelihood* is available for point-wise evaluation, $\Pr(y(t)|x(t))$ based on the measurement model, $C(x(t), u(t), v(t))$; and (v) *resampling* is performed at every time-step. We summarize the algorithm in Table 4.9.

Table 4.9 Bootstrap particle filter (PF) SIR algorithm.

Initialize

$x_i(0) \sim \Pr(x(0))$ $W_i(0) = \frac{1}{N_p}$ $i = 1, \ldots, N_p$ (Sample)

Importance sampling

$x_i(t) \sim \mathcal{A}(x(t)|x_i(t-1)) \Leftarrow A(x(t-1), u(t-1), w_i(t-1));$ (State Transition)
$w_i \sim \Pr(w_i(t))$

Weight update

$W_i(t) = C(y(t)|x_i(t)) \Leftarrow C(x(t), u(t), v(t))$ (Weight/Likelihood)

Weight normalization

$\mathcal{W}_i(t) = \dfrac{W_i(t)}{\sum\limits_{i=1}^{N_p} W_i(t)}$

Resampling decision

$\hat{N}_{\text{eff}} = \dfrac{1}{\sum\limits_{i=1}^{N_p} W_i^2(t)} \cdot \hat{N}_{\text{eff}} = \begin{cases} \text{Resample} \leq N_{\text{thresh}} \\ \text{Accept} > N_{\text{thresh}} \end{cases}$ (Effective Samples)

Resampling

$\hat{x}_i(t) \Rightarrow x_i(t)$

Distribution

$\hat{\Pr}(x(t)|Y_t) \approx \sum\limits_{i=1}^{N_p} \mathcal{W}_i(t)\delta(x(t) - \hat{x}_i(t))$ (Posterior Distribution)

To implement the algorithm, we have the following steps:

- *Generate* the initial state, $x_i(0)$.
- *Generate* the process noise, $w_i(t)$.
- *Generate* the particles, $x_i(t) = A(x_i(t-1), u(t-1), w_i(t-1))$ – the *prediction-step*.
- *Generate* the likelihood, $\Pr(y(t)|x_i(t))$ using the current particle and measurement – the *update step*.
- *Resample* the set of particles retaining and replicating those of highest weight (probability), $\hat{x}_i(t) \Rightarrow x_i(t)$.
- *Generate* the new set, $\{\hat{x}_i(t), \mathcal{W}_i(t)\}$ with $\mathcal{W}_i(t) = \frac{1}{N_p}$.

Next we revisit the nonlinear trajectory estimation problem of Jazwinski [3] and apply the simple bootstrap algorithm to demonstrate the PF solution using the state-space SIR PF algorithm.

Example 4.7 Recall the discrete state-space representation of the nonlinear trajectory problem given by the Markovian model:

$$x(t) = (1 - 0.05\Delta T)x(t-1) + 0.04\Delta T x^2(t-1) + w(t-1)$$

$$y(t) = x^2(t) + x^3(t) + v(t)$$

where $\Delta T = 0.01$, $w \sim \mathcal{N}(0, 10^{-6})$ and $v \sim \mathcal{N}(0, 0.09)$. The initial state is Gaussian distributed with $\bar{x}_i(0) \sim \mathcal{N}(\bar{x}(0), \bar{P}(0))$ and $\bar{x}(0) = 2.3$, $\bar{P}(0) = 10^{-2}$.

We selected the following *simulation* run parameters:

Number of particles: 250
Number of samples: 150
Number of states: 1
Sampling interval: 0.01 seconds
Number of measurements: 1
Process noise covariance: 1×10^{-6}
Measurement noise covariance: 9×10^{-2}
Initial state: 2
Initial state covariance: 10^{-20}.

Thus, the bootstrap SIR algorithm of Table 4.9 for this problem becomes:

1) Draw samples (particles) from the state transition distribution: $x_i(t) \rightarrow \mathcal{N}(x(t) : a[x(t-1)], \mathbf{R}_{ww})$, that is, generate

$$w_i(t) \rightarrow \Pr(w(t)) \sim \mathcal{N}(0, R_{ww})$$

and calculate $\{x_i(t)\}$ using the process model and $w_i(t)$

$$x_i(t) = (1 - 0.05\Delta T)x_i(t-1) + 0.04\Delta T x_i^2(t-1) + w_i(t-1)$$

2) Estimate the weight/likelihood, $W_i(t) = \Pr(y(t)|x(t)) \rightarrow \mathcal{N}(y(t) : c[x(t)], \mathbf{R}_{vv}(t))$

$$c[x_i(t)] = x_i^2(t) + x_i^3(t)$$

$$\ln \Pr(y(t)|x(t)) = 0.5\sqrt{R_{vv}}(y(t) - x_i^2(t) - x_i^3(t))^2$$

3) Update the weight: $W_i(t) = \Pr(y(t)|x_i(t))$
4) Normalize the weight: $\mathcal{W}_i(t) = W_i(t)/\sum_{i=1}^{N_p} W_i(t)$
5) Decide to resample if $N_{\text{eff}} \leq N_{\text{thresh}}$
6) If resample: $\hat{x}_i(t) \Rightarrow x_i(t)$
7) Estimate the instantaneous posterior:

$$\hat{\Pr}(x(t)|Y_t) \approx \sum_{i=1}^{N_p} \mathcal{W}_i \delta(x(t) - \hat{x}_i(t))$$

8) Estimate (inference) the corresponding statistics:

$$\hat{X}_{MAP}(t) = \arg\max_{x_i(t)} \hat{Pr}(x(t)|Y_t)$$

$$\hat{X}_{CM}(t) = E\{x(t)|Y_t\} = \sum_{i=1}^{N_p} \hat{x}_i(t)\hat{Pr}(x(t)|Y_t)$$

The results of the bootstrap PF are shown in Figure 4.9. The usual classical performance metrics are shown: zero-mean ($0.03 < 0.17$), whiteness (0.78% out), and WSSR (below threshold), all indicating (approximately) a tuned processor. The PF tracks the state and measurement after the initial transient error has diminished.

In Figure 4.9a,b, we show the bootstrap state and measurement estimates (inferences), that is, the MAP and CM compared to the UKF. The plots illustrate that the PF can outperform the UKF, which assumes a unimodal Gaussian distribution. The estimated state and predicted measurement posterior distributions are shown in Figure 4.9, demonstrating the capability of the bootstrap PF to characterize the multimodal nature of this problem. □

This completes the development of the most popular and simple PF technique. Next we consider some practical methods to evaluate PF performance.

4.3.7 Practical Bayesian Model-Based Design: Performance Analysis

Pragmatic MC methods, even those that are model-based, are specifically aimed at providing a reasonable estimate of the underlying posterior distribution; therefore, performance testing typically involves estimating just "how close" the estimated posterior is to the "true" posterior. However, within a Bayesian framework, this comparison of posteriors only provides a measure of relative performance, but does not indicate just how well the underlying model embedded in the processor "fits" the measured data. Nevertheless, these closeness methods are usually based on the Kullback–Leibler (KL) divergence measure [39] providing such an answer. However, in many cases, we do not know the true posterior, and therefore, we must resort to other means of assessing performance such as evaluating MSE or more generally checking the validity of the model by evaluating samples generated by the prediction or the likelihood cumulative distribution to determine whether or not the resulting sequences have evolved from a uniform distribution and are i.i.d (see [7]) – analogous to a whiteness test for Gaussian sequences.

Thus, PFs are essentially sequential estimators of the posterior distribution employing a variety of embedded models to achieve meaningful estimates. In contrast to the Kalman filter designs, which are typically based on Gaussian assumptions, the PFs have no such constraints per se. Clearly, when noise is additive Gaussian, then statistical tests can also be performed based on the

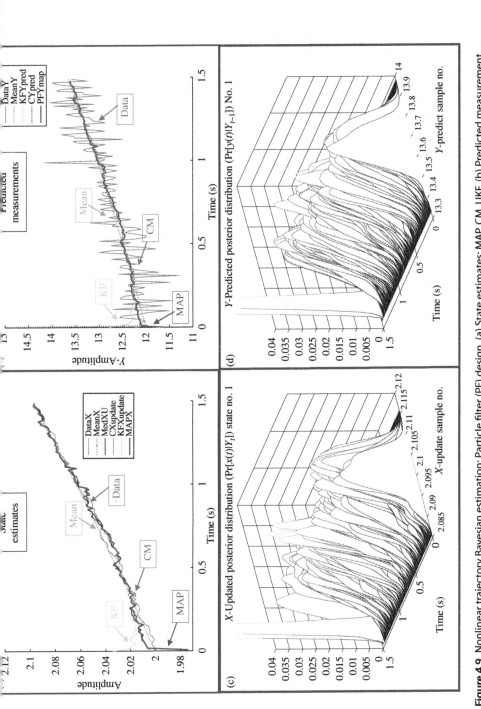

Figure 4.9 Nonlinear trajectory Bayesian estimation: Particle filter (PF) design. (a) State estimates: MAP, CM, UKF. (b) Predicted measurement estimates: MAP, CM, UKF. (c) Updated state instantaneous posterior distribution. (d) Predicted measurement instantaneous posterior distribution.

innovations or residual sequences resulting from the PF estimates (MAP, ML, CM) inferred from the estimated posterior distribution. In what might be termed "sanity testing," the classical methods employ the basic statistical *philosophy* that "if all of the signal information has been removed (explained) from the measurement data, then the residuals (i.e. innovations or prediction error) should be uncorrelated," that is, the model fits the data and all that remains is a white or uncorrelated residual (innovations) sequence. Therefore, performing the zero-mean, whiteness testing as well as checking the *normalized MSE* of the state vector (and/or innovations) given by

$$
\text{NMSE} = E \left\{ \left(\frac{x_{\text{true}}(t) - \hat{x}(t|t)}{x_{\text{true}}(t)} \right)^2 \right\} \approx \sum_{t=1}^{N_t} \left(\frac{x_{\text{true}}(t) - \hat{x}(t|t)}{x_{\text{true}}(t)} \right)^2 \quad (4.109)
$$

are the basic (first) steps that can be undertaken to ensure that the models have been implemented properly and the processors have been "tuned." Here the normalized MSE range is 0 <NMSE< 1.

Also as before, simple ensemble statistics can be employed to provide an average solution enabling us to generate expected results. This approach is viable even for the PF, assuming we have a "truth model" (or data ensemble) available for our designs. As discussed in [7], the performance analysis of the PF is best achieved by combining both simulated and actual measurement data possibly obtained during a calibration phase of the implementation. In this approach, the data are preprocessed and the truth model adjusted to reflect a set of predictive parameters that are then extracted and used to "parameterize" the underlying state-space (or other) representation of the process under investigation. Once the models and parameters are selected, then ensemble runs can be performed; terms such as initial states, covariances, and noise statistics (means, covariances, etc.) can then be adjusted to "tune" the processor. Here classical statistics (e.g. zero-mean/whiteness and WSSR testing) can help in the tuning phase. Once tuned, the processor is then applied to the raw measurement data, and ensemble statistics are estimated to provide a thorough analysis of the processor performance.

Much effort has been devoted to the validation problem with the most significant results evolving from the information theoretical point of view [40]. Following this approach, we start with the basic ideas and converge to a reasonable solution to the *distribution validation* problem [40, 41]. Since many processors are expressed in terms of their "estimated" probability distributions, quality or "goodness" can be evaluated by its similarity to the true underlying probability distribution generating the measured data [42].

Suppose $\Pr(\mathbf{x}(t)|Y_t)$ is the *true posterior* PMF and $\hat{\Pr}(\mathbf{x}_i(t)|Y_t)$ is the estimated (particle) distribution, then the *Kullback–Leibler* information quantity of the true distribution relative to the estimated is defined by [43, 44]

$$\mathcal{I}_{KL}(\Pr(\mathbf{x}(t)|Y_t); \hat{\Pr}(\mathbf{x}_i(t)|Y_t)) := E_\mathbf{x}\left\{\ln\frac{\Pr(\mathbf{x}(t)|Y_t)}{\hat{\Pr}(\mathbf{x}_i(t)|Y_t)}\right\}$$

$$= \sum_{i=1}^{N_p} \ln\frac{\Pr(\mathbf{x}(t)|Y_t)}{\hat{\Pr}(\mathbf{x}_i(t)|Y_t)} \times \Pr(\mathbf{x}(t)|Y_t)$$

(4.110)

The KL satisfies its most important property from a distribution comparison viewpoint – when the true distribution and its estimate are *close* (or identical), then the information quantity is

$$\mathcal{I}_{KL}(\Pr(\mathbf{x}(t)|Y_t); \hat{\Pr}(\mathbf{x}_i(t)|Y_t)) = 0 \Leftrightarrow \Pr(\mathbf{x}(t)|Y_t) = \hat{\Pr}(\mathbf{x}_i(t)|Y_t) \quad \forall i$$

(4.111)

This property infers that as the *estimated* posterior distribution approaches the *true* distribution, then the value of the KL approaches *zero* (minimum).

However, our interest lies in comparing two probability distributions to determine "how close" they are to each other. Even though $\mathcal{I}_{KL}$ does quantify the difference between the true and estimated distributions, unfortunately it is not a distance measure due to its lack of symmetry. However, the *Kullback divergence* (KD) defined by a combination of $\mathcal{I}_{KL}$ for this case is defined by

$$\mathcal{J}_{KD}(\Pr(\mathbf{x}(t)|Y_t); \hat{\Pr}(\mathbf{x}_i(t)|Y_t)) := \mathcal{I}_{KL}(\Pr(\mathbf{x}(t)|Y_t); \hat{\Pr}(\mathbf{x}_i(t)|Y_t))$$

$$+ \mathcal{I}_{KL}(\hat{\Pr}(\mathbf{x}_i(t)|Y_t); \Pr(\mathbf{x}(t)|Y_t)) \quad (4.112)$$

provides the *distance metric* indicating "how close" the estimated posterior is to the true distribution or, from our perspective, "how well does it approximate" the true. Thus, the KD is a very important metric that can be applied to assess the performance of Bayesian processors, providing a measure between the true and estimated posterior distributions.

If we have a "truth model," then its corresponding probability surface is analogous to the PF surface and can be generated through an ensemble simulation with values of the probabilities selected that correspond to the true state values at t (see [7] for details). An even simpler technique for nonanalytic distributions is to synthesize perturbed (low noise covariances) states generated by the model at each point in index (time) and estimate its histogram by kernel density smoothers directly to obtain an estimate of $\Pr(\mathbf{X}_{\text{true}}(t)|Y_t)$ enabling the fact that the KD can now be approximated for a given realization by

$$\mathcal{J}_{KD}(\Pr(\mathbf{X}_{\text{true}}(t)|Y_t); \hat{\Pr}(\hat{\mathbf{X}}_{MAP}(t)|Y_t)) := \mathcal{I}_{KL}(\Pr(\mathbf{X}_{\text{true}}(t)|Y_t); \hat{\Pr}(\hat{\mathbf{X}}_{MAP}(t)|Y_t))$$

$$+ \mathcal{I}_{KL}(\hat{\Pr}(\hat{\mathbf{X}}_{MAP}(t)|Y_t); \Pr(\mathbf{X}_{\text{true}}(t)|Y_t))$$

(4.113)

The implementation of the *Kullback–Leibler divergence approach* (state component-wise) is

- *Generate* samples from the state "truth model": $\{X_{\text{true}}(t)\}$.
- *Estimate* the corresponding state truth model distribution: $\hat{Pr}(X_{\text{true}}(t)|Y_t)$ using the kernel density smoother.
- *Estimate* the corresponding posterior distribution: $\hat{Pr}(\hat{X}_{\text{MAP}}(t)|Y_t)$ from the PF estimates $\{\hat{X}_{\text{MAP}}(t)\}$.
- *Calculate* the Kullback–Leibler divergence: $\mathcal{J}_{\text{KD}}(Pr(X_{\text{true}}(t)|Y_t); \hat{Pr}(\hat{X}_{\text{MAP}}(t)|Y_t))$.
- *Determine* if $\mathcal{J}_{\text{KD}} \approx 0$ for PF performance.

Before we close this section, we briefly mention a similar metric for comparing probability distributions – the *Hellinger distance*. The HD evolved from measure theory and is a true distance metric satisfying all of the required properties [45, 46]. It is a metric that can also be used to compare the similarity of two probability distributions with its values lying between zero and unity, that is, $0 < H_{\text{HD}} < 1$ eliminating the question of "what is the closeness of KD to zero."

It provides a "scaled metric" similar to the "unscaled" Kullback–Leibler divergence providing a reasonable calculation to compare or evaluate overall performance.

For PF design/analysis, the corresponding *Hellinger distance* is specified by

$$H_{\text{HD}}(Pr(x(t)|Y_t); \hat{Pr}(x_i(t)|Y_t)) = \frac{1}{\sqrt{2}}$$

$$\times \left[\sum_{i=1}^{N_p} \left(\sqrt{Pr(x(t)|Y_t)} - \sqrt{\hat{Pr}(x_i(t)|Y_t)} \right)^2 \right]^{\frac{1}{2}} \tag{4.114}$$

Using the truth model as before for the KD and its corresponding probability surface, the values of the probabilities selected correspond to the "true state" probabilities at t can be approximated for a given realization by

$$H_{\text{HD}}(Pr(X_{\text{true}}(t)|Y_t); \hat{Pr}(\hat{X}^i_{\text{MAP}}(t)|Y_t))$$

$$= \frac{1}{\sqrt{2}} \left[\sum_{i=1}^{N_p} \left(\sqrt{Pr(X_{\text{true}}(t)|Y_t)} - \sqrt{\hat{Pr}(\hat{X}^i_{\text{MAP}}(t)|Y_t)} \right)^2 \right]^{\frac{1}{2}} \tag{4.115}$$

Consider the following example illustrating both Kullback–Leibler divergence and Hellinger distance metrics applied to the PF design for the trajectory estimation problem.

Example 4.8   Let us revisit the nonlinear trajectory estimation problem and calculate the Kullback–Leibler divergence using Eq. (4.113). Recall the discrete state-space representation:

$$x(t) = (1 - 0.05\Delta T)x(t - 1) + 0.04\Delta T x^2(t - 1) + w(t - 1)$$

$$y(t) = x^2(t) + x^3(t) + v(t)$$

where $\Delta T = 0.01$, $w \sim \mathcal{N}(0, 10^{-6})$ and $v \sim \mathcal{N}(0, 0.09)$. The initial state is Gaussian distributed with $\bar{x}_i(0) \sim \mathcal{N}(\bar{x}(0), \bar{P}(0))$ and $\bar{x}(0) = 2.3$, $\bar{P}(0) = 10^{-2}$.

Here we see the results of estimating the KD over a 100-member ensemble with the corresponding state and measurement PMFs in Figure 4.10a,b.

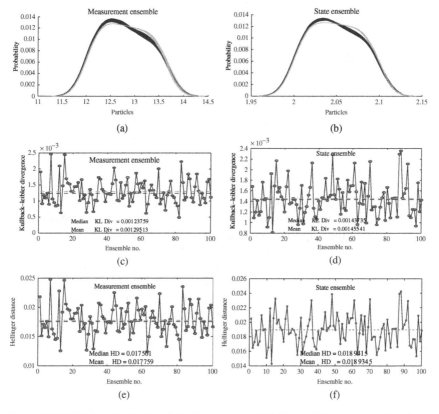

(a)   (b)

(c)   (d)

(e)   (f)

Figure 4.10 PMF data: Kullback–Leibler divergence and Hellinger distance calculations for the nonlinear trajectory estimation problem over 100-member ensemble. (a) Measurement PMF ensemble. (b) State PMF ensemble. (c) KL-divergence measurement median/mean (KD = 0.001 24/0.001 30). (d) KL-divergence state median/mean (KD = 0.001 44/0.001 46) (e) HD-distance measurement median/mean (HD = 0.0176/0.0178). (f) HD-distance state median/mean (HD = 0.0189/0.0189).

The resulting KD over *each* member realization is shown in Figure 4.10c,d, respectively, for the state and measurement ensembles with the HD shown in Figure 4.10e,f. Here the median and mean KD estimates are indicated by the dashed lines in the figure. The corresponding measurement median and average KDs of 0.001 24 and 0.001 30 and state KDs of 0.001 44 and 0.001 46 are shown in (c) and (d). Similarly, the Hellinger distance metric coincides with the KD-metric giving HDs median and average state of 0.0189 and 0.0189 with the corresponding measurement median and average of 0.0176 and 0.0178 as shown in Figure 4.10e,f, respectively. □

This completes the section. Next we discuss an application of nonlinear filtering to a tracking problem.

4.4 Case Study: 2D-Tracking Problem

In this section, we investigate the design of nonlinear MBP to solve a two-dimensional (2D) tracking problem. The hypothetical scenario discussed will demonstrate the applicability of these processors to solve such a problem. It could easily be extrapolated to many other problems in nature (e.g. air control operations, position estimation) and develops the "basic" thinking behind constructing such a problem and solution.

Let us investigate the tracking of a large tanker entering a busy harbor with a prescribed navigation path. In this case, the pilot of the vessel must adhere strictly to the path that has been filed with the harbor master (controller). Here we assume that the ship has a transponder frequently signaling accurate information about its current position. The objective is to safely dock the tanker without any "traffic" incidents. We observe that the ship's path should track the prescribed trajectory (Cartesian coordinates) shown in Figure 4.11 with the corresponding instantaneous XY-positions (vs time) shown as well.

Our fictitious measurement instrument (e.g. low ground clutter phased array radar or a GPS satellite communications receiver) is assumed to instantly report on the tanker position in bearing and range with high accuracy, that is, the measurements are given by

$$\Theta(t) = \arctan\left(\frac{Y(t)}{X(t)}\right) \quad \text{and} \quad R(t) = \sqrt{X^2(t) + Y^2(t)}$$

We use the state-space formulation for a constant velocity model with state vector defined in terms of the physical variables as $x(t) :=$ [$X(t)$ $Y(t)$ $V_x(t)$ $V_y(t)$] along with the incremental velocity input as $u' := [-\Delta V_{x_o} - \Delta V_{y_o}]$.

Using this information (as before), the entire system can be represented as a Gauss–Markov model with the noise sources representing uncertainties in the

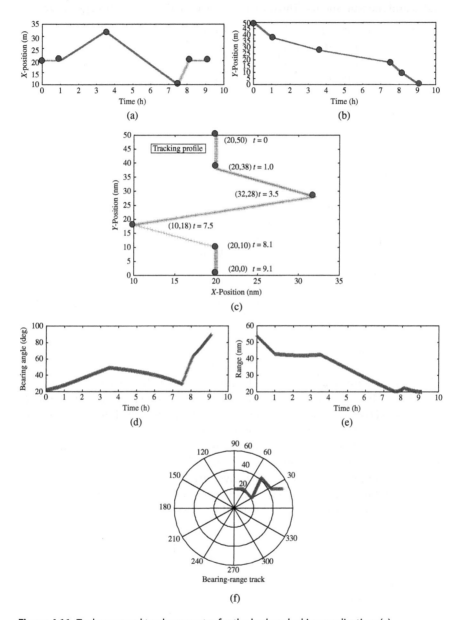

Figure 4.11 Tanker ground track geometry for the harbor docking application: (a) Instantaneous X-position (nm). (b) Instantaneous Y-position (nm). (c) Filed XY-path (nm). (d) Instantaneous bearing (deg). (e) Instantaneous range (nm). (f) Polar bearing–range track from sensor measurement.

states and measurements. Thus, we have the equations of motion given by

$$x(t) = \begin{bmatrix} 1 & 0 & \Delta T & 0 \\ 0 & 1 & 0 & \Delta T \\ 0 & 0 & 1 & 0 \\ 0 & 0 & 0 & 1 \end{bmatrix} x(t-1) + \begin{bmatrix} 0 & 0 \\ 0 & 0 \\ 1 & 0 \\ 0 & 1 \end{bmatrix} \begin{bmatrix} -\Delta V_{x_o}(t-1) \\ -\Delta V_{y_o}(t-1) \end{bmatrix} + w(t-1)$$

with the corresponding measurement model given by

$$y(t) = \begin{bmatrix} \arctan \frac{x_2(t)}{x_1(t)} \\ \sqrt{x_1^2(t) + x_2^2(t)} \end{bmatrix} + v(t)$$

for $w \sim \mathcal{N}(0, R_{ww})$ and $v \sim \mathcal{N}(0, R_{vv})$.

The MATLAB software was used to simulate this Gauss–Markov system for the tanker path, and the results are shown in Figure 4.11. In this scenario, we assume the instrument is capable of making measurements every $\Delta T = 0.02$ hour with a bearing precision of $\pm 0.02°$ and a range precision of ± 0.005 nm (or equivalently $R_{vv} = \mathrm{diag}[4 \times 10^{-4} \quad 1 \times 10^{-4}]$. The model uncertainty was represented by $R_{ww} = \mathrm{diag}(1 \times 10^{-6})$. An impulse-incremental step change in velocity, e.g. V_y going from $-12\ k$ to $-4\ k$ is an incremental change of $+8\ k$ corresponding to $\Delta V_{y_o} = [8 \quad 1.5 \quad -10.83 \quad 3.33]$ knot and $\Delta V_{x_o} = [0 \quad 4.8 \quad -10.3 \quad 22.17]$ knot. These impulses (changes) occur at time fiducials of $t = [0 \quad 1 \quad 3.5 \quad 7.5 \quad 8.1 \quad 9.1]$ hours corresponding to the filed harbor path depicted in the figure. Note that the velocity changes are impulses of height $(\Delta V_x, \Delta V_y)$ corresponding to a known deterministic input, $u(t)$. These changes relate physically to instantaneous direction changes of the tanker and create the path change in the constant velocity model.

The simulated bearing measurements are generated using the initial conditions $x'(0) := [20\ \mathrm{nm} \quad 50\ \mathrm{nm} \quad 0\ k \quad -12\ k\ ]$ and $R_{ww} = \mathrm{diag}\ 1 \times 10^{-6}$ with the corresponding initial covariance given by $\tilde{P}(0) = 1 \times 10^{-6}$. The EKF algorithm of Table 4.6 is implemented using these model parameters and the following Jacobian matrices derived from the Gauss–Markov model above:

$$A[x] = A \quad \text{and} \quad C[x] = \begin{bmatrix} \frac{x_2(t)}{R^2(t)} & \frac{-x_1(t)}{R^2(t)} & 0 & 0 \\ \frac{x_1(t)}{R(t)} & \frac{x_2(t)}{R(t)} & 0 & 0 \end{bmatrix}$$

The EKF, IEKF, and LZKF were run under the constraint that all of the a priori information for the tanker harbor path is "known." Each of the processors performed almost identically with a typical single realization output shown for the EKF in Figure 4.12. In Figure 4.12a,b we observe the estimated states X, Y, V_x, V_y. Note that the velocities are piecewise constant functions with step changes corresponding to the impulsive incremental velocities. The filtered measurements bearing ($\sim 1\%$ out) and range ($\sim 2\%$ out) are shown in Figure 4.12c

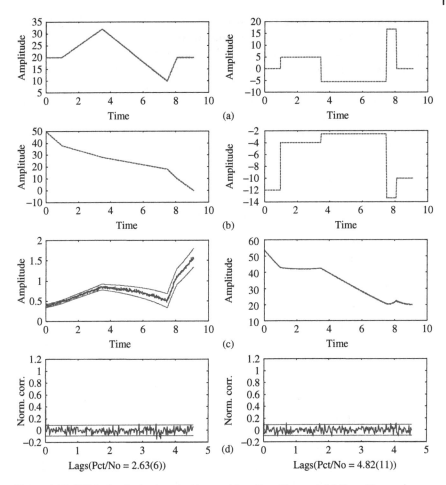

Figure 4.12 EKF design for harbor docking problem (input known). (a) X-position and velocity estimates with bounds (0% out). (b) Y-position and velocity estimates with bounds (0% and 3% out). (c) Bearing and range estimates with bounds (1% and 2% out). (d) Innovations zero-mean/whiteness tests for bearing ($6 \times 10^{-4} < 26 \times 10^{-4}$ and 3% out) and range ($2 \times 10^{-4} < 42 \times 10^{-4}$ and 5% out).

with the resulting innovations zero-mean/whiteness tests depicted in (d). The processor is clearly tuned with bearing and range innovations zero-mean and white ($6 \times 10^{-4} < 26 \times 10^{-4}$/3% out) and ($2 \times 10^{-4} < 42 \times 10^{-4}$/5% out), respectively. This result is not unexpected, since all of the a priori information is given including the precise incremental velocity input, $u(t)$. An ensemble of 101 realizations of the estimator was run by generating random initial condition estimates from the Gaussian assumption. The 101-realization ensemble averaged estimates closely follow the results shown in the figure with the zero-mean/whiteness tests ($2 \times 10^{-4} < 25 \times 10^{-4}/4\%$ out), ($2 \times 10^{-4} < 15 \times 10^{-4}/7\%$ out) slightly worse.

Next, we investigate the usual case where all of the information is known a priori *except* the impulsive incremental velocity changes represented by the deterministic input. Note that *without* the input, the processor cannot respond instantaneously to the velocity changes and therefore will lag (in time) behind in predicting the tanker path. The solution to this problem requires a joint estimation of the states *and* the unknown input, which is really a solution to a deconvolution problem [2]. It is also a problem that is ill-conditioned especially, since $u(t)$ is impulsive.

In any case we ran the nonlinear KF algorithms over the simulated data, and the best results were obtained using the LZKF. This is expected, since we used the exact state reference trajectories, but not the input. Note that the other non-linear KF has no knowledge of this trajectory inhibiting their performance in this problem. The results are shown in Figure 4.13 where we observe the state estimates as before. We note that the position estimates (65% out, 49% out) appear reasonable, primarily because of the reference trajectories. The LZKF is able to compensate for the unknown impulsive input with a slight lag as shown at each of the fiducials. The velocity estimates (4% out, 1% out) are actually low-pass versions of the true velocities caused by the slower LZKF response even with the exact step changes available. These lags are more vividly shown in the bearing estimate of Figure 4.13c, which shows the processor has great difficulty with the instantaneous velocity changes in bearing (0% out). The range (0% out) appears insensitive to this lack of knowledge primarily because the XY-position estimates are good and do not have step changes like the velocity for the LZKF to track. Both processors are *not* optimal and the innovations sequences are zero-mean but *not* white ($75 \times 10^{-3} < 81 \times 10^{-3}/59\%$ out), ($2 \times 10^{-3} < 4 \times 10^{-3}/8\%$ out).

We also designed the UKF to investigate its performance on this problem and its results were quite good[2] as shown in Figure 4.14. The processor does not perform a model linearization but a *statistical linearization* instead; it is clear from the figure that it performs better than any of the other processors for this problem. In Figure 4.14a–d, we see that the XY position estimates "track" the data very well, while the XY-velocity estimates are somewhat nosier due to the abrupt changes (steps) in tuning values of the process noise covariance terms. In order to be able to track the step changes, the process noise covariance could be increased even further at the cost of nosier estimates. The UKF tracks the estimated bearing and range reasonably well as shown in figure with a slight loss of bearing track toward the end of the time sequence. These results are demonstrated by the zero-mean/whiteness test results of the

---

2 We used noisier simulation data for this run than that for the LZKF with
$R_{vv} = \text{diag}[4 \times 10^{-4} \quad 5 \times 10^{-1}]$ providing more realistic measurement uncertainties.

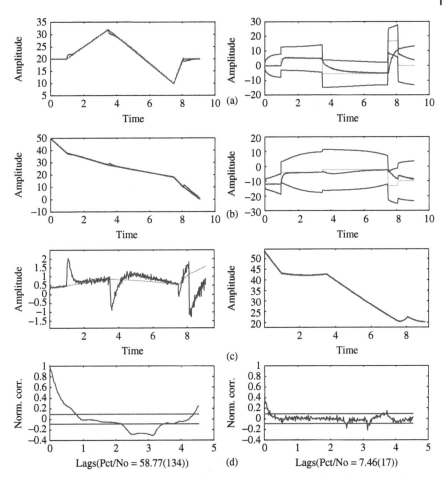

Figure 4.13 LZKF design for harbor docking problem (input known). (a) X-position and velocity estimates with bounds (68% and 4% out). (b) Y-position and velocity estimates with bounds (49% and 1% out). (c) Bearing and range estimates with bounds (0% and 3% out). (d) Innovations zero-mean/whiteness tests for bearing ($75 \times 10^{-3} < 81 \times 10^{-3}$ and 59% out) and range ($2 \times 10^{-3} < 4 \times 10^{-3}$ and 8% out).

corresponding innovations sequences. The bearing innovations statistics are $3.3 \times 10^{-3} < 1.2 \times 10^{-1}$ and 4.7% out and the corresponding range innovations statistics given by $6.5 \times 10^{-3} < 1.2 \times 10^{-1}$ and 4.7% out. Both indicate a tuned processor. These are the best results of all of the nonlinear processors applied. This completes the case study.

It is clear from this study that nonlinear processors can be "tuned" to give reasonable results especially when they are provided with accurate a priori information. If the a priori information is provided in terms of prescribed

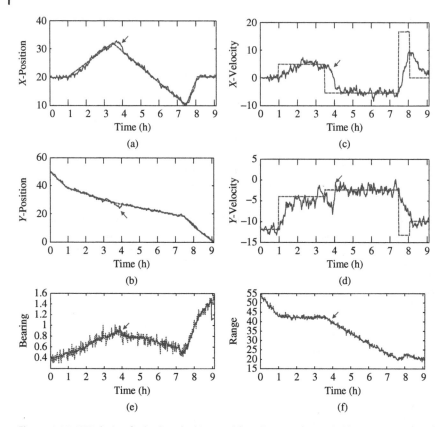

Figure 4.14 UKF design for harbor docking problem (input unknown). (a) X-position estimate. (b) Y-position estimate. (c) X-velocity estimate. (d) Y-velocity estimate. (e) Bearing estimate (zero-mean/whiteness: $3.3 \times 10^{-3} < 1.2 \times 10^{-1}$ and 4.7% out). (f) Range estimate (zero-mean/whiteness: $6.5 \times 10^{-3} < 1.2 \times 10^{-1}$ and 4.7% out).

reference trajectories as in this hypothetical case study, then the LZKF appears to provide superior performance, but in the real-world tracking problem, when this information on the target is not available, then the EKF and IEKF can be tuned to give reasonable results.

There are many variants possible for these processors to improve their performance whether it be in the form of improved coordinate systems [47–49] or in the form of a set of models each with its own independent processor [19]. One might also consider using estimator/smoothers as in the seismic case [14] because of the unknown impulsive input. This continues to be a challenging problem. The interested reader should consult Ref. [19] and the references cited therein for more details.

This completes the chapter on nonlinear state-space-based MBP. Next we consider the performance of parametrically adaptive schemes for MBP.

4.5 Summary

In this chapter, we discussed the concepts of model-based signal processing using state-space models. We first introduced the recursive model-based solution from a motivational perspective. We started with the linear, (time-varying) Gauss–Markov model and developed a solution to the state estimation problem, leading to the well-known LKF from the innovations perspective [3]. The objective was to basically introduce the processor from a pragmatic rather than theoretical approach [2]. Following this development, we investigated properties of the innovations sequence and eventually showed how they can be exploited to analyze processor performance that will be quite useful in subspace identification techniques to follow. With theory now in hand, we analyzed the properties of the "tuned" processor and showed how the MBP parameters can be used to achieve a *minimum variance design*.

Based on this fundamental theme, we progressed to nonlinear estimation problems and to the idea of linearization of the nonlinear state-space model where we derive the LZKF. We then developed the EKF as a special case of the LZKF linearizing about the most currently available estimate. Next we investigated a further enhancement of the EKF by introducing a local iteration of the nonlinear measurement system. Here the processor is called the IEKF and was shown to produce improved estimates at a small computational cost in most cases.

Finally, we introduced the concepts of statistical signal processing from the Bayesian perspective using state-space models. Two different paradigms for state estimation evolved: the UKF and the PF. The UKF employed a statistical rather than model linearization technique that yields superior statistical performance over the Kalman-based methods, while the PF provides a complete Bayesian approach to the problem solution providing a nonparametric estimate of the required posterior distribution. Practical metrics were discussed to evaluate not only the nonlinear processors (LZKF, EKF, IEKF, UKF) but also Bayesian PF processor performance analysis. We summarized the results with a case study implementing a 2D-tracking filter.

MATLAB Notes

MATLAB is a command-oriented vector-matrix package with a simple, yet effective command language featuring a wide variety of embedded C language constructs, making it ideal for signal processing applications and graphics. Linear and nonlinear Kalman filter algorithms are available in the Control Systems (**control**) toolbox. Specifically, the **LKF**, **EKF**, **UKF**, and **PF** algorithms are available as well as state-space simulators (**LSIM**). Also the System Identification (**ident**) provides simulation and analysis tools as well.

MATLAB has a *Statistics Toolbox* that incorporates a large suite of PDFs and CDFs as well as "inverse" CDF functions ideal for simulation-based algorithms. The **mhsample** command incorporate the Metropolis, Metropolis-Hastings, and Metropolis independence samplers in a single command, while the Gibbs sampling approach is adequately represented by the more efficient slice sampler (**slicesample**). There are even specific "tools" for sampling as well as the inverse CDF method captured in the **randsample** command. PDF estimators include the usual histogram (**hist**) as well as the sophisticated kernel density estimator (**ksdensity**) offering a variety of kernel (window) functions (Gaussian, etc.). As yet no sequential algorithms are available.

In terms of statistical testing for particle filtering diagnostics, MATLAB offers the chi-square "goodness-of-fit" test **chi2gof** as well as the Kolmogorov–Smirnov distribution test **kstest**. Residuals can be tested for whiteness using the Durbin–Watson test statistic **dwtest**, while "normality" is easily checked using the **normplot** command indicating the closeness of the test distribution to a Gaussian. Other statistics are also evaluated using the **mean, moment, skewness, std, var,** and **kurtosis** commands. Type *help stats* in MATLAB to get more details or go to the MathWorks website.

References

1 Kalman, R. (1960). A new approach to linear filtering and prediction problems. *Trans. ASME J. Basic Eng.* 82: 34–45.

2 Candy, J. (2006). *Model-Based Signal Processing*. Hoboken, NJ: Wiley/IEEE Press.

3 Jazwinski, A. (1970). *Stochastic Processes and Filtering Theory*. New York: Academic Press.

4 Kailath, T. (1970). The innovations approach to detection and estimation theory. *Proc. IEEE* 58 (5): 680–695.

5 Kailath, T. (1981). *Lectures on Kalman and Wiener Filtering Theory*. New York: Springer.

6 Kailath, T., Sayed, A., and Hassibi, B. (2000). *Linear Estimation*. Englewood Cliffs, NJ: Prentice-Hall.

7 Candy, J. (2016). *Bayesian Signal Processing, Classical, Modern and Particle Filtering*, 2e. Hoboken, NJ: Wiley/IEEE Press.

8 Anderson, B. and Moore, J. (1979). *Optimal Filtering*. Englewood Cliffs, NJ: Prentice-Hall.

9 Wilsky, A.S. (1976). A survey of design methods for failure detection in dynamic systems. *Automatica* 12 (6): 601–611.

10 Mehra, R.K. and Peschon, J. (1971). An innovations approach to fault detection and diagnosis in dynamic systems. *Automatica* 7 (5): 637–640.

11 Martin, W.C. and Stubberud, A.R. (1976). The innovations process with application to identification. In: *Control and Dynamic Systems* (ed. C.T. Leondes), vol. 12, 173–258. New York: Academic Press.

12 Schweppe, F. (1973). *Uncertain Dynamic Systems*. Enegelwood Cliffs, NJ: Prentice-Hall.

13 Maybeck, P. (1979). *Stochastic Models, Estimation, and Control*. New York: Academic Press.

14 Mendel, J. (1995). *Lessons in Estimation Theory for Signal Processing, Communications and Control*. Englewood Cliffs, NJ: Prentice-Hall.

15 Bierman, G. (1977). *Factorization Methods of Discrete Sequential Estimation*. New York: Academic Press.

16 Sage, A. and Melsa, J. (1971). *Estimation Theory with Applications to Communications and Control*. New York: McGraw-Hill.

17 Simon, D. (2006). *Optimal State Estimation: Kalman, H_∞ and Nonlinear Approaches*. Hoboken, NJ: Wiley.

18 Julier, S. and Uhlmann, J. (2004). Unscented filtering and nonlinear estimation. *Proc. IEEE* 92 (3): 401–422.

19 Bar-Shalom, Y. and Li, X. (1993). *Estimation and Tracking: Principles, Techniques and Software*. Boston, MA: Artech House.

20 Candy, J. and Rozsa, R. (1980). Safeguards for a plutonium concentrator–an applied estimation approach. *Automatica* 16: 615–627.

21 Candy, J. and Sullivan, E. (1992). Ocean acoustic signal processing: a model-based approach. *J. Acoust. Soc. Am.* 92 (6): 3185–3201.

22 Mendel, J.M., Kormylo, J., Aminzadeh, F. et al. (1981). A novel approach to seismic signal processing and modeling. *Geophysics* 46 (10): 1398–1414.

23 Sorenson, H. (ed.) (1983). Special issue on applications of Kalman filtering. *IEEE Trans. Autom. Control* AC-28 (3): 253–427.

24 Julier, S., Uhlmann, J., and Durrant-Whyte, H. (2000). A new method for the nonlinear transformation of means and covariances in filters and estimators. *IEEE Trans. Autom. Control* 45 (3): 477–482.

25 Julier, S. and Uhlmann, J. (1996). *A General Method for Approximating Nonlinear Transformations of Probability Distributions*. University of Oxford Report.

26 Haykin, S. and de Freitas, N. (eds.) (2004). Sequential state estimation: from Kalman filters to particle filters. *Proc. IEEE* 92 (3): 399–574.

27 Wan, E. and van der Merwe, R. (2000). The unscented Kalman filter for nonlinear estimation. *Proceedings of IEEE Symposium 2000 on Adaptive Systems for Signal Processing, Communication and Control, Lake Louise, Alberta*.

28 van der Merwe, R., Doucet, A., de Freitas, N., and Wan, E. (2000). The Unscented Particle Filter. Cambridge University Technical Report, CUED/F-INFENG-380.

29 Hogg, R. and Craig, A. (1970). *Introduction to Mathematical Statistics*. New York: MacMillan.

30 Ristic, B., Arulampalam, S., and Gordon, N. (2004). *Beyond the Kalman Filter: Particle Filters for Tracking Applications*. Boston, MA: Artech House.

31 Papoulis, A. (1965). *Probability, Random Variables and Stochastic Processes*. New York: McGraw-Hill.

32 Liu, J. (2001). *Monte Carlo Strategies in Scientific Computing*. New York: Springer.

33 Arulampalam, M., Maskell, S., Gordon,N., and Clapp, T. (2002). A tutorial on particle filters for online nonlinear/non-gaussian Bayesian tracking. *IEEE Trans. Signal Process.* 50 (2): 174–188.

34 Djuric, P., Kotecha, J., Zhang, J. et al. (2003). Particle filtering. *IEEE Signal Process. Mag.* 20 (5): 19–38.

35 Doucet, A., de Freitas, N., and Gordon, N. (2001). *Sequential Monte Carlo Methods in Practice*. New York: Springer.

36 Godsill, S. and Djuric, P. (2002). Special issue: Monte Carlo methods for statistical signal processing. *IEEE Trans. Signal Proc.* 50: 173–499.

37 Gordon, N., Salmond, D., and Smith, A. (1993). A novel approach to nonlinear non-gaussian Bayesian state estimation. *IEE Proc. F.* 140 (2): 107–113.

38 Isard, M. and Blake, A. (1998). Condensation–conditional density propagation for visual tracking. *Int. J. Comput. Vis.* 29 (1): 5–28.

39 Gustafsson, F. (2000). *Adaptive Filtering and Change Detection*. Hoboken, NJ: Wiley.

40 Akaike, H. (1974). A new look at the statistical model identification. *IEEE Trans. Autom. Control* 19 (6): 716–723.

41 Sakamoto, Y., Ishiguro, M., and Kitagawa, G. (1986). *Akaike Information Criterion Statistics* (ed. D. Reidel). Boston, MA: Kluwer Academic.

42 Bhattacharyya, A. (1943). On a measure of divergence between two statistical populations defined by their probability distributions. *Bull. Calcutta Math. Soc.* 35: 99–109.

43 Kullback, S. and Leibler, R. (1951). On information and sufficiency. *Ann. Math. Stat.* 22: 79–86.

44 Kullback, S. (1978). *Information Theory and Statistics*. Gloucester, MA: Peter Smith.

45 Nikulin, M. (2001). Hellinger distance. In: *Encyclopedia of Mathematics* (ed. M. Hazewinkel). New York: Springer.

46 Beran, R. (1977). Minimum Hellinger distance estimates for parametric models. *Ann. Stat.* 5 (3): 445–463.

47 Aidala, V.J. and Hammel, S.M. (1983). Utilization of modified polar-coordinates for bearings-only tracking. *IEEE Trans. Autom. Control* AC-28 (3): 283–294.

48 Aidala, V.J. (1979). Kalman filter behavior in bearings-only velocity and position estimation. *IEEE Trans. Aerosp. Electron. Syst.* AES-15: 29–39.

49 Candy, J.V. and Zicker, J.E. (1982). Deconvolution of Noisy Transient Signals: A Kalman Filtering Application. LLNL Rep., UCID-87432, and Proceedings of CDC Conference, Orlando.

Problems

4.1 Derive the *continuous-time* KF by starting with the discrete equations of Table 4.1 and using the following sampled-data approximations:

$$A = e^{A_c \triangle t} \approx I + A_c \triangle t$$
$$B = B_c \triangle t$$
$$W = W_c \triangle t$$
$$R_{ww} = R_{w_c w_c} \triangle t$$

4.2 Suppose we are given a continuous-time Gauss–Markov model characterized by

$$\dot{x}(t) = A_c(t)x(t) + B_c(t)u(t) + W_c(t)w(t)$$

and discrete (sampled) measurement model such that $t \to t_k$ then

$$y(t_k) = C(t_k)x(t_k) + v(t_k)$$

where the continuous process, $w(t) \sim \mathcal{N}(0, R_{ww})$, and $v(t_k) \sim \mathcal{N}(0, R_{vv})$ with Gaussian initial conditions.

(a) Determine the state mean $(m_x(t))$ and covariance $(P(t))$.
(b) Determine the measurement mean $(m_y(t_k))$ and covariance $(R_{yy}(t_k))$.
(c) Develop the relationship between the continuous and discrete Gauss–Markov models based on the solution of the continuous state equations and approximation using a first-order Taylor series for the state transition matrix, $\Phi(t, t_o)$, and the associated system matrices.
(d) Derive the continuous–discrete KF using first-difference approximations for derivatives and the discrete (sampled) system matrices derived above.

4.3 The covariance correction equation of the KF algorithm is seldom used directly. Numerically, the covariance matrix $\tilde{P}(t|t)$ must be positive semidefinite, but in the correction equation we are subtracting a matrix from a positive semidefinite matrix and cannot guarantee that the result will remain positive semidefinite (as it should be) because of roundoff and truncation errors. A solution to this problem is to replace the standard correction equation with the stabilized *Joseph form*, that is,

$$\tilde{P}(t|t) = [I - K(t)C(t)]\tilde{P}(t|t-1)[I - K(t)C(t)]' + K(t)R_{vv}(t)K'(t)$$

(a) Derive the *Joseph* stabilized form.
(b) Demonstrate that it is equivalent to the standard correction equation.

4.4 Prove that a necessary and sufficient condition for a *linear* KF to be optimal is that the corresponding innovations sequence is zero-mean and white.

4.5 A bird watcher is counting the number of birds migrating to and from a particular nesting area. Suppose the number of migratory birds, $m(t)$, is modeled by a first-order ARMA model:

$$m(t) = -0.5m(t-1) + w(t) \quad \text{for} \quad w \sim \mathcal{N}(10, 75)$$

while the number of resident birds is static, that is,

$$r(t) = r(t-1)$$

The number of resident birds is averaged, leading to the expression

$$y(t) = 0.5r(t) + m(t) + v(t) \quad \text{for} \quad w \sim \mathcal{N}(0, 0.1)$$

(a) Develop the two-state Gauss–Markov model with initial values $r(0) = 20$ birds, $m(0) = 100$ birds, cov $r(0) = 25$.
(b) Use the KF algorithm to estimate the number of resident and migrating birds in the nesting area for the data set $y(t) = \{70, 80\}$, that is, what is $\hat{x}(2|2)$?

4.6 Suppose we are given a measurement device that not only acquires the current state but also the state delayed by one time-step (multipath) such that

$$y(t) = Cx(t) + Ex(t-1) + v(t)$$

Derive the recursive form for this associated KF. (*Hint*: Recall from the properties of the state transition matrix that $x(t) = \Phi(t, \tau)x(\tau)$ and $\Phi^{-1}(t, \tau) = \Phi)\tau, t)$.)

(a) Using the state transition matrix for discrete systems, find the relationship between $x(t)$ and $x(t-1)$ in the Gauss–Markov model.

(b) Substitute this result into the measurement equation to obtain the usual form

$$y(t) = \tilde{C}x(t) + \tilde{v}(t)$$

What are the relations for $\tilde{C}$ and $\tilde{v}(t)$ in the new system?
(c) Derive the new statistics for $\tilde{v}(t)$ ($\mu_{\tilde{v}}$, $R_{\tilde{v}\tilde{v}}$).
(d) Are w and v correlated? If so, use the prediction form to develop the KF algorithm for this system.

4.7 Develop the discrete linearized (perturbation) models for each of the following nonlinear systems [7]:
- *Synchronous (unsteady) motor:* $\ddot{x}(t) + C\dot{x}(t) + p \sin x(t) = L(t)$
- *Duffing equation:* $\ddot{x}(t) + ax(t) + \beta x^3(t) = F \cos \omega t)$
- *Van der Pol equation:* $\ddot{x}(t) + \epsilon \dot{x}(t)[1 - \dot{x}^2(t)] + x(t) = m(t)$
- *Hill equation:* $\ddot{x}(t) - ax(t) + \beta p(t)x(t) = m(t)$
(a) Develop the LZKF.
(b) Develop the EKF.
(c) Develop the IEKF.

4.8 Suppose we are given the following discrete system:

$$x(t) = -\omega^2 x(t-1) + \sin x(t-1) + \alpha u(t-1) + w(t-1)$$
$$y(t) = x(t) + v(t)$$

with w and v zero-mean, white Gaussian with usual covariances, R_{ww} and R_{vv}.
(a) Develop the LZKF for this process. (b) Develop the EKF for this process.
(b) Develop the IEKF for this process.
(c) Develop the UKF for this process.
(d) Suppose the parameters ω and α are unknown. Develop the EKF such that the parameters are jointly estimated along with the states. (*Hint*: Augment the states and parameters to create a new state vector.)

4.9 Suppose we assume that a target is able to maneuver, that is, we assume that the target velocity satisfies a first-order AR model given by

$$v_\tau(t) = -\alpha v_\tau(t-1) + w_\tau(t-1) \quad \text{for} \quad w \sim \mathcal{N}(0, R_{w_\tau w_\tau})$$

(a) Develop the Cartesian tracking model for this process.
(b) Develop the corresponding EKF assuming all parameters are known a priori.
(c) Develop the corresponding EKF assuming α is unknown.
(d) Develop the UKF for this process.

4.10 Consider the problem of estimating a random signal from an AM modulator characterized by

$$s(t) = \sqrt{2P}a(t) \sin \omega_c t$$
$$r(t) = s(t) + v(t)$$

where $a(t)$ is assumed to be a Gaussian random signal with power spectrum

$$S_{aa}(\omega) = \frac{2k_a P_a}{\omega^2 + k_a^2}$$

also assume that the processes are contaminated with the usual additive noise sources: w and v zero-mean, white Gaussian with covariances, R_{ww} and R_{vv}.

(a) Develop the continuous-time Gauss–Markov model for this process.

(b) Develop the corresponding discrete-time Gauss–Markov model for this process using first differences.

(c) Develop the KF.

(d) Assume the carrier frequency, ω_c, is unknown. Develop the EKF for this process.

4.11 Let x_1 and x_2 be i.i.d. with distribution $\mathcal{N}(0,1)$. Suppose $y = x_1^2 + x_2^2$, then

(a) What is the distribution of y, $\Pr_Y(y)$?

(b) Suppose $E\{y\} = 2$ and $\sigma_y^2 = 4$, using the σ-point transformation what are the σ-points for $\mathbf{x} = [x_1 \ x_2]'$?

(c) What are the σ-points for y?

4.12 Suppose $x \sim \mathcal{N}(0,1)$ and $y = x^2$.

(a) What is the distribution of y, $\Pr_Y(y)$?

(b) What is the Gaussian approximation of the mean and variance of $\Pr_Y(y)$? (*Hint*: Use linearization.)

(c) What is the σ-point transformation and corresponding mean and variance estimates of $P_Y(y)$?

4.13 We are given a sequence of data and know it is Gaussian with an unknown mean and variance. The distribution is characterized by $y \sim \mathcal{N}(\mu, \sigma^2)$.

(a) Formulate the problem in terms of a state-space representation. (*Hint*: Assume that the measurements are modeled in the usual manner (scale by standard deviation and add mean to a $\mathcal{N}(0,1)$ "known" sequence, say $v(t)$.)

(b) Using this model, develop the UKF technique to estimate the model parameters.

(c) Synthesize a set of data of 2000 samples at a sampling interval $dt = 0.01$ with process covariance, $R_{ww} = \text{diag}[1 \times 10^{-5}, 1 \times 10^{-6}]$ and measurement noise of $R_{vv} = 1 \times 10^{-6}$ with $\bar{x}(0) = [\sqrt{20} \ 3]'$.

(d) Develop the UKF algorithm and apply it to this data and show the performance results (final parameter estimates, etc.). That is, find the best estimate of the parameters defined by $\Theta := [\mu\sigma]'$ using the UKF approach. Show the mathematical steps in developing the technique and construct a simple UKF to solve.

4.14 Given a sequence of Gaussian data (measurements) characterized by $y \sim \mathcal{N}(\mu, \sigma^2)$, find the best estimate of the parameters defined by $\Theta := [\mu\sigma]'$ using a "sequential" MC approach. Show the mathematical steps in developing the technique and construct a simple PF to solve the problem.

4.15 Consider the following simple model

$$x(t) = ax(t-1) + w(t) \quad \text{for} \ w \sim \mathcal{N}(0, R_{ww}(i)) \quad \text{with} \ \mathcal{I}(t) = i$$

$$y(t) = x(t) + v(t) \quad \text{for} \ v \sim \mathcal{N}(0, R_{vv})$$

with $\Pr(\mathcal{I}(t) = i | \mathcal{I}(t-1), x(t-1))) = \Pr(\mathcal{I}(t) = i) = p_i$

(a) Suppose $i = \{1, 2\}$, what is the distribution, $\Pr(\mathcal{I}(t) = (i_1, i_2), x_1 | y_1, x_0)$?

(b) How would the marginal be estimated using a Kalman filter?

(c) Develop a computational approach to bootstrap PF algorithm for this problem.

4.16 Suppose we have two multivariate Gaussian distributions for the parameter vector, $\Theta \sim \mathcal{N}(\mu_i, \Sigma_i); i = 1, 2$

(a) Calculate the Kullback–Leibler (KL) distance metric, J_{KL}.

(b) Calculate the Hellinger (HD) distance metric, H_{HD}.

(c) Suppose $\Sigma = \Sigma_1 = \Sigma_2$. Recalculate the KL and HD for this case.

4.17 An aircraft flying over a region can use the terrain and an archival digital map to navigate. Measurements of the terrain elevation are collected in real time while the aircraft altitude over mean sea level is measured by a pressure meter with ground clearance measured by a radar altimeter. The measurement differences are used to estimate the terrain elevations and are compared to a digital elevation map to estimate the aircraft's position. The discrete navigation model is given by

$$x(t) = x(t-1) + u(t-1) + w(t) \quad \text{for} \ w \sim \mathcal{N}(0, R_{ww})$$

$$y(t) = c[x(t)] + v(t) \text{for} \ v \sim \mathcal{N}(0, R_{vv})$$

where x is the 2D-position, y is the terrain elevation measurement, the navigation system's output u and w are the respective distance traveled and error drift during one time interval. The nonlinear function $c[\cdot]$ denotes the terrain database yielding terrain elevation outputs with v the associated database errors and measurements. Both noises are assumed zero-mean, Gaussian with known statistics, while the initial state is also Gaussian, $x(0) \sim \mathcal{N}(\bar{x}(0), \bar{P}(0))$.

(a) Based on this generic description, construct the bootstrap PF algorithm for this problem.

(b) Suppose: $P(0) = \text{diag}[10^4 \ 10^4]'$, $R_{ww} = \text{diag}[25 \ 25]'$, $R_{vv} = 16$, $N = 150$ samples, $u(t) = [25 \ 25]'$. Simulate the aircraft measurements and apply the bootstrap algorithm to estimate the aircraft position.

4.18 Develop a suite of particle filters for the RC-circuit problem where the output voltage was given by

$$e(t) = \left(1 - \frac{\Delta T}{RC}\right) e(t-1) + \frac{\Delta T}{C} I_{\text{in}}(t-1)$$

where e_o is the initial voltage and R is the resistance with C the capacitance. The measurement equation for a voltmeter of gain K_e is simply

$$e_{\text{out}}(t) = K_e e(t)$$

Recall that for this circuit the parameters are: $R = 3.3$ kΩ and $C = 1000$ μF, $\Delta T = 100$ ms, $e_o = 2.5$ V, $K_e = 2.0$, and the voltmeter is precise to within ± 4 V. Then transforming the physical circuit model into state-space form by defining $x = e$, $y = e_{\text{out}}$, and $u = I_{in}$, we obtain

$$x(t) = 0.97x(t-1) + 100u(t-1) + w(t-1)$$

$$y(t) = 2x(t) + v(t)$$

The process noise covariance is used to model the circuit parameter uncertainty with $R_{ww} = 0.0001$, since we assume standard deviations, ΔR, ΔC of 1%. Also, $R_{vv} = 4$, since two standard deviations are $\Delta V = 2\left(\frac{1}{2} \ 4V\right)$. We also assume initially that the state is $x(0) \sim \mathcal{N}(2.5, 10^{-12})$, and that the input current is a step function of $u(t) = 300$ μA.

With this in mind, we know that the optimal processor to estimate the state is the linear KF (Kalman filter).

(a) After performing the simulation using the parameters above, construct a bootstrap PF and compare its performance to the optimal. How does it compare? Whiteness? Zero-mean? State estimation error?

(b) Let us assume that the circuit is malfunctioning and we do not precisely know the current values of RC. Construct a parameter estimator for $A = 1/RC$ using the EKF or UKF and compare its performance to the bootstrap and linearized PF.

4.19 We are asked to investigate the possibility of finding the range of a target using a hydrophone sensor array towed by an AUV assuming a near-field target characterized by its spherical wavefront instead of the far-field target of the previous problem using a plane wave model. The near-field processor can be captured by a wavefront curvature scheme (see for more details) with process and measurement models (assuming that the parameters as well as the measurements are contaminated by additive white Gaussian noise). The following set of dynamic relations can be written succinctly as the *Gauss–Markov wavefront curvature model* as

$$\Theta(t_k) = \Theta(t_{k-1}) + w(t_k) \quad \text{for} \quad \Theta(t_k) = [\alpha \ f_o \ \theta_o \ r_o]'$$
$$p_\ell(t_k) = \theta_1(t_k)e^{j2\pi\theta_2(t_k)(t_k - \tau_\ell(\Theta;t_k))} + v_\ell(t_k); \quad \ell = 1, \dots, L$$

where α, f_o, θ_o, and r_o are the respective amplitude, target frequency, bearing, and range. The time delay at the ℓth-sensor and time t_k is given in terms of the unknown parameters of

$$\tau_\ell(\Theta; t_k) := \frac{1}{c}\left(\theta_4(t_k) - \sqrt{\theta_4^2(t_k) + d_\ell^2(t) - 2d_\ell(t)\theta_4(t_k)\sin\theta_3(t_k)}\right)$$

for $d_\ell(t)$ the distance between the ℓth sensor and reference range r_o given by

$$d_\ell(t) = x_\ell + \upsilon \times t_k \quad \text{for} \quad x_\ell \quad \text{the position of sensor } \ell$$

(a) Using this model, develop the bootstrap PF and UKF processors for this problem. Assume we would like to estimate the target bearing, frequency, and range ($\alpha = 1$).

(b) Perform a simulation with initial parameters $r_o = 3$ km, $f_o = 51.1$ Hz, and $\theta_o = 27°$ and true parameters at $r = 2$ km, $f = 51$ Hz, and $\theta = 25°$, $L = 4$.

(c) Apply the bootstrap algorithm with and without "roughening" along with the UKF. Discuss processor performances and compare. Zero-mean? Whiteness?

(d) Implement the "optimal" PF processor using the EKF or UKF linearization. How does its performance compare?

4.20 Consider a bearings-only tracking problem given by the state-space model. The entire system can be represented as an approximate

Gauss–Markov model with the noise sources representing uncertainties in the states and measurements. The equations of motion given by

$$x(t) = \begin{bmatrix} 1 & 0 & \Delta T & 0 \\ 0 & 1 & 0 & \Delta T \\ 0 & 0 & 1 & 0 \\ 0 & 0 & 0 & 1 \end{bmatrix} x(t-1) + \begin{bmatrix} 0 & 0 \\ 0 & 0 \\ 1 & 0 \\ 0 & 1 \end{bmatrix} \begin{bmatrix} -\Delta v_{ox}(t-1) \\ -\Delta v_{oy}(t-1) \end{bmatrix} + w(t-1)$$

with the nonlinear sensor model given by

$$y(t) = \arctan \frac{x_1(t)}{x_2(t)} + v(t)$$

for $w \sim \mathcal{N}(0, R_{ww})$ and $v \sim \mathcal{N}(0, R_{vv})$.

(a) Using this model, develop the bootstrap PF and UKF processors for this problem. Assume we would like to estimate the target bearing, frequency, and range ($\alpha = 1$).

(b) Perform a simulation with the following parameters: an impulse-incremental step change ($\Delta v_{ox} = -24$ knot and $\Delta v_{oy} = +10$ knot) was initiated at 0.5 hour, resulting in a change of observer position and velocity depicted in the figure. The simulated bearing measurements are shown in Figure 5.6d. The initial conditions for the run were $x'(0) := [0 \quad 15 \text{ nm} \quad 20 \text{ knot} - 10 \text{ knot}]$ and $R_{ww} = \text{diag } 10^{-6}$ with the measurement noise covariance given by $R_{vv} = 3.05 \times 10^{-4} \text{ rad}^2$ for $\Delta T = 0.33$ hour.

(c) Apply the bootstrap algorithm along with the UKF. Discuss processor performances and compare. Zero-mean? Whiteness?

(d) Implement the "optimal" PF processor using the EKF or UKF linearization. How does its performance compare?

5

Parametrically Adaptive Processors

5.1 Introduction

The model-based approach to the parameter estimation/system identification problem [1–4] is based on the decomposition of the joint posterior distributions that incorporate *both* dynamic state and parameter variables. From this formulation, the following problems evolve: joint state/parameter estimation; state estimation; and parameter (fixed and/or dynamic) estimation. The state estimation problem was discussed in the previous chapters. However, the most common problem found in the current literature is the parameter estimation problem that can be solved "off-line" using batch approaches (maximum entropy, maximum likelihood, minimum variance, least squares, etc.) or "on-line" using the recursive identification approach, the stochastic Monte Carlo approach and for that matter almost any (deterministic) optimization technique [5, 6]. These on-line approaches follow the classical (EKF), modern (UKF), and the sequential Monte Carlo particle filter (PF). However, it still appears that there is *no* universally accepted technique to solve this problem especially for fixed parameters [7–9].

From the pragmatic perspective, the most useful problem is the *joint* state/parameter estimation problem, since it evolves quite naturally from the fact that a model is developed to solve the basic state estimation problem and it is found that its inherent parameters are poorly specified, just bounded, or even unknown, inhibiting the performance of the processor. We call this problem the "joint" state/parameter estimation, since *both* states and parameters are estimated simultaneously (on-line) and the resulting processor is termed *parametrically adaptive* [10]. This terminology evolves because the inherent model parameters are adjusted sequentially as the measurement data becomes available.

In this chapter, we concentrate primarily on the joint state/parameter estimation problem and refer the interested reader to the wealth of literature available on this subject [7–19]. First, we precisely define the basic problem from the Bayesian perspective and then investigate the classical, modern, and particle

Model-Based Processing: An Applied Subspace Identification Approach, First Edition. James V. Candy.
© 2019 John Wiley & Sons, Inc. Published 2019 by John Wiley & Sons, Inc.

approaches to its solution. We incorporate the nonlinear trajectory estimation problem of Jazwinski [20] used throughout as an example of a parametrically adaptive design. Next we briefly develop the prediction error approach constrained to a linear model (RC-circuit) and then discuss a case study to demonstrate the approach.

5.2 Parametrically Adaptive Processors: Bayesian Approach

To be more precise, we start by defining the joint state/parametric estimation problem. We begin by formulating the Bayesian recursions in terms of the posterior distribution using Bayes' rule for decomposition, that is,

$$\Pr(x(t), \theta(t)|Y_t) = \Pr(x(t)|\theta(t), Y_t) \times \Pr(\theta(t)|Y_t)$$
$$= \Pr(\theta(t)|x(t), Y_t) \times \Pr(x(t)|Y_t) \tag{5.1}$$

From this relation, we begin to "see" just how the variety of state and parameter estimation related problems evolve, that is,

- Optimize the state posterior:

    $\Pr(x(t)|Y_t)$ (State Estimation)

- Optimize the parametric posterior:

    $\Pr(\theta(t)|Y_t)$ (Parameter Estimation)

- Optimize the joint state/parameter posterior:

    $\Pr(x(t), \theta(t)|Y_t)$ (State/Parameter Estimation)

Now if we proceed with the usual factorizations, we obtain the Bayesian decomposition for the *state estimation* problem as

$$\Pr(x(t)|Y_t) = \frac{\Pr(y(t)|x(t))}{\Pr(y(t)|Y_{t-1})} \times \Pr(x(t)|Y_{t-1}) \tag{5.2}$$

Equivalently for the *parameter estimation* problem, we have

$$\Pr(\theta(t)|Y_t) = \frac{\Pr(y(t)|\theta(t))}{\Pr(y(t)|Y_{t-1})} \times \Pr(\theta(t)|Y_{t-1}) \tag{5.3}$$

Now for the *joint state/parameter estimation* problem of Eq. (5.1), we can substitute the above equations to obtain the posterior decomposition of interest, that is,

$$\Pr(x(t), \theta(t)|Y_t) = \frac{\Pr(y(t)|\theta(t))}{\Pr(y(t)|Y_{t-1})} \times \left[\Pr(x(t)|\theta(t), Y_t) \times \Pr(\theta(t)|Y_{t-1})\right] \tag{5.4}$$

This is the most common decomposition found in the literature [7–19] and leads to the maximization of the first term (in brackets) with respect to x and the second with respect to θ [21, 22].

5.3 Parametrically Adaptive Processors: Nonlinear Kalman Filters

In this section, we develop parametrically adaptive processors for nonlinear state-space systems. It is a *joint state/parametric processor*, since it estimates *both* the states and the embedded (unknown) model parameters. It is parametrically adaptive, since it adjusts or "adapts" the model parameters at each time step. The simplified structure of the classical extended Kalman filter state/parameter estimator is shown in Figure 5.1. We see the basic structure of the EKF with the *augmented* state vector that leads to two distinct, yet coupled subprocessors: a parameter estimator and a state estimator (filter). The parameter estimator provides estimates that are corrected by the corresponding innovations during each recursion. These estimates are then provided to the state estimator in order to update the model parameters used in the estimator. After both state and parameter estimates are calculated, a new measurement is processed and the procedure continues. In general, this processor can be considered to be a form of identifier, since system identification is typically concerned with the estimation of a model and its associated parameters from

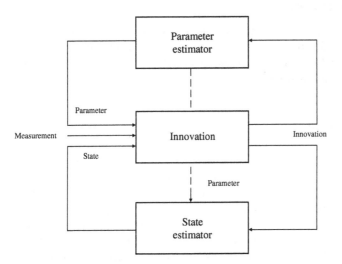

Figure 5.1 Parametrically adaptive processor structure illustrating the coupling between parameter and state estimators through the innovations and measurement sequences.

noisy measurement data. Usually, the model structure is predefined and then a parameter estimator is developed to "fit" parameters according to some error criterion. After completion or during this estimation, the quality of the estimates must be evaluated to decide if the processor performance is satisfactory or equivalently the model adequately represents the data. There are various types (criteria) of identifiers employing many different model (usually linear) structures [2–4]. Here we are primarily concerned with joint estimation in which the models and parameters are nonlinear and discuss the linear case in Section 5.5. Thus, we will concentrate on developing parameter estimators capable of on-line operations with nonlinear dynamics.

The solution to this joint estimation problem involves the augmentation of a parametric model (Θ) along with the dynamic state (x) that embeds the parameters directly. Here the simple solution is to apply the same algorithm (LZKF, EKF, IEKF, UKF) with the original state vector replaced by the new "augmented" state $X \rightarrow [x \mid \Theta]'$. The identical algorithm is then applied directly to the data. This is true for the classical and modern techniques, but not true for the MC-based particle filters, since they must now draw particles from its priors and conditional parametric distributions. However, what happens if we do *not* have a direct physics-based model for the parameters? The nonlinear processors require some characterization or parametric model to be embedded in the joint processor; therefore, we must have some way of providing a parametric representation, if a "physics-based" model is not available.

5.3.1 Parametric Models

Perhaps the most popular approach to this quandary is to use the so-called "random walk" model that has evolved from the control/estimation area. This model essentially assumes a constant parameter (vector) θ with slight variations. In continuous time, a *constant* can be modeled by the differential equation

$$\frac{d\theta_t}{dt} = 0 \quad \text{for} \quad \theta_0 = \text{constant} \tag{5.5}$$

here the solution is just the $\theta_0 \forall t$. Note that this differential equation is a state-space formulation in θ enabling us to incorporate it directly into the overall dynamic model. Now if we artificially assign a Gaussian process noise term w_t to excite the model, then a Wiener process or random walk (pseudo-dynamic) model evolves. Providing a small variance to the noise source approximates small variations in the parameter, enabling it to "walk" in the underlying parameter space and converge to the true value. Since we typically concentrate on a sampled-data system, we can discretize this representation using first differences (with $\Delta t = 1$) and incorporate its corresponding mean/covariance propagation into our Gauss–Markov model, that is,

$$\theta(t) = \theta(t-1) + \mathbf{w}_\theta(t-1) \qquad \text{for} \qquad \mathbf{w}_\theta \sim \mathcal{N}(0, \mathbf{R}_{w_\theta w_\theta})$$

$$\mathbf{m}_\theta(t) = \mathbf{m}_\theta(t-1)$$

$$\mathbf{P}_{\theta\theta}(t) = \mathbf{P}_{\theta\theta}(t-1) + \mathbf{R}_{w_\theta w_\theta}(t-1) \tag{5.6}$$

The process noise covariance $\mathbf{R}_{w_\theta w_\theta}$ controls the excursions of the random walk or equivalently the parameter space.

We are primarily interested in physical systems that have parametric uncertainties that are well modeled by the random walk or other statistical variations. Of course, if the parametric relations are truly dynamic, then the joint approach incorporates these dynamics directly into the augmented state-space model and yields an optimal filtering solution to this joint problem. Using these "artificial dynamics" is the approach employed in the classical LZKF, EKF, IEKF, modern UKF and other nonlinear state-space techniques as well as the purely Bayesian processors (PF).

A popular variation of the random walk is based on incorporating the "forgetting factor" that has evolved from the system identification literature [1] and is called so because it controls the time constant of an inherent exponential window over the data. To see this, consider a squared-error performance function defined by

$$\mathcal{J} = \sum_{t=1}^{k} \alpha(t,k)\epsilon^2(t) \qquad \text{for} \quad \alpha(t,k) = \lambda^{k-t}, \quad 0 < \lambda < 1 \tag{5.7}$$

where the window or weight α is the forgetting factor that creates an exponentially decaying window of length, say T_λ, in which the most recent data are weighted more than the past data. This follows from the following approximation [1, 2]:

$$\alpha(t,k) = e^{(t-k)\ln\lambda} \approx e^{-(t-k)(1-\lambda)} \qquad \text{for} \qquad T_\lambda = \frac{1}{1-\lambda} \tag{5.8}$$

where T_λ is called the memory time constant that remains constant (approximately) over T_λ samples with typical choices between $0.98 < T_\lambda < 0.995$ in practice.

The forgetting factor has been introduced into a variety of sequential algorithms in different roles. For instance, a direct implementation into the random walk model of Eq. (5.6) is [23]

$$\theta(t) = (1 - \lambda(t))\,\theta(t-1) + \lambda(t)\mathbf{w}_\theta(t-1) \tag{5.9}$$

here λ affects the tracking speed of the updates providing a trade-off between speed and noise attenuation. Setting $\lambda = 1$ implies $T_\lambda \to \infty$ an "infinite memory" while $\lambda = 0$ provides instantaneous response forgetting the entire past.

Thus, the forgetting factor introduces a certain amount of flexibility into the joint state/parameter estimation processors by enabling them to be more responsive to new data and therefore yielding improved signal estimates. The same effect is afforded by the process noise covariance matrix of Eq. (5.6) using an equivalent technique called "annealing." The covariance matrix, usually assumed diagonal, is annealed toward zero using exponentially (decay) weighting (see [21, 22] for more details).

Other variations of the parameter estimation models exist besides the statistical random walk model. For instance, similar to the random walk is a process called roughening, which is very similar in concept and is another simple pragmatic method of preventing the sample impoverishment problem for particle filters. It is a method suggested by Gordon et al. [15] and refined in [19], termed particle "roughening," which is similar to adding process noise to constant parameters when constructing a random walk model. It is useful in estimating embedded state-space model parameters and can also be applied to the joint state/parameter estimation problem.

Roughening consists of adding random noise to each particle *after* resampling is accomplished, that is, the a posteriori particles are modified as

$$\tilde{x}_i(t) = \hat{x}_i(t) + \epsilon_i(t) \tag{5.10}$$

where $\epsilon_i \sim \mathcal{N}(0, \text{diag}[\kappa \mathcal{M}_n N_p^{-1/N_x}])$ and κ is a constant "tuning" parameter (e.g. ~ 0.2), $\mathcal{M}_n$ is a vector of the maximum difference between particle components *before* roughening with the nth element of $\mathcal{M}$ given by

$$\mathcal{M}_n = \max_{ij} \left| x_i^{(n)}(t) - x_j^{(n)}(t) \right| \qquad \text{for} \qquad n = 1, \ldots, N_x \tag{5.11}$$

5.3.2 Classical Joint State/Parametric Processors: Augmented Extended Kalman Filter

From our previous discussion in Chapter 4, it is clear that the extended Kalman filter can be applied to this problem directly through *augmentation*. We begin our analysis of the EKF as a joint processor closely following the approach of Ljung [24] for the *linear problem* discussed in Section 4.2. The general nonlinear parameter estimator structure can be derived directly from the EKF algorithm in Table 4.6.

The Bayesian solution to the classical problem is based on solving for the posterior (see Eq. (5.2)) such that each of the required distributions is represented in terms of the EKF estimates. In the joint state/parameter estimation case, these distributions map simply by defining an augmented state vector $X(t) := [x(t)|\theta(t)]'$ to give (see Section 4.2)

$$\Pr(y(t)|x(t)) \sim \mathcal{N}\left(c[x(t)], R_{vv}(t)\right)$$
$$\Updownarrow$$
$$\Pr(y(t)|x(t), \theta(t)) \sim \mathcal{N}\left(c[x(t), \theta(t)], R_{vv}(t)\right)$$

$$\Pr(x(t)|Y_{t-1}) \sim \mathcal{N}\left(\hat{x}(t|t-1), \tilde{P}(t|t-1)\right)$$

$$\Updownarrow$$

$$\Pr(X(t)|Y_{t-1}) \sim \mathcal{N}\left(\hat{X}(t|t-1), \tilde{P}(t|t-1)\right)$$

$$\Pr(y(t)|Y_{t-1}) \sim \mathcal{N}\left(\hat{y}(t|t-1), R_{ee}(t)\right)$$

$$\Updownarrow$$

$$\Pr(y(t)|Y_{t-1}) \sim \mathcal{N}\left(\hat{y}_\theta(t|t-1), \tilde{R}_{e_\theta e_\theta}(t)\right) \tag{5.12}$$

where

$$\hat{X}(t|t-1) := \left[\hat{x}(t|t-1)|\hat{\theta}(t|t-1)\right]'$$

$$\hat{y}_\theta(t|t-1) := c\left[\hat{x}(t|t-1), \hat{\theta}(t|t-1)\right]$$

$$\tilde{P}(t|t-1) := \begin{bmatrix} \tilde{P}_{xx}(t|t-1) & | & \tilde{P}_{x\theta}(t|t-1) \\ - & - & - \\ \tilde{P}_{\theta x}(t|t-1) & | & \tilde{P}_{\theta\theta}(t|t-1) \end{bmatrix}$$

$$\mathcal{R}_{e_\theta e_\theta}(t) := C[\hat{X}(t|t-1)]\tilde{P}(t|t-1)C[\hat{X}(t|t-1)]' + R_{vv}(t)$$

To develop the actual internal structure of the processor, we start with the EKF equations, augment them with the unknown parameters, and then investigate the resulting algorithm. We first define the composite state vector (as before) consisting of the original states, $x(t)$, and the unknown "augmented" parameters represented by $\theta(t)$, that is,

$$X(t) := \begin{bmatrix} x(t) \\ - - - \\ \theta(t) \end{bmatrix} \tag{5.13}$$

where t is the time index, $X \in R^{(N_x+N_\theta)\times 1}$, and $x \in R^{N_x\times 1}$, $\theta \in R^{N_\theta\times 1}$.

Substituting this augmented state vector into the EKF relations of Table 4.6, the following matrix partitions evolve as

$$\tilde{P}(t|t-1) := \begin{bmatrix} \tilde{P}_{xx}(t|t-1) & | & \tilde{P}_{x\theta}(t|t-1) \\ - & - & - \\ \tilde{P}_{\theta x}(t|t-1) & | & \tilde{P}_{\theta\theta}(t|t-1) \end{bmatrix}, \tilde{P}_{\theta x}(t|t-1) = \tilde{P}'_{x\theta}(t|t-1) \tag{5.14}$$

where $\tilde{P} \in R^{(N_x+N_\theta)\times(N_x+N_\theta)}$, $\tilde{P}_{xx} \in R^{N_x\times N_x}$, $\tilde{P}_{\theta\theta} \in R^{N_\theta\times N_\theta}$, and $\tilde{P}_{x\theta} \in R^{N_x\times N_\theta}$. This also leads to the partitioning of the corresponding gain

$$\mathcal{K}(t) := \begin{bmatrix} K_x(t) \\ - - - \\ K_\theta(t) \end{bmatrix} \tag{5.15}$$

for $\mathcal{K} \in R^{(N_x+N_\theta)\times N_y}$, $K_x \in R^{N_x\times N_y}$, and $K_\theta \in R^{N_\theta\times N_y}$.

We must also partition the state and measurement predictions as equations, that is,

$$
\hat{X}(t|t-1) = \begin{bmatrix} \hat{x}(t|t-1) \\ --- \\ \hat{\theta}(t|t-1) \end{bmatrix}
$$

$$
= \begin{bmatrix} \mathbf{a}[\hat{x}(t-1|t-1), \hat{\theta}(t-1|t-1)] \\ + \mathbf{b}[\hat{x}(t-1|t-1), \hat{\theta}(t-1|t-1), u(t-1)] \\ --- \\ \hat{\theta}(t-1|t-1) \end{bmatrix} \tag{5.16}
$$

where the corresponding predicted measurement equation becomes

$$
\hat{y}(t|t-1) = \mathbf{c}[\hat{x}(t|t-1), \hat{\theta}(t|t-1)] \tag{5.17}
$$

Here we have "implicitly" assumed that the parameters can be considered *piecewise constant*, $\Theta(t) = \Theta(t-1)$ or follow a random walk if we add process noise to this representation in the Gauss–Markov sense. However, if we *do* have process dynamics with linear or nonlinear models characterizing the parameters, then they will replace the random walk and the associated Jacobian.

Next, we consider the predicted error covariance

$$
\tilde{P}(t|t-1) = A[\hat{x}(t|t-1), \hat{\theta}(t|t-1)]\tilde{P}(t|t-1)A'[\hat{x}(t|t-1), \hat{\theta}(t|t-1)]
$$

$$
+ R_{ww}(t-1) \tag{5.18}
$$

where R_{ww} is a block diagonal matrix consisting of the state and parameter covariance matrices: $R_{ww} = \text{diag}\left[R_{w_x w_x} R_{w_\theta w_\theta}\right]$. The error covariance can be written in partitioned form using Eq. (5.14) and the following Jacobian process matrix:

$$
A[\hat{x}, \hat{\theta}] = \begin{bmatrix} A_x[\hat{x}(t|t-1), \hat{\theta}(t|t-1)] & | & A_\theta[\hat{x}(t|t-1), \hat{\theta}(t|t-1)] \\ - & - & - \\ O & | & I_{N_\theta} \end{bmatrix} \tag{5.19}
$$

where

$$
A_x[\hat{x}, \hat{\theta}] := \frac{\partial a[\hat{x}, \hat{\theta}]}{\partial x} + \frac{\partial b[\hat{x}, \hat{\theta}]}{\partial x}
$$

$$
A_\theta[\hat{x}, \hat{\theta}] := \frac{\partial a[\hat{x}, \hat{\theta}]}{\partial \theta} + \frac{\partial b[\hat{x}, \hat{\theta}]}{\partial \theta} \tag{5.20}
$$

with $A \in R^{(N_x+N_\theta)\times(N_x+N_\theta)}, A_x \in R^{N_x\times N_x}, A_\theta \in R^{N_x\times N_\theta}$.

Here is where the underlying random walk model enters the structure. The lower block rows of A_x could be replaced by $[A_{\theta x}[\hat{x}, \hat{\theta}] \mid A_{\theta\theta}[\hat{x}, \hat{\theta}]]$, which enables a linear or nonlinear dynamic model to be embedded directly.

Using these partitions, we can develop the parametrically adaptive processor directly from the EKF processor in this joint state and "parameter estimation" form. Substituting Eq. (5.19) into the EKF prediction covariance relation of Table 4.6 and using the partition defined in Eq. (5.14), we obtain (suppressing the $\hat{x}$, $\hat{\theta}$, time index t notation for simplicity)

$$
\tilde{P}(t|t-1)
$$

$$
= \begin{bmatrix} A_x & | & A_\theta \\ - & - & - \\ 0 & | & I_{N_\theta} \end{bmatrix} \tilde{P}(t-1|t-1) \begin{bmatrix} A_x & | & A_\theta \\ - & - & - \\ 0 & | & I_{N_\theta} \end{bmatrix}' + \begin{bmatrix} R_{w_x w_x} & | & 0 \\ - & - & - \\ 0 & | & R_{w_\theta w_\theta} \end{bmatrix}
$$
(5.21)

Expanding these equations, we obtain the following set of predicted covariance relations:

$$
\tilde{P}(t|t-1)
$$

$$
= \begin{bmatrix} A_x \tilde{P}_{xx} A_x' + A_\theta \tilde{P}_{\theta x} A_x' + A_x \tilde{P}_{x\theta} A_\theta' + A_\theta \tilde{P}_{\theta\theta} A_\theta' + R_{w_x w_x} & | & A_x \tilde{P}_{x\theta} + A_\theta \tilde{P}_{\theta\theta} \\ - & - & - \\ \tilde{P}_{\theta x} A_x' + \tilde{P}_{\theta\theta} A_\theta' & | & \tilde{P}_{\theta\theta} + R_{w_\theta w_\theta} \end{bmatrix}
$$
(5.22)

The innovations covariance follows from the EKF as

$$
R_{ee}(t) = C[\hat{x}, \hat{\theta}] \tilde{P}(t|t-1) C'[\hat{x}, \hat{\theta}] + R_{vv}(t)
$$
(5.23)

Now, we must use the partitions of $\tilde{P}$ above along with the measurement Jacobian

$$
C[\hat{x}, \hat{\theta}] = \begin{bmatrix} C_x[\hat{x}(t|t-1), \hat{\theta}(t|t-1)] & | & C_\theta[\hat{x}(t|t-1), \hat{\theta}(t|t-1)] \end{bmatrix}
$$
(5.24)

where

$$
C_x[\hat{x}, \hat{\theta}] := \frac{\partial c[\hat{x}, \hat{\theta}]}{\partial x}
$$
$$
C_\theta[\hat{x}, \hat{\theta}] := \frac{\partial c[\hat{x}, \hat{\theta}]}{\partial \theta}
$$
(5.25)

with $C \in R^{N_y \times (N_x + N_\theta)}$, $C_x \in R^{N_y \times N_x}$, $C_\theta \in R^{N_y \times N_\theta}$.

The corresponding innovations covariance follows from Eqs. (5.23) and (5.24) as

$$
R_{ee}(t) = \begin{bmatrix} C_x & | & C_\theta \end{bmatrix} \begin{bmatrix} \tilde{P}_{xx} & | & \tilde{P}_{x\theta} \\ - & - & - \\ \tilde{P}_{\theta x} & | & \tilde{P}_{\theta\theta} \end{bmatrix} \begin{bmatrix} C_x' \\ - - - \\ C_\theta' \end{bmatrix} + R_{vv}(t)
$$
(5.26)

or expanding

$$
R_{ee}(t) = C_x \tilde{P}_{xx} C_x' + C_\theta \tilde{P}_{\theta x} C_x' + C_x \tilde{P}_{x\theta} C_\theta' + C_\theta \tilde{P}_{\theta\theta} C_\theta' + R_{vv}
$$
(5.27)

$R_{ee} \in R^{N_y \times N_y}$. The gain of the EKF in Table 4.6 is calculated from these partitioned expressions as

$$\mathcal{K}(t) = \begin{bmatrix} K_x(t) \\ --- \\ K_\theta(t) \end{bmatrix} = \begin{bmatrix} \tilde{P}_{xx} & | & \tilde{P}_{x\theta} \\ - & - & - \\ \tilde{P}_{\theta x} & | & \tilde{P}_{\theta\theta} \end{bmatrix} \begin{bmatrix} C_x \\ --- \\ C_\theta \end{bmatrix}' R_{ee}^{-1}(t) \qquad (5.28)$$

or

$$\mathcal{K}(t) = \begin{bmatrix} K_x(t) \\ --- \\ K_\theta(t) \end{bmatrix} = \begin{bmatrix} (\tilde{P}_{xx}C_x' + \tilde{P}_{x\theta}C_\theta')R_{ee}^{-1}(t) \\ --- \\ (\tilde{P}_{\theta x}C_x' + \tilde{P}_{\theta\theta}C_\theta')R_{ee}^{-1}(t) \end{bmatrix} \qquad (5.29)$$

where $K \in {}^{(N_x+N_\theta) \times N_y}$, $K_x \in R^{N_x \times N_y}$, $K_\theta \in R^{N_\theta \times N_y}$. With the gain determined, the corrected state/parameter estimates follow easily, since the innovations remain unchanged, that is,

$$e(t) = y(t) - \hat{y}(t|t-1) = y(t) - c[\hat{x}(t|t-1), \hat{\theta}(t|t-1)] \qquad (5.30)$$

and therefore partitioning the corrected state equations, we have

$$\hat{X}(t|t) = \begin{bmatrix} \hat{x}(t|t) \\ --- \\ \hat{\theta}(t|t) \end{bmatrix} = \begin{bmatrix} \hat{x}(t|t-1) \\ --- \\ \hat{\theta}(t|t-1) \end{bmatrix} + \begin{bmatrix} K_x(t)e(t) \\ --- \\ K_\theta(t)e(t) \end{bmatrix} \qquad (5.31)$$

Finally, the corrected covariance expression is easily derived from the following partitions:

$$\tilde{P}(t|t) = \begin{bmatrix} \tilde{P}_{xx} & | & \tilde{P}_{x\theta} \\ - & - & - \\ \tilde{P}_{\theta x} & | & \tilde{P}_{\theta\theta} \end{bmatrix} - \begin{bmatrix} K_x(t) \\ --- \\ K_\theta(t) \end{bmatrix} \begin{bmatrix} C_x & | & C_\theta \end{bmatrix} \begin{bmatrix} \tilde{P}_{xx} & | & \tilde{P}_{x\theta} \\ - & - & - \\ \tilde{P}_{\theta x} & | & \tilde{P}_{\theta\theta} \end{bmatrix}$$

$$(5.32)$$

Performing the indicated multiplications leads to the final expression

$$\tilde{P}(t|t) = \begin{bmatrix} \tilde{P}_{xx} - K_x C_x \tilde{P}_{xx} - K_x C_\theta \tilde{P}_{\theta x} & | & \tilde{P}_{x\theta} - K_x C_x \tilde{P}_{x\theta} - K_x C_\theta \tilde{P}_{\theta\theta} \\ --- & --- & --- \\ \tilde{P}_{\theta x} - K_\theta C_x \tilde{P}_{xx} - K_\theta C_\theta \tilde{P}_{\theta x} & | & \tilde{P}_{\theta\theta} - K_\theta C_x \tilde{P}_{x\theta} - K_\theta C_\theta \tilde{P}_{\theta\theta} \end{bmatrix}$$

$$(5.33)$$

We summarize the parametrically adaptive EKF in predictor–corrector form in Table 5.1. We note that this algorithm is *not* implemented in this manner – it merely evolves when the augmented state is processed using the "usual" EKF. Here we are just interested in the overall internal structure of the algorithm and the decompositions that evolve. This completes the development of the generic parametrically adaptive EKF processor. Clearly, this applies to any variants of the EKF including the LZKF.

Table 5.1 Augmented extended Kalman filter algorithm.

Prediction

$\hat{x}(t|t-1) = a[\hat{x}(t|t-1), \hat{\theta}(t-1|t-1)] + b[u(t-1)]$ (State)

$\hat{\theta}(t|t-1) = \hat{\theta}(t-1|t-1)$ (Parameter)

$\tilde{P}_{xx}(t|t-1) = A_x[\hat{x}, \hat{\theta}]\tilde{P}_{xx}(t-1|t-1)A_x^T[\hat{x}, \hat{\theta}] + A_\theta[\hat{x}, \hat{\theta}, u]\tilde{P}_{\theta x}(t-1|t-1)A_x^T[\hat{x}, \hat{\theta}]$ (State cov.)
$\quad\quad + A_x[\hat{x}, \hat{\theta}]\tilde{P}_{x\theta}(t-1|t-1)A_\theta^T[\hat{x}, \hat{\theta}, u] + A_\theta[\hat{x}, \hat{\theta}, u]\tilde{P}_{\theta\theta}(t-1|t-1)A_\theta^T[\hat{x}, \hat{\theta}, u] + R_{w_x w_x}(t-1)$

$\tilde{P}_{\theta\theta}(t|t-1) = \tilde{P}_{\theta\theta}(t-1|t-1) + R_{w_\theta w_\theta}(t-1)$ (Param. cov.)

$\tilde{P}_{x\theta}(t|t-1) = A_x[\hat{x}, \hat{\theta}]\tilde{P}_{x\theta}(t-1|t-1) + A_\theta[\hat{x}, \hat{\theta}, u]\tilde{P}_{\theta\theta}(t-1|t-1)$ (Cross cov.)

Innovation

$e(t) = y(t) - \hat{y}(t|t-1) = y(t) - c[\hat{x}(t|t-1), \hat{\theta}(t|t-1)]$ (Innovation)

$R_{ee}(t) = C_x[\hat{x}, \hat{\theta}]\tilde{P}_{xx}(t|t-1)C_x^T[\hat{x}, \hat{\theta}] + C_\theta[\hat{x}, \hat{\theta}, u]\tilde{P}_{\theta x}(t|t-1)C_x^T[\hat{x}, \hat{\theta}]$ (Innovation cov.)
$\quad\quad + C_x[\hat{x}, \hat{\theta}]\tilde{P}_{x\theta}(t|t-1)C_\theta^T[\hat{x}, \hat{\theta}, u] + C_\theta[\hat{x}, \hat{\theta}, u]\tilde{P}_{\theta\theta}(t|t-1)C_\theta^T[\hat{x}, \hat{\theta}, u] + R_v(t)$

Gain

$K_x(t) = \left(\tilde{P}_{xx}(t|t-1)C_x^T[\hat{x}, \hat{\theta}] + \tilde{P}_{x\theta}(t|t-1)C_\theta^T[\hat{x}, \hat{\theta}, u]\right) R_{ee}^{-1}(t)$ (State Gain)

$K_\theta(t) = \left(\tilde{P}_{\theta x}(t|t-1)C_x^T[\hat{x}, \hat{\theta}] + \tilde{P}_{\theta\theta}(t|t-1)C_\theta^T[\hat{x}, \hat{\theta}, u]\right) R_{ee}^{-1}(t)$ (Parameter Gain)

Correction

$\hat{x}(t|t) = \hat{x}(t|t-1) + K_x(t)e(t)$ (State)

$\hat{\theta}(t|t) = \hat{\theta}(t|t-1) + K_\theta(t)e(t)$ (Parameter)

$\tilde{P}_{xx}(t|t) = \tilde{P}_{xx}(t|t-1) - K_x(t)C_x[\hat{x}, \hat{\theta}]\tilde{P}_{xx}(t|t-1) - K_x(t)C_\theta[\hat{x}, \hat{\theta}, u]\tilde{P}_{\theta x}(t|t-1)$ (State cov.)

$\tilde{P}_{\theta\theta}(t|t) = \tilde{P}_{\theta\theta}(t|t-1) - K_\theta(t)C_\theta[\hat{x}, \hat{\theta}, u]\tilde{P}_{\theta\theta}(t|t-1) - K_\theta(t)C_x[\hat{x}, \hat{\theta}, u]\tilde{P}_{x\theta}(t|t-1)$ (Parameter cov.)

$\tilde{P}_{x\theta}(t|t) = \tilde{P}_{x\theta}(t|t-1) - K_x(t)C_x[\hat{x}, \hat{\theta}]\tilde{P}_{x\theta}(t|t-1) - K_x(t)C_\theta[\hat{x}, \hat{\theta}, u]\tilde{P}_{\theta\theta}(t|t-1)$ (Cross cov.)

Initial Conditions

$\hat{x}(0|0)$ $\tilde{P}(0|0)$ $A[\hat{x}, \theta] := \frac{\partial}{\partial x}A[x, \theta]\Big|_{\substack{x=\hat{x}(t|t-1)\\\theta=\hat{\theta}(t|t-1)}}$ $C[\hat{x}, \theta] := \frac{\partial}{\partial x}C[x, \theta]\Big|_{\substack{x=\hat{x}(t|t-1)\\\theta=\hat{\theta}(t|t-1)}}$

It is important to realize that besides its numerical implementation this processor is simply the EKF with an augmented state vector, thereby implicitly creating the partitions developed above. The implementation of these decomposed equations directly is **not** necessary – just augment the state with the unknown parameters, and it evolves naturally from the standard EKF algorithm of Table 4.6. The parametrically adaptive processor of Table 5.1 indicates where to locate the partitions. That is, suppose we would like to extract the submatrix, $\tilde{P}_{\theta\theta}$, but the EKF only provides the overall $(N_x + N_\theta)$ error covariance matrix, $\tilde{P}$. However, locating the lower $N_\theta \times N_\theta$ submatrix of $\tilde{P}$ enables us to extract $\tilde{P}_{\theta\theta}$ directly.

Next, let us reconsider the nonlinear system example given in Chapter 4 and investigate the performance of the parametrically adaptive EKF processor.

Example 5.1 Recall the discrete nonlinear trajectory estimation problem [20] of Chapter 4 given by

$$x(t) = (1 - 0.05\Delta T)x(t-1) + 0.04x^2(t-1) + w(t-1)$$

with the corresponding measurement model

$$y(t) = x^2(t) + x^3(t) + v(t)$$

where $v(t) \sim \mathcal{N}(0, 0.09)$, $x(0) = 2.0$, $P(0) = 0.01$, $\Delta T = 0.01$ second and $R_{ww} = 0$.

Here, we generalize the problem to the case where the coefficients of the process are unknown leading to the parametrically adaptive solution. Therefore, the process equations for this problem become

$$x(t) = (1 - \theta_1 \Delta T)x(t-1) + \theta_2 x^2(t-1) + w(t-1)$$

with the identical measurement and covariances as before. The true parameters are $\Theta_{\text{true}} = [0.05 \ 0.04]'$. The augmented EKF can be applied to this problem by defining the parameter vector as

$$\Theta(t) = \Theta(t-1) + w_\theta(t-1)$$

and augmenting it to form the new state vector $X = [x' \ \theta_1 \ \theta_2]'$. Therefore, the process model becomes

$$X(t) = \begin{bmatrix} (1 - \theta_1(t-1)\Delta T)x(t-1) + \theta_2(t-1)\Delta T x^2(t-1) \\ \theta_1(t-1) \\ \theta_2(t-1) \end{bmatrix}$$
$$+ \begin{bmatrix} w_x(t-1) \\ w_{\theta_1}(t-1) \\ w_{\theta_2}(t-1) \end{bmatrix}$$

$$y(t) = x^2(t) + x^3(t) + v(t)$$

To implement the EKF, the required Jacobians are

$A[X(t-1)]$

$$= \begin{bmatrix} [1 - \theta_1(t-1)\Delta T + 2\Delta T \theta_2(t-1)x(t-1)] & \Delta T)x(t-1) & \Delta T x^2(t-1) \\ 0 & 1 & 0 \\ 0 & 0 & 1 \end{bmatrix}$$

$C[X(t-1)] = [2x(t-1) + 3x^2(t-1) \ 0 \ 0]$

Using MATLAB [10], the augmented EKF processor is applied to solve this problem for 1500 samples with $\Delta T = 0.01$ second. Initially, we used the starting parameters:

$$\tilde{P}(0|0) = \text{diag}\,[100 \ \ 100 \ \ 100] \text{ and } \hat{x}(0|0) = [2 \ \ 0.055 \ \ 0.044]'$$

The results of the parametrically adaptive run are shown in Figure 5.2. We see the estimated state and parameter estimates in Figure 5.2a and b. After a

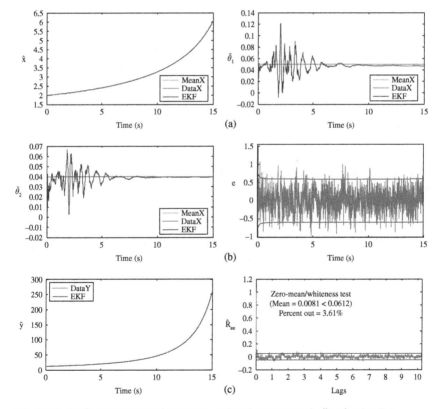

Figure 5.2 Nonlinear (uncertain) trajectory estimation: parametrically adaptive EKF algorithm. (a) Estimated state and parameter no. 1. (b) Estimated parameter no. 2 and innovations. (c) Predicted measurement and zero-mean/whiteness test (0.080 < 0.061 and 3.6% out).

short transient (25 samples), the state estimate begins tracking the true state as evidenced by the innovations sequence in Figure 5.2c. The parameter estimates slowly converge to their true values as evidenced by the plots and insets of the figure. The final estimates are

$$\hat{\theta}_1 = 0.0470$$

$$\hat{\theta}_2 = 0.0395$$

Part of the problem for the slow convergence results stems from the lack of sensitivity of the measurement, or equivalently, innovations to parameter variations in this problem. This is implied from the zero-mean/whiteness tests shown in Figure 5.2c. The innovations are statistically white (3.6% out), but indicate a slight bias (0.080 > 0.061). The filtered measurement is also shown in Figure 5.2c . This completes the example. □

As pointed out by Ljung and Soderstrom [2, 24], it is important to realize that the EKF is suboptimal as a parameter estimator compared to the recursive prediction error (RPE) method based on the Gauss–Newton (stochastic) descent algorithm (see Section 5.5). Comparing the processors in this context, we see that if the gradient term $[\nabla_\theta K(\Theta)]e(t)$ is incorporated into the EKF (add this term to A_θ), its convergence will be improved approaching the performance of the RPE algorithm (see Ljung [24] for details). Next, we consider the development of the "modern" approach using the unscented Kalman filter of Chapter 4.

5.3.3 Modern Joint State/Parametric Processor: Augmented Unscented Kalman Filter

The modern unscented processor offers a similar representation as the extended processor detailed in section 5.3.2. Here, we briefly outline its structure for solution to the *joint* problem and apply it to the trajectory estimation problem for comparison. We again start with the augmented state vector defined initially by sigma-points, that is,

$$\mathcal{X}(t) := \begin{bmatrix} x(t) \\ --- \\ \theta(t) \end{bmatrix} \tag{5.34}$$

$\mathcal{X} \in R^{(N_x+N_\theta)\times 1}$ and $x \in R^{N_x\times 1}$, $\theta \in R^{N_\theta\times 1}$.

Substituting this augmented state vector into the UKF relations of Table 4.8 yields the desired processor. We again draw the equivalences (as before):

$$\Pr(y(t)|x(t),\theta(t)) \sim \mathcal{N}(c[x(t),\theta(t)], R_{vv}(t))$$

$$\Pr(\mathcal{X}(t)|Y_{t-1}) \sim \mathcal{N}(\mathcal{X}(t|t-1), \tilde{P}(t|t-1))$$

$$\Pr(y(t)|Y_{t-1}) \sim \mathcal{N}(\hat{y}_\theta(t|t-1), \tilde{R}_{e_\theta e_\theta}(t)) \tag{5.35}$$

where

$$\mathcal{X}(t|t-1) := [\hat{x}(t|t-1)|\hat{\theta}(t|t-1)]'$$

$$\hat{y}_\theta(t|t-1) := \mathbf{c}[\hat{x}(t|t-1), \hat{\theta}(t|t-1)]$$

$$\tilde{\mathcal{P}}(t|t-1) := \left[\begin{array}{c|c} \tilde{P}_{xx}(t|t-1) & \tilde{P}_{x\theta}(t|t-1) \\ - & - \\ \tilde{P}_{\theta x}(t|t-1) & \tilde{P}_{\theta\theta}(t|t-1) \end{array} \right]$$

$$\mathcal{R}_{e_\theta e_\theta}(t) := C[\hat{X}(t|t-1)]\tilde{P}(t|t-1)C[\hat{X}(t|t-1)]' + R_{vv}(t)$$

With this in mind, it is possible to derive the internal structure of the UKF in a manner similar to that of the EKF. But we will not pursue that derivation here. We just note that the sigma-points are also augmented to give

$$\mathcal{X}_i := \left[\begin{array}{c} \hat{x}(t|t-1) \\ --- \\ \hat{\theta}(t|t-1) \end{array} \right] + \left((N_\mathcal{X} + \kappa) \left[\begin{array}{c|c} \tilde{P}_{xx}(t|t-1) & \tilde{P}_{x\theta}(t|t-1) \\ - & - \\ \tilde{P}_{\theta x}(t|t-1) & \tilde{P}_{\theta\theta}(t|t-1) \end{array} \right] \right)^{\frac{1}{2}}$$

(5.36)

with the corresponding process noise covariance partitioned as

$$\mathcal{R}_{ww}(t) := \left[\begin{array}{c|c} R_{w_x w_x}(t|t-1) & 0 \\ - & - & - \\ 0 & R_{w_\theta w_\theta}(t) \end{array} \right]$$

(5.37)

It also follows that the prediction step becomes

$$\mathcal{X}_i(t|t-1) = \left[\begin{array}{c} \mathbf{a}[\hat{x}(t|t-1), \hat{\theta}(t|t-1)] + \mathbf{b}[\hat{\theta}(t|t-1), u(t-1)] \\ --- \\ \mathbf{a}[\hat{\theta}(t|t-1)] \end{array} \right]$$

(5.38)

and in the multichannel (vector) measurement case we have that

$$\mathcal{Y}_i(t|t-1) = \left[\begin{array}{c} \mathbf{c}[\hat{x}(t|t-1), \hat{\theta}(t|t-1)] \\ --- \\ \mathbf{c}[\hat{\theta}(t|t-1)] \end{array} \right]$$

(5.39)

Using the augmented state vector, we apply the "joint" approach to the trajectory estimation problem [20] and compare its performance to that of the EKF.

Example 5.2 Using the discrete nonlinear trajectory estimation problem of the previous example with unknown coefficients as before, we define the augmented sigma-point vector $\mathcal{X}(t)$ defined above and apply the UKF algorithm of Table 4.8.

The process equations for this problem are as follows:

$$x(t) = (1 - \theta_1 \triangle T)x(t-1) + \theta_2 x^2(t-1) + w_x(t-1)$$

with the identical measurement and covariances as before. The true parameters are $\Theta_{true} = [0.05 \quad 0.04]'$. The UKF can be applied to this problem by defining the parameter vector as a constant characterized by a random walk and augmenting it to form the new sigma-point vector $\mathcal{X} = [x' \quad \theta_1 \quad \theta_2]'$. Therefore, the process model becomes

$$
\begin{bmatrix} x(t) \\ -- \\ \theta(t) \end{bmatrix} = \begin{bmatrix} (1 - \theta_1(t-1)\Delta T)x(t-1) + \theta_2(t-1)\Delta Tx^2(t-1) \\ \theta_1(t-1) \\ \theta_2(t-1) \end{bmatrix}
$$
$$
+ \begin{bmatrix} w_x(t-1) \\ w_{\theta_1}(t-1) \\ w_{\theta_2}(t-1) \end{bmatrix}
$$

$$
y(t) = x^2(t) + x^3(t) + v(t)
$$

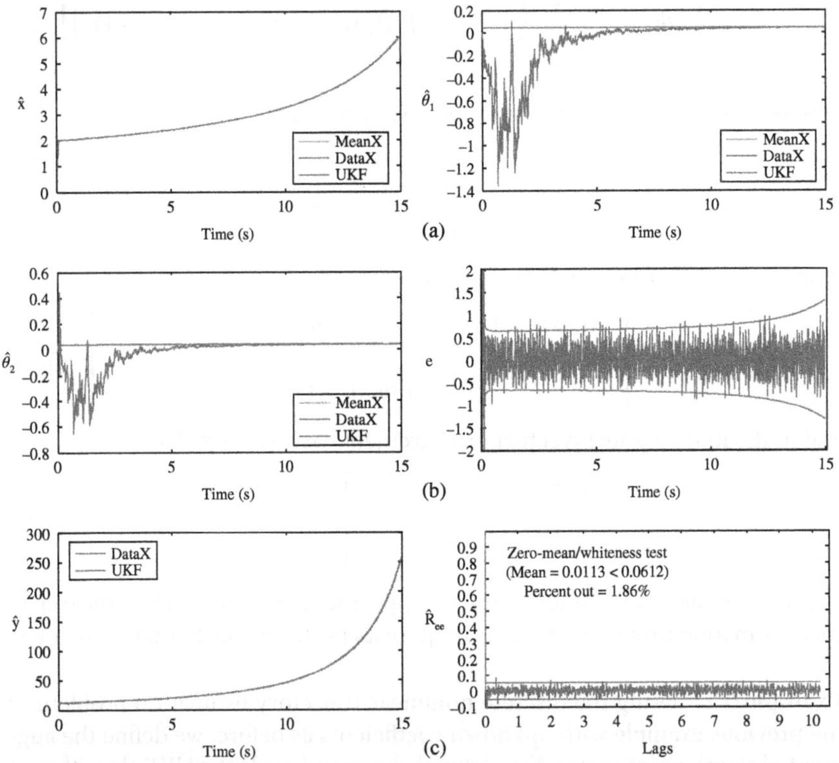

Figure 5.3 Nonlinear (uncertain) trajectory estimation: Parametrically adaptive UKF processor. (a) Estimated state and parameter no. 1. (b) Estimated parameter no. 2 and innovations. (c) Predicted measurement and zero-mean/whiteness test (0.0113 < 0.0612 and 1.86% out)

To implement the UKF, the sigma-points are selected as before for a Gaussian distribution using the scaled transformation with $\alpha = 1$, $\kappa = 0$, and $\beta = 2$ for the same initial conditions as the EKF.

Using MATLAB [10], the augmented UKF is applied to solve this problem for 1500 samples with $\Delta T = 0.01$ second.

The results of the UKF run are shown in Figure 5.3. We see the estimated state and parameter estimates in Figure 5.3a and b. After a short transient, the state estimate begins tracking the true state as evidenced by the predicted measurement and innovations sequence in Figure 5.3c. The parameter estimates converge to their true values as evidenced by the plots. The final estimates are $\hat{\theta}_1 = 0.05$; $\hat{\theta}_2 = 0.04$. The processor appears to converge much faster than the EKF demonstrating its improved capability. This is implied from the zero-mean/whiteness tests shown in Figure 5.3c. The innovations are statistically white (1.86% out) and unbiased (0.0113 < 0.0612). This completes the example.   □

We also note in closing that a "dual" rather than "joint" approach has evolved in the literature. Originally developed as a bootstrap approach, it is constructed by two individual (decoupled) processors: one for the state estimator and one for the parameter estimator, which pass updated estimates back and forth to each other as they become available. This is a suboptimal methodology, but appears to perform quite well in some applications (see [21, 22] for more details). This completes our discussion of the augmented modern UKF processor. Next we investigate the PF approach to solve the joint problem.

5.4 Parametrically Adaptive Processors: Particle Filter

In this section, we briefly develop the sequential Monte Carlo approach to solve the joint state/parametric processing problem. There are a variety of ways to search the feasible space in pursuing solutions to the parameter estimation problem, but when states (or signals) must be extracted along with the parameters, then a solution to the joint estimation problem must follow as discussed before. One of the major problems that evolve is how to "model" the parameter evolution in order to provide an effective way to proceed with the search especially if there does *not* exist a physical model characterizing the parameters. The usual approach is to incorporate a random walk model as discussed previously, if the parameters are slowly varying.

5.4.1 Joint State/Parameter Estimation: Particle Filter

Here we are concerned with the joint estimation problem consisting of setting a prior for θ and augmenting the state vector to solve the joint estimation problem as defined above, thereby converting the parameter estimation problem to one of optimal filtering. Thus, confining our discussion to state-space models and the Bayesian equivalence, we develop the Bayesian approach using the following relations:

$$x \sim \Pr(x(t)|x(t-1), \theta(t-1))$$
$$\theta \sim \Pr(\theta(t)|\theta(t-1), x(t-1))$$
$$y \sim \Pr(y(t)|x(t), \theta(t)) \tag{5.40}$$

Here, we separate the state transition function into the individual vectors for illustrative purposes, but in reality (as we shall observe), they can be jointly coupled. The key idea is to develop the PF technique to estimate the joint posterior $\Pr(x(t), \theta(t)|Y_t)$ relying on the parametric posterior $\Pr(\theta(t)|Y_t)$ in the Bayesian decomposition. We will follow the approach outlined in [17, 25] starting with the full posterior distributions and proceeding to the filtering distributions.

Suppose it is possible to sample N_p-particles, $\{X_t(i), \Theta_t(i)\}$ for $i = 1, \dots, N_p$ from the joint posterior distribution where we define $X_t := \{x(0), \dots, x(t)\}$ and $\Theta_t := \{\theta(0), \dots, \theta(t)\}$. Then the corresponding empirical approximation of the joint posterior is given by

$$\hat{\Pr}(X_t, \Theta_t|Y_t) \approx \frac{1}{N_p} \sum_{i=1}^{N_p} \delta(X_t - X_t(i), \Theta_t - \Theta_t(i)) \tag{5.41}$$

and it follows directly that the filtering posterior is given by

$$\hat{\Pr}(x(t), \theta(t)|Y_t) \approx \frac{1}{N_p} \sum_{i=1}^{N_p} \delta(x(t) - x_i(t), \theta(t) - \theta_i(t)) \tag{5.42}$$

Unfortunately, it is not possible to sample directly from the full joint posterior $\Pr(X_t, \Theta_t|Y_t)$ at any time t. However, one approach to mitigate this problem is by using the importance sampling approach.

Suppose we define a (full) importance distribution, $\mathcal{I}(X_t, \Theta_t|Y_t)$ such that $\Pr(X_t, \Theta_t|Y_t) > 0$ implies $\mathcal{I}(\cdot) > 0$, then we define the corresponding importance weight (as before) by

$$W(X_t, \Theta_t) \propto \frac{\Pr(X_t, \Theta_t|Y_t)}{\mathcal{I}(X_t, \Theta_t|Y_t)} \tag{5.43}$$

From Bayes' rule, we have that the posterior can be expressed as

$$\Pr(X_t, \Theta_t|Y_t) = \frac{\Pr(Y_t|X_t, \Theta_t) \times \Pr(X_t, \Theta_t)}{\Pr(Y_t)} \tag{5.44}$$

Thus, if N_p-particles, $\{X_t(i), \Theta_t(i)\}; i = 1, \dots, N_p$, can be generated from the importance distribution

$$\{X_t(i), \Theta_t(i)\} \rightarrow \mathcal{I}(X_t, \Theta_t|Y_t) \tag{5.45}$$

then the empirical distribution can be estimated and the resulting normalized weights specified by

$$\mathcal{W}_t(i) = \frac{W(X_t(i), \Theta_t(i))}{\sum_{k=1}^{N_p} W(X_t(k), \Theta_t(k))} \quad \text{for} \quad i = 1, \ldots, N_p \tag{5.46}$$

to give the desired empirical distribution of Eq. (5.41) leading to the corresponding filtering distribution of Eq. (5.42).

If we have a state transition model available, then for a fixed parameter estimate, the state estimation problem is easily solved as in Chapter 4. Therefore, we will confine our discussion to the parameter posterior distribution estimation problem, that is, marginalizing the joint distribution with respect to the states (that have already been estimated) gives

$$\Pr(\Theta_t|Y_t) = \int \Pr(X_t, \Theta_t|Y_y) dX_t \tag{5.47}$$

and it follows that

$$\mathcal{W}(\Theta_t) \propto \frac{\Pr(\Theta_t|Y_t)}{\mathcal{I}(\Theta_t|Y_t)} \tag{5.48}$$

Assuming that a set of particles can be generated from the importance distribution as $\Theta_t \sim \mathcal{I}(\Theta_t|Y_t)$, then we have the set of normalized weights

$$\mathcal{W}_t(\Theta(i)) = \frac{W(\Theta_t(i))}{\sum_{k=1}^{N_p} W(\Theta_t(k))} \quad \text{for} \quad i = 1, \ldots, N_p \tag{5.49}$$

which is the joint "batch" importance sampling solution when coupled with the dynamic state vectors.

The sequential form of this joint formulation follows directly as in Chapter 4. We start with the factored form of the importance distribution and focus on the full posterior $\Pr(\Theta_t|Y_t)$, that is,

$$\mathcal{I}(\Theta_t|Y_t) = \prod_{k=0}^{t} \mathcal{I}(\theta(k)|\Theta_{k-1}, Y_k) \tag{5.50}$$

with $\mathcal{I}(\theta(0)|\Theta_{-1}, Y_t) \to \mathcal{I}(\theta(0)|Y_t)$.

Assuming that this factorization can be expressed recursively in terms of the previous step, $\mathcal{I}(\Theta_{t-1}|Y_{t-1})$ and extracting the tth term gives

$$\mathcal{I}(\Theta_t|Y_t) = \mathcal{I}(\theta(t)|\Theta_{t-1}, Y_t) \times \prod_{k=0}^{t-1} \mathcal{I}(\theta(k)|\Theta_{k-1}, Y_k) \tag{5.51}$$

or simply

$$\mathcal{I}(\Theta_t|Y_t) = \mathcal{I}(\theta(t)|\Theta_{t-1}, Y_t) \times \mathcal{I}(\Theta_{t-1}|Y_{t-1}) \tag{5.52}$$

With this in mind, the weight recursion becomes

$$W(\Theta_t) = W(t) \times W(\Theta_{t-1}) \tag{5.53}$$

Applying Bayes' rule to the posterior, we define

$$
\begin{aligned}
W(t) &:= \frac{\Pr(y(t)|\Theta_t, Y_{t-1}) \times \Pr(\theta(t)|\Theta(t-1))}{\Pr(y(t)|Y_{t-1}) \times \mathcal{I}(\theta(t)|\Theta_{t-1}, Y_t)} \\
&\propto \frac{\Pr(y(t)|\Theta_t, Y_{t-1}) \times \Pr(\theta(t)|\theta(t-1))}{\mathcal{I}(\theta(t)|\Theta_{t-1}, Y_t)}
\end{aligned}
\tag{5.54}
$$

As before, in the state estimation problem, we must choose an importance distribution before we can construct the algorithm. The bootstrap approach can also be selected as the importance distribution leading to a simpler alternative with $\mathcal{I}(\theta(t)|\Theta_{t-1}, Y_t) \to \Pr(\theta(t)|\theta(t-1))$ and the weight of Eq. (5.54) becomes the likelihood

$$W_{BS}(t) = \Pr(y(t)|\Theta_t, Y_{t-1}) \tag{5.55}$$

From the pragmatic perspective, we must consider some practical approaches to implement the processor for the joint problem. The first approach, when applicable, is to incorporate the random walk model when reasonable [17]. Another variant is to use the roughening model of Eq. (5.10) that "moves" the particles (after resampling) by adding a Gaussian sequence of specified variance.

This completes the discussion of the joint state/parameter estimation problem using the PF approach. Next we consider an example to illustrate the Bayesian approach.

Example 5.3 Again we consider the trajectory estimation problem [20] using the particle filter technique. At first, we applied the usual bootstrap technique and found what was expected, a collapse of the parameter particles, giving an unsatisfactory result. Next we tried the "roughening" approach, and the results were much more reasonable. We used a roughening factor or $\kappa = 5 \times 10^{-5}$ along with $N_p = 350$. The results are shown in Figure 5.4. We see both the estimated states and measurement along with the associated zero-mean/whiteness test. The result, although not as good as the modern approach, is reasonable with the final estimates converging to the static parameter values of true parameters: $\theta_1 = 0.05$ (0.034) and $\theta_2 = 0.04$ (0.039). The state and measurement estimates are quite good as evidenced by the zero-mean (0.002 < 0.061) and whiteness (1.76% out). We show the estimated posterior distributions for the states and parameters in Figure 5.5, again demonstrating a reasonable solution. Note how the distributions are initially multimodal and become unimodal as the parameter estimates converge to their true values as depicted in Figure 5.5.

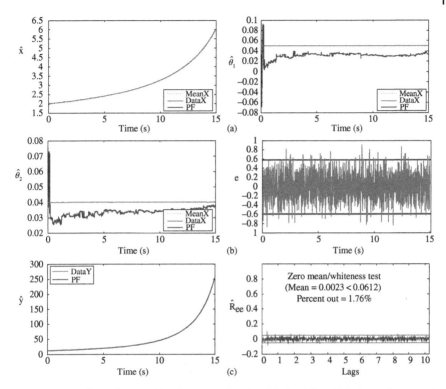

Figure 5.4 Nonlinear (uncertain) trajectory estimation: PF algorithm. (a) Estimated state and parameter no. 1. (b) Estimated parameter no. 2 and innovations. (c) Predicted measurement and zero-mean/whiteness test (0.002 < 0.061 and 1.76% out).

Next we generate an ensemble consisting of a set of a 100-member realizations, execute the PF and obtain the ensemble estimates shown in Figure 5.6. In Figure 5.6a, we see the ensemble state estimate and the corresponding joint parameter estimates in Figure 5.6b and c. Comparing these with those of Figure 5.4, we see that after the initial transients die out, they converge to reasonable estimates for both state and parameters. The ensemble processor gives reasonable zero-mean/whiteness (sanity) test results: (Z-M: 0.023 < 0.123/W-test: 1.2% out) along with relative root mean-squared errors (RMSE: state = 0.0010; parameter no. 1 = 0.0013; and parameter no. 2 = 0.0010). The corresponding *median* Kullback–Leibler divergence/Hellinger distance statistics are: (KL/HD: state = 0.024/0.077; parameter no. 1 = 1.794/0.629; parameter no. 2 = 2.201/0.642 and measurement = 0.025/0.079). These metrics also indicate a reasonable performance for both state and measurement estimates and a mediocre at best performance for the parameter estimates. □

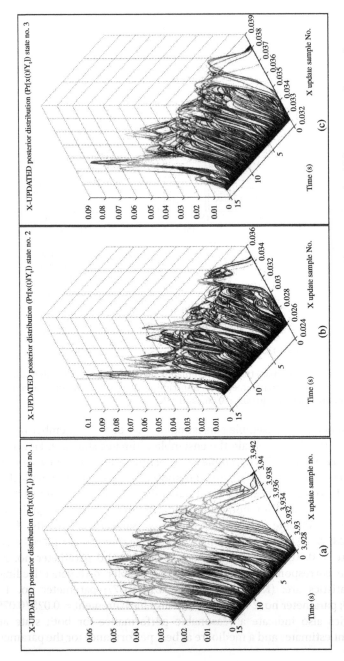

Figure 5.5 PF posterior distribution estimation. (a) Estimated state posterior. (b) Parameter no. 1 posterior. (c) Parameter no. 2 posterior.

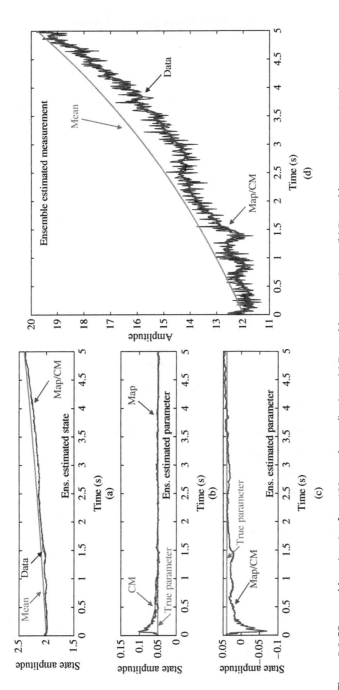

Figure 5.6 PF ensemble estimation for a 100-member realizations. (a) Ensemble state estimate. (b) Ensemble parameter no. 1 estimate. (c) Ensemble parameter no. 2 estimate. (d) Ensemble measurement estimate.

5.5 Parametrically Adaptive Processors: Linear Kalman Filter

In this section, we briefly describe the recursive prediction error (RPE) approach to model-based identification using the innovations model of Section 4.3 in parametric form. We follow the numerical optimization approach using the recursive Gauss–Newton algorithm developed by Ljung and Soderstrom [2] and available in MATLAB. This approach can also be considered "parametrically adaptive," since it consists of both embedded state and parameter estimators as shown in Figure 5.1.

Our task is to develop a parametrically adaptive version of the linear innovations state-space model. The time-invariant *innovations model* with parameter-dependent (Θ) system, input, output, and gain matrices: $A(\Theta)$, $B(\Theta)$, $C(\Theta)$, $K(\Theta)$ is given by

$$\hat{x}_\theta(t+1|t) = A(\Theta)\hat{x}_\theta(t|t-1) + B(\Theta)u(t) + K(\Theta)e_\theta(t)$$
$$\hat{y}_\theta(t|t-1) = C(\Theta)\hat{x}_\theta(t|t-1)$$
$$e_\theta(t) = y(t) - \hat{y}_\theta(t|t-1) \tag{5.56}$$

For the parametrically adaptive algorithm, we transform the basic innovations structure to its equivalent prediction form.

Substituting for the innovations and collecting like terms, we obtain the *prediction error model* directly from the representation above, that is,

$$\hat{x}_\theta(t+1|t) = [A(\Theta) - K(\Theta)C(\Theta)]\hat{x}_\theta(t|t-1) + B(\Theta)u(t) + K(\Theta)y(t)$$
$$\hat{y}_\theta(t|t-1) = C(\Theta)\hat{x}_\theta(t|t-1) \tag{5.57}$$

Comparing these models, we see that the following mappings occur:

$$A(\Theta) \longrightarrow A(\Theta) - K(\Theta)C(\Theta)$$
$$B(\Theta) \longrightarrow [B(\Theta), K(\Theta)]$$
$$C(\Theta) \longrightarrow C(\Theta) \tag{5.58}$$

Prediction error algorithms are developed from weighted quadratic criteria that evolve by minimizing the weighted sum-squared error:

$$\min_\theta \quad \mathcal{J}(\Theta) = \frac{1}{2}\sum_{k=1}^{t} e'_\theta(k)\Lambda^{-1}(\Theta)e_\theta(k) \tag{5.59}$$

for the prediction error (innovations), $e_\theta \in R^{N_y \times 1}$ and the innovations covariance (weighting) matrix, $\Lambda(\Theta) \in R^{N_y \times N_y}$. The algorithm is developed by performing a second-order Taylor series expansion about the parameter estimate, $\hat{\Theta}$, and applying the Gauss–Newton assumptions (e.g. $\hat{\Theta} \to \Theta_{\text{true}}$) leading to the basic *parametric recursion*:

$$\hat{\Theta}(t+1) = \hat{\Theta}(t) - [\nabla_{\theta\theta}\mathcal{J}(\hat{\Theta}(t))]^{-1}\nabla_\theta\mathcal{J}(\hat{\Theta}(t)) \tag{5.60}$$

where $\nabla_\theta J(\Theta)$ is the $N_\theta \times 1$ *gradient vector* and $\nabla_{\theta\theta} J(\Theta)$ is the $N_\theta \times N_\theta$ *Hessian matrix*. As part of the RPE algorithm, we must perform these operations on the predictor; therefore, we define the *gradient matrix* by

$$\Psi_\theta(t) := [\psi_{\theta_1}(t) \cdots \psi_{\theta_{N_\theta}}(t)]' \quad \text{for} \quad \Psi \in \mathcal{R}^{N_\theta \times N_y}; \quad \psi \in \mathcal{R}^{N_y \times 1} \quad (5.61)$$

Differentiating Eq. (5.59) using the chain rule,[1] we obtain the gradient

$$\nabla_\theta J(\Theta) = \nabla_\theta e'_\theta(t) \Lambda^{-1}(t) e_\theta(t) = -\Psi_\theta(t) \Lambda^{-1}(t) e_\theta(t) \quad (5.62)$$

using the gradient matrix from the definition above.

Defining the Hessian matrix as $R(t) := \nabla_{\theta\theta} J(\Theta)$ and substituting for the gradient in Eq. (5.60), we obtain the parameter vector recursion

$$\hat{\Theta}(t+1) = \hat{\Theta}(t) + \gamma(t)[R(t)]^{-1}\Psi_\theta(t)\Lambda^{-1}(t)e_\theta(t) \quad (5.63)$$

where $\gamma(t)$ is an N_θ-weighting vector sequence. However, this does give us a recursive form, but it can be transformed to yield the desired result, that is, under the Gauss–Newton approximation, it has been shown [2] that the Hessian can be approximated by a weighted sample variance estimator of the form

$$R(t) = \frac{1}{t} \sum_{k=1}^{t} \gamma(k)\psi(k)\Lambda^{-1}(k)\psi'(k) \quad (5.64)$$

which is easily placed in recursive form to give

$$R(t) = R(t-1) + \gamma(t)[\psi(t)\Lambda^{-1}(t)\psi'(t) - R(t-1)] \quad (5.65)$$

Following the same arguments, the prediction error covariance can be placed in recursive form as well to give

$$\Lambda(t) = \Lambda(t-1) + \gamma(t)[e_\theta(t)e'_\theta(t) - \Lambda(t-1)] \quad (5.66)$$

The $N_\theta \times N_y$-*gradient matrix* in terms of our innovations model is given by

$$\Psi_\theta(t) = \nabla_\theta e_\theta(t) = \nabla_\theta \hat{y}'_\theta(t|t-1)$$
$$= \nabla_\theta [C(\Theta)\hat{x}_\theta(t|t-1)]' \text{ and } \Psi_\theta \in \mathcal{R}^{N_\theta \times N_y} \quad (5.67)$$

A matrix expansion of the gradient matrix with operations developed component-wise is

$$\Psi_\theta(t) = \nabla_\theta [C(\Theta)\hat{x}_\theta(t|t-1)]'$$

$$= \begin{bmatrix} \frac{\partial}{\partial\theta_1} \\ \vdots \\ \frac{\partial}{\partial\theta_{N_\theta}} \end{bmatrix} [c'_1(\theta)\hat{x}_\theta(t|t-1) \cdots c'_{N_y}(\theta)x_\theta(t|t-1)]$$

---

[1] The chain rule from vector calculus is defined by $\nabla_\Theta(a'b) := (\nabla_\Theta a')b + (\nabla_\Theta b')a \quad a, b \in R^{N_\theta \times 1}$ with a, b functions of Θ.

for $c_i' \in \mathcal{R}^{1 \times N_x}$ and $\hat{x}_\theta \in \mathcal{R}^{N_x \times 1}$ which can be expressed in a succinct notation as (see [10] for details)

$$
\begin{aligned}
\Psi_\theta(t) &= \nabla_\theta [C(\Theta)\hat{x}_\theta(t|t-1)]' \\
&= [\nabla_\theta C(\theta)]\hat{x}_\theta(t|t-1) + C(\theta)[\nabla_\theta \hat{x}_\theta(t|t-1)]
\end{aligned}
\tag{5.68}
$$

Define the following terms, which will ease the notational burden:

$$
\mathcal{W}_\theta(t) := \nabla_\theta \hat{x}_\theta(t|t-1) \text{ and } \mathcal{Z}_\theta(\hat{x}) := \nabla_\theta [C(\Theta)]\hat{x}_\theta(t|t-1)
\tag{5.69}
$$

for $\mathcal{W}_\theta \in \mathcal{R}^{N_x \times N_\theta}$, the *state predictor gradient weighting matrix* and $\mathcal{Z}_\theta \in \mathcal{R}^{N_y \times N_\theta}$, the *measurement gradient weighting matrix.*

Thus, in terms of these definitions, we have

$$
\Psi_\theta'(t) = C(\Theta)\mathcal{W}_\theta(t) + \mathcal{Z}_\theta(\hat{x})
\tag{5.70}
$$

and therefore

$$
\nabla_\theta e_\theta(t) = \nabla_\theta [y(t) - C(\Theta)\hat{x}_\theta(t|t-1)] = -\Psi_\theta'(t)
\tag{5.71}
$$

Next we must develop an expression for $\mathcal{W}_\theta$ by differentiating the predictor of Eq. (5.56), that is,

$$
\begin{aligned}
\mathcal{W}_\theta(t+1) &= \nabla_\theta \hat{x}_\theta(t+1|t) \\
&= \nabla_\theta [A(\Theta)\hat{x}_\theta(t|t-1) + B(\Theta)u(t) + K(\Theta)e_\theta(t)]
\end{aligned}
\tag{5.72}
$$

Performing component-wise derivatives, we differentiate the predictor as

$$
\begin{aligned}
\mathcal{W}_{\theta_i}(t+1) &:= \frac{\partial}{\partial \theta_i} \hat{x}_\theta(t+1|t) \\
&= \frac{\partial}{\partial \theta_i} [A(\theta)\hat{x}_\theta(t|t-1) + B(\theta)u(t) + K(\theta)e_\theta(t)] \\
&= \frac{\partial}{\partial \theta_i} [A(\theta)\hat{x}_\theta(t|t-1) + B(\theta)u(t) + K(\theta)y(t) - K(\theta)C(\theta)\hat{x}_\theta(t|t-1)]
\end{aligned}
\tag{5.73}
$$

Using our definitions for $\mathcal{W}_\theta$ and e_θ, we have

$$
\begin{aligned}
\mathcal{W}_{\theta_i}(t+1) = {}& [A(\theta) - K(\theta)C(\theta)]\mathcal{W}_{\theta_i}(t) - K(\theta)\left(\left[\frac{\partial c_i'(\theta)}{\partial \theta_i} \right] \hat{x}_\theta(t|t-1) \right) \\
& + \left(\left[\frac{\partial A(\theta)}{\partial \theta_i} \right] \hat{x}_\theta(t|t-1) + \left[\frac{\partial B(\theta)}{\partial \theta_i} \right] u(t) + \left[\frac{\partial K(\theta)}{\partial \theta_i} \right] e_\theta(t) \right)
\end{aligned}
\tag{5.74}
$$

Define the $N_x \times N_\theta$-matrix,

$$
\mathcal{U}_\theta(\hat{x}_\theta, u, e_\theta) := [\nabla_\theta A(\Theta)]\hat{x}_\theta(t|t-1) + [\nabla_\theta B(\Theta)]u(t) + [\nabla_\theta K(\Theta)]e_\theta(t)
\tag{5.75}
$$

Substituting component-wise into Eq. (5.74) for each of the definitions ($\mathcal{W}_{\theta_i}$, $\mathcal{Z}_{\theta_i}$, $\mathcal{U}_{\theta_i}$), we obtain

$$\mathcal{W}_{\theta_i}(t+1) = [A(\theta) - K(\theta)C(\theta)]\mathcal{W}_{\theta_i}(t) + \mathcal{U}_{\theta_i}(\hat{x}_\theta, u, e_\theta)$$

$$- K(\theta)\mathcal{Z}_{\theta_i}(\hat{x}_\theta(t|t-1)) \tag{5.76}$$

Now expanding over $i = 1, \dots, N_\theta$, we obtain the *recursion* for the state predictor gradient weighting matrix as

$$\mathcal{W}_\theta(t+1) = [A(\Theta) - K(\Theta)C(\Theta)]\mathcal{W}_\theta(t) + \mathcal{U}_\theta(\hat{x}_\theta(t|t-1), u, e_\theta)$$

$$- K(\Theta)\mathcal{Z}_\theta(\hat{x}_\theta(t|t-1)) \tag{5.77}$$

which completes the RPE algorithm applied to the innovations model. We summarize the algorithm in Table 5.2 using the simplified notation for clarity. That is, we use the Gauss–Newton approximations as before: $x_\theta(t) \approx x(t)$,

Table 5.2 Parametrically adaptive innovations model (RPE) algorithm.

Prediction Error

$e(t) = y(t) - \hat{y}(t|t-1)$	(Prediction Error)
$\Lambda(t) = \Lambda(t-1) + \gamma(t)[e(t)e'(t) - \Lambda(t-1)]$	(Prediction Error Covariance)
$R(t) = R(t-1) + \gamma(t)[\psi(t)\Lambda^{-1}(t)\psi'(t) - R(t-1)]$	(Hessian)

Parameter Estimation

$\hat{\Theta}(t+1) = \hat{\Theta}(t) + \gamma(t)R^{-1}(t)\psi(t)\Lambda^{-1}(t)e(t)$	(Parameter Update)

State Prediction

$\hat{x}(t+1|t) = A(\Theta)\hat{x}(t|t-1) + B(\Theta)u(t) + K(\Theta)e(t)$	(State)
$\hat{y}(t|t-1) = C(\Theta)\hat{x}(t|t-1)$	(Measurement)

Gradient Prediction

$\mathcal{W}(t) = [A(\Theta) - K(\Theta)C(\Theta)]\mathcal{W}(t) + \mathcal{U}(\hat{x}, u, e) - K(\Theta)\mathcal{Z}(\hat{x})$	(Weight)
$\psi'(t) = C(\Theta)\mathcal{W}(t) + \mathcal{Z}(\hat{x})$	(Gradient)

for $\mathcal{W}(t) := \nabla_\theta \hat{x}(t|t-1)$, $\mathcal{Z}(\hat{x}) := \nabla_\theta[C(\Theta)\hat{x}(t|t-1)]$ and
$\mathcal{U}(\hat{x}, u, e) := [\nabla_\theta A(\Theta)]\hat{x}(t|t-1) + [\nabla_\theta B(\Theta)]u(t) + [\nabla_\theta K(\Theta)]e(t)$

Weighting Vector Sequence

$\gamma(t)$

Initial Conditions

$\hat{x}(0|0)$ $\Theta(0)$ $\Lambda(0)$ $R(0)$

$\hat{y}_\theta(t) \approx \hat{y}(t)$, $\Psi_\theta(t) \approx \psi(t)$ and $e_\theta(t) \approx e(t)$. We now apply this algorithm to the RC-circuit problem.

Example 5.4 Consider the RC-circuit problem of Section 4.3 with unknown parameters, $[A, C]$, that is, the underlying Gauss–Markov model has the form

$$x(t) = \Theta_1 x(t-1) + u(t-1) + w(t-1)$$

$$y(t) = \Theta_2 x(t) + v(t)$$

The corresponding innovations model is therefore

$$\hat{x}(t+1|t) = \Theta_1 \hat{x}(t|t-1) + u(t) + \Theta_3 e(t)$$

$$\hat{y}(t|t-1) = \Theta_2 \hat{x}(t|t-1)$$

We seek the adaptive processor that provides joint estimates of both the unknown state and parameters. We applied the RPEM algorithm available in MATLAB [10] to the simulated data with $R_{ww} = 10^{-4}, R_{vv} = 4, x(0) = 2.5, P(0) = 10^{-6}$ to obtain the adaptive solution. We show the noisy measurement data (thin dashed line), true measurement (dashed line), and predicted (filtered) measurement. The processor is able to produce a good estimate of the true measurement tracking it closely (see Figure 5.7a). The corresponding innovations sequence is shown in Figure 5.7b with 5.5% of the samples exceeding the predicted bounds. The performance is validated by investigating the statistics of the innovations sequence. The zero-mean/whiteness test ($0.04 < 0.17/2.35\%$ out) indicates a converged processor as shown in Figure 5.7c.

Thus, we see that the adaptive innovations processor is capable of estimating the unknown parameters (5% initial parametric error) with the following final parameter estimates:

$$\Theta_1 = 0.970 \pm 0.003 \quad \text{(Process Matrix)}$$

$$\Theta_2 = 1.987 \pm 0.140 \quad \text{(Measurement Matrix)}$$

$$\Theta_3 = -0.015 \pm 0.011 \quad \text{(Gain Matrix)}$$

$$\hat{R}_{ee} = 3.371$$

This completes the example. □

In practice, the innovations model is the usual approach taken for the linear parametrically adaptive state-space processors. This completes the section on parametrically adaptive innovations model. Next we investigate a case study on the synthetic aperture tracking of an underwater source.

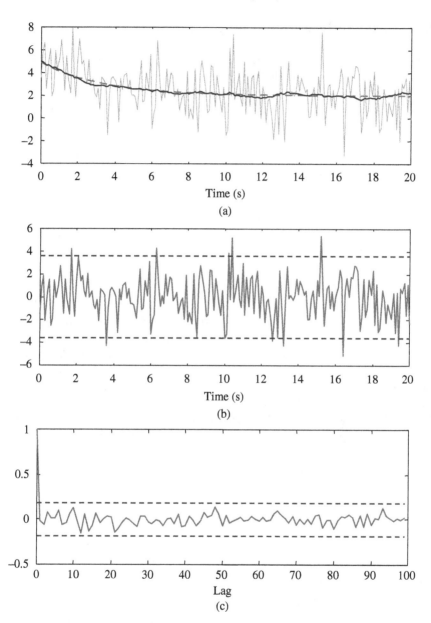

Figure 5.7 Parametrically adaptive innovations state-space processor: RC-circuit problem. (a) Noisy (dashed), true (thick dashed), and predicted (solid) measurements. (b) Innovations sequence (5.5% out). (c) Zero-mean/whiteness test (0.04 < 0.17/2.35% out).

5.6   Case Study: Random Target Tracking

Synthetic aperture processing is well known in airborne radar, but not as familiar in sonar [26–31]. The underlying idea in creating a synthetic aperture is to increase the array length by motion, thereby increasing the spatial resolution (bearing) and gain in SNR. It has been shown that for stationary targets the motion-induced bearing estimates have smaller variances than that of a stationary array [29, 32]. Here we investigate the case of *both* array and target motion. We define the acoustic *array space–time processing problem* as follows:

GIVEN a set of noisy pressure-field measurements from a horizontally towed array of L-sensors in motion, FIND the "best" (minimum error variance) estimate of the target bearings.

We use the following nonlinear pressure-field measurement model for M-monochromatic plane wave targets characterized by a corresponding set of temporal frequencies, bearings, and amplitudes, $[\{\omega_m\}, \{\theta_m\}, \{a_m\}]$ given by

$$p(x, t_k) = \sum_{m=1}^{M} a_m e^{j\omega_m t_k - j\beta(x,t_k)\sin\theta_m} + n(t_k) \tag{5.78}$$

where $\beta(x, t_k) := k_o x(t_o) + v t_k$, $k_o = \frac{2\pi}{\lambda_o}$ is the wavenumber, $x(t_k)$ is the current spatial position along the x-axis in meters, v is the tow speed in meter per second, and $n(t_k)$ is the additive random noise. The inclusion of motion in the *generalized* wavenumber, β, is critical to the improvement of the processing since the synthetic aperture effect is actually created through the motion itself and not simply the displacement.

If we further assume that the single sensor equation above is expanded to include an array of L-sensors, $x \to x_\ell$, $\ell = 1, \ldots, L$, then we obtain

$$p(x_\ell, t_k) = \sum_{m=1}^{M} a_m e^{j\omega_m t_k - j\beta(x_\ell, t_k)\sin\theta_m} + n_\ell(t_k)$$

$$= \sum_{m=1}^{M} a_m \cos(\omega_m t_k - \beta(x_\ell, t_k)\sin\theta_m) + n_\ell(t_k) \tag{5.79}$$

Since our hydrophone sensors measure the real part of the complex pressure-field, the final nonlinear measurement model of the system can be written in compact vector form as

$$\mathbf{p}(t_k) = \mathbf{c}[t_k; \Theta] + \mathbf{n}(t_k) \tag{5.80}$$

where $\mathbf{p}, \mathbf{c}, \mathbf{n} \in C^{L\times 1}$, are the respective pressure-field, measurement, and noise vectors and $\Theta \in \mathcal{R}^{M\times 1}$ represents the target bearings. The corresponding vector measurement model

$$\mathbf{c}_\ell(t_k; \Theta) = \sum_{m=1}^{M} a_m \cos(\omega_m t_k - \beta(x_\ell, t_k)\sin\theta_m) \quad \text{for } \ell = 1, \ldots, L$$

Since we model the bearings as a random walk emulating random target motion, then the Markovian state-space model evolves from first differences as

$$\Theta(t_k) = \Theta(t_{k-1}) + \mathbf{w}_\theta(t_{k-1}) \tag{5.81}$$

Thus, the state-space model is linear with no explicit dynamics; therefore, the process matrix $A = I$ (identity) and the relations are greatly simplified.

Now let us see how a particle filter using the bootstrap approach can be constructed according to the generic algorithm of Table 4.9. For this problem, we assume the additive noise sources are Gaussian, so we can compare results to the performance of the approximate processor. We define the discrete notation, $t_{k+1} \to t+1$ for the sampled-data representation.

Let us cast this problem into the sequential Bayesian framework; that is, we would like to estimate the *instantaneous* posterior filtering distribution, $\hat{\mathrm{Pr}}(x(t)|Y_t)$, using the PF representation to be able to perform inferences and extract the target bearing estimates. Therefore, we have that the transition probability is given by $(\Theta(t) \to x(t))$

$$\mathrm{Pr}(\theta(t)|\theta(t-1)) \sim \mathcal{N}(\Theta(t) : \mathbf{a}[\Theta(t-1)], \mathbf{R}_{w_\theta,w_\theta})$$

or in terms of our state transition (bearings) model, we have

$$\Theta(t) = \mathbf{a}[\Theta(t-1)] + w_\theta(t-1) = \Theta(t-1)$$
$$+ \mathbf{w}_\theta(t-1) \text{ for } \mathrm{Pr}(\mathbf{w}_\theta(t)) \sim \mathcal{N}(0, \mathbf{R}_{w_\theta w_\theta})$$

The corresponding likelihood is specified in terms of the measurement model $(y(t) \to p(t))$ as

$$\mathrm{Pr}(y(t)|x(t)) \sim \mathcal{N}(\mathbf{y}(t) : \mathbf{c}[\Theta(t)], \mathbf{R}_{vv}(t))$$

where we have used the notation: $z \sim \mathcal{N}(z : m_z, R_{zz})$ to specify the Gaussian distribution in random vector z. For our problem, we have that

$$\ln \mathrm{Pr}(y(t)|x(t)) = \kappa - \frac{1}{2}(\mathbf{y}(t) - \sum_{m=1}^{M} a_m \cos(\omega_m t - \beta(t) \sin \theta_m))' \mathbf{R}_{vv}^{-1}$$

$$(\mathbf{y}(t) - \sum_{m=1}^{M} a_m \cos(\omega_m t - \beta(t) \sin \theta_m))$$

with κ a constant, $\beta \in \mathcal{R}^{L\times1}$ and $\beta(t) := [\beta(x_1, t) \mid \cdots \mid \beta(x_L, t)]'$, the dynamic wavenumber expanded over the array. Thus, the PF SIR algorithm becomes

- Draw samples (particles) from the state transition distribution:

$$\Theta_i(t) \sim \mathcal{N}(\Theta(t) : \mathbf{a}[\Theta(t-1)], \mathbf{R}_{w_\theta,w_\theta})$$

$$\Theta_i(t) = \Theta_i(t-1) + w_{\theta_i}(t-1), \, w_{\theta_i}(t) \sim \mathrm{Pr}(w(t)) \sim \mathcal{N}(0, R_{w_{\theta_i} w_{\theta_i}});$$

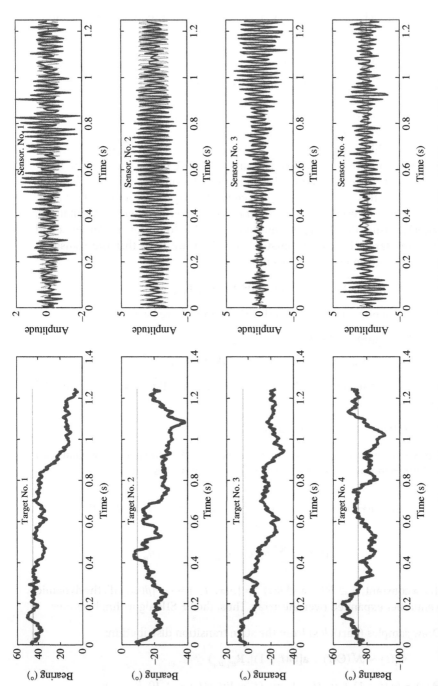

Figure 5.8 Synthetic aperture sonar tracking problem: simulated target motion from initial bearings of 45°, −10°, 5° and −75° and array measurements (−10 dB SNR).

- Estimate the likelihood, $\Pr(\mathbf{y}(t)|\Theta(t)) \sim \mathcal{N}(\mathbf{y}(t) : \mathbf{c}[\Theta(t)], \mathbf{R}_{vv}(t))$ with $c_\ell(t;\Theta_i) = \sum_{m=1}^{M} a_m \cos(\omega_m t_k - \beta(x_\ell, t) \sin \Theta_{m,i}(t))$ for $\ell = 1, \cdots, L$ and $\Theta_{m,i}$ the ith particle at the mth bearing angle.
- Update and normalize the weight: $\mathcal{W}_i(t) = W_i(t) / \sum_{i=1}^{N_p} W_i(t)$
- Resample: $\hat{N}_{\text{eff}}(t) \leq N_{\text{thresh}}$.
- Estimate the instantaneous posterior: $\hat{\Pr}(\Theta(t)|Y_t) \approx \sum_{i=1}^{N_p} \mathcal{W}_i(t)\delta(\Theta(t) - \Theta_i(t))$.
- Perform the inference by estimating the corresponding statistics:

$$\hat{\Theta}_{\text{map}}(t) = \arg \max \hat{\Pr}(\Theta(t)|Y_t);$$

$$\hat{\Theta}_{\text{mmse}}(t) = \hat{\Theta}_{\text{cm}}(t) = E\{\Theta(t)|Y_t\} = \sum_{i=1}^{N_p} \Theta_i(t)\hat{\Pr}(\Theta(t)|Y_t);$$

$$\hat{\Theta}_{\text{median}}(t) = \text{median}(\hat{\Pr}(\Theta(t)|Y_t))).$$

Consider the following simulation of the synthetic aperture using a four-element, linear towed array with "moving" targets using the following parameters:

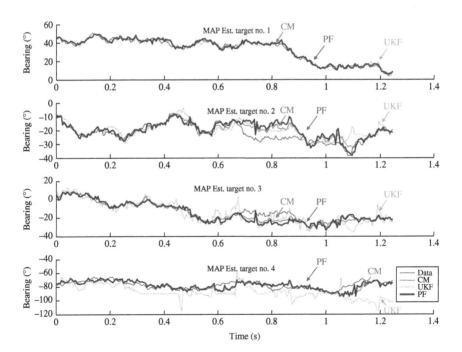

Figure 5.9 Particle filter bearing estimates for four targets in random motion: PF bearing (state) estimates and simulated target tracks (UKF, conditional mean, MAP).

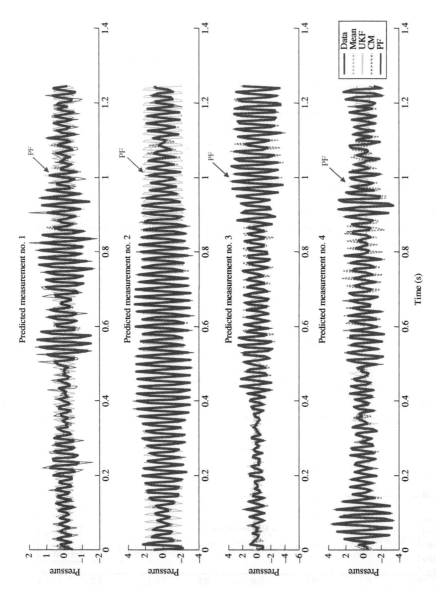

Figure 5.10 Particle filter predicted measurement estimates for four channel hydrophone sensor array.

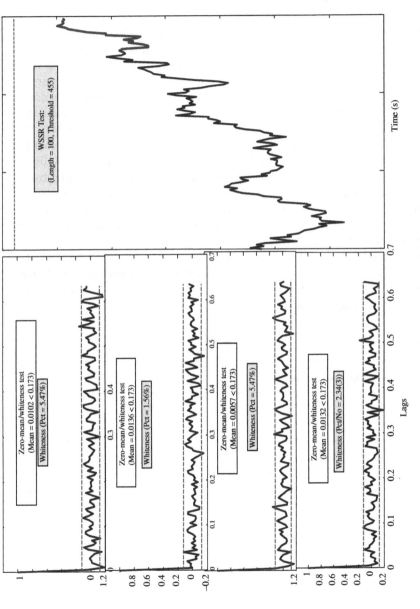

Figure 5.11 Particle filter classical "sanity" performance metrics: zero-mean/whiteness tests for 45°, −10°, 5° and 75° targets as well as the corresponding $\rho(k)$ test.

Target: Unity amplitudes with temporal frequency is 50 Hz, wavelength = 30 m, tow speed = 5 m s^{-1} *Array*: four (4) element linear towed array with 15 m spacing; *Particle filter*: N_θ = 4 states (bearings), N_y = 4 sensors, N = 250 samples, N_p = 250 particles/weights; SNR: is -10 dB; *Noise*: white Gaussian with R_{ww} = diag [2.5], R_{vv} = diag [0.1414]; *Sampling interval* is 0.005 second; *Initial conditions*: *Bearings*: $\Theta_0 = [45° - 10°5° - 75°]'$; *Covariance*: P_0 = diag (10^{-10}).

The array simulation was executed and the targets moved according to a random walk specified by the process noise and sensor array measurements with -10 dB SNR. The results are shown in Figure 5.8, where we see the noisy synthesized bearings (left) and four (4) noisy sensor measurements at the moving array. The bearing (state) estimates are shown in Figure 5.9 where we observe the targets making a variety of course alterations. The PF is able to track the target motions quite well while we observe the unscented Kalman filter (UKF) [10] is unable to respond quickly enough and finally losing track completely for target no. 4. It should be noted that target no. 2 and target no. 4 "crossover" between 0.8 and 1.0 second. The PF loses these tracks during this time period getting them confused, but recovers by the one-second time step. Both the MAP and CM estimates using the estimated posterior provide excellent tracking. Note that these bearing inputs would provide the raw data for an XY-tracker [10]. The PF estimated or filtered measurements are shown in Figure 5.10. As expected the PF tracks the measurement data quite well, while the UKF is again in small error.

Table 5.3 PF Performance towed array problem.

Particle filter performance results

Parameter	RMSE	Median KLD	Median HD
Bearing no. 1	0.117	0.528	0.342
Bearing no. 2	0.292	0.180	0.208
Bearing no. 3	0.251	0.179	0.209
Bearing no. 4	0.684	0.406	0.309
Measurements		Median KLD	Median HD
Sensor no. 1		0.066	0.139
Sensor no. 2		0.109	0.166
Sensor no. 3		0.259	0.275
Sensor no. 4		0.312	0.327
Innovations	Zero-mean < 0.173	Whiteness	WSSR < 455
Sensor no. 1	0.0102	5.47%	Below
Sensor no. 2	0.0136	1.56%	Below
Sensor no. 3	0.0057	5.47%	Below
Sensor no. 4	0.0132	2.34%	Below

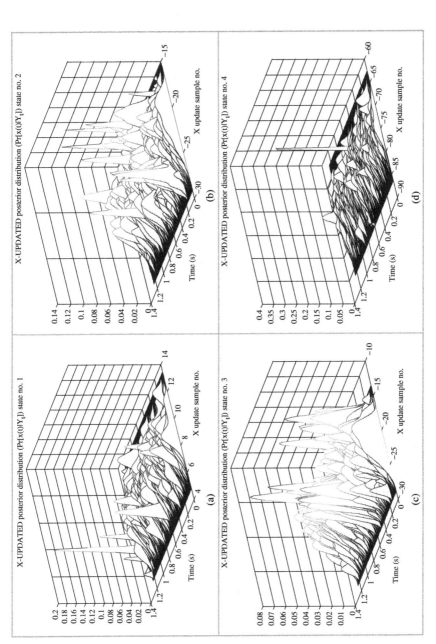

Figure 5.12 Particle filter instantaneous posterior bearing estimates: (a) 45° target no. 1 posterior; (b) −10° target no. 2 posterior; (c) 5° target no. 3; and (d) −75° target no. 4.

For performance analysis, we applied the usual "sanity tests." From this perspective, the PF processor works well, since each measurement channel is zero-mean and white with the WSSR lying below the threshold, indicating white innovations sequences demonstrating the tracking ability of the PF processor at least in a classical sense [10] (see Figure 5.11 and Table 5.3). The corresponding normalized root mean-squared errors RMSE, *median* Kullback–Leibler divergence/Hellinger distance statistics for the bearing estimates are also shown in the table along with the measurements at each sensor. Estimating the KD/HD metrics for the measurements indicate a reasonable performance for both bearings and measurements.

The instantaneous posterior distributions for the bearing estimates are shown in Figure 5.12. Here we see the Gaussian nature of the bearing estimates generated by the random walk. Clearly, the PF performs quite well for this problem. Note also the capability of using the synthetic aperture, since we have only a four-element sensor array, yet we are able to track four targets. Linear array theory implies with a static array that we should only be able to track $L - 1 =$ three targets!

In this case study, we have applied the unscented Kalman filter and the bootstrap PF to an ocean acoustic, synthetic aperture, towed array, target tracking problem to evaluate the performance of both the parametrically adaptive UKF and particle filtering techniques. The results are quite reasonable on this simulated data set.

5.7 Summary

In this chapter, we have discussed the development of joint Bayesian state/parametric processors. Starting with a brief introduction, we defined a variety of problems based on the joint posterior distribution $\Pr(x(t), \theta(t)|Y_t)$ and its decomposition. We focused on the joint problem of estimating *both* states and parameters simultaneously (on-line) – a common problem of high interest. We then briefly showed that all that is necessary for this problem is to define an "augmented" state consisting of the original state variables along with the unknown parameters typically modeled by a random walk, when a dynamic parametric model is not available. This casts the joint problem into an optimal filtering framework. We then showed how this augmentation leads to a decomposition of the classical EKF processor and developed the "decomposed" form for illustrative purposes. The algorithm was implemented by executing the usual processor with the new augmented state vector embedded. We also extended this approach to both modern "unscented" and "particle-based" processors, again only requiring the state augmentation procedure to implement. All of the processors required a random walk parametric model to function, while the particle filters could be implemented using the "roughening" (particle random walks) to track the parameters effectively. Next we discussed the recursive prediction error approach (RPE) and developed the solution for a specific linear, time-invariant, state-space model – the innovations model that will prove useful in subsequent subspace realization chapters to follow. Besides applying the nonlinear processors to the parametrically uncertain nonlinear trajectory estimation problem, we developed a case study for a synthetic aperture towed array and compared the modern UKF to the PF processors to complete the chapter [34–46].

MATLAB Notes

MATLAB has a variety of parametrically adaptive algorithms available in the Controls/Identification toolboxes. The linear Kalman filter (**Kalman**), extended Kalman filter (**extendedKalmanFilter**), and unscented Kalman filter (**unscentedKalmanFilter**) algorithms are available or as a set of commands: (**predict**) and (**correct**). The Identification toolbox has an implementation of the recursive prediction error method for a variety of model sets. The state-space algorithms available are **N4SID** in the **ident** toolbox as well as the **ekf**, **ukf**, and **pf** of the **control** toolbox.

References

1 Ljung, L.J. (1999). *System Identification: Theory for the User*, 2e. Englewood Cliffs, NJ: Prentice-Hall.

2 Ljung, L.J. and Soderstrom, T. (1983). *Theory and Practice of Recursive Identification*. Boston, MA: MIT Press.

3 Soderstrom, T. and Stoica, P. (1989). *System Identification*. Englewood Cliffs, NJ: Prentice-Hall.

4 Norton, J.P. (1986). *An Introduction to Identification*. New York: Academic Press.

5 Liu, J. (2001). *Monte Carlo Strategies in Scientific Computing*. New York: Springer.

6 Cappe, O., Moulines, E., and Ryden, T. (2005). *Inference in Hidden Markov Models*. New York: Springer.

7 Liu, J. and West, M. (2001). Combined parameter and state estimation in simulation-based filtering. In: *Sequential Monte Carlo Methods in Practice*, Chapter 10 (ed. A. Doucet, N. de Freitas, and N. Gordon), 197–223. New York: Springer.

8 Doucet, A., de Freitas, N., and Gordon, N. (2001). *Sequential Monte Carlo Methods in Practice*. New York: Springer.

9 Godsill, S. and Djuric, P. (2002). Special issue: Monte Carlo methods for statistical signal processing. *IEEE Trans. Signal Process.* 50: 173–499.

10 Candy, J. (2016). *Bayesian Signal Processing: Classical, Modern and Particle Filtering Methods*, 2e. Englewood Cliffs, NJ: Wiley/IEEE Press.

11 Cappe, O., Godsill, S., and Moulines, E. (2007). An overview of existing methods and recent advances in sequential Monte Carlo. *Proc. IEEE* 95 (5): 899–924.

12 Kitagawa, G. and Gersch, W. (1997). *Smoothness Priors Analysis of Time Series*. New York: Springer.

13 Kitagawa, G. (1998). Self-organizing state space model. *J. Am. Stat. Assoc.* 97 (447): 1207–1215.

14 van der Merwe, R., Doucet, A., de Freitas, N., and Wan, E. (2000). The unscented particle filter. In: *Advances in Neural Information Processing Systems*, vol. 16. Cambridge, MA: MIT Press, Cambridge University Engineering Technical Report, CUED/F-INFENG/TR 380.

15 Gordon, N., Salmond, D., and Smith, A. (1993). A novel approach to non-linear non-Gaussian Bayesian state estimation. *IEE Proc. F* 140: 107–113.

16 Haykin, S. and de Freitas, N. (2004). Special issue: sequential state estimation: from Kalman filters to particle filters. *Proc. IEEE* 92 (3): 399–574.

17 Andrieu, C., Doucet, A., Singh, S., and Tadic, V. (2004). Particle methods for change detection, system identification and control. *Proc. IEEE* 92 (6): 423–468.

18 Haykin, S. (2001). *Kalman Filtering and Neural Networks*. New York: Wiley.

19 Simon, D. (2006). *Optimal State Estimation: Kalman H_∞ and Nonlinear Approaches*. Englewood Cliffs, NJ: Wiley/IEEE Press.

20 Jazwinski, A. (1970). *Stochastic Processes and Filtering Theory*. New York: Academic Press.

21 van der Merwe, R. (2004). Sigma-point Kalman filters for probabilistic inference in dynamic state-space models. PhD dissertation. OGI School of Science & Engineering, Oregon Health & Science University.

22 Nelson, A. (2000). Nonlinear estimation and modeling of noisy time-series by Dual Kalman filtering methods. PhD dissertation. OGI School of Science & Engineering, Oregon Health & Science University.

23 Gustafsson, F. (2001). *Adaptive Filtering and Change Detection*. Englewood Cliffs, NJ: Wiley/IEEE Press.

24 Ljung, L. (1979). Asymptotic behavior of the extended Kalman filter as a parameter estimator for linear systems. *IEEE Trans. Autom. Control* AC-24: 36–50.

25 Vermaak, J., Andrieu, C., Doucet, A., and Godsill, S. (2002). Particle methods for Bayesian modeling and enhancement of speech signals. *IEEE Trans. Speech Audio Process.* 10 (3): 173–185.

26 Williams, R. (1976). Creating an acoustic synthetic aperture in the ocean. *J. Acoust. Soc. Am.* 60: 60–73.

27 Yen, N. and Carey, W. (1976). Applications of synthetic aperture processing to towed array data. *J. Acoust. Soc. Am.* 60: 764–775.

28 Stergiopoulus, S. and Sullivan, E. (1976). Extended towed array processing by an overlap correlator. *J. Acoust. Soc. Am.* 86: 764–775.

29 Sullivan, E., Carey, W., and Stergiopoulus, S. (1992). Editorial in special issue on acoustic synthetic aperture processing. *IEEE J. Ocean. Eng.* 17: 1–7.

30 Ward, D., Lehmann, E., and Williamson, R. (2003). Particle filtering algorithm for tracking and acoustic source in a reverberant environment. *IEEE Trans. Speech Audio Process.* 11 (6): 826–836.

31 Orton, M. and Fitzgerald, W. (2002). Bayesian approach to tracking multiple targets using sensor arrays and particle filters. *IEEE Trans. Signal Process.* 50 (2): 216–223.

32 Sullivan, E. and Candy, J. (1997). Space-time array processing: the model-based approach. *J. Acoust. Soc. Am.* 102 (5): 2809–2820.

33 Gelb, A. (1975). *Applied Optimal Estimation*. Boston, MA: MIT Press.

34 Candy, J. (2007). Bootstrap particle filtering. *IEEE Signal Process. Mag.* 24 (4): 73–85.

35 Rajan, J., Rayner, P., and Godsill, S. (1997). Bayesian approach to parameter estimation and interpolation of time-varying autoregressive processes using the Gibbs sampler. *IEE Proc.-Vis. Image Signal Process.* 144 (4): 249–256.

36 Polson, N., Stroud, J., and Muller, P. (2002). Practical Filtering with Sequential Parameter Learning. University Chicago Technical Report, 1–18.

37 Andrieu, C., Doucet, A., Singh, S., and Tadic, V. (2004). Particle methods for change detection, system identification and control. *Proc. IEEE* 92 (3): 423–438.

38 Storvik, G. (2002). Particle filters in state space models with the presence of unknown static parameters. *IEEE Trans. Signal Process.* 50 (2): 281–289.

39 Djuric, P. (2001). Sequential estimation of signals under model uncertainty. In: *Sequential Monte Carlo Methods in Practice*, Chapter 18 (ed. A. Doucet, N. de Freitas, and N. Gordon), 381–400. New York: Springer.

40 Lee, D. and Chia, N. (2002). A particle algorithm for sequential Bayesian parameter estimation and model selection. *IEEE Trans. Signal Process.* 50 (2): 326–336.

41 Doucet, A. and Tadic, V. (2003). Parameter estimation in general state-space models using particle methods. *Ann. Inst. Stat. Math.* 55 (2): 409–422.

42 Andrieu, C., Doucet, A., and Tadic, V. (2005). On-line parameter estimation in general state-space models. In: *Proceedings of the 44th IEEE Conference on Decision and Control*, 332–337.

43 Schoen, T. and Gustafsson, F. (2003). Particle Filters for System Identification of State-Space Models Linear in Either Parameters or States. Linkoping University Report, LITH-ISY-R-2518.

44 Faurre, P.L. (1976). Stochastic realization algorithms. In: *System Identification: Advances and Case Studies* (ed. R. Mehra and D. Lainiotis), 1–23. New York: Academic Press.

45 Tse, E. and Wiennert, H. (1979). Structure identification for multivariable systems. *IEEE Trans. Autom. Control* AC-24: 36–50.

46 Bierman, G. (1977). *Factorization Methods of Discrete Sequential Estimation*. New York: Academic Press.

Problems

5.1 Suppose we are given the following innovations model (in steady state):

$$\hat{x}(t) = a\hat{x}(t-1) + ke(t-1)$$

$$y(t) = c\hat{x}(t) + e(t)$$

where $e(t)$ is the zero-mean, white innovations sequence with covariance, R_{ee}.

(a) Derive the Wiener solution using the spectral factorization method of Section 2.4.

(b) Develop the linear steady-state KF for this model.

(c) Develop the parametrically adaptive processor to estimate k and R_{ee}.

5.2 As stated in the chapter, the EKF convergence can be improved by incorporating a gain gradient term in the system Jacobian matrices, that is,

$$A_\theta^*[x, \theta] := A_\theta[x, \theta] + [\nabla_\theta K_x(\Theta)]e(t) \quad \text{for} \quad \mathcal{K} := [K_x | K_\theta]$$

(a) By partitioning the original $N_x \times N_\theta$ Jacobian matrix, $A_\theta[x, \theta]$, derive the general "elemental" recursion, that is, show that

$$A_\theta^*[i, \ell] = \nabla_{\theta_\ell} a_i[x, \theta] + \sum_{j=1}^{N_y} \nabla_{\theta_\ell} k_x(i, j)e_j(t); i = 1, \dots, N_x;$$

$$\ell = 1, \dots, N_\theta$$

(b) Suppose we would like to implement this modification, does there exist a numerical solution that could be used? If so, describe it.

5.3 Using the following *scalar* Gauss–Markov model

$$x(t) = Ax(t-1) + w(t-1)$$

$$y(t) = C\hat{x}(t) + v(t)$$

with the usual zero-mean, R_{ww} and R_{vv} covariances.

(a) Let $\{A, C, K, R_{ee}\}$ be scalars. Develop the EKF solution to estimate A from noisy data.

(b) Can these algorithms be combined to "tune" the resulting hybrid processor?

5.4 Suppose we are given the following structural model:

$$m\ddot{x}(t) + c\dot{x} + kx(t) = p(t) + w(t)$$
$$y(t) = \beta x(t) + v(t)$$

with the usual zero-mean, R_{ww} and R_{vv} covariances.

(a) Convert this model into discrete-time using first differences. Using central difference, create the discrete Gauss–Markov model. (*Hint:* $\ddot{x}(t) \approx \frac{x(t) - 2x(t-1) + x(t-2)}{\Delta_t^2}$.)

(b) Suppose we would like to estimate the spring constant k from noisy displacement measurements, develop the EKF to solve this problem.

(c) Transform the discrete Gauss–Markov model to the innovations representation.

(d) Solve the parameter estimation problem using the innovations model; that is, develop the estimator of the spring constant.

5.5 Given the ARMAX model

$$y(t) = -ay(t-1) + bu(t-1) + e(t)$$

with innovations covariance, R_{ee}:

(a) Write the expressions for the EKF in terms of the ARMAX model.

(b) Write the expressions for the EKF in terms of the state-space model.

5.6 Consider tracking a body falling freely through the atmosphere [33]. We assume it is falling down in a straight line toward a radar. The state vector is defined by $x := [z \quad \dot{z} \quad \beta]$ where $\beta \sim \mathcal{N}(\mu_\beta, R_{\beta\beta}) = (2000, 2.5 \times 10^5)$ is the ballistic coefficient. The dynamics are defined by the state equations

$$\dot{x}_1(t) = x_2(t)$$
$$\dot{x}_2(t) = \frac{\rho x_2^2(t)}{2x_3(t)} - g$$
$$\dot{x}_3(t) = 0$$
$$\rho = \rho_0 e^{-\frac{x_1(t)}{k_\rho}}$$

where d is the drag deceleration, g is the acceleration of gravity (32.2), ρ is the atmospheric density (with ρ_0 (3.4×10^{-3}) density at sea level), and k_ρ a decay constant (2.2×10^4). The corresponding measurement is given by

$$y(t) = x_1(t) + v(t)$$

for $v \sim \mathcal{N}(0, R_{vv}) = \mathcal{N}(0, 100)$. Initial values are $x(0) = \mu \sim \mathcal{N}(1065, 500)$, $\dot{x}(0) \sim \mathcal{N}(-6000, 2 \times 10^4)$ and $P(0) = \text{diag}[p_0(1, 1), p_0(2, 2), p_0(3, 3)] = [500, 2 \times 10^4, 2.5 \times 10^5]$.

(a) Is this an EKF If so, write out the explicit algorithm in terms of the parametrically adaptive algorithm of this chapter.

(b) Develop the EKF for this problem and perform the discrete simulation using MATLAB.

(c) Develop the LZKF for this problem and perform the discrete simulation using MATLAB.

(d) Develop the PF for this problem and perform the discrete simulation using MATLAB.

5.7 Parameter estimation can be performed directly when we are given a nonlinear measurement system such that

$$\mathbf{y} = \mathbf{h}(\theta) + \mathbf{v}$$

where $\mathbf{y}, \mathbf{h} \in \mathcal{R}^{N_y \times 1}$ and $\theta \sim \mathcal{N}(\mathbf{m}_\theta, R_{\theta\theta})$ and $\mathbf{v} \sim \mathcal{N}(0, R_{vv})$.

(a) From the a posteriori density, $\Pr(\theta|\mathbf{y})$, derive the MAP estimator for θ.

(b) Expand $\mathbf{y} = \mathbf{h}(\theta)$ in a Taylor series about θ_o and incorporate the first-order approximation into the MAP estimator (approximate).

(c) Expand $\mathbf{y} = \mathbf{h}(\theta)$ in a Taylor series about θ_o and incorporate the second-order approximation into the MAP estimator (approximate).

(c) Develop an iterated version of both estimators in (b) and (c). How do they compare?

(d) Use the parametrically adaptive formulation of this problem assuming the measurement model is time-varying. Construct the EKF assuming that θ is modeled by a random walk. How does this processor compare to the iterated versions?

5.8 Suppose we are asked to solve a detection problem, that is, we must "decide" whether a signal is present or not according to the following binary hypothesis test:

$$\mathcal{H}_o : y(t) = v(t) \quad \text{for} \quad v \sim \mathcal{N}(0, R_{vv})$$
$$\mathcal{H}_1 : y(t) = s(t) + v(t)$$

The signal is modeled by a Gauss–Markov model

$$s(t) = a[s(t-1)] + w(t-1) \quad \text{for} \quad w \sim \mathcal{N}(0, R_{ww})$$

(a) Calculate the *likelihood-ratio* defined by

$$\mathcal{L}(Y(N)) := \frac{\Pr(Y(N)|\mathcal{H}_1)}{\Pr(Y(N)|\mathcal{H}_o)}$$

where the measurement data set is defined by $Y(N); = \{y(0), y(1), \ldots, y(N)\}$. Calculate the corresponding threshold and construct the detector (binary hypothesis test).

(b) Suppose there is an unknown but deterministic parameter in the signal model, that is,

$$s(t) = a[s(t-1); \theta(t-1)] + w(t-1)$$

Construct the "composite" likelihood ratio for this case. Calculate the corresponding threshold and construct the detector (binary hypothesis test). (*Hint*: Use the EKF to jointly estimate the signal and parameter.)

(c) Calculate a sequential form of the likelihood ratio above by letting the batch of measurements, $N \to t$. Calculate the corresponding threshold and construct the detector (binary hypothesis test). Note that there are two thresholds for this type of detector.

5.9 Angle modulated communications including both frequency modulation (FM) and phase modulation (PM) are basically nonlinear systems from the model-based perspective. They are characterized by high bandwidth requirements, and their performance is outstanding in noisy environments. Both can be captured by the *transmitted* measurement model:

$$s(t) = \sqrt{2P} \sin[\omega_c t + k_p m(t)] \quad (PM)$$

or

$$s(t) = \sqrt{2P} \sin\left[\omega_c t + 2\pi k_f \int_{-\infty}^{t} m(\tau)d\tau\right] \quad (FM)$$

where P is a constant, ω_c is the carrier frequency, k_p and k_f are the deviation constants for the respective modulation systems, and of course, $m(t)$, is the message model. *Demodulation* to extract the message from the transmission is accomplished by estimating the phase of $s(t)$. For FM, the recovered phase is differentiated and scaled to extract the message, while PM only requires the scaling.

Suppose the message signal is given by the Gauss–Markov representation

$$m(t) = -\alpha m(t-1) + w(t-1)$$
$$y(t) = s(t) + v(t)$$

with both w and v zero-mean, Gaussian with variances, R_{ww} and R_{vv}.

(a) Construct a receiver for the PM system using the EKF design.
(b) Construct an equivalent receiver for the FM system.
(c) Assume that the message amplitude parameter α is unknown. Construct the EKF receiver for the PM system to jointly estimate the message and parameter.
(d) Under the same assumptions as (c), construct the EKF receiver for the FM system to jointly estimate the message and parameter.
(e) Compare the receivers for both systems. What are their similarities and differences?

5.10 We are given the population model below and would like to "parameterize" it to design a parametrically adaptive processor, since we know that the parameters are not very well known. The state transition and corresponding measurement model are given by

$$x(t) = \frac{1}{2}x(t-1) + \frac{25x(t-1)}{1+x^2(t-1)} + 8\cos(1.2(t-1)) + w(t-1)$$

$$y(t) = \frac{x^2(t)}{20} + v(t)$$

where $\Delta t = 1.0$, $w \sim \mathcal{N}(0, 10)$ and $v \sim \mathcal{N}(0, 1)$. The initial state is Gaussian distributed with $\bar{x}(0) \sim \mathcal{N}(0.1, 5)$.

In terms of the nonlinear state-space representation, we have

$$a[x(t-1)] = \frac{1}{2}x(t-10) + \frac{25x(t-1)}{1+x^2(t-1)}$$

$$b[u(t-1)] = 8\cos(1.2(t-1))$$

$$c[x(t)] = \frac{x^2(t)}{20}$$

(a) Choose the model constants: 25, 8, 0.5, and $\frac{1}{20}$ as the unknown parameters, reformulate the state estimation problem as a parameter estimation problem with unknown parameter vector, Θ and a random walk model with the corresponding process noise variance, $R_{ww} = \text{diag}[1 \times 10^{-6}]$.
(b) Develop the joint UKF algorithm to solve this problem. Run the UKF algorithm and discuss the performance results.
(c) Develop the joint PF algorithm to solve this problem. Run the PF algorithm and discuss the performance results.
(d) Choose to "move" the particles using the roughening approach. How do these results compare to the standard bootstrap algorithm?
(e) Develop the joint linearized (UKF) PF algorithm to solve this problem. Run this PF algorithm and discuss the performance results.

6

Deterministic Subspace Identification

6.1 Introduction

The basic problem in this text is to identify a state-space model, linear or nonlinear, from data. *Subspace IDentification* (SID) is a technique to extract a black-box model in generic state-space form from uncertain input/output data using robust, numerically stable, linear algebraic methods. SID performs a black-box identification of a linear, time-invariant, state-space model $\Sigma_{ABCD} := \{A, B, C, D\}$ from uncertain measurements. Recall that a *subspace* is a space or subset of a parent space that inherits some or all of its characteristics. Unlike the MBID discussed in the previous chapter, it does *not* require any specific structure of the state-space model only its order (number of states) that can be derived directly from the embedded linear algebraic decomposition. In this chapter, subspace refers primarily to a vector (state) space in which its essential characteristics can be captured by a "smaller" space – the subspace. For example, an Nth order state-space dynamic system can be approximated by an Mth order ($M < N$) system capturing the majority of its properties. Here we focus on the *deterministic* problem ignoring the uncertainties created by noisy measurements that will be developed in the next chapter for stochastic systems.

This chapter focuses on the foundation of the system identification problem for state-space systems that leads to subspace identification techniques. The original basis has evolved from systems theory and the work of Kalman for control system design [1]. The fundamental problem is called the "realization problem" – the extraction of a linear state-space model from impulse response data or equivalently transfer function representation. The basic paper by Ho and Kalman has been cited as the seminal publication showing how to extract the state-space model [2, 3]. Initially, we discuss this problem for two distinct data sets: impulse sequences of an infinite length (number of samples) data leading to the realization problem and input/output data leading to the subspace identification problem.

Model-Based Processing: An Applied Subspace Identification Approach, First Edition. James V. Candy.
© 2019 John Wiley & Sons, Inc. Published 2019 by John Wiley & Sons, Inc.

Compared to the MBID processors that iteratively estimate parameters of the previous section [4, 5], we investigate subspace identification techniques [6–9]. These techniques enable the capability to extract a large number of parameters compared to the prediction error (P-E) methods that are quite time-consuming and not practical for large parametric problems coupled with the need for potential real-time applications. Subspace identification techniques are robust and reliable and capable of real-time applications, but they are *not* as accurate as the optimization-based P-E methods [5]. The SID methods are typically used as "starters" for P-E processors enabling a reasonable starting point for the numerical optimizers.

6.2 Deterministic Realization Problem

In this section, we develop the fundamental realization problem that provides the foundation for the subspace identification problems to follow. Here we review the basic realization (systems) theory (without proof) of the results developed by Kalman [1, 2] and show how they lead directly to the deterministic realization problem and its solution.

The deterministic realization problem has evolved in a twofold manner: (i) the classical approach and (ii) the subspace approach as illustrated in Figure 6.1. The classical approach essentially evolved from system theoretic developments based on the Hankel matrix composed of impulse response matrices (Markov parameters) and its decomposition into the observability and controllability matrices, while the subspace approach is based on projection theory (orthogonal and oblique) leading to the decomposition of a block input/output Hankel matrix. Both of the modern approaches use a singular-value decomposition of their respective Hankel matrix to estimate its rank or equivalently system order leading to the extraction of the underlying system Σ_{ABCD}. In fact, besides the critical system theoretical properties of the Hankel matrix, the singular-value decomposition (SVD) is the common thread or tool critical to a solution of the deterministic realization problem. The SVD, besides its superior numerical properties, enables a reliable and robust approach to extracting the underlying system. In this chapter, we first develop the important system theoretic results essential for comprehending solutions to the deterministic realization problem and then develop a set of classical numerical techniques based on the Ho–Kalman [3] approach leading initially to extract a system or canonical form – critical for understanding the system along with its subsystems. Although quite important, these algorithms are not numerically reliable. The evolution of the SVD has changed that fact; however, unfortunately, the underlying internal structure is difficult to unravel, since no particular coordinate system (e.g. canonical form) evolves – this is a drawback of both SVD classical and subspace approaches,

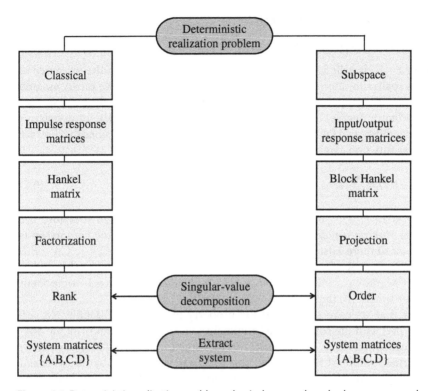

Figure 6.1 Deterministic realization problem: classical approach and subspace approach.

leading to the idea of a black-box identification followed by a similarity transformation to the desired coordinate system – a necessary and reasonable trade-off.

The subspace approach evolves from projection theory leading to a factorization of the block Hankel matrix, similar to the classical approach leading to its decomposition using the SVD and eventual extraction of the system model Σ_{ABCD}. This approach follows leading to the two most popular SID algorithms (MOESP, N4SID). Performance metrics are discussed and applied to illustrate their utility. Examples are then developed to demonstrate all of the algorithms, classical and subspace along with a detailed structural vibration response case study.

6.2.1 Realization Theory

A *realization* is defined in terms of the system transfer function and/or a state-space model, that is, a matrix transfer function $H(z)$ is defined as "realizable," if there exists a finite-dimensional state-space description,

$\Sigma_{ABCD} = \{A, B, C, D\}$, such that [10]

$$H(z) = C(zI - A)^{-1}B + D \tag{6.1}$$

where Σ_{ABCD} is called the *realization* of $H(z)$. The question of the existence of such a realization, termed "realizability," is a complex issue. We tacitly assume that $H(z)$ is a proper rational matrix (numerator polynomial degree less than denominator degree) to ensure such realizability exists (see [10] for details).

Realizability also follows from the system impulse response, since the discrete transfer function can be represented by the power series [10]

$$H(z) = \sum_{t=0}^{\infty} H_t z^{-t} \quad \text{for} \quad H_t = CA^{t-1}B + D\delta(t) \tag{6.2}$$

and H_t is the $N_y \times N_u$ unit impulse response matrix with $H_0 = D$. Therefore, the infinite *Markov sequence* $\{H_t\}_0^{\infty}$ composed of the set of *Markov parameters* is *realizable* if there exists a finite-dimensional Σ_{ABCD} that satisfies $H_t = CA^{t-1}B + D\delta(t); t = 0, \dots, \infty$. Recall that the transfer function matrix $H(z)$ can be used to extract Markov parameter sequence $\{H_t\}$ by *long division* of each component transfer functions, that is, $H_{ij}(z) = N_{ij}(z)/D_{ij}(z)$.

Now that we have established the property of system realizability, we are concerned with the concept of a "minimality." By definition, out of all the possible realizations Σ_{ABCD}, those that have the "smallest" order (number of states) are termed *minimal realizations*. This property is specified by the so-called *McMillan degree* of the matrix transfer function $H(z)$ that is equivalent to the dimension (order) of the system matrix, that is,

$$\dim A = \deg H(z) \tag{6.3}$$

Since the transfer function is the external description of the system from an input/output perspective, only the controllable and observable components are available (see [1] for details). With this in mind, it is clear that system minimality must be connected to this property. In fact, it is based on the fact proven in [2] that "Σ_{ABCD} is a minimal realization of $H(z)$ or equivalently H_t, if and only if, the pairs (A, B) and (A, C) are *completely* controllable and observable, respectively, which is equivalent to Eq. (6.3). So we see that controllability and observability of Σ_{ABCD} is a necessary and sufficient condition for minimality!

This minimality concept leads to a powerful theoretical result proved by Kalman [3]. The Markov sequence $\{H_t\}$ has a *finite*-dimensional realization, if and only if, there exists an integer N and coefficients $\alpha_1, \dots, \alpha_N$ such that

$$H_N = -\sum_{t=1}^{N} \alpha_t H_{N-t} = -\sum_{t=1}^{N} \alpha_t (CA^{N-t-1}B) \tag{6.4}$$

This result follows from the Cayley–Hamilton theorem of Section 3.6 and is based on the Nth-order characteristic equation $\alpha(A)$. With these system theoretic properties in mind, we can now define the deterministic realization problem.

The fundamental problem of extracting the state-space system Σ_{ABCD} is called the realization problem. Simply stated, the deterministic *realization problem* is GIVEN a transfer function matrix $H(z)$ or equivalently a set of impulse response matrices $\{H_t\}$ with corresponding Markov parameters, $H_t = CA^{t-1}B + D\delta(t);\ t = 0,\dots,\infty$, FIND the underlying minimal state-space system, $\Sigma_{ABCD} := \{A,B,C,D\}$.

These properties evolve completely from the development of the Hankel matrix of systems theory [2], which plays an intrinsic role in the solution of the deterministic realization problem. Note that the Hankel matrix is to state-space identification as the Toeplitz matrix is to signal processing identification (e.g. spectral estimation [11]). Note that they are related through an *exchange matrix* that is cross diagonal or antidiagonal made up of ones in the scalar case and identity matrices in the multivariable case [9].

We define the $(K \times N_y N_u) \times (K \times N_y N_u)$ block *Hankel matrix* $(K \to \infty)$ by

$$
\mathcal{H}_{K,K} := \begin{bmatrix} H_1 & H_2 & \cdots & H_K \\ H_2 & H_3 & \cdots & H_{K+1} \\ \vdots & \vdots & \cdots & \vdots \\ H_K & H_{K+1} & \cdots & H_{2K-1} \end{bmatrix} \tag{6.5}
$$

If Σ_{ABCD} is a realization of $\{H_t\}$, then based on the Markov parameters, the Hankel matrix can be *factored* as the product of the controllability and observability matrices:

$$
\begin{aligned}
\mathcal{H}_{K,K} &= \begin{bmatrix} CB & CAB & \cdots & CA^{K-1}B \\ CAB & CA^2B & \cdots & CA^KB \\ \vdots & \vdots & \cdots & \vdots \\ CA^{K-1}B & CA^{K-2}B & \cdots & CA^{2K-2}B \end{bmatrix} \\
&= \begin{bmatrix} C \\ CA \\ \vdots \\ CA^{K-1} \end{bmatrix} \begin{bmatrix} B \mid AB \mid \cdots \mid A^{K-1}B \end{bmatrix}
\end{aligned} \tag{6.6}
$$

or simply

$$
\mathcal{H}_{K,K} = \mathcal{O}_K \times \mathcal{C}_K \tag{6.7}
$$

Another interesting property of the Hankel matrix utilized by realization algorithms is its inherent *shift-invariant property*, that is, multiplication by the system matrix A shifts the Hankel matrix "up" ($\uparrow$) or "left" ($\leftarrow$) since

$$
\mathcal{O}_K^{\uparrow} = \underbrace{\begin{bmatrix} C \\ CA \\ \vdots \\ CA^{K-1} \end{bmatrix}}_{\mathcal{O}_K} \times A = \underbrace{\begin{bmatrix} CA \\ CA^2 \\ \vdots \\ CA^K \end{bmatrix}}_{\mathcal{O}_K^{\uparrow}}
\tag{6.8}
$$

$$
C_K^{\leftarrow} = A \times \underbrace{\begin{bmatrix} B & AB & \cdots & A^{K-1}B \end{bmatrix}}_{C_K} = \underbrace{\begin{bmatrix} AB & A^2B & \cdots & A^K B \end{bmatrix}}_{C_K^{\leftarrow}}
$$

$$
\tag{6.9}
$$

therefore

$$
\mathcal{H}_{K,K}^{\uparrow} := \mathcal{O}_K^{\uparrow} C_K = \underbrace{(\mathcal{O}_K \times A)}_{\mathcal{O}_K^{\uparrow}} \times C_K \iff \mathcal{H}_{K,K}^{\leftarrow} := \mathcal{O}_K C_K^{\leftarrow} = \mathcal{O}_K \times \underbrace{(A \times C_K)}_{C_K^{\leftarrow}}
$$

$$
\tag{6.10}
$$

Suppose the dimension of the minimal realization is given from a matrix transfer function as $N = N_x$, then the Hankel matrix could be constructed with $K = N_x$ using $2N_x$ impulse response matrices. Knowledge of the order N_x indicates the minimum number of terms required to exactly "match" the Markov sequence and extract the Markov parameters. It follows that the rank of the Hankel matrix must be N_x, that is, $\rho(\mathcal{H}_{K,K}) = N_x$, which is the dimension of the *minimal realization*. This leads to a very important property of the Hankel matrix. If we did *not* know the dimension of the system, then we would let (in theory) $K \to \infty$ and determine the rank of $\mathcal{H}_{\infty,\infty}$. Thus, the minimal dimension of an "unknown" system is given by the *rank* of the Hankel matrix.

This rank property of the Hankel matrix is fundamental to linear systems theory [2]. In order for a system to be *minimal*, it must be *completely controllable* and *completely observable*; therefore, it follows that the *rank condition*

$$
\rho(\mathcal{H}_{K,K}) = \min[\rho(\mathcal{O}_K), \rho(C_K)] = N_x \qquad \text{(Rank Condition)}
$$

$$
\tag{6.11}
$$

Therefore, we see that the properties of controllability and observability are carefully woven into that of minimality. Testing the rank of the Hankel matrix yields the dimensionality of the underlying dynamic system. This fact will prove crucial when we must "identify" a system, $\Sigma_{ABCD} = \{A, B, C, D\}$, from noisy measurement data.

It is also important to realize that "shifting" the Hankel matrix preserves its rank N_x. If $H_t = CA^{t-1}B; t = 1, \ldots, 2N_x$, then it follows that the corresponding rank

$$\rho(\mathcal{H}_{N_x,N_x}) = \rho(\mathcal{H}_{N_x+1,N_x}) = \rho(\mathcal{H}_{N_x,N_x+1}) = N_x \tag{6.12}$$

is satisfied [9].

Finally, the realization problem based on an infinite Markov sequence $\{H_t\}_0^{\infty}$ leads to $\mathcal{H}_{\infty,\infty}$ with rank preserved

$$\rho(\mathcal{H}_{\infty,\infty}) = \sup_t \rho(\mathcal{H}_{K,K})$$

that is satisfied by a minimal realization with $K = N_x$.

Another interesting system theoretic property that evolves from the observability, controllability, and Hankel matrices is the concept of a balanced realization. Simply stated, a balanced realization is one in which "every state is as controllable as it is observable" [12]. It is a critical property in model order reduction characterizing certain features that uniquely capture the foundation of the underlying system.

We know that if a system is completely observable and controllable, then it is minimal and the Hankel matrix factors as in Eq. (6.7). Therefore, for a stable system, both observability and controllability Gramians must satisfy the following positivity constraints, that is, the Gramian matrices Σ_{AC}, Σ_{AB} are positive definite satisfying the *Lyapunov equations*:

$$\Sigma_{AC} = A'\Sigma_{AC}A + C'C \geq 0 \qquad \text{(Observability Gramian)}$$

$$\Sigma_{AB} = A\Sigma_{AB}A' + BB' \geq 0 \qquad \text{(Controllability Gramian)}$$
$$\tag{6.13}$$

Alternatively, we can express these constraints in terms of the observability and controllability matrices as follows:

$$\Sigma_{AC}(K) = \sum_{k=0}^{K} (A^k)'C'CA^k = \mathcal{O}_K'\mathcal{O}_K \geq 0 \qquad \text{(Observability Gramian)}$$

$$\Sigma_{AB}(K) = \sum_{k=0}^{K} A^kBB'(A^k)' = \mathcal{C}_K\mathcal{C}_K' \geq 0 \qquad \text{(Controllability Gramian)}$$
$$\tag{6.14}$$

Performing a singular-value decomposition (SVD) of the observability and controllability matrices, we obtain

$$\mathcal{O}_K = U_{\mathcal{O}}(K)\Sigma_{\mathcal{O}}(K)V_{\mathcal{O}}'(K) \quad \text{and} \quad \mathcal{C}_K = U_C(K)\Sigma_C(K)V_C'(K) \tag{6.15}$$

Note that applying the orthogonality properties of the SVD ($U'U = I, V'V = I$), we obtain the *Gramians* as

$$\Sigma_{AC}(K) = \mathcal{O}'_K \mathcal{O}_K = V_{\mathcal{O}}(K)\Sigma^2_{\mathcal{O}}(K)V'_{\mathcal{O}}(K)$$

$$\Sigma_{AB}(K) = C_K C'_K = U_C(K)\Sigma^2_C(K)U'_C(K) \tag{6.16}$$

From the factorization of the Hankel matrix and its singular-value decomposition, we have

$$\mathcal{H}_{K,K} = \mathcal{U}_K \times \Sigma_K \times V'_K$$
$$= \underbrace{(U_{\mathcal{O}}(K)\Sigma_{\mathcal{O}}(K)V'_{\mathcal{O}}(K))}_{\mathcal{O}_K} \times \underbrace{(U_C(K)\Sigma_C(K)V'_C(K))}_{C_K}$$

$$\tag{6.17}$$

establishing the relationship between the singular values of the Hankel matrix and those of the so-called "second-order modes" of the observability/controllability product values (see [12] or [13] for details).

Formally, the Hankel singular values are taken as the square root of the observability and controllability Gramian product values, that is, the *Hankel singular values* are given by

$$\sigma_{ABC} = \sqrt{\Sigma_{AC} \times \Sigma_{AB}}$$

It is important to realize that while the system eigenvalues are invariant under a similarity transformation those of the Gramians are *not*, but the eigenvalues of the product of the Gramians *are* similarity invariant [14]. It is these values that are the invariants of the system and are used in model order reduction [15].

6.2.2 Balanced Realizations

With these system theoretic properties in mind, we can now formally express the concept of a "balanced realization" in terms of the observability, controllability, and the process of simultaneously diagonalizing these matrices. Therefore, we define a *balanced realization* for a stable system as one in which its Gramians are equal and diagonal, that is, [13]

$$\Sigma_{AB}(K) = \Sigma_{AC}(K) = \Sigma_K \qquad \text{(Balanced Realization)}$$

where Σ_K is a diagonal matrix obtained from the SVD of the Hankel matrix of Eq. (6.17) such that

$$\Sigma_K = \text{diag}[\sigma_1, \sigma_2, \dots, \sigma_K] \quad \text{for} \quad \sigma_1 \geq \sigma_2 \geq \cdots \geq \sigma_K$$

and σ_i are the *Hankel singular values* of the system based on the product of the Gramian matrices. These values are *invariant* under similarity transformation (see [12–15] for details).

The maximum Hankel singular value defines the so-called *Hankel norm* of the system, a metric of system input/output energy as [13, 15]

$$\| \mathcal{H}_{K,K} \|_2 = \max\{\sigma_i\} \quad \text{for} \quad i = 1, \cdots , K \tag{6.18}$$

The resulting realization is said to be a *balanced realization* and is *scale invariant*, that is, the balanced system is robust to amplitude scaling [12–14]. It is also interesting that the balancing (similarity) transformation for any system is given by

$$T_{\text{bal}} = \Sigma_{AB}^{\frac{1}{2}} \times V_P \times \Sigma_P^{-\frac{1}{2}} \quad \text{and} \quad T_{\text{bal}}^{-1} = \Sigma_P^{-\frac{1}{2}} \times U_P' \times \Sigma_{AC}^{\frac{1}{2}} \tag{6.19}$$

where we define the product matrix in terms of its Cholesky square-root decomposition as well as its corresponding SVD as

$$\mathcal{P} = \Sigma_{AC}^{\frac{1}{2}} \times \Sigma_{AB}^{\frac{1}{2}} = U_P \times \Sigma_P \times V_P' \tag{6.20}$$

with $\Sigma_{\bullet}^{\frac{1}{2}}$ the respective square-root matrices of the observability and controllability Gramians.

6.2.3 Systems Theory Summary

We summarize these important system theoretic properties of Hankel matrices as follows:

- The system defined by Σ_{ABCD} is *minimal*, if and only if, it is *completely controllable* and *completely observable*   (Minimal Order)
- For a minimal realization Σ_{ABCD}, the rank of the Hankel matrix satisfies:

$$\rho(\mathcal{H}_{K,K}) = \min[\rho(\mathcal{O}_K), \rho(C_K)] = N_x \qquad \text{(Rank Property)}$$

- The constant rank under *shifted* Hankel matrix is

$$\rho[\mathcal{H}_{K,K}] = \rho[\mathcal{H}_{K+1,K}] = \rho[\mathcal{H}_{K,K+1}] = N_x \qquad \text{(Shift Property)}$$

- The Hankel matrix is *shift invariant* such that

$$\mathcal{H}_{K,K}^{\uparrow} := \mathcal{O}_K^{\uparrow} C_K \quad \Leftrightarrow \quad \mathcal{H}_{K,K}^{\leftarrow} := \mathcal{O}_K C_K^{\leftarrow} \qquad \text{(Shift Invariant)}$$

- The eigenvalues of the system matrix A lie within the unit circle:

$$\lambda[A] \leq 1 \qquad \text{(Stability)}$$

or if and only if the *Lyapunov equation*

$$P = A'PA + C'C \qquad \text{(Lyapunov Stability)}$$

has a *unique positive definite* solution, $P > 0$.

- A *balanced realization* is defined by the equality of the diagonalized Gramians: $\Sigma_{AB}(K) = \Sigma_{AC}(K) = \Sigma_K$ such that

$$\Sigma_K = \mathrm{diag}[\sigma_1, \sigma_2, \ldots, \sigma_K] \quad \text{for} \quad \sigma_1 \geq \sigma_2 \geq \ldots \geq \sigma_K$$

where Σ_K is given by the singular-value decomposition (SVD) of the Hankel matrix:

$$\mathcal{H}_{K,K} = \mathcal{U}_K \Sigma_K \mathcal{V}_K$$

- The resulting realization is said to be *scale invariant* [12], that is, the balanced system realized is robust to amplitude scaling.
- The *Hankel singular values* used in model order reduction are obtained as the root of the observability and controllability Gramian product, that is,

$$\sigma_{ABC} = \sqrt{\Sigma_{AC} \times \Sigma_{AB}}$$

and are the underlying systemsimilarity invariants.

- The *Hankel norm* is a measure of system input/output energy and is given by the *maximum* Hankel singular value: $\| \mathcal{H}_{K,K} \|_2 = \max\{\sigma_i\}$ for $i = 1, \ldots, K$
- The *balanced similarity transformation* is given in terms of the product matrix and its decompositions (Cholesky and SVD) as

$$T_{\mathrm{bal}} = \Sigma_{AB}^{\frac{1}{2}} \times V_p \times \Sigma_p^{-\frac{1}{2}} \quad \text{and} \quad T_{\mathrm{bal}}^{-1} = \Sigma_p^{-\frac{1}{2}} \times U_p' \times \Sigma_{AC}^{\frac{1}{2}}$$

This completes the fundamental system theoretic properties of a realization Σ_{ABCD}, next we consider further properties that can lead to more robust realizations.

Before we depart from this discussion, we mention the corresponding partial realization problem that is based on a finite-length data sequence $\{H_t\}_0^K$. With *only* K-data samples, the Hankel matrix is incomplete, that is,

$$\mathcal{H}_{K,K} := \begin{bmatrix} H_1 & H_2 & \cdots & H_K \\ H_2 & H_3 & \cdots & * \\ \vdots & \vdots & \cdots & \vdots \\ H_K & * & \cdots & * \end{bmatrix} \tag{6.21}$$

where the * indicates the absence of data samples. The *partial realization problem* is then defined as

GIVEN a set of finite-length impulse response matrices $\{H_t\}_0^K$ with corresponding Markov parameters, $H_t = CA^{t-1}B + D\delta(t)$; $t = 0, \ldots, K$, FIND the underlying minimal state-space system, $\Sigma_{ABCD} := \{A(K), B(K), C(K), D(K)\}$ based on the K available samples. Solutions to this important problem have been investigated extensively, but are beyond the scope of this text [16, 17].

6.3   Classical Realization

The solution to the deterministic realization problem is based primarily on one fundamental property of the Hankel matrix – the *minimal order factorization* into the full-rank observability and controllability matrices. It has been shown that operations directly on the rows or columns of the Hankel matrix are identical to independently operating directly on the rows of the observability matrix or equivalently to operating directly on the columns of the controllability matrix [18, 19]. Therefore, minimality implies this unique decomposition of the Hankel matrix into its full-rank factors. Most realization algorithms assume this property (realizability) leading to a variety of operations. This fact led to the development of not only solutions of the so-called "canonical realization problem" but also to inherent techniques of similarity transformations to canonical forms [19]. We start with the basic classical algorithm developed by Ho and Kalman [3] and show its extension using the SVD of the Hankel matrix [20, 21].

6.3.1   Ho–Kalman Realization Algorithm

The fundamental approach to solving the deterministic realization problem evolved from [3] and is called *Ho–Kalman algorithm*. Assuming the order is known, it proceeds in the following steps starting with performing transformations on the block Hankel matrix developing the elementary matrices,[1] $P \in \mathcal{R}^{KN_y \times KN_y}$ and $Q \in \mathcal{R}^{KN_u \times KN_u}$, that is,

$$P \times \mathcal{H}_{K,K} \times Q = \begin{bmatrix} I_{N_x} & | & 0 \\ - & - & - \\ 0 & | & 0 \end{bmatrix} \in \mathcal{R}^{KN_y \times KN_u} \tag{6.22}$$

After performing this decomposition, the realization Σ_{ABCD} is obtained by extracting the individual state-space matrices from

$$\mathcal{P} := P \times \mathcal{H}_{K,K} = \begin{bmatrix} \overbrace{B}^{N_u} & |AB| \cdots |A^{K-1}B \end{bmatrix} \Rightarrow B = \mathcal{P}(1 : N_x, 1 : N_u) \tag{6.23}$$

$$\mathcal{Q} := \mathcal{H}_{K,K} \times Q = \begin{bmatrix} N_y \{ C \\ CA \\ \vdots \\ CA^{K-1} \end{bmatrix} \Rightarrow C = \mathcal{Q}(1 : N_y, 1 : N_x) \tag{6.24}$$

---

1 An elementary transformation is a nonsingular matrix, constructed by a sequence elementary operations of interchanging and scaling the rows and columns of the identity matrix such that, $P = P_1 \times P_2 \times \cdots \times P_M$ [22].

$$\mathcal{A} := P \times \mathcal{H}^{\uparrow}_{K,K} \times Q \Rightarrow A = \mathcal{A}(1:N_x, 1:N_x) \tag{6.25}$$

where recall that $\mathcal{H}^{\uparrow}_{K,K}$ is the shifted Hankel matrix of Eq. (6.10) and N_u is the dimension of the input vector **u** with N_y that of the output or measurement vector **y** with the notation $1:N \rightarrow 1, 2, \ldots, N$ or equivalently *select* the corresponding rows (columns) 1-to-N of the matrix.

Consider the following example to demonstrate this approach.

Example 6.1 Suppose we have a known $N_y \times N_u$ transfer function matrix (unstable) of Example 3.1 with $N_y = N_u = 2$ as before. The impulse response sequence is given by

$$H(z) = \begin{bmatrix} 1 & 1 \\ 1 & 1 \end{bmatrix} z^{-1} + \begin{bmatrix} -2 & -4 \\ -3 & -1 \end{bmatrix} z^{-2} + \begin{bmatrix} 4 & 10 \\ 7 & 1 \end{bmatrix} z^{-3}$$
$$+ \begin{bmatrix} -8 & -22 \\ -15 & -1 \end{bmatrix} z^{-4} + \begin{bmatrix} 16 & 46 \\ 31 & 1 \end{bmatrix} z^{-5}$$

Constructing the corresponding Hankel and shifted Hankel matrices for this problem, we have

$$H_{2,2} = \begin{bmatrix} 1 & 1 & | & -2 & -4 \\ 1 & 1 & | & -3 & -1 \\ - & - & - & - & - \\ -2 & -4 & | & 4 & 10 \\ -3 & -1 & | & 7 & 1 \end{bmatrix}; \quad H^{\uparrow}_{2,2} = \begin{bmatrix} -2 & -4 & | & 4 & 10 \\ -3 & -1 & | & 7 & 1 \\ - & - & - & - & - \\ 4 & 10 & | & -8 & -22 \\ 7 & 1 & | & -15 & -1 \end{bmatrix}$$

Performing elementary row and column operations on $\mathcal{H}_{2,2}$, we obtain

$$P = \begin{bmatrix} 1 & 0 & 0 & 0 \\ -1 & 0 & -1/2 & 0 \\ 1 & -1 & 0 & 0 \\ -2/3 & -1/6 & -1/6 & -1/6 \end{bmatrix}; \quad Q = \begin{bmatrix} 1 & -1 & 2 & 9 \\ 0 & 1 & 0 & 1 \\ 0 & 0 & 1 & 3 \\ 0 & 0 & 0 & 1 \end{bmatrix}$$

The input $B \Rightarrow P(1:4, 1:2)$ and output $C \Rightarrow Q(1:2, 1:4)$ matrices are found as

$$PH_{2,2} = B \left\{ \begin{bmatrix} 1 & 1 & | & 2 & 9 \\ 0 & 1 & | & 0 & 1 \\ 0 & 0 & | & 1 & 3 \\ 0 & 0 & | & 0 & 1 \end{bmatrix} = [B \mid AB]; \quad H_{2,2}Q = \begin{bmatrix} \overbrace{\begin{matrix} 1 & 0 & 0 & 0 \\ 1 & 0 & -1 & 0 \\ - & - & - & - \\ -2 & -2 & 0 & 0 \\ -3 & -2 & 1 & -5 \end{matrix}}^{C} \end{bmatrix} = \begin{bmatrix} C \\ -- \\ CA \end{bmatrix} \right.$$

Finally, the system matrix $A \Rightarrow \mathcal{A}(1:4, 1:4)$ is found from

$$
P\mathcal{H}_{2,2}^{\uparrow}Q = \begin{bmatrix} 1 & 0 & 0 & 0 \\ -1 & 0 & -1/2 & 0 \\ 1 & -1 & 0 & 0 \\ -2/3 & -1/6 & -1/6 & -1/6 \end{bmatrix} \times \left[\begin{array}{cc|cc} -2 & -4 & 4 & 10 \\ -3 & -1 & 7 & 1 \\ \hline - & - & - & - \\ 4 & 10 & -8 & -22 \\ 7 & 1 & -15 & -1 \end{array}\right]
$$

$$
\times \begin{bmatrix} 1 & -1 & 2 & 9 \\ 0 & 1 & 0 & 1 \\ 0 & 0 & 1 & 3 \\ 0 & 0 & 0 & 1 \end{bmatrix}
$$

$$
A = \begin{bmatrix} -2 & -2 & 0 & 0 \\ 0 & -1 & 0 & 0 \\ 1 & -4 & -1 & 6 \\ 0 & 1 & 0 & -2 \end{bmatrix}
$$

completing the application of the Ho–Kalman algorithm to this MIMO problem. As a check of the realization, we reconstructed the impulse response sequence from the extracted model Σ_{ABCD} using $H_t = CA^{t-1}B; t = 1, \ldots, 5$ with a "perfect match," since the calculations were accomplished manually with perfect precision. □

Note that if the minimal order was *not* known a priori as from the transfer function in this example, then a larger Hankel matrix would be processed until its rank was found using the elementary transformations to determine N_x as expressed in Eq. (6.22).

6.3.2 SVD Realization Algorithm

A more efficient solution to the *deterministic realization problem* is obtained by performing the SVD of the Hankel matrix [20, 21] $\mathcal{H}_{m,K}$, where $m \geq N_x$ and $K \geq N_x$ with N_x the dimension of the minimum realization (system) or equivalently the underlying *true* number of states; therefore,

$$
\mathcal{H}_{m,K} = \begin{bmatrix} U_{N_x} \mid U_N \end{bmatrix} \begin{bmatrix} \Sigma_{N_x} & \mid & 0 \\ - & - & - \\ 0 & \mid & 0 \end{bmatrix} \begin{bmatrix} V'_{N_x} \\ -- \\ V'_N \end{bmatrix} = U_{N_x} \Sigma_{N_x} V'_{N_x} \tag{6.26}
$$

for $\Sigma_{N_x} = \text{diag}[\sigma_1, \sigma_2, \ldots, \sigma_{N_x}]$

From the factorization of the Hankel matrix and its SVD, we have that

$$
\mathcal{H}_{m,K} = \mathcal{O}_m \times C_K = \underbrace{(U_{N_x}\Sigma_{N_x}^{1/2})}_{\mathcal{O}_{N_x}}\underbrace{((\Sigma'_{N_x})^{1/2}V'_{N_x})}_{C_{N_x}} \tag{6.27}
$$

where $\Sigma_{N_x}^{1/2}$ is the matrix square root, that is, $R = S^{1/2} \times (S')^{1/2}$ obtained by a Cholesky decomposition [22].

Now from shift-invariant property (see Eq. (6.10)) of the observability or controllability matrices, we can obtain the system matrices from either

$$\mathcal{O}_{N_x} \times A = \mathcal{O}_{N_x}^{\uparrow}$$

or

$$A \times C_{N_x} = C_{N_x}^{\leftarrow}$$

which yields the unique system matrices by applying the pseudo-inverse[2] $(Z^{\#} := (Z'Z)^{-1}Z')$, since minimality guarantees full-rank observability and controllability matrices

$$A = \mathcal{O}_{N_x}^{\#} \times \mathcal{O}_{N_x}^{\uparrow} \quad \text{or} \quad A = C_{N_x}^{\leftarrow} \times C_{N_x}^{\#} \tag{6.28}$$

with the input and output matrices extracted directly from the controllability and observability matrices as

$$B = C(1 : N_x, 1 : N_u) \quad \text{and} \quad C = \mathcal{O}(1 : N_y, 1 : N_x) \tag{6.29}$$

However, from the SVD of Eq. (6.27), we have

$$A = \mathcal{O}_{N_x}^{\#} \mathcal{O}_{N_x}^{\uparrow} = (\Sigma_{N_x}^{-1/2} U_{N_x}') \times \mathcal{O}_{N_x}^{\uparrow} \text{ or } A = C_{N_x}^{\leftarrow} C_{N_x}^{\#} = C_{N_x}^{\leftarrow} \times (V_{N_x}(\Sigma_{N_x}^{-1/2})') \tag{6.30}$$

which is a very efficient method of calculating the pseudo-inverse and extracting the model.

It is also interesting to note that with the SVD extraction of Σ_{ABCD}, both for the A and C-matrices all lower-order models are "nested" within these structures and can easily be extracted as submatrices (see [23] and Problem 6.6 for more details).

Example 6.2 Consider Example 6.1 and assume that we do not know the order N_x. We would create an extended Hankel matrix in this case with $K = 5$ leading to

$$H_{3,3} = \begin{bmatrix} 1 & 1 & | & -2 & -4 & | & 4 & 10 \\ 1 & 1 & | & -3 & -1 & | & 7 & 1 \\ - & - & - & - & - & - & - & - \\ -2 & -4 & | & 4 & 10 & | & -8 & -22 \\ -3 & -1 & | & 7 & 1 & | & -15 & -1 \\ - & - & - & - & - & - & - & - \\ 4 & 10 & | & -8 & -22 & | & 16 & 46 \\ 7 & 1 & | & -15 & -1 & | & 31 & 1 \end{bmatrix}$$

---

2 Recall the SVD of $Z = U\Sigma V'$ efficiently provides $Z^{\#} = V\Sigma^{-1}U' = V(\Sigma^{-1/2})'\Sigma^{-1/2}U'$.

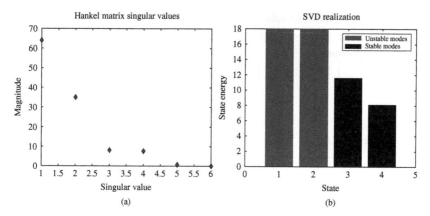

Figure 6.2 SVD realization. (a) Estimated singular values of the Hankel matrix. (b) Estimated Hankel singular values.

Investigating the Hankel matrix, we calculated its SVD to observe the rank in Figure 6.2 along with the corresponding *Hankel singular values* discussed previously. From the figure, it is clear that the order (rank) is four (4), since the first four singular values are large relative to the final two indicating a fourth-order system. The Hankel singular values indicate two unstable poles (eigenvalues); however; there should be four $(-1, -1, -2, -2)$, but inaccuracies in the identified model cause variations as indicated by the eigenvalues:

$$\lambda(A_{\text{true}}) = \{-1, -1, -2, -2\} \rightarrow \lambda(\hat{A}) = \{-1, 0.474, -2, -1.606\}$$

indicating a nonperfect realization.

We applied the SVD approach to this set of impulse response matrices and extracted the following system Σ_{ABCD}:

$$\hat{A} = \begin{bmatrix} -2.095 & 0.376 & -0.448 & 0.004 \\ -0.057 & -1.884 & -0.303 & 0.064 \\ 0.260 & 0.756 & -0.745 & 0.344 \\ 0.140 & -0.147 & -0.418 & 5.92 \end{bmatrix};$$

$$\hat{B} = \begin{bmatrix} -0.873 & -0.943 \\ 0.940 & -0.235 \\ 0.0526 & -2.008 \\ -0.005 & 1.930 \end{bmatrix}; \quad \hat{C} = \begin{bmatrix} -1.100 & -0.0248 & 1.934 & 1.989 \\ -0.657 & 0.9780 & -0.262 & 0.0463 \end{bmatrix}$$

Table 6.1 Ho–Kalman (SVD) realization algorithm

Singular value decomposition

$$\mathcal{H}_{m,K} = [U_{N_x} \mid U_N] \begin{bmatrix} \Sigma_{N_x} \mid 0 \\ -- \quad - \\ 0 \mid \quad 0 \end{bmatrix} \begin{bmatrix} V'_{N_x} \\ -- \\ V'_N \end{bmatrix} = U_{N_x} \Sigma_{N_x} V'_{N_x}$$ (Rank Matrix)

Factorization

$$\mathcal{H}_{m,K} = \mathcal{O}_m \times C_K = \underbrace{(U_{N_x} \Sigma_{N_x}^{1/2})}_{\mathcal{O}_{N_x}} \underbrace{((\Sigma'_{N_x})^{1/2} V'_{N_x})}_{C_{N_x}}$$ (Observability/ Controllability Matrices)

System matrix

$$A = \mathcal{O}_{N_x}^\# \times \mathcal{O}_{N_x}^\uparrow = (\mathcal{O}'_{N_x} \mathcal{O}_{N_x})^{-1} \mathcal{O}'_{N_x} \times \mathcal{O}_{N_x}^\uparrow = (\Sigma_{N_x}^{-1/2} U'_{N_x}) \times \mathcal{O}_{N_x}^\uparrow$$ (System/Process Matrix)

or

$$A = \quad C_{N_x}^\leftarrow \times C_{N_x}^\# = \quad C_{N_x}^\leftarrow \times C'_{N_x} (C'_{N_x} C_{N_x})^{-1} = C_{N_x}^\leftarrow \times (V_{N_x} (\Sigma_{N_x}^{-1/2})')$$

Measurement (output) matrix

$C = \mathcal{O}(1 : N_y, 1 : N_x)$ (Output Matrix)

Input transmission matrices

$B = C(1 : N_x, 1 : N_u)$ (Input Matrix)

Input/output transmission matrices

$D = H_0$ (Input/Output Matrix)

As a check, we estimated the impulse response (as before) with $\hat{H}_t = \hat{C}\hat{A}^{t-1}\hat{B}$

$$H(z) = \begin{bmatrix} 1.08 & 1.00 \\ 1.68 & 1.03 \end{bmatrix} z^{-1} + \begin{bmatrix} -2.09 & 3.99 \\ -3.43 & -1.03 \end{bmatrix} z^{-2} + \begin{bmatrix} 3.97 & 9.96 \\ 6.98 & 0.76 \end{bmatrix} z^{-3}$$
$$+ \begin{bmatrix} -7.91 & -22.01 \\ -14.12 & -1.14 \end{bmatrix} z^{-4} + \begin{bmatrix} 15.83 & 46.18 \\ 28.73 & 1.85 \end{bmatrix} z^{-5}$$

Examining the estimated impulse response, we see that errors have evolved but the results are still quite reasonable. □

We summarize the Ho–Kalman/SVD algorithm in Table 6.1.

6.3.2.1 Realization: Linear Time-Invariant Mechanical Systems

Mechanical systems are important in many applications, especially when considering vibrational responses of critical components such as turbine–generator pairs in nuclear systems on ships or aircraft structures that transport people throughout the world. Next we briefly present the generic multivariable mechanical system representation that will be employed in examples and case studies to follow.

Linear, time-invariant multiple input/output (MIMO) mechanical systems are characterized by the vector–matrix differential equations that can be expressed as

$$M\ddot{\mathbf{d}}(t) + C_d \dot{\mathbf{d}}(t) + K\mathbf{d}(t) = B_p \mathbf{p}(t) \tag{6.31}$$

where $\mathbf{d}$ is the $N_d \times 1$ displacement vector, $\mathbf{p}$ is the $N_p \times 1$ excitation force, and M, C_d, K, are the $N_d \times N_d$ lumped mass, damping, and spring constant matrices characterizing the vibrational process model, respectively. The structure of these matrices, typically, takes the form as

$$M = \begin{bmatrix} M_1 & 0 & 0 & 0 & 0 \\ 0 & M_2 & 0 & 0 & 0 \\ \vdots & \vdots & \ddots & \vdots & \vdots \\ 0 & 0 & 0 & M_{N_d-1} & 0 \\ 0 & 0 & 0 & 0 & M_{N_d} \end{bmatrix}, \quad C_d = [C_{d_{ij}}], \text{ and}$$

$$K = \begin{bmatrix} (K_1 + K_2) & -K_2 & 0 & 0 & 0 \\ -K_2 & (K_2 + K_3) & \ddots & 0 & 0 \\ 0 & \ddots & \ddots & -K_{N_d-1} & 0 \\ 0 & 0 & -K_{N_d-1} & (K_{N_d-1} + K_{N_d}) & -K_{N_d} \\ 0 & 0 & 0 & -K_{N_d} & K_{N_d} \end{bmatrix}$$

If we define the $2N_d$-state vector as $\mathbf{x}(t) := [\mathbf{d}(t) \mid \dot{\mathbf{d}}(t)]'$, then the continuous-time state-space representation of this process can be expressed as

$$\dot{\mathbf{x}}(t) = \underbrace{\begin{bmatrix} 0 & | & I \\ --- & | & --- \\ -M^{-1}K & | & -M^{-1}C_d \end{bmatrix}}_{A} \mathbf{x}(t) + \underbrace{\begin{bmatrix} 0 \\ --- \\ M^{-1}B_p \end{bmatrix}}_{B} \mathbf{p}(t) \tag{6.32}$$

The corresponding measurement or output vector relation can be characterized by

$$\mathbf{y}(t) = \mathbf{C_a}\ddot{\mathbf{d}}(t) + \mathbf{C_v}\dot{\mathbf{d}}(t) + \mathbf{C_d}\mathbf{d}(t) \tag{6.33}$$

where the constant matrices: $\mathbf{C_a}$, $\mathbf{C_v}$, $\mathbf{C_d}$ are the respective acceleration, velocity, and displacement weighting matrices of appropriate dimension.

In terms of the state vector relations of Eq. (6.32), we can express the acceleration vector as

$$\ddot{\mathbf{d}}(t) = -M^{-1}K\mathbf{d}(t) - M^{-1}C_d\dot{\mathbf{d}}(t) + M^{-1}B_p\mathbf{p}(t) \tag{6.34}$$

Substituting for the acceleration term in Eq. (6.33), we have that

$$\mathbf{y}(t) = -\mathbf{C_a}M^{-1}[B_p\mathbf{p}(t) - C_d\dot{\mathbf{d}}(t) - K\mathbf{d}(t)] + \mathbf{C_v}\dot{\mathbf{d}}(t) + \mathbf{C_d}\mathbf{d}(t)$$

or

$$\mathbf{y}(t) = \underbrace{[\mathbf{C}_d - \mathbf{C_a}M^{-1}K \mid \mathbf{C_v} - \mathbf{C_a}M^{-1}\mathbf{C_d}]}_{C} \begin{bmatrix} \mathbf{d}(t) \\ - - - \\ \dot{\mathbf{d}}(t) \end{bmatrix} + \underbrace{\mathbf{C_a}M^{-1}B_p\mathbf{p}(t)}_{D} \quad (6.35)$$

to yield the vibrational measurement as

$$\mathbf{y}(t) = C\mathbf{x}(t) + D\mathbf{u}(t) \quad (6.36)$$

where the output or measurement vector is $\mathbf{y} \in \mathcal{R}^{N_y \times 1}$ completing the MIMO-mechanical system model.

When the SNR is high, an SVD realization can be performed and reasonable results can be achieved – especially in the deterministic case. Next, we investigate a simple mechanical system developed previously in [15].

Example 6.3 Consider the simple three-mass (spring–damper) mechanical system[3] illustrated in Figure 6.3 from [15] where $m_i = 1; k_i = 3$, $k_4 = 0$; and $d_i = 0.01k_i, d_4 = 0; i = 1, \dots, 3$. These parameters along with the input/output transmission matrices lead to the following state-space representation ($N_d = 3, N_x = 6$).

$$\dot{\mathbf{x}}(t) = A\mathbf{x}(t) + B\mathbf{u}(t)$$
$$\mathbf{y}(t) = C\mathbf{x}(t) + D\mathbf{u}(t)$$

$$A = \begin{bmatrix} 0 & 0 & 0 & \mid & 1 & 0 & 0 \\ 0 & 0 & 0 & \mid & 0 & 1 & 0 \\ 0 & 0 & 0 & \mid & 0 & 0 & 1 \\ - & - & - & \mid & - & - & - \\ -6 & 3 & 0 & \mid & -0.06 & 0.03 & 0 \\ 3 & -6 & 3 & \mid & 0.03 & -0.06 & 0.03 \\ 0 & 3 & -3 & \mid & 0 & 0.03 & -0.03 \end{bmatrix}; \ B = \begin{bmatrix} 0 \\ 0 \\ 0 \\ 0 \\ 0 \\ 0 \\ 1 \end{bmatrix}; \ C = \begin{bmatrix} 1 & 0 & 0 & \mid & 0 & 0 & 0 \\ 0 & 0 & 0 & \mid & 1 & 0 & 0 \\ 0 & 0 & 0 & \mid & 0 & 0 & 1 \end{bmatrix}$$

with $D = 0$.

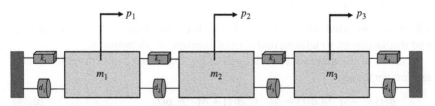

Figure 6.3 The three-output mechanical system under investigation.

---

3 See [15] for a detailed analysis of this mechanical system.

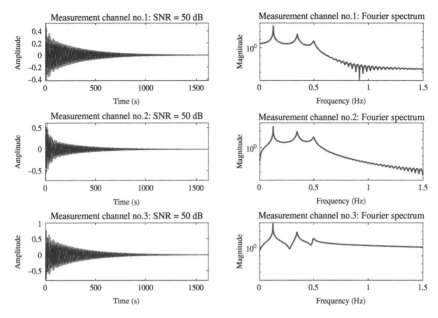

Figure 6.4 Impulse response and spectrum of three-output mechanical system. (a) Synthesized channel impulse responses. (b) Fourier transforms of channel responses with spectral peaks at 0.497 Hz, 0.343 Hz, 0.123 Hz.

Exciting this system with an impulse yields the three-output response shown in Figure 6.4 along with its accompanying Fourier spectra. Discretizing the response with a sampling interval of $\Delta t = 0.01$ seconds, creating the Hankel matrix and decomposing it as in Eq. (6.26) using the SVD reveals a sixth-order system $(N_d = 3, N_x = 6)$. The log-plot of the singular values is shown in Figure 6.5a. The modal frequencies of the system estimated from the identified "discrete" system matrix can be transformed from the discrete (z-domain) to the continuous (s-domain) – $(\mathcal{Z} \to \mathcal{S})$, that is,

$$ s = \frac{1}{\Delta t} \ln z $$

yielding the modal frequencies at: $\{p_i = 0.497 \text{ Hz}, 0.343 \text{ Hz}, 0.123 \text{ Hz}\}$ that correspond to the peaks of the Fourier transform, indicating a reasonable realization. We will investigate this structural problem even further applying the subspace identification techniques subsequently.

The deterministic SVD realization technique was performed using the impulse response data at a high SNR of 50 dB. The rank of the Hankel matrix from the SVD decomposition indicates a sixth-order system (three-modes) as observed in Figure 6.5a. The realization results are shown in Figure 6.5b–d where we see the modal frequency estimates that match those of the system

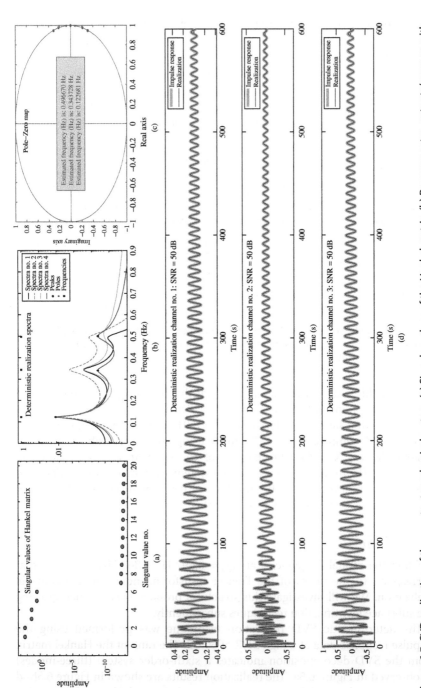

Figure 6.5 SVD realization of three-output mechanical system. (a) Singular values of the Hankel matrix. (b) Power spectra: outputs/average with poles (boxes). (c) Frequencies with poles/zeros. (d) Realization output channels (data/realization) overlays.

closely along with the estimated channel outputs overlaid onto the raw impulse response data – again indicating an accurate realization due to the high SNR. In Figure 6.5b, we compare the estimated average power spectra from the SVD realization of the model to the three-output spectra and show the location of the extracted modal frequencies. Note how all of the peaks align indicating a valid realization of the underlying mechanical system. The pole–zero representation along with the extracted modal frequencies are shown in Figure 6.5c. This completes the example. □

Next we consider another approach of estimating realizations in unique structural descriptions that have evolved from algebraic group theory – canonical forms.

6.3.3 Canonical Realization

In this section, we consider solving the realization problem in canonical coordinates of the state-space. Here, specific "invariant" entities will be extracted directly from the Hankel matrix impulse response data enabling considerable insights into the inherent structure of the identified system. There are advantages in this approach, since the minimal number of parameters required are utilized to represent a complex system in this state-space form. Canonical realizations characterize a unique representation of an unknown system. However, the disadvantage is that the error in realizing the model in this form can represent an entirely different system than the true system due to its uniqueness.

6.3.3.1 Invariant System Descriptions

A unique parametric representation of a system Σ_{ABCD} is called a canonical form [24]. Formally, this characterization evolves from algebraic module theory through the action of a transformation group (similarity transformation) acting on an equivalence class (transfer function/Markov sequence) [2]. The result of this action is a complete set of independent "invariants" that uniquely characterize the system under investigation and form the essential entities that represent that system. These invariants are the essentially critical parameters of the structure of a system revealing its internal interconnections and couplings at the subsystem level. From these invariants, we can construct a "canonical form" [18, 25] that uniquely represents Σ_{ABCD} revealing its internal structure.

For an observable pair (A, C), define v_j as the jth *observability index*. There exists exactly one set of ordered scalars $\beta_{jk\ell}$ such that

$$c_j' A^{v_j} = \sum_{k=1}^{j-1} \sum_{\ell=0}^{\min(v_j,v_k-1)} \beta_{jk\ell} \, c_k' A^\ell + \sum_{k=j}^{N_y} \sum_{\ell=0}^{\min(v_j,v_k)-1} \beta_{jk\ell} \, c_k' A^\ell \tag{6.37}$$

where the set $[\{v_j\}, \{\beta_{jk\ell}\},]; j, k \in N_y, \ell = 0, \ldots, v_j - 1$ are defined as the *observability invariants* for the observable pair (A, C).

This relation provides a unique set of basis vectors, $c'_j A^{v_j}$, and the corresponding coefficients of linear dependence, $\beta_{jk\ell}$. Here, the jth observability index v_j is the smallest positive integer such that the vector $c'_j A^{v_j}$ is a linear combination of its basis vectors or *predecessors*, where a predecessor of $c'_j A^{v_j}$ is any vector $c'_r A^s$ such that $rN_y + s < kN_y + j$.

With this in mind, we define *predecessor independence* as a row vector of a matrix as independent, if it is not a linear combination of its predecessors in a *chain* as given in Eq. (6.37). For instance, for our observability invariants, the "observability" chain is assembled as

$$[c'_1 \; c'_1 A \cdots c'_1 A^{v_1-1} \mid c'_2 \; c'_2 A \cdots c'_2 A^{v_2-1} \mid \cdots \mid c'_{N_y} \cdots c'_{N_y} A^{v_{N_y}-1}]'$$

$$\underbrace{\qquad\qquad\qquad\qquad\qquad\qquad\qquad\qquad\qquad\qquad\qquad\qquad}$$

$$\text{Observability Chain 1} \quad \text{Observability Chain 2} \quad \cdots \quad \text{Observability Chain}$$

$$\times N_y(\text{Row Predecessors}) \tag{6.38}$$

The significance of this definition is that sequentially examining rows of the *Hankel matrix* for dependencies in "row" predecessor order of the chain $\{c'_j \; c'_j A \cdots c'_j A^{v_j-1}\}$ is equivalent to examining the rows of the observability matrix in identical order. When these rows are examined for predecessor independence, then the corresponding indices $\{v_j\}$ and coefficients of linear dependence $\{\beta_{jk\ell}\}$ have special meaning – they are the set of observability invariants.

Analogously, for a controllable pair (A, B), define μ_i as the ith *controllability index*, then there exists exactly one set of ordered scalars α_{imn} such that

$$A^{\mu_i} b_i = \sum_{m=1}^{i-1} \sum_{n=0}^{\min(\mu_i,\mu_m-1)} \alpha_{imn} A^n b_m + \sum_{m=i}^{N_u} \sum_{n=0}^{\min(\mu_i,\mu_m)-1} \alpha_{imn} A^n b_m \tag{6.39}$$

where the set $[\{\mu_i\}, \{\alpha_{imn}\},]; i, m \in N_u, n = 0, \cdots, \mu_i - 1$ are defined as the *controllability invariants* for the pair (A, B).

This relation provides a unique set of basis vectors, $A^{\mu_i} b_i$, and corresponding coefficients of linear dependence, α_{imn}. Here the ith controllability index μ_i is the smallest positive integer such that the vector $A^{\mu_i} b_i$ is a linear combination of its predecessors, where a predecessor of $A^{\mu_i} b_i$ is any vector $A^s b_r$ such that $rN_u + s < mN_u + i$.

As before, we define *predecessor independence* as a column vector of a matrix as independent, if it is not a linear combination of its predecessors in a column *chain* as given in Eq. (6.39). For instance, for our controllability invariants, the "controllability" chain is assembled as

$$[b_1 \; Ab_1 \; \cdots \; A^{\mu_1-1} b_1 \mid b_2 \; Ab_2 \; \cdots \; A^{\mu_2-1} b_2 \mid \cdots \mid b_{N_u} \; \cdots \; A^{\mu_{N_u}-1} b_{N_u}]$$

$$\underbrace{\qquad\qquad\qquad\qquad\qquad\qquad\qquad\qquad\qquad\qquad\qquad\qquad}$$

$$\text{Controllability Chain 1} \quad \text{Controllability Chain 2} \quad \cdots \quad \text{Controllability Chain}$$

$$\times N_y(\text{Column Predecessors}) \tag{6.40}$$

Also the significance of this definition is that sequentially examining columns of the *Hankel matrix* for dependencies in "column" predecessor order of the chain $\{b_i \ Ab_i \ \cdots \ A^{\mu_i-1}b_i\}$ is equivalent to examining the columns of the controllability matrix in identical order. When these columns are examined for predecessor independence, then the corresponding indices $\{\mu_i\}$ and coefficients of linear dependence $\{\alpha_{imn}\}$ are the set of controllability invariants.

The observability and controllability "indices" are related to the overall system *observability index* v and *controllability index* μ defined as the *minimum* integer value such that Σ_{ABCD} is completely observable and completely controllable, respectively. Therefore, we have

$$v = \sum_{\ell=1}^{N_y} v_\ell = N_x \quad \text{and} \quad \mu = \sum_{n=1}^{N_u} \mu_n = N_x \quad (6.41)$$

With these definitions in mind, we can reformulate the *rank condition* in terms of the observability and controllability indices, that is, for a minimal realization Σ_{ABCD}, the rank of the Hankel matrix satisfies

$$\rho(\mathcal{H}_{v,\mu}) = \min[\rho(\mathcal{O}_v), \rho(C_\mu)] = N_x \quad \text{(Rank Property)} \quad (6.42)$$

Using the sets of observability and controllability invariants, we can construct the corresponding canonical forms (Luenberger) under the similarity transformation group [24, 25]. From the dependence relations of Eqs. (6.37) and (6.39), we see that the corresponding vectors $(\{c_j'A^{v_j-1}\}; j,k \in N_y)$ span the rows of $\mathcal{O}_{v+1}$ and analogously, the vectors $(\{A^{\mu_i-1}b_i\}; i,m \in N_u)$ span the columns of $C_{\mu+1}$. Therefore, the canonical forms (Luenberger) are defined by the row pair (A_R, C_R) and the column pair (A_C, C_C).

We have the *Luenberger row canonical form* as

$$A_R = \begin{bmatrix} e_2' \\ \vdots \\ e_{m_1}' \\ \beta_1' \\ \hline \vdots \\ \hline e_{m_{n-1}+2}' \\ \vdots \\ e_{m_n}' \\ \beta_m' \end{bmatrix} ; \ C_R = \begin{bmatrix} e_1' \\ e_{m_1+1}' \\ \vdots \\ e_{m_{n-1}+1}' \end{bmatrix} \quad \text{for} \quad m_i = \sum_{k=1}^{i} v_k; i = 1, \ldots, n \quad (6.43)$$

and the corresponding *Luenberger column canonical form* as

$$A_C = [e_2 \; \cdots \; e_{q_1}\alpha_1 \; | \; \cdots \; | \; e_{q_{p-1}+2} \; \cdots \; e_{q_p}\alpha_p]$$

$$B_C = [e_1 \, e_{q_1+1} \; \cdots \; e_{q_{p-1}+1}]; \quad \text{for} \quad q_i = \sum_{k=1}^{i} \mu_k; i = 1, \ldots, p \qquad (6.44)$$

where e'_m is a unit row vector with a 1 in the mth row, while e_q is a unit column vector with a 1 in the qth column. The invariants are contained in the respective row and column vectors β_m and α_q.

The key to understanding the row coefficient indices evolves directly from the corresponding observability dependence relations of Eq. (6.37). Therefore, for the jth row vector, $c'_j A^{v_j}$, we have the following coefficient mapping (vector → coefficient) for the jth chain with $k = 1, \ldots, v_j - 1$; $\ell = 0, \ldots, v_j - 1$. That is, the mapping from Eq. (6.37), through the indices, onto the coefficients can be seen as, first, the observability invariant chain mapping into the β-coefficients as

$$[c'_1 \; c'_1 A \; \cdots \; c'_1 A^{v_1-1} \; | \; c'_2 \; c'_2 A \; \cdots \; c'_2 A^{v_2-1} \; | \; \cdots \; | \; c'_{N_y} \; c'_{N_y} A \; \cdots \; c'_{N_y} A^{v_{N_y}-1}]'$$

$$\Downarrow \quad \Downarrow \quad \Downarrow \quad \Downarrow \quad \Downarrow \quad \Downarrow \quad \Downarrow \quad \Downarrow \quad \Downarrow$$

$$[\beta_{j10} \; \beta_{j11} \; \cdots \; \beta_{j1v_1-1} \; | \; \beta_{j20} \; \beta_{j21} \; \cdots \; \beta_{j2v_2-1} \; | \; \cdots \; | \; \beta_{jN,0} \; \beta_{jN,1} \; \cdots \; \beta_{jN_y v_{N_y}-1}]'$$

Similarly from the corresponding controllability dependence relations of Eq. (6.39), for the ith column vector, $A^{\mu_i} b_i$, we have the following mapping for the ith controllability chain with $m = 1, \ldots, \mu_i - 1$; $n = 0, \ldots, \mu_i - 1$ can be seen as, first, the controllability chain mapping into the α-coefficients as

$$[b_1 \; Ab_1 \; \cdots \; A^{\mu_1-1}b_1 \; | \; b_2 \; Ab_2 \; \cdots \; A^{\mu_2-1}b_2 \; | \; \cdots \; | \; b_{N_u} \; Ab_{N_u} \; \cdots \; A^{\mu_{N_u}-1}b_{N_u}]$$

$$\Downarrow \quad \Downarrow \quad \Downarrow \quad \Downarrow \quad \Downarrow \quad \Downarrow \quad \Downarrow \quad \Downarrow \quad \Downarrow$$

$$[\alpha_{i10} \; \alpha_{i11} \; \cdots \; \alpha_{i1\mu_1-1} \; | \; \alpha_{i20} \; \alpha_{i21} \; \cdots \; \alpha_{i2\mu_2-1} \; | \; \cdots \; | \; \alpha_{iN_u0} \; \alpha_{iN_u1} \; \cdots \; \alpha_{iN_u\mu_{N_u}-1}]$$

The following example illustrates these constructs.

Example 6.4 Suppose we are given two sets of invariants $[\{v_j\}, \{\beta_{jk\ell}\},]; j, k \in N_y, \ell = 0, \ldots, v_j - 1$ and $[\{\mu_i\}, \{\alpha_{imn}\},]; i, m \in N_u, n = 0, \ldots, \mu_i - 1$ for $N_y = 3; N_u = 2$ with observability invariants $v_j = \{1, 2, 1\}$ and $\{\beta_{jk\ell}\}$ and controllability invariants $\mu_i = \{3, 1\}$ and $\{\alpha_{imn}\}$ and we are asked to construct both of the corresponding Luenberger canonical forms.

The row canonical form is constructed from each of the following observability chains:

$[c_1' \mid c_2' \; c_2'A \mid c_3']'$

$\Downarrow \quad \Downarrow \quad \Downarrow \quad \Downarrow$

$[\beta_{j10} \mid \beta_{j20} \; \beta_{j21} \mid \beta_{j30}]'$

for $j = 1, 2, 3$. Therefore, we have the row canonical form as follows:

$$A_R = \begin{bmatrix} \beta_1' \\ e_2' \\ \beta_2' \\ \beta_3' \end{bmatrix} = \begin{bmatrix} \beta_{110} & \mid \beta_{120} & \beta_{121} & \mid \beta_{130} \\ 0 & \mid 1 & 0 & \mid 0 \\ \beta_{210} & \mid \beta_{220} & \beta_{221} & \mid \beta_{230} \\ \beta_{310} & \mid \beta_{320} & \beta_{321} & \mid \beta_{330} \end{bmatrix}; \quad C_R = \begin{bmatrix} e_1' \\ e_2' \\ e_4' \end{bmatrix} = \begin{bmatrix} 1 & \mid 0 & 0 & \mid 0 \\ 0 & \mid 1 & 0 & \mid 0 \\ 0 & \mid 0 & 0 & \mid 1 \end{bmatrix}$$

The column canonical form is constructed from each of the following controllability chains:

$[b_1 \; Ab_1 \; A^2 b_1 \mid b_2]'$

$\Downarrow \quad \Downarrow \quad \Downarrow \quad \Downarrow$

$[\alpha_{i10} \; \alpha_{i11} \; \alpha_{i12} \mid \alpha_{20}]'$

for $i = 1, 2$ and the corresponding column canonical form is

$$A_C = [e_2 \; e_3 \; \alpha_1 \mid \alpha_2] = \begin{bmatrix} 0 & 0 & \alpha_{110} & \mid \alpha_{210} \\ 1 & 0 & \alpha_{111} & \mid \alpha_{211} \\ 0 & 1 & \alpha_{112} & \mid \alpha_{212} \\ 0 & 0 & \alpha_{120} & \mid \alpha_{220} \end{bmatrix}; \quad B_C = [e_1 \; e_4] = \begin{bmatrix} 1 & 0 \\ 0 & 0 \\ 0 & 0 \\ 0 & 1 \end{bmatrix}$$

This completes the example of mapping the indices and coefficients into the state-space canonical forms. □

It is also important to understand that the development of the realization algorithms to follow evolve directly from the construction of the state-space similarity transformations [24]. Therefore, the similarity transformation matrices for the row and column canonical forms are given by [24]

$$T_R = \begin{bmatrix} T_1^R \\ --- \\ T_2^R \\ --- \\ \vdots \\ --- \\ T_{N_y}^R \end{bmatrix} \quad \text{and} \quad T_C = [T_1^C \mid T_2^C \mid \cdots \mid T_{N_u}^C] \tag{6.45}$$

where we have (see Problem 6.10)

$$
\overbrace{}^{j\text{th chain}}
$$

$$
T_j^R = \begin{bmatrix} c_j' \\ c_j' A \\ \vdots \\ c_j' A^{v_j-1} \end{bmatrix} \quad \text{and} \quad T_i^C = \overbrace{[b_i \mid b_i A \mid \cdots \mid b_i A^{\mu_{i-1}}]}^{i\text{th chain}} \tag{6.46}
$$

Now that we have the canonical forms for the matrix pairs, (A_R, C_R) and (A_C, B_C), we still must capture the "canonical triples" $A_\bullet, B_\bullet, C_\bullet$.

It has been shown that a canonical form for an input/output (transfer function or Markov sequence) equivalent system, Σ_{ABCD}, can be obtained from the Hankel matrix [18]. Define $\mathcal{H}_{K,K'}$ in terms of its "block" *Hankel row vectors* $\mathbf{h}'_{j_\bullet}$; $j = 1, \ldots, K$ and block *Hankel column vectors* $\mathbf{h}_{\bullet_i}$; $i = 1, \ldots, K'$, the set of observability invariants *and* block Hankel row vectors $[\{v_j\}, \{\beta_{jk\ell}\}, \{\mathbf{h}'_{j_\bullet}\}]$ form the set of "row realizability invariants". The block Hankel row vectors satisfy the relation analogous to Eq. (6.37) with $\mathbf{h}'_{j_\bullet} \to c_j' A^{v_j}$, that is, there exists exactly one set of ordered scalars $\beta_{jk\ell}$ with $j = i + v_i N_y$; $i = 1, \ldots, N_y$ such that

$$
\mathbf{h}'_{j_\bullet} = \sum_{k=1}^{j-1} \sum_{\ell=0}^{\min(v_j, v_k-1)} \beta_{jk\ell} \, \mathbf{h}'_{k+\ell N_{y_\bullet}} + \sum_{k=j}^{N_y} \sum_{\ell=0}^{\min(v_j, v_k)-1} \beta_{jk\ell} \, \mathbf{h}'_{k+\ell N_{y_\bullet}} \tag{6.47}
$$

where the set $[\{v_j\}, \{\beta_{jk\ell}\},]$ are the *observability invariants* and $\mathbf{h}'_{n_\bullet}$, $n \in KN_y$ is the nth row vector of $\mathcal{H}_{K,K'}$.

Similar results hold for the column vectors of the Hankel matrix leading to the set of "column controllability invariants," that is, set $[\{\mu_i\}, \{\alpha_{imn}\}, \{\mathbf{h}'_{m_\bullet}\}]$ are the *controllability invariants* with $\mathbf{h}'_{m_\bullet}$, $m \in K'N_u$ is the mth column vector of $\mathcal{H}_{K,K'}$.

$$
\mathbf{h}_{\bullet_i} = \sum_{m=1}^{i-1} \sum_{n=0}^{\min(\mu_i, \mu_m-1)} \alpha_{imn} \, \mathbf{h}_{\bullet_{m+nN_u}} + \sum_{m=i}^{N_u} \sum_{n=0}^{\min(\mu_i, \mu_m)-1} \alpha_{imn} \, \mathbf{h}_{\bullet_{m+nN_u}} \tag{6.48}
$$

Thus, all of the dependent block Hankel rows and dependent Hankel columns can be directly generated from this unique set of invariants. With these results available, we can now complete the realization triple with the corresponding

B_R and C_C canonical matrices as

$$
B_R = \begin{bmatrix}
\mathbf{h}'_{1\bullet} \\
\vdots \\
\mathbf{h}'_{(\nu_1-1)N_y+1\bullet} \\
-\ -\ - \\
\vdots \\
-\ -\ - \\
\mathbf{h}'_{N_{y\bullet}} \\
\vdots \\
\mathbf{h}'_{(\nu_{N_y}-1)N_{y\bullet}}
\end{bmatrix};
$$

$$
C_C = [\mathbf{h}_{\bullet 1} \ \cdots \ \mathbf{h}_{\bullet(\mu_1-1)N_u+1} \quad | \quad \cdots \quad | \quad \mathbf{h}_{\bullet N_u} \ \cdots \ \mathbf{h}_{\bullet(\mu_{N_u}-1)N_u)}] \quad (6.49)
$$

and the canonical systems are denoted by $\Sigma_R = \{A_R, B_R, C_R, D_R\}$ and $\Sigma_C = \{A_C, B_C, C_C, D_C\}$ with $D_R = D_C = D$, since D is not included in the similarity transformation.

We mention in closing that Rissanen has also shown that canonical forms for MIMO transfer functions can be constructed directly from these invariants (see [18] for more details). The so-called (left) *matrix fraction description* (MFD) is defined as

$$
H(z) = B^{-1}(z)D(z) \quad \text{for} \quad B(z) = \sum_{i=0}^{\nu} B_i z^i, \ |B_i| \neq 0; D(z) = \sum_{i=0}^{\nu-1} D_i z^i
$$

$$(6.50)$$

The canonical forms for both the left and right MFD are defined by the polynomial pairs $H_R(z) \to (B_R(z), D_R(z))$ and $H_C(z) \to (B_C(z), D_C(z))$ respectively (see both [10] and [18] for detailed constructions).

This completes the section on canonical forms. Next we see how their construction leads to canonical realization techniques.

6.3.3.2 Canonical Realization Algorithm

Perhaps the most significant aspect and the key to canonical realizations is the primary fact that was obtained when developing the canonical triples Σ_R in Eq. (6.47) and Σ_C in Eq. (6.48) – "row or column operations on the Hankel matrix are identical to row or column operations on the respective observability or controllability matrices."

With this in mind, a canonical realization algorithm and equivalently the corresponding similarity transformation for a particular canonical form has already been prescribed by its predecessor construction. That is, the "formula" has already been developed [24, 25]. Here, we will concentrate on the Luenberger canonical forms, but be aware that other forms such as the Bucy canonical form are as easily extracted from the Hankel matrix as well [26].

Identifying a canonical realization from the Markov sequence proceeds like the Ho–Kalman algorithm by performing respective elementary row (P) and/or column (Q) operations on the Hankel matrix. Performing row operations on the observability matrix such that the predecessor dependencies of $P\mathcal{O}_K$ are identical to $\mathcal{O}_K$ and performing column operations on the controllability matrix so that $\mathcal{C}_{K'}E$ and $\mathcal{C}_{K'}$ have the identical dependencies, then it follows that

$$\mathcal{H}^*_{K,K'} = P \times \mathcal{H}_{K,K'} \times E \quad \Longleftrightarrow \quad \mathcal{H}_{K,K'} \tag{6.51}$$

where $\mathcal{H}^*_{K,K'}$ is defined as the *structural array* of $\mathcal{H}_{K,K'}$ revealing the internal coupling of the system, since its nonzero elements occur at (row, column)-locations corresponding to the observability and controllability indices with the remaining elements 0, that is,

$$\mathcal{H}^*_{K,K'}(i,j) = \begin{cases} H(i,j) = H(v_i, \mu_j) \text{ for } i = 1, \dots, N_y; j = 1, \dots, N_u \\ 0 \qquad \text{elsewhere} \end{cases} \tag{6.52}$$

or simply

(observability invariants, controllability invariants) $\rightarrow$ (row/column indices).

There are even more remarkable results from a "predecessor-based" canonical realization algorithm. It can be shown that the structure of the final elementary row transformation matrix P becomes a lower triangular matrix with *1*s on the diagonal with the coefficient dependencies occurring in each row corresponding to the observability indices $\{v_j\}$, while the final elementary column transformation in each column corresponding to the controllability indices μ_i. In fact, it has been shown [19] sets of invariants $\{\beta_{jk\ell}\}$, $\{\alpha_{imn}\}$ or more compactly the sets of N_x-vectors $\{\boldsymbol{\beta}'_i\}$, $\{\boldsymbol{\alpha}_j\}$ are given by the rows of P and the columns of E such that

$$\boldsymbol{\beta}' = -[p_{qr} \; p_{qr+N_y} \; \cdots \; p_{qr+(v_j-1)N_y}]; \text{ for } q = j + v_j N_y, \; j, r = 1, \dots, N_y$$

$$\boldsymbol{\alpha} = -[e_{st} \; e_{s+tN_u} \; \cdots \; e_{s+t(\mu_i-1)N_u}]'; \text{ for } t = i + \mu_i N_y, \; i, s = 1, \dots, N_u$$

where

$$p_{qr}, e_{st} = \begin{cases} 1 & q = r; s = t \\ 0 & q < r; s > t \end{cases} \tag{6.53}$$

Tracking the tedium of these invariant indices is quite cumbersome, but fortunately it is not necessary in order to extract the canonical realization, since the steps of the algorithm are much more straightforward. It is just the canonical structures Σ_R and Σ_C requiring that knowledge in order to construct these canonical forms from the identified invariants. There exist a number of mathematical proofs of each of these results that lead to the subsequent canonical realization algorithm, and they can be found in [16]. Next we present the algorithm.

Initially, we have the subsequent transformation sequence $P \to Q$; $Q \to E$ resulting in the structural array $\mathcal{H}_{K,K'} \to \mathcal{H}_{K,K'}^*$ as

| Initialize | Row Operations | Column Operations |

$$[I_{KN_y} \mid \mathcal{H}_{K,K'} \mid I_{K'N_u}] \longrightarrow [P \mid Q \mid I_{K'N_u}] \longrightarrow [P \mid \mathcal{H}_{K,K'}^* \mid E]$$

The canonical realization algorithm proceeds in the following steps, by first performing the row operations on $\mathcal{H}_{K,K'}$ to obtain

$$[P \mid Q \mid I_{K'N_u}]$$

1. Examine the rows of $\mathcal{H}_{K,K'}$ *column-by-column* for *independence* (0-elements) to extract the row predecessor chains of Eq. (6.38).
2. Continue to perform the predecessor chain row operations to search for dependencies (0-row elements) until the last N_y-rows of Q are zero ensuring the *rank condition* is satisfied.
3. Obtain the *observability* and *controllability indices* $[\{v_j\}, \{\mu_i\}]$ from their location (*row/column indices*) in the Q-matrix.
4. Extract the set $\{\beta_j\}$; $j = 1, \dots, N_y$ from the appropriate rows $(q = j + v_j N_y)$ of the P-matrix as in Eq. (6.53) for A_R.
5. Construct C_R selecting the unit row vectors e'_p based on the extracted observability indices.
6. Extract B_R from the corresponding rows of $\mathcal{H}_{K,K'}$ as in Eq. (6.49).

We now have enough information to construct Σ_R using the extracted "row" invariants. If the column canonical forms are also desired, then the following steps can be performed on the intermediate Q-matrix to obtain the structural matrix $\mathcal{H}_{K,K'}^*$ such that

$$[P \mid \mathcal{H}_{K,K'}^* \mid E]$$

1. Perform column operations on Q to obtain the column predecessor chains for 0-elements in the remaining rows ensuring that the remaining elements are the nonzero *leading elements* corresponding to the observability and controllability index positions (as before).
2. Extract the set $\{\alpha_i\}$; $i = 1, \dots, N_u$ from the appropriate columns $(q = i + \mu_i N_u)$ of the E-matrix as in Eq. (6.53) for A_C.
3. Construct B_C selecting the unit column vectors e'_p based on the extracted controllability indices.
4. Extract C_C from the corresponding columns $\mathcal{H}_{K,K'}$ as in Eq. (6.49).

This completes the canonical realization algorithm. Next we illustrate the algorithm with the following example.

Example 6.5 Consider Example 6.1 and assume that we do not know the order N_x. We would create an extended Hankel matrix in this case with $K = 7$ leading to

$$H_{4,4} = \begin{bmatrix}
1 & 2 & | & 2 & 4 & | & 4 & 8 & | & 8 & 16 \\
1 & 2 & | & 2 & 4 & | & 6 & 10 & | & 13 & 22 \\
1 & 0 & | & 1 & 0 & | & 3 & 2 & | & 6 & 6 \\
- & - & - & - & - & - & - & - & - & - & - \\
2 & 4 & | & 4 & 8 & | & 8 & 16 & | & 16 & 32 \\
2 & 4 & | & 6 & 10 & | & 13 & 22 & | & 28 & 48 \\
1 & 0 & | & 3 & 2 & | & 6 & 6 & | & 13 & 16 \\
- & - & - & - & - & - & - & - & - & - & - \\
4 & 8 & | & 8 & 16 & | & 16 & 32 & | & 32 & 64 \\
6 & 10 & | & 13 & 22 & | & 28 & 48 & | & 58 & 102 \\
3 & 2 & | & 6 & 6 & | & 13 & 16 & | & 27 & 38 \\
- & - & - & - & - & - & - & - & - & - & - \\
8 & 16 & | & 16 & 32 & | & 32 & 64 & | & 64 & 128 \\
13 & 22 & | & 28 & 48 & | & 58 & 102 & | & 119 & 214 \\
6 & 6 & | & 13 & 16 & | & 27 & 38 & | & 56 & 86
\end{bmatrix}$$

(1) Start with $[I_{12} \mid \mathcal{H}_{4,4} \mid I_8]$ and perform elementary row operations to obtain $[P \mid Q \mid I_8]$

$$\left[\begin{array}{cccccccccccccccccccc}
1 & 0 & 0 & 0 & 0 & 0 & 0 & 0 & 0 & 0 & 0 & 0 & | & \underline{1} & 2 & 2 & 4 & 4 & 8 & 8 & 16 \\
-1 & 1 & 0 & 0 & 0 & 0 & 0 & 0 & 0 & 0 & 0 & 0 & | & 0 & 0 & 0 & 0 & \underline{2} & 2 & 5 & 6 \\
-3/2 & 1/2 & 1 & 0 & 0 & 0 & 0 & 0 & 0 & 0 & 0 & 0 & | & 0 & \underline{-2} & -1 & -4 & 0 & -5 & 1/2 & -7 \\
-2 & 0 & 0 & 1 & 0 & 0 & 0 & 0 & 0 & 0 & 0 & 0 & | & 0 & 0 & 0 & 0 & 0 & 0 & 0 & 0 \\
1/2 & -5/2 & 0 & 0 & 1 & 0 & 0 & 0 & 0 & 0 & 0 & 0 & | & 0 & 0 & \underline{2} & 2 & 0 & 1 & -1/2 & 1 \\
1 & 1 & -1 & 0 & -1 & 1 & 0 & 0 & 0 & 0 & 0 & 0 & | & 0 & 0 & 0 & 0 & 0 & 0 & 0 & 0 \\
-4 & 0 & 0 & 0 & 0 & 1 & 0 & 0 & 0 & 0 & 0 & 0 & | & 0 & 0 & 0 & 0 & 0 & 0 & 0 & 0 \\
-3 & 0 & -1 & 0 & -1 & 0 & 0 & 1 & 0 & 0 & 0 & 0 & | & 0 & 0 & 0 & 0 & 0 & 0 & 0 & 0 \\
0 & 1 & -2 & 0 & -1 & 0 & 0 & 0 & 1 & 0 & 0 & 0 & | & 0 & 0 & 0 & 0 & 0 & 0 & 0 & 0 \\
-8 & 0 & 0 & 0 & 0 & 0 & 0 & 0 & 0 & 1 & 0 & 0 & | & 0 & 0 & 0 & 0 & 0 & 0 & 0 & 0 \\
-8 & 1 & -2 & 0 & -2 & 0 & 0 & 0 & 0 & 1 & 0 & 0 & | & 0 & 0 & 0 & 0 & 0 & 0 & 0 & 0 \\
-1 & 2 & -3 & 0 & -2 & 0 & 0 & 0 & 0 & 0 & 0 & 1 & | & 0 & 0 & 0 & 0 & 0 & 0 & 0 & 0
\end{array}\right]$$

$$\underbrace{}_{P} \qquad\qquad\qquad \underbrace{}_{Q}$$

(2) Determine the observability and controllability invariants using the row and column predecessor chain constructs of Eqs. (6.38) and (6.40), that is, from the underlined entries of the Q-matrix we see that (independent rows/columns annotated in parentheses).

Observability Invariants:

Chain No. 1 : $c_1' \to$ nonzero (1); $c_1'A \equiv 0$ (4th row); $\Rightarrow$ $\nu_1 = 1$
Chain No. 2 : $c_2' \to$ nonzero (1); $c_2'A \to$ nonzero (2);
$$c_2'A^2 \equiv 0 \text{ (8th row); } \Rightarrow \nu_2 = 2$$
Chain No. 3 : $c_3' \to$ nonzero (1); $c_3'A \equiv 0$ (6th row); $\Rightarrow$ $\nu_3 = 1$

Controllability Invariants:

Chain 1 : $b_1 \to$ nonzero (1); $Ab_1 \to$ nonzero (2);
$$A^2b_1 \to \text{nonzero (3); } A^3b_1 \equiv 0 \text{ (7th col)} \Rightarrow \mu_1 = 3$$
Chain 2 : $b_2' \to$ nonzero (1); $Ab_2 \equiv 0$ (4th col) $\Rightarrow$ $\mu_2 = 1$

and check the rank condition

$$\rho \mathcal{H}_{2,3} = \rho \mathcal{H}_{3,3} = \rho \mathcal{H}_{2,4} = 4$$

(3) *Extract the coefficients from the rows and columns of P as*
$\mathbf{A}_R$:

$\beta_1' = [\beta_{110} \mid \beta_{120}\ \beta_{121} \mid \beta_{130}] = -[p_{41} \mid p_{42}\ p_{45} \mid p_{43}] = -[2 \quad \mid \quad 0 \quad 0 \quad \mid \quad 0]$
$\beta_2' = [\beta_{210} \mid \beta_{220}\ \beta_{221} \mid \beta_{230}] = -[p_{81} \mid p_{82}\ p_{85} \mid p_{83}] = -[3 \quad \mid \quad 0 \quad 1 \quad \mid \quad 1]$
$\beta_3' = [\beta_{310} \mid \beta_{320}\ \beta_{321} \mid \beta_{330}] = -[p_{61} \mid p_{62}\ p_{65} \mid p_{63}] = -[-1 \mid -1 \quad 1 \quad \mid \quad 1]$

$\mathbf{C}_R$:

$$\mathbf{e}_1' = [1 \quad \mid \quad 0 \quad 0 \quad \mid \quad 0]$$
$$\mathbf{e}_2' = [0 \quad \mid \quad 1 \quad 0 \quad \mid \quad 0]$$
$$\mathbf{e}_4' = [0 \quad \mid \quad 0 \quad 0 \quad \mid \quad 1]$$

$\mathbf{B}_R$:

$$\mathbf{h}_{1\bullet}' = [1 \quad 2]$$
$$\mathbf{h}_{2\bullet}' = [1 \quad 2]$$
$$\mathbf{h}_{5\bullet}' = [2 \quad 4]$$
$$\mathbf{h}_{3\bullet}' = [1 \quad 0]$$

The *Luenberger row canonical forms* are constructed from the observability invariants as

$$A_R = \begin{bmatrix} \beta_1' \\ \mathbf{e}_1' \\ \mathbf{e}_2' \\ \beta_2' \\ \beta_3' \end{bmatrix} = \begin{bmatrix} -2 \mid 0 & 0 \mid 0 \\ 0 \mid 1 & 0 \mid 0 \\ -3 \mid 0 & -1 \mid -1 \\ 1 \mid 1 & -1 \mid -1 \end{bmatrix} ; \quad C_R = \begin{bmatrix} \mathbf{e}_1' \\ \mathbf{e}_2' \\ \mathbf{e}_4' \end{bmatrix} = \begin{bmatrix} 1 \mid 0 & 0 \mid 0 \\ 0 \mid 1 & 0 \mid 0 \\ 0 \mid 0 & 0 \mid 1 \end{bmatrix}$$

$$B_R = \begin{bmatrix} h'_{1\bullet} \\ h'_{2\bullet} \\ h'_{5\bullet} \\ h'_{2\bullet} \end{bmatrix} = \begin{bmatrix} 1 & 2 \\ 1 & 2 \\ 2 & 4 \\ 1 & 0 \end{bmatrix}$$

(4) *Perform the elementary column operations to obtain* $[P \mid \mathcal{H}^*_{4,4} \mid E]$

$$\begin{bmatrix}
\underline{1} & 0 & 0\,0\,0\,0\,0\,0 & | & 1 & -2 & -1 & 1 & -4 & 3/2 & 5/4 & 7/2 \\
0 & 0 & 0\,0\,\underline{2}\,0\,0\,0 & | & 0 & 1 & -1/2 & -3/2 & 0 & -9/4 & 1/8 & -13/4 \\
0 & -\underline{2} & 0\,0\,0\,0\,0\,0 & | & 0 & 0 & 1 & -1 & 0 & -1/2 & 1/4 & -1/2 \\
0 & 0 & 0\,0\,0\,0\,0\,0 & | & 0 & 0 & 0 & 1 & 0 & 0 & 0 & 0 \\
0 & 0 & \underline{2}\,0\,0\,0\,0\,0 & | & 0 & 0 & 0 & 0 & 1 & -1 & -5/2 & -3 \\
0 & 0 & 0\,0\,0\,0\,0\,0 & | & 0 & 0 & 0 & 0 & 0 & 1 & 0 & 0 \\
0 & 0 & 0\,0\,0\,0\,0\,0 & | & 0 & 0 & 0 & 0 & 0 & 0 & 1 & 0 \\
0 & 0 & 0\,0\,0\,0\,0\,0 & | & 0 & 0 & 0 & 0 & 0 & 0 & 0 & 1 \\
0 & 0 & 0\,0\,0\,0\,0\,0 & | & - & - & - & - & - & - & - & - \\
0 & 0 & 0\,0\,0\,0\,0\,0 & | & & & & & & & & \\
0 & 0 & 0\,0\,0\,0\,0\,0 & | & & & & & & & & \\
0 & 0 & 0\,0\,0\,0\,0\,0 & | & & & & & & & &
\end{bmatrix}$$

$$\underbrace{\hspace{3cm}}_{\mathcal{H}^*_{4,4}} \qquad\qquad \underbrace{\hspace{3cm}}_{E}$$

(5) *Extract the coefficients from the columns and rows of E as*

$$\mathbf{A}_C :$$

$$\alpha_1 = [\alpha_{110}\ \alpha_{111}\ \alpha_{112} \mid \alpha_{120}]' = -[e_{17}\ e_{37}\ e_{57} \mid e_{27}]' = -[5/4 \quad 1/4 - 5/2 \mid 1/8]'$$

$$\alpha_2 = [\alpha_{210}\ \alpha_{211}\ \alpha_{212} \mid \alpha_{220}]' = -[e_{14}\ e_{34}\ e_{54} \mid e_{24}]' = -[1 \ -1 \quad 0 \mid -3/2]'$$

$$\mathbf{B}_C :$$

$$\mathbf{e}_1 = [1 \qquad 0 \qquad 0 \qquad 0]'$$

$$\mathbf{e}_4 = [0 \qquad 0 \qquad 0 \qquad 1]'$$

$$\mathbf{C}_C :$$

$$\mathbf{h}_{\bullet 1} = [1 \qquad 1 \qquad 1]'$$

$$\mathbf{h}_{\bullet 3} = [2 \qquad 2 \qquad 1]'$$

$$\mathbf{h}_{\bullet 5} = [4 \qquad 6 \qquad 3]'$$

$$\mathbf{h}_{\bullet 2} = [2 \qquad 2 \qquad 0]'$$

The *Luenberger column canonical form* is constructed from the controllability invariants as

$$A_C = [\mathbf{e}_2 \ \mathbf{e}_3 \ \boldsymbol{\alpha}_1 \mid \boldsymbol{\alpha}_2] = \begin{bmatrix} 0 & 0 & -5/4 & | -1 \\ 1 & 0 & -1/4 & | \ 1 \\ 0 & 1 & 5/2 & | \ 0 \\ 0 & 0 & -1/8 & | \ 3/2 \end{bmatrix}; \ B_C = [\mathbf{e}_1 \quad \mathbf{e}_4] = \begin{bmatrix} 1 & 0 \\ 0 & 0 \\ 0 & 0 \\ 0 & 1 \end{bmatrix}$$

$$C_C = \begin{bmatrix} \mathbf{h}_{\bullet 1} \\ \mathbf{h}_{\bullet 3} \\ \mathbf{h}_{\bullet 5} \\ \mathbf{h}_{\bullet 2} \end{bmatrix} = \begin{bmatrix} 1 & 2 & 4 & | \ 2 \\ 1 & 2 & 6 & | \ 2 \\ 1 & 1 & 3 & | \ 0 \end{bmatrix}$$

This completes the canonical realization example.

□

We summarize the canonical realization algorithm in Table 6.2.

Table 6.2 Canonical realization algorithm.

Hankel Matrix Elementary Row (Predecessor) Operations
Observability/Controllability Indices (Rank)

$$[I_{KN_y} \mid \mathcal{H}_{K,K'} \mid I_{K'N_u}] \rightarrow \underbrace{[P \mid Q \mid I_{K'N_u}]}_{\text{Row Operations}}$$

- Perform the predecessor chain row operations until the last N_y-rows of Q are zero *rank condition satisfied*
- Extract the *observability* and *controllability indices* $[\{v_j\}, \{\mu_i\}]$ from row/column indices of the Q-matrix

Extract Row Invariants

- Extract the set $\{\beta_j\}$; $j = 1, \cdots, N_y$ from the appropriate rows $(q = j + v_j N_y)$

Construct Row Canonical System: $\Sigma_{A_R, B_R, C_R, D_R}$

- A_R from rows of P-matrix based on the extracted observability indices
- C_R selecting the unit row vectors e'_p based on the extracted observability indices
- B_R from the corresponding rows of $\mathcal{H}_{K,K'}$
- D_R from H_0

$$[P \mid Q \mid I_{K'N_u}] \rightarrow \underbrace{[P \mid \mathcal{H}^*_{K,K'} \mid E]}_{\text{Column Operations}}$$

- Perform column operations on Q to obtain the column predecessor chains for 0-elements in the remaining rows.
- Extract the set $\{\alpha_i\}$; $i = 1, \cdots, N_u$ from the appropriate columns $(q = i + \mu_i N_u)$ of the E-matrix to construct for A_C.
- Construct B_C selecting the unit column vectors e'_p based on the extracted controllability indices.
- Extract C_C from the corresponding columns $\mathcal{H}_{K,K'}$.
- Set $D_C = H_0$.

In this subsection, we have shown how to use canonical state-space structures to extract unique models from the Markov sequences or impulse response matrices evolving from MIMO systems. Before we conclude this section, it is important to understand that input/output data could be used directly to perform these operations as detailed in [27], rather than estimating the impulse responses first and then applying a similarity transformation. In the next section, we develop the subspace techniques that extract the state-space model directly from MIMO data, but in contrast to the canonical structures developed here – they are "black-box" models that could be transformed to the canonical structures through a similarity transformation such as those of Eq. (6.45) for the Luenberger canonical forms. This is the basic philosophy of subspace realizations, "fit a black-box state-space model and transform it to the desired coordinate system" (e.g. canonical form) for processing. Next we investigate these subspace methods.

6.4 Deterministic Subspace Realization: Orthogonal Projections

In this subsection, we develop the fundamental subspace realization approach to identifying the state-space representation from input/output data extending the realization from impulse response to input/output data – still assumed *deterministic*. In order to comprehend just how the deterministic realization problem can be solved when input/output sequences rather than impulse response sequences evolve, it is important to understand that most subspace realization techniques require the extraction of the extended observability matrix directly from acquired data first, followed by the system model extraction, second. The existing methods differ in the manner in which the observability matrix is estimated and their methodology for extracting the system model. To extract the state-space model, Σ_{ABCD}, it is only required to obtain either the extended column space of the observability matrix or the row space of the state sequence both can easily be obtained using the SVD, that is, the *key* elements of subspace identifications are based on extracting:

$$
\mathcal{O}_K = \begin{bmatrix} C \\ CA \\ \vdots \\ CA^{K-1} \end{bmatrix} ; \qquad \mathcal{X}_k = \underbrace{[\mathbf{x}(k) \quad \mathbf{x}(k+1) \quad \cdots \quad \mathbf{x}(K+k-1) \quad]}_{\text{Row space}}
$$

$\underbrace{\phantom{\mathcal{O}_K = \begin{bmatrix} C \\ CA \end{bmatrix}}}_{\text{Column space}}$

Perhaps the major "link" between classical realization theory employing impulse response sequences followed by the Hankel matrix factorization and subspace realization theory from input/output sequences lies within the inherent system theoretic "rank condition" defining the minimal dimension of the state-space. Subspace realization theory relies primarily on the concept of the persistent excitation of the input sequences as well as the minimal state-space dimension for solution [9], that is, the observability matrix spans the N_x-dimensional state-space that contains that part of the output due to the states exclusive of the inputs. In order to see this, we first construct input/output data matrices (as before).

Recall that the discrete LTI solution of Eq. (3.33) given by

$$x(t) = A^t x(0) + \sum_{\ell=0}^{t-1} A^{t-\ell-1} Bu(\ell); \qquad t = 0, 1, \dots, K \tag{6.54}$$

with the measurement or output system given by

$$y(t) = CA^t x(0) + \sum_{\ell=0}^{t-1} CA^{t-\ell-1} Bu(\ell) + D\delta(t) \tag{6.55}$$

Expanding this relation further over k-samples and collecting terms, we obtain

$$\begin{bmatrix} y(0) \\ y(1) \\ \vdots \\ y(k-1) \end{bmatrix} = \underbrace{\begin{bmatrix} C \\ CA \\ \vdots \\ CA^{k-1} \end{bmatrix}}_{\mathcal{O}_k} x(0) + \underbrace{\begin{bmatrix} D & \cdots & 0 \\ CB & D & \cdots & 0 \\ \vdots & \vdots & \vdots & \vdots \\ CA^{k-1}B & \cdots & CB & D \end{bmatrix}}_{\mathcal{T}_k} \begin{bmatrix} u(0) \\ u(1) \\ \vdots \\ u(k-1) \end{bmatrix} \tag{6.56}$$

where $\mathcal{O}_k \in \mathcal{R}^{kN_y \times N_x}$ is the *observability matrix* and $\mathcal{T}_k \in \mathcal{R}^{kN_y \times kN_u}$ is *impulse response matrix* – a Toeplitz matrix [4].

Shifting these relations in time $(0 \rightarrow t)$ yields

$$\begin{bmatrix} y(t) \\ y(t+1) \\ \vdots \\ y(t+k-1) \end{bmatrix} = \mathcal{O}_k x(t) + \mathcal{T}_k \begin{bmatrix} u(t) \\ u(t+1) \\ \vdots \\ u(t+k-1) \end{bmatrix} \tag{6.57}$$

leads to the *vector* input/output relation

$$\mathbf{y}_k(t) = \mathcal{O}_k \mathbf{x}(t) + \mathcal{T}_k \mathbf{u}_k(t) \tag{6.58}$$

where $\mathbf{y} \in \mathcal{R}^{kN_y \times 1}$, $\mathbf{x} \in \mathcal{R}^{N_x \times 1}$ and $\mathbf{u} \in \mathcal{R}^{kN_u \times 1}$.

We stack these vectors to create batch data (*block Hankel*) matrices to obtain the subsequent "data equation." That is, defining the block output data matrix $(\mathcal{Y}_{0|k-1})$ and block input data matrix $(\mathcal{U}_{0|k-1})^4$ over K-samples are

$$\mathcal{Y}_{0|k-1} := \begin{bmatrix} \mathbf{y}(0) & \mathbf{y}(1) & \cdots & \mathbf{y}(K-1) \\ \mathbf{y}(1) & \mathbf{y}(2) & \cdots & \mathbf{y}(K) \\ \vdots & \vdots & \vdots & \vdots \\ \mathbf{y}(k-1) & \mathbf{y}(k) & \cdots & \mathbf{y}(K+k-2) \end{bmatrix} \tag{6.59}$$

---

4 The notation "$0|k-1$" defines the first and last row elements of the K-column data matrix.

and

$$
\mathcal{U}_{0|k-1} := \begin{bmatrix} \mathbf{u}(0) & \mathbf{u}(1) & \cdots & \mathbf{u}(K-1) \\ \mathbf{u}(1) & \mathbf{u}(2) & \cdots & \mathbf{u}(K) \\ \vdots & \vdots & \vdots & \vdots \\ \mathbf{u}(k-1) & \mathbf{u}(k) & \cdots & \mathbf{u}(K+k-2) \end{bmatrix} \tag{6.60}
$$

or more compactly

$$
\mathcal{Y}_{0|k-1} = [\mathbf{y}_k(0) \ \cdots \ \mathbf{y}_k(K-1)] \in \mathcal{R}^{kN_y \times K}
$$
$$
\mathcal{U}_{0|k-1} = [\mathbf{u}_k(t) \ \cdots \ \mathbf{u}_k(K-1)] \in \mathcal{R}^{kN_u \times K} \tag{6.61}
$$

with the corresponding state vector defined by

$$
\mathcal{X}_0 := [\mathbf{x}(0) \ \mathbf{x}(1) \ \cdots \ \mathbf{x}(K-1)] \in \mathcal{R}^{N_x \times K} \tag{6.62}
$$

Therefore, we have the *vector–matrix* input/output equation or the *data equation* that relates the system model to the data (input, state, and output matrices)

$$
\mathcal{Y}_{0|k-1} = \mathcal{O}_k \mathcal{X}_0 + \mathcal{T}_k \mathcal{U}_{0|k-1} \tag{6.63}
$$

The overall objective is to extract the observability matrix and subsequently the system matrices Σ_{ABCD} from the deterministic input/output data. That is, we seek a solution that possesses the column space (range) equivalent to $\mathcal{O}_K$ (e.g. range($\mathcal{Y}_{0|k-1}$) $\equiv$ range($\mathcal{O}_K$)). Clearly, if the impulse response matrix $\mathcal{T}_K$ is *known* besides the data and input matrices, then we can merely subtract the last term from the data equation and we would have

$$
\mathcal{Y}_{0|k-1} = \mathcal{O}_k \mathcal{X}_0
$$

providing the desired result enabling the extraction of Σ_{ABCD} after performing an SVD to obtain the equivalent column space, but unfortunately $\mathcal{T}_K$ is *unknown*! Thus, we must somehow estimate the impulse response matrix, first, in order to remove it from the data equation.

6.4.1 Subspace Realization: Orthogonal Projections

One approach to extract the observability matrix is *least-squares* [7, 28–31], that is, to formulate the problem as minimizing a cost function $\mathcal{J}$ with respect to $\mathcal{T}_k$

$$
\mathcal{J} = \min_{\mathcal{T}_k} \ \| \mathcal{Y}_{0|k-1} - \mathcal{T}_k \mathcal{U}_{0|k-1} \| \tag{6.64}
$$

in order to remove the effect of the input using the idea of developing a projection operator $\mathcal{P}$ that is orthogonal to the input space $\mathcal{U}$ such that

$< \mathcal{U}, \mathcal{P}^{\perp} > = \mathbf{0}$. Well-known methods exist to accomplish this estimation such as setting the gradient operator to zero ($\nabla_{\mathcal{T}} J = \mathbf{0}$) and solving for $\hat{\mathcal{T}}_k$ or completing the square [7, 31]. Therefore, the least-squares estimate for this problem is obtained as

$$\hat{\mathcal{T}}_k := \mathcal{Y}_{0|k-1} \mathcal{U}_{0|k-1}' (\mathcal{U}_{0|k-1} \mathcal{U}_{0|k-1}')^{-1} \tag{6.65}$$

Removing the input from the data equation gives

$$\mathcal{Y}_{0|k-1} - \hat{\mathcal{T}}_k \times \mathcal{U}_{0|k-1} = \mathcal{Y}_{0|k-1} - \mathcal{Y}_{0|k-1} \mathcal{U}_{0|k-1}' (\mathcal{U}_{0|k-1} \mathcal{U}_{0|k-1}')^{-1} \times \mathcal{U}_{0|k-1}$$

and factoring we have

$$\mathcal{Y}_{0|k-1} - \hat{\mathcal{T}}_k \mathcal{U}_{0|k-1} = \mathcal{Y}_{0|k-1} \underbrace{(I - \mathcal{U}_{0|k-1}' (\mathcal{U}_{0|k-1} \mathcal{U}_{0|k-1}')^{-1} \mathcal{U}_{0|k-1})}_{\mathcal{P}^{\perp}_{\mathcal{U}_{0|k-1}}} = \mathcal{Y}_{0|k-1} \mathcal{P}^{\perp}_{\mathcal{U}_{0|k-1}}$$

$$\tag{6.66}$$

where

$$\mathcal{P}^{\perp}_{\mathcal{U}_{0|k-1}} = I - \mathcal{P}_{\mathcal{U}_{0|k-1}} = I - \mathcal{U}_{0|k-1}' (\mathcal{U}_{0|k-1} \mathcal{U}_{0|k-1}')^{-1} \mathcal{U}_{0|k-1} \tag{6.67}$$

is the orthogonal projection $< \mathcal{U}_{0|k-1}, \mathcal{P}^{\perp}_{\mathcal{U}_{0|k-1}} >$ of the (column) space of $\mathcal{U}_{0|k-1}$ onto the null space of $\mathcal{U}$ such that

$$\mathcal{U}_{0|k-1} \times \mathcal{P}^{\perp}_{\mathcal{U}_{0|k-1}} = \mathcal{U}_{0|k-1} - (\mathcal{U}_{0|k-1} \mathcal{U}_{0|k-1}')(\mathcal{U}_{0|k-1} \mathcal{U}_{0|k-1}')^{-1} \mathcal{U}_{0|k-1} = \mathbf{0} \quad (6.68)$$

Here the solution depends on assuming that $\mathcal{U}_{0|k-1}$ is "persistently exciting" or equivalently it is full rank, $\rho(\mathcal{U}) = kN_u$.

Applying the projection operator to the data equation of Eq. (6.63) projects the "past data" onto (" | ") the null space of $\mathcal{U}$ defined by $\mathcal{P}^{\perp}_{\mathcal{Y}_{0|k-1}|\mathcal{U}_{0|k-1}}$ yielding the desired results

$$\mathcal{P}^{\perp}_{\mathcal{Y}_{0|k-1}|\mathcal{U}_{0|k-1}} := \mathcal{Y}_{0|k-1} \times \mathcal{P}^{\perp}_{\mathcal{U}_{0|k-1}} = \mathcal{O}_k \mathcal{X}_0 \mathcal{P}^{\perp}_{\mathcal{U}_{0|k-1}} + \mathcal{T}_k \underbrace{\mathcal{U}_{0|k-1} \mathcal{P}^{\perp}_{\mathcal{U}_{0|k-1}}}_{0} = \mathcal{O}_k \mathcal{X}_0 \mathcal{P}^{\perp}_{\mathcal{U}_{0|k-1}}$$

$$\tag{6.69}$$

of eliminating the input and extracting the observability matrix (see Figure 6.6).

Next, we must be assured that the column spaces of the data and the observability matrix are identical in order to extract Σ_{ABCD}.

From the data equation of Eq. (6.63), we can construct the *augmented* (input/output) *data matrix*, $\mathcal{D} \in \mathcal{R}^{k(N_u + N_y) \times K}$

$$\mathcal{D}_{0|k-1} := \begin{bmatrix} \mathcal{U}_{0|k-1} \\ --- \\ \mathcal{Y}_{0|k-1} \end{bmatrix} = \begin{bmatrix} I_{kN_u} & | & 0 \\ - & & - \\ \mathcal{T}_k & | & \mathcal{O}_k \end{bmatrix} \begin{bmatrix} \mathcal{U}_{0|k-1} \\ -- \\ \mathcal{X}_0 \end{bmatrix} \tag{6.70}$$

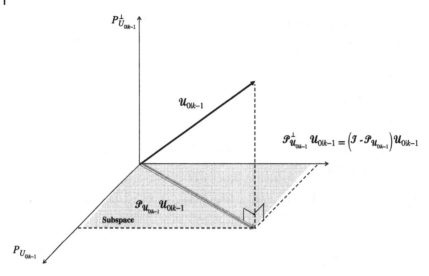

Figure 6.6 Orthogonal projection operator: orthogonal complement operation.

and

$$\rho\left(\begin{bmatrix} \mathcal{U}_{0|k-1} \\ -- \\ \mathcal{X}_0 \end{bmatrix}\right) = N_x + kN_u$$

leads to the *rank condition*

$$\rho(\mathcal{Y}_{0|k-1}P^{\perp}_{\mathcal{U}_{0|k-1}}) = \rho(\mathcal{O}_K\mathcal{X}_0 P^{\perp}_{\mathcal{U}_{0|k-1}}) = N_x \qquad \text{(Rank Condition)}$$

therefore, it follows immediately that

$$\text{range}(\mathcal{Y}_{0|k-1}) \equiv \text{range}(\mathcal{O}_K)$$

the desired condition (see [7] or [9] for proof).

This "rank condition" provides a powerful result for the identification of the system model Σ_{ABCD} from deterministic input/output data just as the rank condition for impulse response data enabling the Hankel matrix factorization provides for the identification. It leads to pragmatic numerically efficient linear algebraic methods (to follow) providing a robust solution to solve the problem.

However, before we close this discussion, the system model must be extracted from the output data, which can be accomplished using a truncated SVD decomposition – similar to the method of Section 6.3.2, that is, the SVD partitions the projected data, as

$$P^{\perp}_{\mathcal{Y}_{0|k-1}|\mathcal{U}_{0|k-1}} = \mathcal{Y}_{0|k-1} P^{\perp}_{\mathcal{U}_{0|k-1}} = \mathcal{O}_k \mathcal{X}_0 \, P^{\perp}_{\mathcal{U}_{0|k-1}} = [U_S \mid U_N] \begin{bmatrix} \Sigma_S & | & 0 \\ - & - & - \\ 0 & | & \Sigma_N \end{bmatrix} \begin{bmatrix} V'_S \\ -- \\ V'_N \end{bmatrix}$$

$$= \underbrace{U_S \Sigma_S V'_S}_{\text{Signal Subspace}} + \underbrace{U_N \Sigma_N V'_N}_{\text{Noise Subspace}}$$

(6.71)

Note that an added benefit of this decomposition is that the system order N_x can be estimated directly from its singular values enabling the partition into signal and noise subspaces (e.g. see Example 6.2). There exists a variety of *order estimation* criteria to accomplish this choice (see Section 6.6 and [4] for more details).

For a *balanced realization* choice (see Section 6.2), the observability matrix can be extracted directly through this decomposition as

$$\underbrace{\mathcal{O}_k \mathcal{X}_0 \, P^{\perp}_{\mathcal{U}_{0|k-1}} \approx (U_S \Sigma_S^{1/2})}_{\mathcal{O}_k} \times ((\Sigma'_S)^{1/2} V'_S)$$

(6.72)

assuming the noise subspace is negligible ($\Sigma_N \ll \Sigma_S$).

Here unique system matrices can be obtained by applying the pseudo-inverse ($Z^{\#} = (Z'Z)^{-1}Z'$ or $Z^{\#} = Z'(ZZ')^{-1}$), since the input/output rank condition guarantees full rank of the observability matrix to extract[5]

$$A = \mathcal{O}^{\#}_{N_x} \times \mathcal{O}^{\uparrow}_{N_x} = \Sigma_S^{-1/2} U'_S \times \mathcal{O}^{\uparrow}_{N_x} \quad \text{and} \quad C = \mathcal{O}_{N_x}(1 : N_y, 1 : N_x)$$

(6.73)

To obtain the input transmission matrix B and the input/output transmission matrix D, we can use the orthogonality of the left-singular matrix, U_N, from the SVD of Eq. (6.71) to isolate (extract) the impulse response matrix $\mathcal{T}_k$, that is, premultiplying the data equation by U_N gives [9, 32, 33]

$$U'_N \times \mathcal{Y}_{0|k-1} = \underbrace{U'_N \times \mathcal{O}_k \mathcal{X}_0}_{0} + U'_N \times \mathcal{T}_k \mathcal{U}_{0|k-1} = U'_N \times \mathcal{T}_k \mathcal{U}_{0|k-1}$$

(6.74)

since the first term is null, that is, from the SVD we have

$$\underbrace{U'_N \times (\mathcal{O}_k \mathcal{X}_0 \, P^{\perp}_{\mathcal{U}_{0|k-1}})}_{0} = \underbrace{U'_N (U_S \Sigma_S V'_S)}_{0} = 0$$

Postmultiplying by the pseudo-inverse $\mathcal{U}^{\#}_{0|k-1} = \mathcal{U}_{0|k-1}(\mathcal{U}_{0|k-1}\mathcal{U}_{0|k-1})^{-1}$

---

5 Recall the shift-invariant property of the observability matrix in Eq. (6.8).

yields the relation

$$U'_N \mathcal{Y}_{0|k-1} \times \mathcal{U}^\#_{0|k-1} = U'_N \mathcal{T}_k \mathcal{U}_{0|k-1} \times \mathcal{U}^\#_{0|k-1} = U'_N \mathcal{T}_k \tag{6.75}$$

to isolate and extract the impulse response matrix.
Define the following matrices:

$$\mathcal{Z} := U'_N \mathcal{Y}_{0|k-1} \mathcal{U}^\#_{0|k-1} := [\mathbf{z}_1 \; \mathbf{z}_1 \; \cdots \; \mathbf{z}_k] \in \mathcal{R}^{(kN_y - N_x) \times kN_u}$$
$$U'_N := [\mathbf{u}^N_1 \; \mathbf{u}^N_2 \; \cdots \; \mathbf{u}^N_k] \in \mathcal{R}^{(kN_y - N_x) \times kN_y}$$

$$\tag{6.76}$$

for $\mathcal{T}_k \in \mathcal{R}^{kN_y \times kN_u}$ and $\mathbf{z}_i \in \mathcal{R}^{(kN_y - N_x) \times N_u}$, $\mathbf{u}_i \in \mathcal{R}^{(kN_y - N_x) \times N_y}$, $i = 1, \dots, k$.
Therefore, with these definitions, we have

$$\mathcal{Z} = U'_N \times \mathcal{T}_k \tag{6.77}$$

expanding gives

$$\mathcal{Z} = [\mathbf{z}_1 \quad \mathbf{z}_1 \quad \cdots \quad \mathbf{z}_k] = [\mathbf{u}^N_1 \; \mathbf{u}^N_2 \; \cdots \; \mathbf{u}^N_k] \begin{bmatrix} D & & \cdots & 0 \\ CB & D & \cdots & 0 \\ \vdots & \vdots & \vdots & \vdots \\ CA^{k-1}B & \cdots & CB & D \end{bmatrix}$$

$$\tag{6.78}$$

Performing the multiplications to expand, we obtain the following (vector) relations:

$$\mathbf{z}_1 = \mathbf{u}^N_1 D + \mathbf{u}^N_2 CB + \cdots + \mathbf{u}^N_k CA^{k-2}B$$
$$\mathbf{z}_2 = \mathbf{u}^N_2 D + \mathbf{u}^N_3 B + \cdots + \mathbf{u}^N_k CA^{k-3}B$$
$$\vdots = \vdots \qquad \vdots$$
$$\mathbf{z}_{k-1} = \mathbf{u}^N_{k-1} D + \mathbf{u}^N_k CB$$
$$\mathbf{z}_k = \mathbf{u}^N_k D \tag{6.79}$$

resulting in a stacking (reordering) of the $\mathbf{z}_i$ into a new structure defined by $\tilde{\mathcal{Z}} \in \mathcal{R}^{k(kN_y - N_x) \times N_u}$

$$\tilde{\mathcal{Z}} = \begin{bmatrix} \mathbf{z}_1 \\ \mathbf{z}_2 \\ \vdots \\ \mathbf{z}_{k-1} \\ \mathbf{z}_k \end{bmatrix} = \begin{bmatrix} \mathbf{u}^N_1 \mid [\mathbf{u}^N_2 \cdots \mathbf{u}^N_k] \times \mathcal{O}_{k-1} \\ \mathbf{u}^N_2 \mid [\mathbf{u}^N_3 \cdots \mathbf{u}^N_k] \times \mathcal{O}_{k-2} \\ \vdots \qquad\qquad \vdots \\ \mathbf{u}^N_{k-1} \mid [\quad \mathbf{u}^N_k \quad] \times \mathcal{O}_1 \\ \mathbf{u}^N_k \mid \qquad [\,0\,] \end{bmatrix} \times \begin{bmatrix} D \\ -- \\ B \end{bmatrix}$$

or defining $\overline{\mathbf{u}}_i^N := [\mathbf{u}_i^N \; \mathbf{u}_{i+1}^N \; \cdots \mathbf{u}_k^N]$, we have

$$
\tilde{\mathcal{Z}} = \underbrace{\begin{bmatrix} \mathbf{u}_1^N & | & \overline{\mathbf{u}}_2^N \times \mathcal{O}_{k-1} \\ \mathbf{u}_2^N & | & \overline{\mathbf{u}}_3^N \times \mathcal{O}_{k-2} \\ \vdots & & \vdots \\ \mathbf{u}_{k-1}^N & | & \overline{\mathbf{u}}_k^N \times \mathcal{O}_1 \\ \mathbf{u}_k^N & | & [\,0\,] \end{bmatrix}}_{U} \times \begin{bmatrix} D \\ -- \\ B \end{bmatrix} := U \times \begin{bmatrix} D \\ -- \\ B \end{bmatrix} \tag{6.80}
$$

Since U is a full-rank matrix, we obtain a unique *least-squares* solution to extract the $\hat{B}$ and $\hat{D}$ input transmission matrices as

$$
\begin{bmatrix} \hat{D} \\ -- \\ \hat{B} \end{bmatrix} = U^{\#} \times \tilde{\mathcal{Z}} = (U'U)^{-1}U' \times \tilde{\mathcal{Z}} \tag{6.81}
$$

Even though this result is theoretically pleasing and provides insights into the motivation of the subsequent algorithms, it is *not* numerically efficient, leading us to a more robust linear algebraic solution for this state-space identification problem.

6.4.2 Multivariable Output Error State-Space (MOESP) Algorithm

A numerically efficient linear algebraic approach to solving these identification problems can be achieved by applying the LQ-decomposition[6] method to numerically calculate orthogonal projections enabling a reliable "separation" of data into subspaces in order to eliminate disturbances and noise. Therefore, to extract the state-space model, Σ_{ABCD}, it is required to obtain the extended column space of the observability matrix that can be obtained using the SVD, that is, one of the key elements of subspace identification is based on extracting:

$$
\mathcal{O}_K = \begin{bmatrix} C \\ CA \\ \vdots \\ CA^{K-1} \end{bmatrix}
$$

Column space

The MOESP[7] technique is based on performing an orthogonal LQ-decomposition on the data matrix (see Appendix C), that is, $\mathcal{D} = L \times Q$

---

6 The LQ-decomposition is the *dual* of the well-known Gram–Schmidt orthogonalization method using the QR-decomposition of a matrix, that is, $LQ = (QR)' = R'Q'$ with L now lower triangular and replacing $Q' \rightarrow Q$ [9].

7 This method is called the PI-MOESP technique with instrument variable $\mathcal{U}_{0|k-1}$ [28].

with L a lower block triangular matrix, $L \in \mathcal{R}^{k(N_u+N_y) \times k(N_u+N_y)}$ and Q an orthogonal matrix, $Q \in \mathcal{R}^{k(N_u+N_y) \times K}$ such that

$$\mathcal{D}_{0|k-1} = \begin{bmatrix} I_{kN_u} & | & O_{kN_u \times N_x} \\ -- & | & -- \\ \mathcal{T}_k & | & \mathcal{O}_k \end{bmatrix} \begin{bmatrix} \mathcal{U}_{0|k-1} \\ ---- \\ \mathcal{X}_0 \end{bmatrix} = \begin{bmatrix} L_{11} & | & O_{kN_u \times kN_y} \\ -- & | & -- \\ L_{21} & | & L_{22} \end{bmatrix} \begin{bmatrix} Q_1' \\ -- \\ Q_2' \end{bmatrix} \tag{6.82}$$

with the $L_{11} \in \mathcal{R}^{kN_u \times kN_u}, L_{21} \in \mathcal{R}^{kN_y \times kN_u}, L_{22} \in \mathcal{R}^{kN_y \times kN_y}; Q_1' \in \mathcal{R}^{kN_u \times K}$ and $Q_2' \in \mathcal{R}^{kN_y \times K}$.

This operation produces an efficient orthogonal decomposition of the augmented data matrix, $\mathcal{D}_{0|k-1}$, that is, the LQ-decomposition is a numerical (matrix) version of the *Gram–Schmidt orthogonalization* procedure [22, 29, 30]. Recall that the LQ-decomposition of a matrix contains the set of orthonormal basis vectors in Q ($Q'Q = I$) with $\mathbf{q}_i \in \mathcal{R}^{K \times 1}$ and L contains the coefficients of linear dependence establishing the relationship the basis vectors of $\mathcal{D}$ to those of Q. In fact, from this viewpoint, we can think of partitioning $Q = [Q_D \mid Q_D^\perp]$ such that a rank-deficient data matrix $\mathcal{D}$ can (hypothetically) be decomposed as

$$\mathcal{D} = LQ = [L_D \mid 0] \begin{bmatrix} Q_D \\ -- \\ Q_D^\perp \end{bmatrix} = L_D \times Q_D$$

The *projection matrix* associated with the decomposition is specified by [30]

$$P = Q \times Q' \qquad \text{(Projection Matrix)}$$

From Eq. (6.70) (assuming full-rank matrices), we have that

$$\mathcal{D}_{0|k-1} = \begin{bmatrix} L_{11} & | & O \\ -- & | & -- \\ L_{21} & | & L_{22} \end{bmatrix} \begin{bmatrix} Q_1' \\ -- \\ Q_2' \end{bmatrix} = \begin{bmatrix} \mathcal{U}_{0|k-1} \\ ---- \\ \mathcal{Y}_{0|k-1} \end{bmatrix} = \begin{bmatrix} L_{11} \times Q_1' \\ --- \\ L_{21} Q_1' + L_{22} Q_2' \end{bmatrix} \tag{6.83}$$

Orthogonally projecting $\mathcal{Y}_{0|k-1}$ onto $\mathcal{U}_{0|k-1}$ and $\mathcal{U}_{0|k-1}^\perp$ enables the decomposition by applying the projection matrices

$$P_{\mathcal{Y}_{0|k-1} \mid \mathcal{U}_{0|k-1}} = \mathcal{Y}_{0|k-1}(\underbrace{Q_1 Q_1'}_{P_{\mathcal{U}_{0|k-1}}}) = (L_{21}Q_1' + L_{22}Q_2')Q_1 \times Q_1' = L_{21} \times Q_1'$$

$$P_{\mathcal{Y}_{0|k-1} \mid \mathcal{U}_{0|k-1}^\perp} = \mathcal{Y}_{0|k-1}(\underbrace{Q_2 Q_2'}_{P_{\mathcal{U}_{0|k-1}^\perp}^\perp}) = (L_{21}Q_1' + L_{22}Q_2')Q_2 \times Q_2' = L_{22} \times Q_2'$$

$$\tag{6.84}$$

where we have applied the orthonormal properties of Q such that $Q_i'Q_i = I$ and $Q_i'Q_j = 0'$.

Returning to the original formulation of Eq. (6.70) and equating with these expressions, we see that

$$\mathcal{U}_{0|k-1} = L_{11}Q_1'$$
$$\mathcal{Y}_{0|k-1} = \mathcal{O}_k\mathcal{X}_0 + \mathcal{T}_k\mathcal{U}_{0|k-1} = \mathcal{O}_k\mathcal{X}_0 + \mathcal{T}_k(\underbrace{L_{11}Q_1'}_{\mathcal{U}_{0|k-1}}) = L_{21}Q_1' + L_{22}Q_2' \quad (6.85)$$

Postmultiplying Eq. (6.85) by Q_2 and imposing the orthogonality conditions of the LQ-decomposition as $Q_2'\,Q_2 = I$ and $Q_1'\,Q_2 = 0$ enables the extraction of the observability matrix

$$\mathcal{Y}_{0|k-1}Q_2 = \mathcal{O}_k\mathcal{X}_0Q_2 + \mathcal{T}_kL_{11}\underbrace{Q_1'Q_2}_{0} = L_{21}\underbrace{Q_1'Q_2}_{0} + L_{22}\underbrace{Q_2'Q_2}_{I} = L_{22}$$

or simply

$$\mathcal{O}_k\mathcal{X}_0Q_2 = L_{22} \quad (6.86)$$

with $\rho(L_{22}) = N_x$ from the rank condition for input/output sequences. Performing the SVD of L_{22} is given by

$$L_{22} = [U_S \mid U_N]\begin{bmatrix} \Sigma_S & | & 0 \\ - & - & - \\ 0 & | & \Sigma_N \end{bmatrix}\begin{bmatrix} V_S' \\ -- \\ V_N' \end{bmatrix} = \underbrace{U_S\Sigma_S V_S'}_{\text{Signal subspace}} + \underbrace{U_N\Sigma_N V_N'}_{\text{Noise subspace}} \quad (6.87)$$

Assuming the uncertainty or equivalently noise component is negligible compared to the signal component and selecting the balanced realization (as before) yields

$$\mathcal{O}_k\mathcal{X}_0Q_2 \approx U_S \times \Sigma_S \times V_S' \approx \underbrace{(U_S\Sigma_S^{1/2})}_{\mathcal{O}_k} \times ((\Sigma_S')^{1/2}V_S') \quad (6.88)$$

extracting the observability matrix – the desired result.

The system matrices A, B, C, D can now be determined as before by

$$A = \mathcal{O}_{N_x}^{\#} \times \mathcal{O}_{N_x}^{\uparrow} = \Sigma_S^{-1/2}U_S'; \qquad C = \mathcal{O}(1 : N_y, 1 : N_x) \quad (6.89)$$

with B and D obtained by solving a least-squares problem directly, since premultiplying Eq. (6.85) by U_N' gives

$$U_N'\mathcal{O}_k\mathcal{X}_0 + U_N'\mathcal{T}_k\mathcal{U}_{0|k-1} = U_N'L_{21}Q_1' + U_N'L_{22}Q_2'$$

but after substituting for $\mathcal{U}_{0|k-1}$ and applying the orthogonality conditions $U'_N \mathcal{O}_k = 0$ and $U'_N L_{22} = 0$, we have

$$U'_N \mathcal{T}_k(L_{11}Q'_1) = U'_N L_{21}Q'_1$$

postmultiply by Q_1 and using its orthonormal property ($Q'_1 Q_1 = I$) gives

$$Z := U'_N \mathcal{T}_k = U'_N L_{21} \times L_{11}^{-1} \tag{6.90}$$

analogous to Eq. (6.77) with $Z \to \mathcal{Z}$. Stacking and reordering the vectors (as before) leads to an overdetermined set of linear equations and a least-squares solution with $\tilde{Z} \to \tilde{\mathcal{Z}}$ to give

$$\left[\begin{array}{c} \hat{D} \\ -- \\ \hat{B} \end{array} \right] = \mathcal{U}^\# \times \tilde{Z} \tag{6.91}$$

where $\mathcal{U}$ is the block coefficient matrix equivalent to Eq. (6.81) ($\mathcal{U} \to \mathbf{U}$).

Therefore, we have the *Multivariable Output Error State-Space* (MOESP[8]) algorithm given by

- Compute the LQ-decomposition of $\mathcal{D}$ of Eq. (6.82).
- Perform the SVD of L_{22} in Eq. (6.87) to extract $\mathcal{O}_k$.
- Obtain A and C from Eq. (6.89).
- Solve the least-squares problem to obtain B and D from Eq. (6.91).

We summarize the deterministic MOESP algorithm in Table 6.3.

6.5 Deterministic Subspace Realization: Oblique Projections

Finally, for the deterministic realization problem from input/output data, we develop an alternative technique, the *parallel* or *oblique* projection method based on both past and future data. This approach differs from the MOESP-method in that it first estimates the state sequence directly from the data through a projection of the row space of block Hankel matrices. That is, in this approach, the row space of the state sequence can be obtained using the SVD; here the *key* element of oblique subspace identification is based on extracting:

$$\mathcal{X}_k = \underbrace{[\mathbf{x}(k)\ \mathbf{x}(k+1)\ \cdots\ \mathbf{x}(K+k-1)\,]}_{\text{State-Space Sequence}}$$

---

8 Note that if the instrument variable is selected as both $\mathcal{U}_{0|k-1}$ and $\mathcal{Y}_{0|k-1}$, the past inputs and outputs, then the PO-MOESP technique evolves rather than the PI-MOESP [28].

Table 6.3 Multivariable output error state-space algorithm (MOESP).

$$
\mathcal{D}_{0|k-1} = \begin{bmatrix} I_{kN_u}| & 0 \\ -- & - \\ \mathcal{T}_k| & \mathcal{O}_k \end{bmatrix} \begin{bmatrix} \mathcal{U}_{0|k-1} \\ -- \\ \mathcal{X}_0 \end{bmatrix} = \begin{bmatrix} L_{11}| & 0 \\ -- & - \\ L_{21}| & L_{22} \end{bmatrix} \begin{bmatrix} Q'_1 \\ -- \\ Q'_2 \end{bmatrix}
$$

LQ-Decomposition

(Data Matrix)

Singular-Value Decomposition

$$
L_{22} = [U_S \mid U_N] \begin{bmatrix} \Sigma_S| & 0 \\ -- & - \\ 0| & \Sigma_N \end{bmatrix} \begin{bmatrix} V'_S \\ -- \\ V'_N \end{bmatrix} \approx \underbrace{(U_S\Sigma_S^{1/2})}_{\mathcal{O}_{N_x}} \times ((\Sigma'_S)^{1/2}V'_S)
$$

(Observability Matrix)

System Matrix

$$
A = \mathcal{O}_{N_x}^{\#} \times \mathcal{O}_{N_x}^{\uparrow} = (\mathcal{O}'_{N_x}\mathcal{O}_{N_x})^{-1}\mathcal{O}'_{N_x} \times \mathcal{O}_{N_x}^{\uparrow} = \Sigma_S^{-1/2}U'_S \times \mathcal{O}_{N_x}^{\uparrow}
$$

(System/Process Matrix)

Measurement (Output) Matrix

$$
C = \mathcal{O}(1 : N_y, 1 : N_x)
$$

(Output Matrix)

Input/Input/Output Transmission Matrices

$$
\begin{bmatrix} \hat{D} \\ -- \\ \hat{B} \end{bmatrix} = \mathcal{U}^{\#} \times \tilde{Z}
$$

(Input Matrices)

for $\mathcal{U}^{\#} := (\mathcal{U}'\mathcal{U})^{-1}\mathcal{U}'$

State estimation is then followed by a linear least-squares solution extracting the desired system matrices, Σ_{ABCD}. It is based on the concept that the projection of the "future outputs" depend solely on the past inputs/outputs as well as on the "future" inputs that evolve directly from the available data. Since the state sequence is available from the input/output data, it is used to establish a solution to the least-squares problem enabling subspace identification to be accomplished [6–9].

We begin by defining the data matrices for samples ranging from k-to-$(2k - 1)$ – the *future* data as [9]

$$
\mathcal{Y}_{k|2k-1} := \begin{bmatrix} \mathbf{y}(k) & \mathbf{y}(k+1) & \cdots & \mathbf{y}(K+k-1) \\ \mathbf{y}(k+1) & \mathbf{y}(k+2) & \cdots & \mathbf{y}(K+k) \\ \vdots & \vdots & \vdots & \vdots \\ \mathbf{y}(2k-1) & \mathbf{y}(2k) & \cdots & \mathbf{y}(K+2k-2) \end{bmatrix}
\tag{6.92}
$$

and

$$
\mathcal{U}_{k|2k-1} := \begin{bmatrix} \mathbf{u}(k) & \mathbf{u}(k+1) & \cdots & \mathbf{u}(K+k-1) \\ \mathbf{u}(k+1) & \mathbf{u}(k+2) & \cdots & \mathbf{u}(K+k) \\ \vdots & \vdots & \vdots & \vdots \\ \mathbf{u}(2k-1) & \mathbf{u}(2k) & \cdots & \mathbf{u}(K+2k-2) \end{bmatrix} \tag{6.93}
$$

Leading to the corresponding *data equation* given by

$$
\mathcal{Y}_{k|2k-1} = \mathcal{O}_k \mathcal{X}_k + \mathcal{T}_k \mathcal{U}_{k|2k-1} \tag{6.94}
$$

where the block (Hankel) matrices are defined as before to give

$$
\mathcal{Y}_{k|2k-1} = [\mathbf{y}_k(k) \quad \mathbf{y}_k(k+1) \quad \cdots \quad \mathbf{y}_k(K+k-1) \quad] \in \mathcal{R}^{kN_y \times K}
$$
$$
\mathcal{U}_{k|2k-1} = [\mathbf{u}_k(k) \quad \mathbf{u}_k(k+1) \quad \cdots \quad \mathbf{u}_k(K+k-1) \quad] \in \mathcal{R}^{kN_u \times K}
$$
$$
\mathcal{X}_k = [\mathbf{x}(k) \quad \mathbf{x}(k+1) \quad \cdots \quad \mathbf{x}(K+k-1) \quad] \in \mathcal{R}^{N_x \times K}
$$

Here $\mathcal{U}_{0|k-1}$, $\mathcal{Y}_{0|k-1}$, $\mathcal{X}_0$ are defined as the *past* inputs, outputs, and states ($\mathcal{U}_p$, $\mathcal{Y}_p$, $\mathcal{X}_0$), while $\mathcal{U}_{k|2k-1}$, $\mathcal{Y}_{k|2k-1}$, $\mathcal{X}_k$ are the *future* inputs, outputs, and states (U_f, Y_f, $\mathcal{X}_f$), which are all *block Hankel* matrices as before, that is,

$$
\mathcal{U}_p := \mathcal{U}_{0|k-1}; \qquad \mathcal{Y}_p := \mathcal{Y}_{0|k-1}; \qquad \mathcal{X}_p := \mathcal{X}_0
$$
$$
\mathcal{U}_f := \mathcal{U}_{k|2k-1}; \qquad \mathcal{Y}_f := \mathcal{Y}_{k|2k-1}; \qquad \mathcal{X}_f := \mathcal{X}_k \tag{6.95}
$$

Similar to the input/output relations developed previously in Eq. (6.63), the corresponding state equations can be developed by expanding the solution of the deterministic state-space model:

$$
\mathbf{x}(k+j) = A^k \mathbf{x}(j) + \sum_{\ell=0}^{k-1} A^{k-\ell-1} Bu(\ell+j)
$$
$$
= A^k \mathbf{x}(j) + A^{k-1} Bu(j) + A^{k-2} Bu(j+1) + \cdots + Bu(k+j-1)
$$

or

$$
\mathbf{x}(k+j) = A^k \mathbf{x}(j) + \underbrace{[A^{k-1}B \; A^{k-2}B \; \cdots \; AB \; B \,]}_{C_{AB}^{\leftarrow}} \underbrace{\begin{bmatrix} u(j) \\ u(j+1) \\ \vdots \\ u(j+k-1) \end{bmatrix}}_{\mathbf{u}}
$$

simplifying and grouping, we have

$$
\mathbf{x}(k+j) = A^k \mathbf{x}(j) + C_{AB}^{\leftarrow}(k) \, \mathbf{u}(j) \tag{6.96}
$$

where the "reversed" (input) controllability matrix $C_{AB}^{\leftarrow} \in \mathcal{R}^{N_x \times kN_u}$ is defined by

$$
C_{AB}^{\leftarrow}(k) := [A^{k-1}B \; A^{k-2}B \; \cdots \; AB \; B \,] \; [\text{reversed input controllability}] \tag{6.97}
$$

Expanding this relation over K-columns with $j = 0$, then

$$\underbrace{[\mathbf{x}(k) \; \cdots \; \mathbf{x}(K + k - 1)]}_{\mathcal{X}_k} = A^k \underbrace{[\mathbf{x}(0) \; \cdots \; \mathbf{x}(K - 1)]}_{\mathcal{X}_0}$$

$$+ C_{AB}^{\leftarrow}(k) \underbrace{[\mathbf{u}_k(0) \; \cdots \; \mathbf{u}_k(K - 1)]}_{\mathcal{U}_{0|k-1}}$$

that is, we now have

$$\mathcal{X}_k = A^k \mathcal{X}_0 + C_{AB}^{\leftarrow}(k) \, \mathcal{U}_{0|k-1} \tag{6.98}$$

or in terms of our definitions (above), the set of future states in terms of past states $\mathcal{X}_p$, inputs $\mathcal{U}_p$ becomes

$$\mathcal{X}_f = A^k \mathcal{X}_p + C_{AB}^{\leftarrow}(k) \, \mathcal{U}_p \tag{6.99}$$

Therefore, the set of "deterministic" data equations for the past/future inputs/outputs/states are as follows:

$$\mathcal{Y}_p = \mathcal{O}_k \mathcal{X}_p + \mathcal{T}_k \mathcal{U}_p \qquad \text{(Past Outputs)}$$

$$\mathcal{Y}_f = \mathcal{O}_k \mathcal{X}_f + \mathcal{T}_k \mathcal{U}_f \qquad \text{(Future Outputs)}$$

$$\mathcal{X}_f = A^k \mathcal{X}_p + C_{AB}^{\leftarrow}(k) \mathcal{U}_p \qquad \text{(Future States)} \tag{6.100}$$

with the corresponding data matrices, past and future, defined by $\mathcal{D}_p := \mathcal{D}_{0|k-1}$ and $\mathcal{D}_f := \mathcal{D}_{k|2k-1}$ where

$$\mathcal{D}_p = \begin{bmatrix} \mathcal{U}_p \\ --- \\ \mathcal{Y}_p \end{bmatrix} \text{ and } \mathcal{D}_{0|k-1} = \begin{bmatrix} \mathcal{U}_{0|k-1} \\ --- \\ \mathcal{Y}_{0|k-1} \end{bmatrix};$$

$$\mathcal{D}_f = \begin{bmatrix} \mathcal{U}_f \\ --- \\ \mathcal{Y}_f \end{bmatrix} \text{ and } \mathcal{D}_{k|2k-1} = \begin{bmatrix} \mathcal{U}_{k|2k-1} \\ --- \\ \mathcal{Y}_{k|2k-1} \end{bmatrix} \tag{6.101}$$

With these input/state/output relations available, we see the relationship between the future state vector with the corresponding past and future data as

$$\text{span}\{\mathcal{X}_f\} = \text{span}\{\mathcal{D}_p\} \cap \text{span}\{\mathcal{D}_f\} \tag{6.102}$$

under the assumption of a persistently excited, completely controllable and observable system ($\rho(\mathcal{O}_k) = \rho(C_k) = N_x$) along with the constraint on the initial state matrix being independent of the past inputs (span$\{\mathcal{X}_0\}$ ∩ span$\{\mathcal{U}_{0|k-1}\}$ = $\{0\}$).

Therefore, solving Eq. (6.100) for the past and future states employing the pseudo-inverse ($^\#$) of the full-rank observability matrix ($\mathcal{O}_k^\#$) provides

$$\mathcal{X}_p = \mathcal{O}_k^\# \times (\mathcal{Y}_p - \mathcal{T}_k \mathcal{U}_p) \quad \text{for} \quad \mathcal{X}_p \in \text{span}\{\mathcal{D}_p\} \tag{6.103}$$

and

$$\mathcal{X}_f = \mathcal{O}_k^\# \times (\mathcal{Y}_f - \mathcal{T}_k \mathcal{U}_f) \quad \text{for} \quad \mathcal{X}_f \in \text{span}\{\mathcal{D}_f\} \tag{6.104}$$

Substituting for $\mathcal{X}_p$ into Eq. (6.99), we have

$$\mathcal{X}_f = A^k (\mathcal{O}_k^\# \mathcal{Y}_p - \mathcal{O}_k^\# \mathcal{T}_k \mathcal{U}_p) + C_{AB}^\leftarrow(k)\, \mathcal{U}_p \tag{6.105}$$

expanding this expression and grouping like-terms, we obtain a vector relation in terms of $\mathcal{D}_p$

$$\mathcal{X}_f = \underbrace{\left[C_{AB}^\leftarrow(k) - A^k \mathcal{O}_k^\# \mathcal{T}_k \mid A^k \mathcal{O}_k^\# \right]}_{\Theta_p} \begin{bmatrix} \mathcal{U}_p \\ -- \\ \mathcal{Y}_p \end{bmatrix} = \Theta_p \times \mathcal{D}_p \tag{6.106}$$

demonstrating that the future states *depend only* on the past inputs and outputs and not on the future inputs $\mathcal{U}_f$ indicating that Eq. (6.102) is satisfied; that is, the future states provide the set of basis vectors evolving from the intersection of the past and future data subspaces. Therefore, intuitively, $\mathcal{X}_f$ enables the exchange of information between the past and future.

In terms of Eq (6.106), it also follows that the future output can be written as

$$\mathcal{Y}_f = \mathcal{O}_k\, \Theta_p\, \mathcal{D}_p + \mathcal{T}_k\, \mathcal{U}_f \tag{6.107}$$

With this information available, we can now establish the basis of N4SID approach.

6.5.1 Subspace Realization: Oblique Projections

Another approach to decomposing or separating the data spaces is to apply an *oblique projection operator* enabling the extraction of the state-space model and eliminating the past input/output data. The parallel (oblique) projection ("$\|$") of the future output data $\mathcal{Y}_f$ projected ("$|$") *onto* the past data subspace $\mathcal{D}_p$ *along* ("$\circ$") the future $\mathcal{U}_f$ is depicted in Figure 6.7. As illustrated in the figure, the orthogonal projector can be constructed from the sum of two oblique projectors and one orthogonal complement projector. Here, the orthogonal projection is performed first enabling the projection onto the joint space ($\mathcal{D}_p \vee \mathcal{U}_f$) that is followed by the decomposition along either $\mathcal{D}_f$ or $\mathcal{U}_p$ [9]

$$P_{\mathcal{Y}_f|\mathcal{D}_p \vee \mathcal{U}_f} = P_{\mathcal{Y}_f|\mathcal{D}_p \circ \, \mathcal{U}_f}^{\|} + P_{\mathcal{Y}_f|\mathcal{U}_f \circ \, \mathcal{D}_p}^{\|} + P_{\mathcal{Y}_f|\mathcal{U}_f \vee \mathcal{D}_p}^{\perp} \tag{6.108}$$

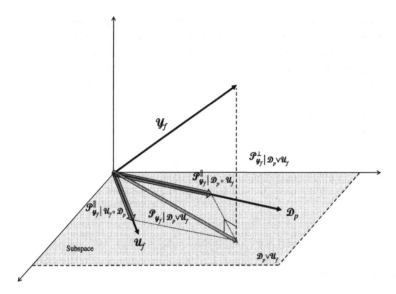

Figure 6.7 Oblique projections: Projection $(\mathcal{P}^{\parallel}_{\mathcal{Y}_f | \mathcal{D}_p \,\circ\, \mathcal{U}_f})$ of future outputs $(\mathcal{Y}_f)$ onto past data $(\mathcal{D}_p)$ along the future inputs $(\mathcal{U}_f)$.

Another way to express this idea from an algebraic rather than geometric perspective is that the "vectors in the row space of $\mathcal{Y}_f$ are the sum of linear combinations of vectors in the row space of the state $\mathcal{X}_f$ and with those in the row space of the future inputs $\mathcal{U}_f$."

In contrast, the oblique projection matrix can also be defined in terms of orthogonal projections as well [6, 8]

$$
\mathcal{P}^{\parallel}_{\mathcal{Y}_f | \mathcal{D}_p \circ \, \mathcal{U}_f} := \left[\mathcal{P}^{\perp}_{\mathcal{Y}_f | \mathcal{U}_f} \right] \times \left[\mathcal{P}^{\perp}_{\mathcal{D}_p | \mathcal{U}_f} \right]^{\#} \mathcal{D}_p = \left[\mathcal{Y}_f \mathcal{P}^{\perp}_{\mathcal{U}_f} \right] \times \left[\mathcal{D}_p \mathcal{P}^{\perp}_{\mathcal{U}_f} \right]^{\#} \mathcal{D}_p \quad (6.109)
$$

Substituting Eq. (6.107) for the "future" output into this expression, we have

$$
\mathcal{P}^{\parallel}_{\mathcal{Y}_f | \mathcal{D}_p \circ \, \mathcal{U}_f} = \left[(\mathcal{O}_k \Theta_p \mathcal{D}_p + \underbrace{\mathcal{T}_k \mathcal{U}_f}_{0}\,) \mathcal{P}^{\perp}_{\mathcal{U}_f} \right] \left[\mathcal{D}_p \mathcal{P}^{\perp}_{\mathcal{U}_f} \right]^{\#} \mathcal{D}_p
$$

$$
= \mathcal{O}_k \Theta_p \underbrace{\left[\mathcal{D}_p \mathcal{P}^{\perp}_{\mathcal{U}_f} \right] \left[\mathcal{D}_p \mathcal{P}^{\perp}_{\mathcal{U}_f} \right]^{\#} \mathcal{D}_p}_{I} \qquad (6.110)
$$

or simply from Eq. (6.106), we have

$$
\mathcal{P}^{\parallel}_{\mathcal{Y}_f | \mathcal{D}_p \circ \, \mathcal{U}_f} = \mathcal{O}_k \Theta_p \mathcal{D}_p = \mathcal{O}_k \mathcal{X}_f \qquad (6.111)
$$

This is the underlying concept of the N4SID-approach when applied to the deterministic realization problem. This relationship shows that obliquely projecting the future data nullifies the inputs, while enabling the extraction of the observability and future states directly from the given data. Therefore, the *state vector* provides a basis for the *intersection* of the past and future subspaces and can be extracted from this projection.

Performing the SVD on Eq. (6.111), we obtain the partitioned signal and noise (uncertainty) relationship as

$$
\mathcal{P}^{\parallel}_{\mathcal{Y}_f|\mathcal{D}_p} \circ \mathcal{U}_f = [U_S \mid U_N] \begin{bmatrix} \Sigma_S & \mid & 0 \\ -- & -- & -- \\ 0 & \mid & \Sigma_N \end{bmatrix} \begin{bmatrix} V'_S \\ -- \\ V'_N \end{bmatrix} = \underbrace{U_S \Sigma_S V'_S}_{\text{Signal Subspace}} + \underbrace{U_N \Sigma_N V'_N}_{\text{Noise Subspace}}
$$

(6.112)

Assuming the uncertainties are "negligible" ($\Sigma_N \ll \Sigma_S$) and selecting a *balanced deterministic realization* ($\Sigma_S = \Sigma_S^{1/2} \times (\Sigma'_S)^{1/2}$), we obtain the critical N4SID subspace relations:

$$
\mathcal{P}^{\parallel}_{\mathcal{Y}_f|\mathcal{D}_p} \circ \mathcal{U}_f \approx \underbrace{U_S \Sigma_S^{1/2}}_{\mathcal{O}_k} \underbrace{(\Sigma'_S)^{1/2} V'_S}_{\mathcal{X}_f} \in \mathcal{R}^{kN_y \times K} \qquad \text{(Projection)}
$$

$$
\mathcal{O}_k = U_S \Sigma_S^{1/2} \in \mathcal{R}^{kN_y \times N_x} \qquad \text{(Observability)}
$$

$$
\mathcal{X}_f = (\Sigma'_S)^{1/2} V'_S \in \mathcal{R}^{N_x \times K} \qquad \text{(States)} \qquad (6.113)
$$

6.5.2 Numerical Algorithms for Subspace State-Space System Identification (N4SID) Algorithm

Again, a numerically efficient approach to extract the system Σ_{ABCD} from input/output data is to employ the LQ-decomposition of the block Hankel data matrices of the past $\mathcal{D}_p$ and future input/output data matrices $\{\mathcal{U}_f, \mathcal{Y}_f\}$ that will lead us to the oblique projection of $\mathcal{Y}_f$ onto $\mathcal{D}_p$ along $\mathcal{U}_f$ [9, 34]

$$
\mathcal{D} := \begin{bmatrix} \mathcal{U}_f \\ -- \\ \mathcal{D}_p \\ -- \\ \mathcal{Y}_f \end{bmatrix} = \underbrace{\begin{bmatrix} L_{11} & \mid & 0 & \mid & 0 \\ -- & \mid & -- & \mid & -- \\ L_{21} & \mid & L_{22} & \mid & 0 \\ -- & \mid & -- & \mid & -- \\ L_{31} & \mid & L_{32} & \mid & L_{33} \end{bmatrix}}_{L} \times \underbrace{\begin{bmatrix} Q'_1 \\ -- \\ Q'_2 \\ -- \\ Q'_3 \end{bmatrix}}_{Q} \qquad (6.114)
$$

with $L_{11} \in \mathcal{R}^{kN_u \times kN_u}$, $L_{21} \in \mathcal{R}^{k(N_u+N_y) \times kN_u}$, $L_{22} \in \mathcal{R}^{k(N_u+N_y) \times k(N_u+N_y)}$, $L_{31} \in \mathcal{R}^{kN_y \times kN_u}$, $L_{32} \in \mathcal{R}^{kN_y \times k(N_u+N_y)}$, $L_{33} \in \mathcal{R}^{kN_y \times kN_y}$; and $Q'_1 \in \mathcal{R}^{kN_u \times K}$, $Q'_2 \in \mathcal{R}^{k(N_u+N_y) \times K}$, $Q'_3 \in \mathcal{R}^{kN_y \times K}$.

As illustrated in Figure 6.7, the oblique projection of the future data $\mathcal{Y}_f$ onto the "joint" space of future inputs $\mathcal{U}_f$ and past inputs/outputs data $\mathcal{D}_p$ can be expressed in terms of the summation of two oblique projections: one onto $\mathcal{U}_f$ along $\mathcal{D}_p$ ($\mathcal{P}^{\parallel}_{\mathcal{Y}_f \mid \mathcal{U}_f \circ \mathcal{D}_p}$) and one onto $\mathcal{D}_p$ along $\mathcal{U}_f$ ($\mathcal{P}^{\parallel}_{\mathcal{Y}_f \mid \mathcal{D}_p \circ \mathcal{U}_f}$) combined with complement of the orthogonal projection of $\mathcal{Y}_f$ onto the joint space of future inputs and past inputs/outputs ($\mathcal{P}^{\perp}_{\mathcal{Y}_f \mid \mathcal{U}_f \vee \mathcal{D}_p}$), that is,

$$
\mathcal{P}^{\parallel}_{\mathcal{Y}_f \mid \mathcal{U}_f \vee \mathcal{D}_p} = \underbrace{\mathcal{P}^{\parallel}_{\mathcal{Y}_f \mid \mathcal{U}_f \circ \mathcal{D}_p} + \mathcal{P}^{\parallel}_{\mathcal{Y}_f \mid \mathcal{D}_p \circ \mathcal{U}_f}}_{\text{Oblique Projections}} + \underbrace{\mathcal{P}^{\perp}_{\mathcal{Y}_f \mid \mathcal{U}_f \vee \mathcal{D}_p}}_{\text{Orthogonal Complement Projection}}
$$

(6.115)

The orthogonal projection of $\mathcal{Y}_f$ onto the joint space of $\mathcal{U}_f \vee \mathcal{D}_p$ is the subspace spanned by the orthonormal basis vectors of Q_1' and Q_2'; therefore, we have that

$$
\mathcal{P}_{\mathcal{Y}_f \mid \mathcal{U}_f \vee \mathcal{D}_p} = [L_{31} \mid L_{32}] \begin{bmatrix} Q_1' \\ -- \\ Q_2' \end{bmatrix} \quad \text{(Orthogonal Projection)} \quad (6.116)
$$

Alternatively, this projection can also be expressed in terms of *oblique projection operators* corresponding to the summation of the two oblique projections as in Eq. (6.115)

$$
\mathcal{P}_{\mathcal{Y}_f \mid \mathcal{U}_f \vee \mathcal{D}_p} = [\mathcal{P}^{\parallel}_{\bullet \mid \mathcal{U}_f \circ \mathcal{D}_p} \mid \mathcal{P}^{\parallel}_{\bullet \mid \mathcal{D}_p \circ \mathcal{U}_f}] \begin{bmatrix} L_{11} \mid 0 \\ -- \mid -- \\ L_{21} \mid L_{22} \end{bmatrix} \begin{bmatrix} Q_1' \\ -- \\ Q_2' \end{bmatrix}
$$

[Orthogonal Projection]

(6.117)

since Q_1', Q_2' span the joint subspace of $\mathcal{U}_f \vee \mathcal{D}_p$.

Equating these relations for the "joint" orthogonal projection ($\mathcal{P}_{\mathcal{Y}_f \mid \mathcal{U}_f \vee \mathcal{D}_p}$) and solving in terms of the oblique projection operators gives

$$
[\mathcal{P}^{\parallel}_{\bullet \mid \mathcal{U}_f \circ \mathcal{D}_p} \mid \mathcal{P}^{\parallel}_{\bullet \mid \mathcal{D}_p \circ \mathcal{U}_f}] = [L_{31} \mid L_{32}] \begin{bmatrix} L_{11} \mid 0 \\ --- \mid --- \\ L_{21} \mid L_{22} \end{bmatrix}^{-1}
$$

$$
= [L_{31} \mid L_{32}] \begin{bmatrix} L_{11}^{-1} \mid 0 \\ --- \mid --- \\ -L_{22}^{-1} L_{21} L_{11}^{-1} \mid L_{22}^{-1} \end{bmatrix}
$$

(6.118)

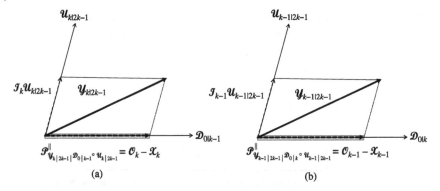

Figure 6.8 Extraction of future state and shifted state sequences: (a) Oblique projection for future state sequence $\mathcal{X}_k$. (b) Oblique projection for shifted state sequence $\mathcal{X}_{k+1}$.

or simply

$$\mathcal{P}^{\parallel}_{\bullet|\mathcal{U}_f \circ \mathcal{D}_p} = L_{31}L_{11}^{-1} - L_{32}L_{22}^{-1}L_{21}L_{11}^{-1}$$

$$\mathcal{P}^{\parallel}_{\bullet|\mathcal{D}_p \circ \mathcal{U}_f} = L_{32}L_{22}^{-1} \tag{6.119}$$

Obliquely projecting $\mathcal{Y}_f$ onto $\mathcal{D}_p$ along $\mathcal{U}_f$ and substituting for the past data from Eq. (6.114) as illustrated in Figure 6.8a, we obtain the *oblique projection* in terms of the reduced LQ-decomposition and perform the corresponding SVD as in Eq. (6.113) to obtain the set of relations:

$$\mathcal{P}^{\parallel}_{\mathcal{Y}_f|\mathcal{D}_p \circ \mathcal{U}_f} = \mathcal{O}_k\mathcal{X}_f = L_{32}L_{22}^{-1}\,\mathcal{D}_p \qquad \text{[Oblique Projection]}$$

$$\mathcal{O}_k = U_S\Sigma_S^{1/2} \qquad\qquad\qquad \text{[Observability]}$$

$$\mathcal{X}_f = \Sigma_S^{1/2}V_S' \qquad\qquad\qquad \text{[States]} \tag{6.120}$$

For the N4SID approach, both state sequences $\mathcal{X}_k$ and $\mathcal{X}_{k+1}$ will be required[9] subsequently to extract the model from the input/output data [25]. Therefore, another oblique projection based on projecting the future (shifted) output $\mathcal{Y}_{k+1|2k-1}$ onto the (shifted) data $\mathcal{D}_{0|k}$ along the (shifted) future inputs $\mathcal{U}_{k+1|2k-1}$, that is, $\mathcal{P}^{\parallel}_{\mathcal{Y}_{k+1|2k-1}|\mathcal{D}_{0|k} \circ \mathcal{U}_{k+1|2k-1}}$ will be required (see Figure 6.8b).

---

9 Some approaches use the extracted states available from $\mathcal{X}_f \to \{\mathcal{X}_k\}; k \to k + K - 1$ for K large enabling a reasonable approximation for $\mathcal{X}_{k+1}$ [9, 35].

These oblique projections are [10]

$$\mathcal{P}^{\|}_{\mathcal{Y}_f|\mathcal{D}_p} \circ \mathcal{U}_f = \mathcal{P}^{\|}_{\mathcal{Y}_{k|2k-1}|\mathcal{D}_{0|k-1}} \circ \mathcal{U}_{k|2k-1} = \mathcal{O}_k \mathcal{X}_k \quad \text{[Oblique Projection]}$$

$$\mathcal{P}^{\|}_{\mathcal{Y}_{k+1|2k-1}|\mathcal{D}_{0|k}} \circ \mathcal{U}_{k+1|2k-1} = \mathcal{O}_{k-1} \mathcal{X}_{k+1} \quad \text{[Shifted Projection]}$$

$$(6.121)$$

where $\mathcal{O}_{k-1} \in \mathcal{O}_k$ is the extended observability matrix with the *last block row eliminated*.[11]

From these projections, we can estimate *both* required state sequences, $\overline{\mathcal{X}}_{k+1}$ and $\overline{\mathcal{X}}_k$ in order to solve a subsequent least-squares problem and extract the model Σ_{ABCD}. It is clear from the oblique projection of Eq. (6.120) that this can easily be accomplished from the SVD of Eq. (6.112) ($\overline{\mathcal{X}}_k \to \mathcal{X}_k \equiv \mathcal{X}_f$) by applying the pseudo-inverse of the full-rank observability matrix as

$$\overline{\mathcal{X}}_k = \mathcal{O}_k^{\#} \times \mathcal{P}^{\|}_{\mathcal{Y}_{k|2k-1}|\mathcal{D}_{0|k-1}} \circ \mathcal{U}_{k|2k-1} = \Sigma_{S_k}^{-1/2} U'_{S_k} \times \mathcal{P}^{\|}_{\mathcal{Y}_{k|2k-1}|\mathcal{D}_{0|k-1}} \circ \mathcal{U}_{k|2k-1}$$

$$\overline{\mathcal{X}}_{k+1} = \mathcal{O}_{k-1}^{\#} \times \mathcal{P}^{\|}_{\mathcal{Y}_{k+1|2k-1}|\mathcal{D}_{0|k}} \circ \mathcal{U}_{k+1|2k-1} = \Sigma_{S_{k-1}}^{-1/2} U'_{S_{k-1}} \times \mathcal{P}^{\|}_{\mathcal{Y}_{k+1|2k-1}|\mathcal{D}_{0|k}} \circ \mathcal{U}_{k+1|2k-1}$$

$$(6.122)$$

where $\Sigma_{S_{k-1}}^{-1/2} \in \Sigma_{S_k}^{-1/2}$ and $U_{S_{k-1}} \in U_{S_k}$ are the singular values and orthogonal matrices (respectively) with the last block row eliminated (as above for the observability matrix).

With these states now available, we can define the underlying "batch" state-space system $\overline{\Sigma}_{ABCD}$ as

$$\overline{\mathcal{X}}_{k+1} = A\overline{\mathcal{X}}_k + B\overline{\mathcal{U}}_k$$
$$\overline{\mathcal{Y}}_k = C\overline{\mathcal{X}}_k + D\overline{\mathcal{U}}_k$$

$$(6.123)$$

with

$$\overline{\mathcal{X}}_{k+1} := [x(k+1) \ \cdots \ x(k+K-1)]$$
$$\overline{\mathcal{X}}_k := [x(k) \ \cdots \ x(k+K-2)]$$
$$\overline{\mathcal{Y}}_k := [y(k) \ \cdots \ y(k+K-2)]$$
$$\overline{\mathcal{U}}_k := [u(k) \ \cdots \ u(k+K-2)]$$

---

10 Recall that $\mathcal{X}_k \equiv \mathcal{X}_f$ and $\mathcal{X}_0 \equiv \mathcal{X}_p$.

11 The shifted projection relation follows from the same steps as Eq. (6.107)(6.111) replacing $\mathcal{Y}_f \to \mathcal{Y}_{k+1|2k-1}, \mathcal{U}_f \to \mathcal{U}_{k+1|2k-1}, \mathcal{O}_k \to \mathcal{O}_{k-1}$ and $\mathcal{X}_k \to \mathcal{X}_{k+1}$ in the derivation [25].

Table 6.4 Numerical algorithm for subspace state-space system identification (N4SID).

$$D = \begin{bmatrix} U_f \\ D_p \\ Y_f \end{bmatrix} = \begin{bmatrix} L_{11} & 0 & 0 \\ L_{21} & L_{22} & 0 \\ L_{31} & L_{32} & L_{33} \end{bmatrix} \begin{bmatrix} \tilde{Q}_1 \\ \tilde{Q}_2 \\ \tilde{Q}_3 \end{bmatrix}$$	*Oblique Projection (LQ-Decomposition)* (Data Matrix)
$$P^{\parallel}_{Y_f\mid D_p} \circ U_f = [U_S \ \ U_N] \begin{bmatrix} \Sigma_S & | & 0 \\ - & - & - \\ 0 & | & \Sigma_N \end{bmatrix} \begin{bmatrix} V'_S \\ V'_N \end{bmatrix} \approx U_S \Sigma_S V'_S$$	*Singular-Value Decomposition* (Projection Matrix)
$$P^{\parallel}_{Y_{k\mid 2k-1}\mid D_{0\mid k-1} \circ U_{k\mid 2k-1}} = \mathcal{O}_k \mathcal{X}_k = L_{32} L_{22}^{-1} D_{0\mid k-1}$$ $$\mathcal{O}_k = U_S \Sigma_S^{1/2}$$ $$\mathcal{X}_k = \Sigma_S^{1/2} V'_S$$	*Observability/State Extraction* (Oblique Projection) (Observability Matrix) (State Matrix)
$$\overline{\mathcal{X}}_k = \mathcal{O}_k^{\#} \times P^{\parallel}_{Y_{k\mid 2k-1}\mid D_{0\mid k-1} \circ U_{k\mid 2k-1}} = \Sigma_{S_k}^{-1/2} U'_{S_k} \times P^{\parallel}_{Y_{k\mid 2k-1}\mid D_{0\mid k-1} \circ U_{k\mid 2k-1}}$$ $$\overline{\mathcal{X}}_{k+1} = \mathcal{O}_{k-1}^{\#} \times P^{\parallel}_{Y_{k+1\mid 2k-1}\mid D_{0\mid k} \circ U_{k+1\mid 2k-1}} = \Sigma_{S_{k-1}}^{-1/2} U'_{S_{k-1}} \times P^{\parallel}_{Y_{k+1\mid 2k-1}\mid D_{0\mid k} \circ U_{k+1\mid 2k-1}}$$	*State Matrices* (State Matrices)
$$\begin{bmatrix} \hat{A} & | & \hat{B} \\ - & - & - \\ \hat{C} & | & \hat{D} \end{bmatrix} = \left(\begin{bmatrix} \overline{\mathcal{X}}_{k+1} \\ - - - \\ \overline{Y}_k \end{bmatrix} \begin{bmatrix} \overline{\mathcal{X}}_k \\ - - - \\ \overline{U}_k \end{bmatrix}' \right) \left(\begin{bmatrix} \overline{\mathcal{X}}_k \\ - - - \\ \overline{U}_k \end{bmatrix} \begin{bmatrix} \overline{\mathcal{X}}_k \\ - - - \\ \overline{U}_k \end{bmatrix}' \right)^{-1}$$	*System Matrices* (Σ_{ABCD})

with

$$\overline{\mathcal{X}}_{k+1} = [x(k+1) \cdots x(k+K-1)]$$
$$\overline{\mathcal{X}}_k = [x(k) \cdots x(k+K-2)]$$
$$\overline{Y}_k = [y(k) \cdots y(k+K-2)]$$
$$\overline{U}_k = [u(k) \cdots u(k+K-2)]$$

or more compactly

$$
\begin{bmatrix} \overline{\mathcal{X}}_{k+1} \\ --- \\ \overline{\mathcal{Y}}_k \end{bmatrix} = \begin{bmatrix} A & | & B \\ - & - & - \\ C & | & D \end{bmatrix} \begin{bmatrix} \overline{\mathcal{X}}_k \\ --- \\ \overline{\mathcal{U}}_k \end{bmatrix}
\tag{6.124}
$$

$$\underbrace{}_{\text{known}} \qquad\qquad \underbrace{}_{\text{known}}$$

which can be solved as a least-squares problem, that is,

$$
\begin{bmatrix} \hat{A} & | & \hat{B} \\ - & - & - \\ \hat{C} & | & \hat{D} \end{bmatrix} = \min_{ABCD} \left[\begin{bmatrix} \overline{\mathcal{X}}_{k+1} \\ --- \\ \overline{\mathcal{Y}}_k \end{bmatrix} - \begin{bmatrix} A & | & B \\ - & - & - \\ C & | & D \end{bmatrix} \begin{bmatrix} \overline{\mathcal{X}}_k \\ --- \\ \overline{\mathcal{U}}_k \end{bmatrix} \right]^2
\tag{6.125}
$$

where $\| \ \|$ is the Frobenius matrix norm.
Therefore, the *least-squares solution* follows as

$$
\begin{bmatrix} \hat{A} & | & \hat{B} \\ - & - & - \\ \hat{C} & | & \hat{D} \end{bmatrix} = \left(\begin{bmatrix} \overline{\mathcal{X}}_{k+1} \\ --- \\ \overline{\mathcal{Y}}_k \end{bmatrix} \begin{bmatrix} \overline{\mathcal{X}}_k \\ --- \\ \overline{\mathcal{U}}_k \end{bmatrix}' \right) \left(\begin{bmatrix} \overline{\mathcal{X}}_k \\ --- \\ \overline{\mathcal{U}}_k \end{bmatrix} \begin{bmatrix} \overline{\mathcal{X}}_k \\ --- \\ \overline{\mathcal{U}}_k \end{bmatrix}' \right)^{-1}
\tag{6.126}
$$

This *oblique* projection algorithm is termed Numerical algorithms 4 Subspace IDentification (N4SID) [9] and can be summarized by the following steps:

- Compute the LQ-decomposition of the data matrix (augmented) $\mathcal{D}$ of Eq. (6.114);
- Perform the SVD of $\mathcal{P}^{\|}_{\mathcal{Y}_f | \mathcal{D}_p} \circ \mathcal{U}_f$ in Eq. (6.112) to extract $\mathcal{O}_k$ and $\mathcal{O}_{k-1}$.
- Compute $\overline{\mathcal{X}}_{k+1}$ and $\overline{\mathcal{X}}_k$ of Eq. (6.122).
- Construct the "batch" state-space system of Eq. (6.124).
- Obtain A, B, C, D by solving the least-squares problem as in Eq. (6.126).

We summarize the N4SID algorithm in Table 6.4. Next we consider the important questions of order estimation and model validation techniques that are embedded in the subspace identification process.

6.6 Model Order Estimation and Validation

Once the subspace identification technique is selected, then a sequence of questions arise that must be answered prior to application of the approach to measured data. The primary question is that of the overall *system order* required by the subspace identifier: how many states are needed to adequately capture the underlying data sets? Once the order or equivalently the number of states (N_x) is obtained or estimated based on sound physical insight of the problem at hand, we select a data length K such that $K \gg N_x$ to ensure adequate coverage of the data space. In any case with this information available, we can perform some preliminary model "fits" and estimate the order based on some criteria (to follow) leading to the next pertinent question that must be answered – one of *validation*: how well does the model predict the data? Validation is based on generating an ensemble of data gathered from a sophisticated simulation of the process under investigation or a controlled experiment. If an ensemble is not possible, then partitioning of the data is required to ensure validation from data that was *not* employed for the identification. Once the ensemble is generated, subspace identification is performed on each realization to estimate the model $\hat{\Sigma}_{ABCD}$ and evaluate how well it predicts the data.

In any case, we must

- *estimate* the system order or equivalently the number of states such that $K \gg N_x$; and
- *validate* the performance of the estimated models $\hat{\Sigma}_{ABCD}(n)$; $n = 1, \dots, N$ over the N-ensemble of data.

We start with the order estimation problem and then proceed to evaluating the ensemble performance of the identified models.

6.6.1 Order Estimation: SVD Approach

An important problem in any of these subspace identification schemes is that of estimating the order of the underlying system model. In fact, most schemes require the system order to be "known" prior to estimating the model parameters. Therefore, we must first determine the correct number of states to estimate from the data directly. Order estimation has evolved specifically from the identification problem [36] where the first step is to estimate or guess the system order and then attempt to validate prediction error testing of the identified model. Much effort has been devoted to this problem [36, 37] with the most significant results evolving from the information theoretical point of view [38, 39] in the statistical case.

We consider some equivalent order estimation methods based on the singular-value decomposition method applied to the data. The method is deterministic based on the Hankel singular values or the best rank approximation. These approaches are convenient for both classical and subspace identification, since the SVD has become an integral part of these algorithms and the singular values are available.

We know from "theoretical" systems (realization) theory that the key element is determining the order or number of states of the underlying system under investigation. Classically, order estimation is based on the rank condition that evolves from the Hankel matrix and its observability/controllability factors. However, the root of achieving a robust and reliable estimate lies in the singular-value decomposition of the Hankel matrix, that is,

$$\mathcal{H}_{K,K'} = U \, \Sigma \, V' = [U_S \mid U_N] \begin{bmatrix} \Sigma_S & \mid & 0 \\ - & - & - \\ 0 & \mid & \Sigma_N \end{bmatrix} \begin{bmatrix} V'_S \\ -- \\ V'_N \end{bmatrix} = \underbrace{U_S \Sigma_S V'_S}_{\text{Signal}} + \underbrace{U_N \Sigma_N V'_N}_{\text{Noise}}$$

(6.127)

Ideally, $\Sigma_N \ll \Sigma_S$ and the number of nonzero singular values residing in Σ_S reveals the order or number of states of the underlying system. Therefore, in theory, it is a simple matter to perform the SVD of $\mathcal{H}_{KK'}$ and count the number of nonzero singular values to obtain N_x, since $\rho(\mathcal{H}_{K,K'}) = \rho(\Sigma) = \rho(\Sigma_S) = N_x$. Unfortunately, zero singular values do not occur even in high-precision simulated data due to roundoff and truncation errors (e.g. see Example 6.2 of Section 6.3). Canonical realization theory presents an even more difficult problem, since the rank based on a particular canonical form desired dictates the manner in which the Hankel matrix is to be decomposed. For instance, the predecessor evaluation scheme of Section 6.3 leading to the Luenberger canonical form is *not* a robust and reliable numerical approach for estimating the system order based on extracting the observability and controllability invariants (indices), since $N_x = \sum_{i=1}^{N_y} v_i = \sum_{i=1}^{N_u} \mu_i$. Therefore, this provides a strong argument to first perform a numerically reliable approach to extract a generic "black-box" state-space system and then transform it to the desired canonical coordinate system (see Eq. (6.45)).

With this in mind, we understand that the key *first step* in subspace identification lies in the extraction of the singular values either employing the classical Hankel matrix of the SVD-based realization algorithm of Section 6.3 or the SVD-based projection algorithms of Section 6.4. It will be shown in the next chapter that this philosophy holds also for the stochastic case where impulse response matrix sequences are replaced by covariance matrix sequences and deterministic input/output sequences replaced by noisy, uncertain sequences.

Next let us consider properties of the Hankel matrix that aid in the order estimation problem. The Hankel matrix singular values enable us to estimate its rank and therefore the minimal number of states of the underlying system. However, when noise or computational error occurs, then it is necessary to apply judgment when evaluating these singular values. One approach is simply to search for a meaningful change in these values, referred to as a "gap" or "jump," selecting the order corresponding to this abrupt change as illustrated in Figures 6.1 and 6.2 for a fourth-order or sixth-order system, respectively.

Another approach is to employ the so-called *best rank approximation* that is calculated by the ratio of the summation of singular values up to the selected order over all the values, that is, β (percentage) based on the ratio of singular values, [29]

$$\beta(\hat{N}_x) := \left(\frac{\sum\limits_{k=1}^{\hat{N}_x} \sigma_k}{\sum\limits_{k=1}^{K} \sigma_k} \right) \times 100 \tag{6.128}$$

Here β (%) is the percentage of the original matrix (Hankel) approximated by choosing $\hat{N}_x$. A threshold, τ_x (%) can be selected to search over k for that particular $\hat{N}_x$ that "best" approximates $\mathcal{H}_{K,K}$ up to the threshold, that is,

$$\beta_n(\hat{N}_x) \geq \tau_x(\%) \quad \text{for } n = 1, \dots, \hat{N}_x \tag{6.129}$$

Once the order is selected, then the error in this choice can be specified by the $\hat{N}_{x+1}$ singular value σ_{N_x+1} as a particular metric or the so-called truncation or SVD *model error* metric defined by the summation over all of the remaining singular values as [12–15]

$$\mathcal{M}_\epsilon := 2 \sum_{k=N_x+1}^{K} \sigma_k \tag{6.130}$$

Other properties of these Hankel singular values are based on norms of the Hankel matrix and are used in intensity measures and model reduction [13, 15]. For instance, the *Hankel norm* is a measure of the energy stored in a system and is specified by the "largest" singular value of the Hankel matrix:

$$\| \mathcal{H}_{K,K} \| = \max\{\sigma_k\} = \sigma_1; \quad k = 1, \dots, K$$

while the $\| \mathcal{H} \|_2$ norm is the *rms* sum of the impulse responses [15]. In any case, it is clear that the Hankel matrix is a critical entity in identifying a system from measured data.

Another popular SVD metric that is applied is the *model order estimator* (MOE)[12] [40] given by

$$\hat{N}_x = \max\{N : \log \hat{\sigma}_N > (\log \hat{\sigma}_{max} + \log \hat{\sigma}_{min})\} \tag{6.131}$$

is specified as the largest integer corresponding to the singular value (σ_{N_x}) *greater than* the geometric mean of the largest and smallest singular values extracted. Here the idea is to formally search for the "gap" or "jump" in singular values.

We summarize the metrics available from the singular-value decomposition of the Hankel matrix as follows:

---

12 This is the order estimator used in the identification toolbox of MATLAB.

1. *Hankel order estimation* (SVD, best rank) of (6.135) $\Rightarrow \hat{N}_x$;
2. *Hankel singular values*: $\sigma_{ABC} = \sqrt{\Sigma_{AC} \times \Sigma_{AB}}$;
3. *Hankel norm* (energy): $\| \mathcal{H}_{K,K'} \| = \max\{\sigma_i\}$; $i = 1, \cdots, N_x$; and
4. SVD *model error*: $\mathcal{M}_e = 2 \sum_{k=\hat{N}_x+1}^{K} \sigma_k$.
5. SVD *model order estimation*: $\hat{N}_x = \max\{N : \log \hat{\sigma}_N > (\log \hat{\sigma}_{\max} + \log \hat{\sigma}_{\min})\}$.

This completes the discussion of order estimation for a deterministic system based on the SVD of the Hankel matrix. These results are valid when we replace the Hankel matrix with the projection matrices of the subspace realization techniques as well. These resulting singular values hold the same credence and are evaluated and analyzed identically.

6.6.2 Model Validation

Once the order is selected, the subspace identification completed and the "black-box" model Σ_{ABCD} identified, the question of its validity must be answered to decide just "how good" the model fits the data. For prediction error methods, a wealth of statistical techniques exist that we will discuss in the next chapter; however, for the deterministic subspace identification case, we are more limited and rely on the identified model response compared to the system response data to calculate the fitting error.

The response of the identified model $\Sigma_M := \{A_M, B_M, C_M, D_M\}$ is easily calculated using the available deterministic inputs that were used to extract it from the data. We define the *model response*, $\mathbf{y}_M(t)$ in state-space form as follows:

$$x(t) = A_M x(t-1) + B_M u(t-1) \qquad \text{(State Response)}$$

$$y_M(t) = C_M x(t) + D_M u(t) \qquad \text{(Measurement Response)}$$
$$(6.132)$$

or equivalently as a *convolution response*

$$y_M(t) = C_M A_M^t x(0) + \sum_{k=0}^{t-1} C_M A_M^{t-k-1} B_M u(k-1) + D_M \delta(t)$$

$$\text{(Convolution Response)} \quad (6.133)$$

or directly from the *impulse response*

$$y_M(t) = H_t = C_M A_M^{t-1} B_M + D_M \delta(t); \qquad t = 1, \ldots, K \quad \text{(Impulse Response)}$$
$$(6.134)$$

We can now use any of these representations to estimate the *model (residual) error* or simulation error defined by [5] as

$$\epsilon(t) := \mathbf{y}(t) - \mathbf{y}_M(t) \qquad \text{(Model Residual Error)} \qquad (6.135)$$

and define the *model error metric* for the *i*th-measurement channel as

$$\varepsilon_i = \sqrt{\frac{\sum\limits_{t=1}^{K} \epsilon_i^2(t)}{\sum\limits_{t=1}^{K} y_i^2(t)}}; \quad i = 1, \ldots, N_y \qquad \text{(Model Error Metric)} \qquad (6.136)$$

to evaluate the performance of the model response $\{y_M(t)\}$ relative to the response data $\{y(t)\}$ [6, 7].

This metric can also be calculated over an N-ensemble of data for *each* measurement channel as well to provide statistics enabling a reasonable estimate of model quality. Here we have generated a data ensemble $\{y_n(t)\}; n = 1, \ldots, N$ with the corresponding *ensemble model error* for the *n*th realization defined by

$$\epsilon_n(t) := y_n(t) - y_M^n(t) \qquad \text{(Ensemble Model Error)} \qquad (6.137)$$

Ensemble statistics and confidence intervals can be estimated from the sample mean and variance estimates across the realizations $\{\epsilon_n(t)\}; n = 1, \ldots, N$ as

$$\mu_\epsilon(t) := \hat{E}\{\epsilon_n(t)\} = \frac{1}{N} \sum_{n=1}^{N} \epsilon_n(t)$$

$$\sigma_\epsilon(t) := \hat{E}\{\epsilon_n(t) - \mu_\epsilon(t)\} = \frac{1}{N} \sum_{n=1}^{N} \epsilon_n^2(t) - \mu_\epsilon^2(t) \qquad (6.138)$$

enabling a confidence interval estimate

$$I_\epsilon(t) = \mu_\epsilon(t) \pm 1.96\sigma_\epsilon(t) \qquad \text{for} \qquad N > 30$$

for *each* measurement channel realization providing another means of validating the identified model.

Another statistic that can be applied to validate the performance of the identified model is the *median absolute deviation* (MAD) statistic because of its inherent robustness property relative to the usual ensemble mean/standard deviation approach [41]:

$$MAD(t) := \gamma \mathcal{M}_t(|\epsilon_n(t) - \mathcal{M}_i(\epsilon_i)|) \qquad (6.139)$$

where $\epsilon_n(t) := \{\epsilon_n(0) \cdots \epsilon_n(t)\}$ is the *n*th realization of the set of model errors up to time t, $\mathcal{M}_t$ is the median of the data, and $\gamma = 1.4826$ is a constant based on the normalized (assumed Gaussian) data. A confidence limit that can be used for outlier detection can also be calculated as

$$\mathcal{M}_t - \beta \times MAD(t) < \epsilon_n(t) < \mathcal{M}_t + \beta \times MAD(t) \qquad (6.140)$$

for β is a threshold equivalent to a confidence limit (e.g. $\beta = 1, 2$).

If it is not possible to create a deterministic data ensemble through simulation or experiment, then the original measurement can be partitioned to create an ensemble or bootstrap data resampling techniques may be applied as well [42]. We summarize the steps in order estimation and model validation:

- *Generate* ensemble data: through experiments or simulation, if not possible, use partitioning or bootstrap techniques.
- *Create* Hankel impulse or input/output data matrices.
- *Perform* an SVD of the Hankel matrix extracting the observability matrix (classical) or extended observability matrix (subspace).
- *Estimate* the order from the extracted singular values and extract the corresponding statistics (Hankel order, Hankel singular values, Hankel norm, Hankel SVD-model error, SVD-model order estimator).
- *Identify* the model Σ_{ABCD} using classical realization or subspace identification techniques.
- *Estimate* model (simulation) error(s) and its statistics over each realization of the ensemble.
- *Validate* model performance using ensemble statistics and error metrics.

We now have the tools in place to perform system identification and evaluate the identified model performance. Consider the following example.

Example 6.6 Consider the simple three-mass (spring–damper) mechanical system illustrated in Figure 6.2 from [15] where $m_i = 1; k_i = 3, k_4 = 0$; and $d_i = 0.01k_i, d_4 = 0; i = 1, \ldots, 3$. These parameters along with the input/output transmission matrices led to the state-space representation ($N_d = 3, N_x = 6$) developed in Example 6.3.

$$\ddot{\mathbf{x}}(t) = A\mathbf{x}(t) + B\mathbf{u}(t)$$
$$\mathbf{y}(t) = C\mathbf{x}(t) + D\mathbf{u}(t)$$

Since the system is mechanical, it is common practice in operational modal analysis to excite the system "randomly" with the idea of a persistent excitation [43]. We chose a pseudo-random sequence to be applied as a single input. Exciting the system with this excitation yields the three-output response shown in Figure 6.9 along with its accompanying Fourier spectra. The deterministic SVD realization technique was performed as before at a high SNR of 50 dB. Discretizing the response with a sampling interval of $\Delta t = 0.01$ seconds, creating the input/output Hankel matrices and decomposing it with the SVD as in Eq. (6.26) using the MOESP or N4SID subspace realization technique reveals a sixth-order system as before. The log-plots of the singular values are shown in

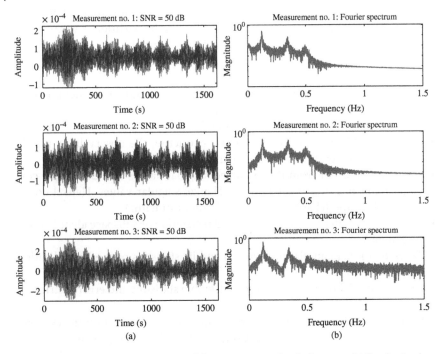

Figure 6.9 Pseudo-random response of three-output mechanical system. (a) Synthesized channel pseudo-random responses. (b) Fourier transforms of channel responses.

Figure 6.10b. The corresponding Hankel metrics are as follows:

Hankel order estimation : $\hat{N}_x = 6$

Hankel singular values : $\sigma_i = \{56.0,\ 54.7,\ 7.5,\ 13.0,\ 2.5,\ 4.4\}$

Hankel norm (energy) : $\|\mathcal{H}_{K,K'}\| = 8.67 \times 10^{-4}$

SVD model error : $\mathcal{M}_\epsilon = 8.3 \times 10^{-6}$

The corresponding output power spectra of each channel and the average power spectrum of the identified model are compared in Figure 6.10a indicating a reasonable realization of the data. Again the "identified" modal frequencies of the system estimated from the "discrete" identified system matrix and transformation from the discrete ($\mathcal{Z}$-domain) to the continuous ($\mathcal{S}$-domain) given by ($s = \frac{1}{\Delta t} \ln z$) yields $p_i = \{0.497\ \text{Hz}, 0.343\ \text{Hz}, 0.123\ \text{Hz}\}$ and are shown in Figure 6.10c.

To confirm that the realization *matches* the modal frequency estimates of the system closely, the estimated channel outputs are overlaid onto the raw impulse response data as well as the random measurement data in Figure 6.11a and b – again indicating an accurate realization due to the high SNR.

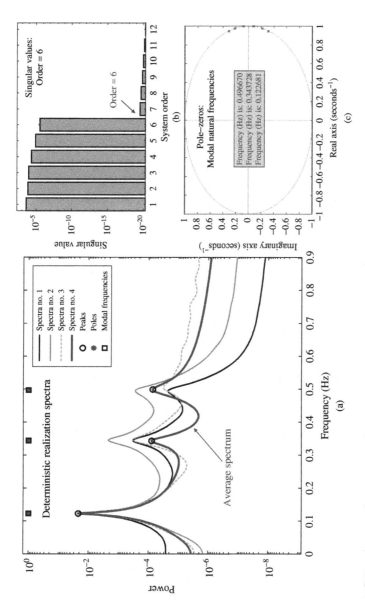

Figure 6.10 SVD realization of three-output mechanical system. (a) Power spectra: outputs/average spectra indicated with poles (boxes). (b) Singular values of the Hankel matrix. (c) Frequencies with poles/zeros.

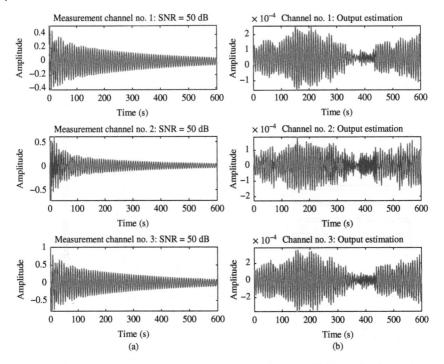

Figure 6.11 SVD realization of three-output mechanical system. (a) Raw and estimated impulse response. (b) Raw and estimated channel measurements.

Table 6.5 Deterministic subspace identification: three-mass mechanical system.

		Model error/validation results		
Channel no.	Residual error	Average model error	Deviation model error	MAD model error
1	9.7×10^{-10}	5.4×10^{-4}	7.8×10^{-3}	5.9×10^{-3}
2	2.7×10^{-10}	4.8×10^{-4}	6.6×10^{-3}	5.2×10^{-3}
3	6.6×10^{-10}	5.4×10^{-4}	7.8×10^{-3}	5.9×10^{-3}
Average	6.3×10^{-10}	5.2×10^{-3}	7.4×10^{-3}	5.6×10^{-3}

Next we calculate the various validation metrics to observe how well they predict/validate a reasonable realization of this data. The results are shown in Table 6.5. From the table, we examine the average results of each of the statistics. Over an ensemble of 100-realizations and deterministic subspace identifications, the average residual error is 6.3×10^{-10} as expected, since the synthesized signal is essentially deterministic (50 dB SNR). The calculated *model errors* are reasonable for the corresponding average, standard deviation, and MAD statistics on the order of 10^{-3} corresponding to less than 1% error. This completes the example. □

This completes the section on order estimation and deterministic model validation techniques. Next we consider a case study that applies these methods to a synthesized data set of a more complex mechanical system – a vibrating structure.

6.7 Case Study: Structural Vibration Response

In this section, we study the application of the deterministic subspace identification of a structure represented by a LTI, MIMO, mass–spring–damper mechanical system consisting of 8-modes or 16-states (see [44] for details) measured by 3-output accelerometers. The structure is excited by a random input. Structurally, the system mass ($\mathcal{M}$) is characterized by an identity matrix while the coupled spring constants in (N/m) are given by the tridiagonal matrix

$$
\mathcal{K} = \begin{bmatrix}
2400 & -1600 & 0 & 0 & 0 & 0 & 0 & 0 \\
-1600 & 4000 & -2400 & 0 & 0 & 0 & 0 & 0 \\
0 & -2400 & 5600 & -3200 & 0 & 0 & 0 & 0 \\
0 & 0 & -3200 & 7200 & -4000 & 0 & 0 & 0 \\
0 & 0 & 0 & -4000 & 8800 & -4800 & 0 & 0 \\
0 & 0 & 0 & 0 & -4800 & 10\,400 & -5600 & 0 \\
0 & 0 & 0 & 0 & 0 & -5600 & 12\,000 & -6400 \\
0 & 0 & 0 & 0 & 0 & 0 & -6400 & 13\,600
\end{bmatrix}
$$

the damping matrix is constructed using the relation (Rayleigh damping)

$$
C_d = 0.680\mathcal{M} + 1.743 \times 10^{-4}\mathcal{K} \quad (\text{N s})\text{m}^{-1}
$$

The measurement system consisted of three accelerometers placed to measure the modes at the 1, 4, and 8 locations on the structure. The accelerometer data is acquired and digitized at a sampling frequency of 50 Hz ($\Delta t = 0.02$ second). The input signal from a randomly excited stinger rod is applied at a specified spatial location such that

$$
\mathcal{M}\ddot{\mathbf{d}}(t) + C_d\dot{\mathbf{d}}(t) + \mathcal{K}\mathbf{d}(t) = B_p\mathbf{p}(t)
$$

The structural system model is transformed to state-space such that

$$
\dot{\mathbf{x}}(t) = \begin{bmatrix} 0 & | & I \\ --- & | & --- \\ -\mathcal{M}^{-1}\mathcal{K} & | & -\mathcal{M}^{-1}C_d \end{bmatrix}\mathbf{x}(t) + \begin{bmatrix} 0 \\ --- \\ \mathcal{M}^{-1}B_p \end{bmatrix}\mathbf{p}(t)
$$

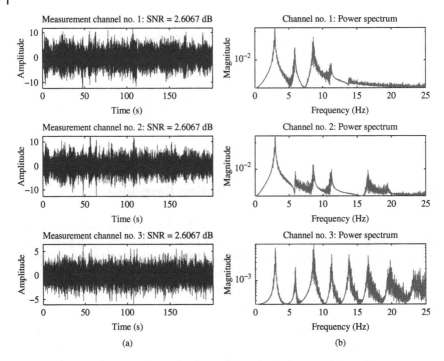

Figure 6.12 Structural vibrations of eight-mode mechanical system: (a) accelerometer responses of three-output system. (b) Fourier power spectra of channel responses.

and

$$\mathcal{B}_p = \mathbf{e}_{N_x}; \mathcal{C}_d = \begin{bmatrix} \mathbf{e}'_1 \\ \mathbf{e}'_4 \\ \mathbf{e}'_8 \end{bmatrix}$$

where $\mathbf{e}_{N_x}$ is a N_x-column vector of all 1s and $\mathbf{e}'_i$ is a unit row vector with a 1 in the ith column and the excitation $\mathbf{p}$ is a unit variance, pseudo-random sequence. Three accelerometer outputs of the synthesized vibrating structure were recorded for 200 seconds with the vibrational responses shown in Figure 6.12 along with their corresponding power spectra where we see a persistently excited system ideal for subspace identification.

The deterministic subspace identification algorithm (N4SID) was applied to the response data, and the results are summarized in Figure 6.13. We observe the estimated power spectra (thin lines) of each channel measurement along with the "average" identified spectra (thick line) in Figure 6.13a as well as the extracted modal frequencies (squares) from the identified model system matrix (A). The SVD-order test is shown in Figure 6.13b. This test reveals a 16th-order (8-mode) representation as adequate with the large gap shown to indicate the proper order. The poles/zeros (discrete) populating the unit circle

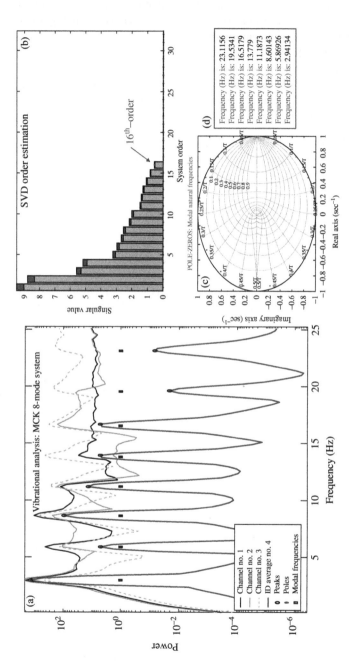

Figure 6.13 Structural vibration subspace identification eight-mode mechanical system. (a) Power spectra: outputs/average with poles (boxes). (b) Singular values of the Hankel matrix. (c) Poles/zeros. (d) Estimated modal frequencies.

are shown in Figure 6.13c with the estimated modal frequencies obtained from the discrete-to-continuous transform ($\mathcal{Z} \rightarrow \mathcal{S}$) listed in Figure 6.13d corresponding to those of the simulation.

We show both synthesized data along with their corresponding estimates in Figure 6.14, indicating quite a reasonable "fit" of both the underlying impulse response, synthesized data and their corresponding estimates. The subspace identifier has clearly captured the essence of the signals.

The corresponding Hankel metrics are:

Hankel order estimation : $\hat{N}_x = 16$

Hankel singular values : $\sigma_i = \{18.8,\ 17.9,\ 10.4,\ 10.6,\ 6.4,\ \ldots, 2.3\}$

Hankel norm (energy) : $\| \mathcal{H}_{K,K'} \| = 1.2 \times 10^{-5}$

SVD model error : $\mathcal{M}_\epsilon = 3.0 \times 10^{-5}$

Next we calculate the various validation metrics to observe how well they predict/validate a reasonable realization of this data. The results are shown in Table 6.6. From the table, we examine the average results of each of the statistics. Over an ensemble of 100-realizations and deterministic subspace identifications, the average residual error is 1.3×10^{-11} as expected since the

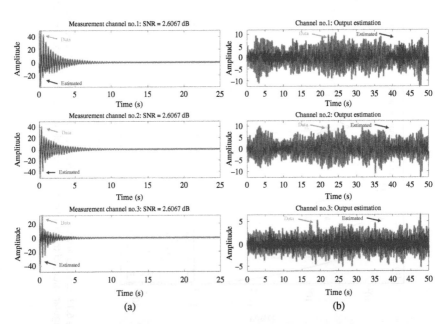

Figure 6.14 Subspace identification of eight-mode mechanical system. (a) Raw data and estimated impulse responses. (b) Raw and estimated channel measurements.

Table 6.6 Deterministic subspace identification: eight-mass mechanical system.

	Model error/validation results			
Channel no.	Residual model error	Average model error	Deviation model error	MAD model error
1	1.3×10^{-11}	3.3×10^{-3}	4.2×10^{-3}	3.3×10^{-3}
2	1.3×10^{-11}	3.3×10^{-3}	4.1×10^{-3}	3.3×10^{-3}
3	1.3×10^{-11}	3.1×10^{-3}	3.8×10^{-3}	3.1×10^{-3}
Average	1.3×10^{-11}	3.2×10^{-3}	4.0×10^{-3}	3.2×10^{-3}

synthesized signal is essentially deterministic (100 dB SNR). The calculated *model errors* are reasonable for the corresponding average, standard deviation, and MAD statistics on the order of 10^{-3} corresponding to less than 1% error. This completes the case study of a mechanical structure.

6.8 Summary

In this chapter, we have discussed solutions to the deterministic realization problem ranging from the classical system theoretic approaches to the modern subspace projection approaches. We have shown that both modern techniques rely on the singular-value decomposition of the Hankel and block Hankel matrices, each with underlying theoretical implications. Once the appropriate Hankel matrix is decomposed, the system order is estimated and the underlying system matrices Σ_{ABCD} are extracted. This is essentially a "black-box" approach using MIMO embedded state-space models as its inherent model set. The philosophy is that once Σ_{ABCD} is extracted, it can be transformed to any coordinate system for analysis (e.g. canonical or modal coordinates). Using examples throughout, these approaches were developed from a pragmatic rather than theoretical perspective. It was demonstrated that the numerically superior singular-value decomposition for both classical and subspace methods provides a robust and reliable solution to the deterministic realization problem. SVD-based order estimation was shown to reveal simple metrics (model error, ensemble model error) that can be used to extract the underlying order and state-space system. Validation techniques based on estimated model error evolved leading to pragmatic design solutions. A case study from a well-known structural dynamics problem was investigated to illustrate the approach and solution.

In the next chapter, we discuss the all-important "stochastic" realization problem that accounts not only for noisy, uncertain data, but also inherent parameter uncertainties.

MATLAB Notes

MATLAB offers a wealth of linear algebraic algorithms that can be applied to the subspace realization problem including commands to create the **Hankel** and **Toeplitz** matrices and their decompositions using the singular-value decomposition (**svd**), the Gram Schmidt, or **QR** decomposition that is easily transformed to the lower triangular/orthogonal LQ-transformation used throughout. Other techniques such as the **Cholesky** factorization for covariance matrices as well as the **LDU** decomposition methods extend these methods even further. The system identification toolbox (**ident**) offers a large variety of both subspace (**n4sid**) and prediction-error methods for state-space models including *block-box* as well as *gray-box* representations. **Analysis** command enables the performance analysis of the identified models as well as well-founded validation methods. Using the embedded GUI allows the user to easily import the input/output data, perform preliminary preprocessing (e.g. mean and trend removal), partition the data for both identification and validation as well as enable instantaneous analysis of results performing signal estimation plots, residual analysis, frequency domain plots, and transient response analysis. This is quite a powerful toolbox for this problem sets. Other implementations of subspace algorithms exist and can be extracted from the texts of [6] (disk) and [9].

References

1 Kalman, R. (1963). Mathematical description of linear dynamical systems. *SIAM J. Control* 82: 152–192.

2 Kalman, R., Falb, P., and Arbib, M. (1969). *Topics in Mathematical System Theory*. New York: McGraw-Hill.

3 Ho, B. and Kalman, R. (1966). Effective reconstruction of linear state variable models from input/output data. *Regelungstechnik* 14: 545–548.

4 Candy, J. (2006). *Model-Based Signal Processing*. Hoboken, NJ: Wiley/IEEE Press.

5 Ljung, L. (1999). *System Identification: Theory for the User*, 2e. Englewood Cliffs, NJ: Prentice-Hall.

6 van Overschee, P. and De Moor, B. (1996). *Subspace Identification for Linear Systems: Theory, Implementation, Applications*. Boston, MA: Kluwer Academic Publishers.

7 Verhaegen, M. and Verdult, V. (2007). *Filtering and System Identification: A Least-Squares Approach*. Cambridge: Cambridge University Press.

8 Tangirala, A. (2015). *Principles of System Identification: Theory and Practice*. Boca Raton, FL: CRC Press.

9 Katayama, T. (2005). *Subspace Methods for System Identification*. London: Springer.

10 Chen, C. (1984). *Introduction to Linear System Theory*. New York: Holt, Rhinehart, and Winston.

11 Rao, B. and Arun, K. (1992). Model based processing of signals: a state space approach. *Proc. IEEE* 80 (2): 283–309.

12 Moore, B. (1981). Principal component analysis in linear systems: controllability, observability and model reduction. *IEEE Trans. Autom. Control* AC-26: 17–32.

13 Glover, K. (1984). All optimal Hankel-norm approximations of linear multivariable systems and their L∞-error bounds. *Int. J. Control* 39: 1115–1193.

14 Laub, A., Heath, M., Paige, C., and Ward, R. (1987). Computation of system balancing transformations and other applications of simultaneous diagonalization algorithms. *IEEE Trans. Autom. Control* AC-32: 115–122.

15 Gawronski, W. (2004). *Advanced Structural Dynamics and Active Control of Structures*. New York: Springer.

16 Candy, J. (1976). Realization of invariant system descriptions from Markov sequences. PhD dissertation. Gainesville, FL: University Florida.

17 Candy, J., Warren, M., and Bullock, T. (1978). Partial realization of invariant system descriptions. *Int. J. Control* 28 (1): 113–127.

18 Rissanen, J. (1974). Basis of invariants and canonical forms for linear dynamic systems. *Automatica* 10: 175–182.

19 Candy, J., Warren, M., and Bullock, T. (1977). Realization of an invariant system description from Markov sequences. *IEEE Trans. Autom. Control* AC-23 (12): 93–96.

20 Zeiger, H. and McEwen, A. (1974). Approximate linear realization of given dimension via Ho's algorithm. *IEEE Trans. Autom. Control* AC-19 (2): 153.

21 Kung, S. (1978). A new identification and model reduction algorithm via the singular value decomposition. In: *Proceedings of the 12th Asilomar Conference Circuits, Systems, and Computers*, 305–314.

22 Noble, B. and Daniel, J. (1977). *Applied Linear Algebra*, 2e. Englewood Cliffs, NJ: Prentice-Hall.

23 Aoki, M. (1990). *State Space Modeling of Time Series*, 2e. London: Springer.

24 Luenberger, D. (1967). Canonical forms for linear multivariable systems. *IEEE Trans. Autom. Control* AC-12: 290–293.

25 Popov, V. (1972). Invariant system description of linear, time invariant controllable systems. *SIAM J. Control* 10: 254–264.

26 Ackermann, J. and Bucy, R. (1971). Canonical minimal realization of a matrix of impulse response sequences. *Inf. Control* 19: 224–231.

27 Guidorzi, R. (1975). Canonical structures in the identification of multivariable systems. *Automatica* 11: 361–374.

28 Viberg, M. (1995). Subspace-based methods fro the identification of linear time-invariant systems. *Automatica* 31 (12): 1835–1851.

29 Golub, G. and van Loan, C. (1996). *Matrix Computations*, 3e. Baltimore, MD: Johns Hopkins University Press.

30 Bretscher, O. (2015). *Linaer Algebra with Applications*, 5e. Upper Saddle River, NJ: Pearson Prentice-Hall.

31 Sharf, L. (1991). *Statistical Signal Processing: Detection, Estimation and Time Series Analysis*. Reading, MA: Addison-Wesley.

32 Swevers, J., Adams, M., and De Moor, B. (1987). A new direct time domain identification method for linear systems using singular value decomposition. In: *Proceedings of the 12th International Seminar on Modal Analysis, 12-3*, 1–20.

33 deMoor, B., Vandewalle, J., Vandenberghe, L., and Mieghem, P. (1988). A geometrical strategy for the identification of state space models of linear multivariable systems with singular value decomposition. *Proc. IFAC 88*: 700–704.

34 Van Der Veen, A., Deprettere, E.F., and Swindlehurst, A.L. (1993). Subspace-based methods for the identification of linear time-invariant systems. *Proc. IEEE* 81 (9): 1277–1308.

35 Katayama, T. and Picci, G. (1994). Realization of stochastic systems with exogenous inputs and subspace identification methods. *Automatica* 35 (10): 75–93.

36 Astrom, K. and Eykhoff, P. (1971). System identification–a survey. *Automatica* 7 (2): 123–162.

37 Ljung, L. and Soderstrom, T. (1983). *Theory and Practice of Recursive Identification*. Boston, MA: MIT Press.

38 Akaike, H. (1974). A new look at the statistical model identification. *IEEE Trans. Autom. Control* 19 (6): 716–723.

39 Sakamoto, Y., Ishiguro, M., and Kitagawa, G. (1986). *Akaike Information Criterion Statistics*. Boston, MA: D. Reidel/Kluwer Academic.

40 Bauer, D. (2001). Order estimation for subspace methods. *Automatica* 37: 1561–11573.

41 Pham Gia, T. and Hung, T. (2001). The mean and median absolute deviations. *Math. Comput. Model.* 34: 921–936.

42 Zoubir, A. and Iskander, D. (2004). *Bootstrap Techniques for Signal Processing*. Cambridge: Cambridge University Press.

43 Reynders, E. (2012). System identification methods for (operational) modal analysis: review and comparison. *Arch. Comput. Methods Eng.* 19 (1): 51–124.

44 Cara, F., Juan, J., Alarcon, E. et al. (2013). Modal contribution and state space order selection in operational modal analysis. *Mech. Syst. Sig. Process.* 38: 276–298.

Problems

6.1 Classical realization theory is based on the factorization of the Hankel matrix of MIMO impulse response matrices into the full-rank, observability, and controllability matrices, $\mathcal{H}_{K,K'} = \mathcal{O}_K \times C_{K'}$, show that each row (column) of $\mathcal{H}_{K,K'}$ is dependent if and only if it is a dependent row (column) of $\mathcal{O}_K$ ($C_{K'}$). (Proof of this result forms the foundation for canonical realization theory.)

6.2 Suppose we are given the following state-space description of a continuous-time system with [10]

$$\dot{x}(t) = \begin{bmatrix} -1 & -4/\theta \\ 4\theta & -2 \end{bmatrix} x(t) + \begin{bmatrix} 1 \\ 2\theta \end{bmatrix} u(t)$$

$$y(t) = [-1 \quad 2/\theta] \, x(t)$$

The continuous-time (subscript "c") observability and controllability Gramians satisfy the following Lyapunov equations:

$$A'_c \mathcal{O}_c + \mathcal{O}_c A_c = -C'_c C_c \qquad \text{(Continuous Gramians)}$$
$$A_c C_c + C_c A'_c = -B_c B'_c$$

(a) Calculate the observability and controllability Gramians.
(b) Is this a balanced realization? Show results.
(c) If so, what values of θ are required?

6.3 Suppose we are given the transfer function

$$H(s) = \frac{4s^2 - 2s - 6}{2s^4 + 2s^3 + 2s^2 + 3s + 1}$$

(a) What order is this system? Verify your results.
(b) Find a balanced realization using the Ho–Kalman approach?

6.4 We are given the following continuous system:

$$\dot{x}(t) = Ax(t)$$

with

$$A = \begin{bmatrix} -4 & -4 \\ 2 & -6 \end{bmatrix}$$

(a) Develop the solution to the corresponding *Lyapunov* equation:

$$A'P + PA = -Q \qquad \text{(Lyapunov Equation)}$$

with $Q = I$.

(b) Is this continuous system stable? Show results.

(c) Validate stability using the calculated eigenvalues.

6.5 Consider the following scalar example with impulse response, $H(k) = \{1, 1/2, 1/4, 1/8\ 1/16\}$ with $N = 5$. Using the SVD approach, we would like to extract the realization $\Sigma = (A, b, c)$. Develop the Hankel matrix and apply the SVD–Ho algorithm to extract the system.

6.6 We choose $K > N_x$, then perform an SVD of a given $\mathcal{H}_{K,K}$ and select a *balanced realization* such that

$$\mathcal{H}_{K,K} = U\Sigma V' = (U(\Sigma_{N_x}^{1/2})) \times ((\Sigma_{N_x}^{1/2})'V') = \mathcal{O}_K \times \mathcal{C}_K$$

where N_x is the rank of the Hankel matrix. Using the shift-invariant property of $\mathcal{H}_{K,K} \rightarrow \mathcal{H}_{K,K}^{\uparrow}$ show that if we select the first N_1 of the N_x-singular values ($N_1 < N_x$), then we can extract the N_1-dimensional realization $\Sigma_{A_1 B_1 C_1 D_1}$ from the N_xth. Does this imply that we can extract *all* lower-order systems ($N < N_x$) from a single N_x-realization as submatrices? Show this? Are these realizations "nested" ($A_1 \in A_{N_x}$)?

6.7 Suppose we have the following discrete transfer function:

$$H(z) = \frac{z + 3}{z^2 + 3z + 2}$$

(a) What is impulse response of this system? (*Hint*: Long division)

(b) Create the Hankel matrix from the impulse response.

(c) Using the original Ho–Kalman algorithm, extract the system Σ_{ABCD}.

(d) Using the SVD–Ho–Kalman algorithm, also extract the system Σ_{ABCD}.

(e) Are the realizations equivalent? (*Hint*: Eigenvalues? Impulse responses?)

6.8 We are given the following discrete system represented by the transfer function

$$H(z) = \begin{bmatrix} \frac{z-0.5}{z^2-1} & \frac{1}{z-1} \\ \frac{z^2}{z^2-1} & \frac{2z}{z-1} \end{bmatrix}$$

(a) Calculate the impulse response of this system.

(b) Create the Hankel matrix.

(c) Apply the original Ho–Kalman algorithm, extract the system Σ_{ABCD}.

(d) Apply the SVD–Ho–Kalman algorithm, also extract the system Σ_{ABCD}.

(e) Perform the canonical realization algorithm of Section 6.3 on the Hankel matrix. What are the observability and controllability invariants?

(f) Extract both row and column realizations Σ_R and Σ_C.

(g) Are the realizations equivalent? (*Hint*: Eigenvalues? Impulse responses?)

6.9 Revisit the problem of Example 6.1 where we have a known $N_y \times N_u$ transfer function matrix with $N_y = N_u = 2$ as before. The impulse response sequence is given by

$$H(z) = \begin{bmatrix} 1 & 1 \\ 1 & 1 \end{bmatrix} z^{-1} + \begin{bmatrix} -2 & -4 \\ -3 & -1 \end{bmatrix} z^{-2} + \begin{bmatrix} 4 & 10 \\ 7 & 1 \end{bmatrix} z^{-3}$$
$$+ \begin{bmatrix} -8 & -22 \\ -15 & -1 \end{bmatrix} z^{-4} + \begin{bmatrix} 16 & 46 \\ 31 & 1 \end{bmatrix} z^{-5}$$

(a) Show the elementary row and column operations that can be performed, that is, develop the P (row) and Q (column) elementary matrices leading to the resulting realization Σ_{ABCD} using the Ho–Kalman algorithm. Verify the results with the example.

(b) Perform the SVD–Ho algorithm and compare with the Σ_{ABCD} of the example. Are they equivalent?

(c) Perform the canonical realization algorithm on this impulse response data and show the steps.

6.10 From the extracted model Σ_{ABCD} of the previous problem, determine the canonical similarity transformations T_R and T_C of Eq. (6.45). Transform the extracted system matrices $\{A, B, C, D\}$ to the Luenberger canonical forms $\{A_R, B_R, C_R, D_R\}$ and $\{A_C, B_C, C_C, D_C\}$. How do these realizations compare to those obtained above in (c)?

6.11 Suppose we are given the following MIMO transfer function matrix:

$$H(s) = \begin{bmatrix} \frac{s+1}{(s+2)(s+4)} & \frac{s+4}{s(s+2)} \\ \frac{s+1}{s^2(s+4)} & \frac{s+1}{(s+2)(s+4)} \end{bmatrix}$$

(a) What is impulse response of this system? (*Hint*: Long division)

(b) Create the Hankel matrix from the impulse response.

(c) Using the original SVD–Ho–Kalman algorithm, also extract the system Σ_{ABCD}.

(d) Is this system stable?

(e) What is the model error? (*Hint*: Compare impulse responses.)

6.12 Given the following transfer function matrix,

$$H(s) = \begin{bmatrix} \frac{4s-10}{2s+1} & \frac{3}{s+2} \\ \frac{1}{2s^2+5s+2} & \frac{1}{s^2+4s+4} \end{bmatrix}$$

(a) What is impulse response of this system?
(b) Create the Hankel matrix from the impulse response.
(c) Using the original SVD–Ho–Kalman algorithm, also extract the system Σ_{ABCD}.
(d) Is this system stable?
(e) What is the model error? (*Hint*: Compare impulse responses.)

6.13 We have a deterministic system characterized by sinusoids given by

$$y(t) = \sum_{m=1}^{M} \alpha_m \sin(2\pi f_i t + \phi_i)$$

(a) Simulate this system for the set of frequencies, $f_i = \{1, 2, 5, 10, 20\}$ Hz with $dt = 0.01$ second for $N = 5000$ samples with $\alpha_i = 1$ and $\phi_i = 0, \ \forall i$.
(b) Generate the required Hankel matrices and perform a subspace identification using the MOESP algorithm (MATLAB).
(c) What is the order of the system based on the SVD decomposition? Show the results.
(d) Repeat the order estimation using the AIC and MDL as well as the best rank approach. How do they compare to the SVD result above?
(e) Extract the underlying state-space model Σ_{ABCD}.
(f) Calculate the corresponding model error statistics for a single realization.
(g) Add a small amount of Gaussian noise (SNR = 100 dB) and generate an ensemble of 100-realizations. Calculate the corresponding error statistics as in (f) and compare the results. Are they close?
(h) Repeat the subspace identification for the N4SID algorithm. How do these results compare to that of the MOESP approach? Are the realizations equivalent?

6.14 Consider the following discrete Gauss–Markov system:

$$x(t) = 0.97x(t-1) + 0.03u(t-1) + w(t) \quad \text{for } w \sim \mathcal{N}(0, R_{ww}(i))$$
$$y(t) = 2.0x(t) + v(t) \quad \text{for } v \sim \mathcal{N}(0, R_{vv})$$

with $x(0) = 2.5$, $R_{ww} = 1 \times 10^{-12}$ and $R_{vv} = R_{ww} = 1^10^{-10}$ – essentially deterministic.

(a) Perform a simulation of this process with a sampling interval of $dt = 0.1sec$ and $u(t)$ a unit step function.

(b) Identify the system Σ_{ABCD} using the MOESP subspace technique and calculate the model error statistics for a single realization.

(c) Generate an ensemble from the system Σ_{ABCD} and again apply the MOESP subspace technique and calculate the model error statistics over the ensemble. How do the statistics compare to that of (b)?

(d) Repeat (b) and (c) identifying the system Σ_{ABCD} now using the N4SID subspace technique and calculate the model error statistics for both a single realization and an ensemble? How do the statistics compare? How do the statistics compare with those of the MOESP approach?

(e) For the system Σ_{ABCD} let $u(t)$ be a unit impulse and simulate the system response. Identify the model using the SVD–Ho algorithm and calculate the corresponding model error statistic for both single and ensemble realizations? How do these statistics compare?

(f) From the model error statistics calculated, summarize the identification results. Which of these techniques performed best?

6.15 An RLC circuit is characterized by the following continuous-time Gauss–Markov system:

$$\dot{x}(t) = \begin{bmatrix} 0 & 1 \\ -4 & -2 \end{bmatrix} x(t) + \begin{bmatrix} 0 \\ -4\theta \end{bmatrix} u(t) + \begin{bmatrix} 0 \\ -4\theta \end{bmatrix} w(t)$$

$$y(t) = \begin{bmatrix} 1 & 0 \end{bmatrix} x(t) + v(t)$$

with $x(0) = 0$, $R_{ww} = 1 \times 10^{-12}$ and $R_{vv} = R_{ww} = 1^{1}0^{-12}$ – essentially deterministic.

(a) Perform a simulation of this process with a sampling interval of $dt = 0.1\ msec$ and $u(t)$ a unit pulse train with a period of 7 milliseconds. (*Hint*: Convert to a sampled-data system.)

(b) Identify the system Σ_{ABCD} using the MOESP subspace technique and calculate the model error statistics for a single realization.

(c) Generate an ensemble from the system Σ_{ABCD} and again apply the MOESP subspace technique and calculate the model error statistics over the ensemble. How do the statistics compare to that of (b)?

(d) Repeat (b) and (c) identifying the system Σ_{ABCD} now using the N4SID subspace technique and calculate the model error statistics for both a single realization and an ensemble? How do the statistics compare? How do the statistics compare with those of the MOESP approach?

(e) For the system Σ_{ABCD}, let $u(t)$ be a unit impulse and simulate the system response. Identify the model using the SVD–Ho algorithm and calculate the corresponding model error statistic for both single and ensemble realizations. How do these statistics compare?

(f) From the model error statistics calculated, summarize the identification results. Which of these techniques performed best?

7

Stochastic Subspace Identification

7.1 Introduction

In this chapter, we discuss the stochastic realization problem from both the classical and the subspace perspectives as before in the previous chapter. We start with the classical problem that mimics the deterministic approach with covariance matrices replacing impulse response matrices [1–10]. The underlying Hankel matrix, now populated with covariance rather than impulse response matrices, admits a factorization leading to the fundamental rank condition (as before); however, the problem expands somewhat with the set of additional relations, termed the Kalman–Szego-Popov (KSP) equations, that must be solved to provide the desired identification [11–16]. All of these results follow from the innovations (INV) model of Chapter 3 or equivalently the steady-state Kalman filter of Chapter 4. We again begin the development of this model and discuss the underlying system theoretical properties that lead to a "stochastic" realization.

The subspace approach also follows in a development similar to the deterministic case. Starting with the multivariable output error state-space (MOESP) formulation using orthogonal projection theory, both the "past input multivariable output error state-space" (PI-MOESP) and "past input/output multivariable output error state-space" (PO-MOESP) techniques evolve [17–19]. Next the numerical algorithms for state-space system identification (N4SID) approach follow based on oblique projection theory [19–27]. We concentrate primarily on a solution to the "combined problem" that is the identification of *both* the deterministic and stochastic systems directly (without separation) through the estimated state vector embedded in the INV representation of the system. Here the input/output and noise data are incorporated into the "stochastic" data model.

In this section, we develop the underlying relations that are necessary to precisely define the stochastic realization problem evolving from a

Model-Based Processing: An Applied Subspace Identification Approach, First Edition. James V. Candy.
© 2019 John Wiley & Sons, Inc. Published 2019 by John Wiley & Sons, Inc.

Gauss–Markov (GM) model with correlated noise sources. We must develop equations that describe the output covariance and corresponding power spectrum of this representation in order to evolve into a set of Markov parameters or power spectra needed to recover the true stochastic realization of the process under investigation. Generally, the basic concept of a stochastic realization is that of producing a random process that captures its underlying statistics. For the Gauss–Markov process model, we need to be concerned only with its first and second moments. Recall from the simple development of Chapter 2 that in order to accomplish a stochastic realization we had to perform a factorization of the power spectrum from its sum decomposition (see Eq. (2.29)). This concept is still valid for correlated, multiple input/multiple output (MIMO) Gauss–Markov models as well. Therefore, we investigate the correlated GM model from its recursive state-space model and then examine the detailed representations to show the relationships with the previous development of Chapter 2. All of this effort leads to the measurement or output covariance relations yielding the Markov parameters of the linear, time-invariant (LTI) state-space representation of a stationary, correlated Gauss–Markov process as

$$\Lambda_{yy}(\ell) := \begin{cases} CA^{\ell-1}(A\Pi C' + R_{wv}) & \ell > 0 \\ C\Pi C' + R_{vv} & \ell = 0 \end{cases} \tag{7.1}$$

where the "stationary" *state variance*, $\Pi := P_{xx}(t,t) = \text{cov}(x(t))$ satisfies the Lyapunov recursion:

$$\Pi(t) = A\Pi(t-1)A' + R_{ww} \quad \text{for} \quad t = 0,\ldots,\infty \tag{7.2}$$

which is the "starting" point of classical realization theory.

The underlying correlated GM model becomes more complex than the model of Chapter 2 due to the fact that we now have correlated noise sources $\text{cov}(w(t)v(t)) \neq 0$ introducing more terms into the output covariance of Eq. (7.1). Just as the GM model was developed previously, we can easily extract the recursions of the mean and variance, but it is the closed-form solutions of these relations that provide the required insight to define the stochastic realization problem.

We summarize the classical approaches to solve this problem as well as the modern subspace approaches in Figure 7.1. Here we see that the classical approach not only requires the extraction of the Markov parameters, but also the solution of a nonlinear algebraic equation with certain constraints (positivity) to yield a proper stochastic realization. The subspace approach offers an alternative, but also requires such an underlying algebraic solution as shown in the figure. Next we start with the development of the relations above in order to *define* the stochastic realization problem.

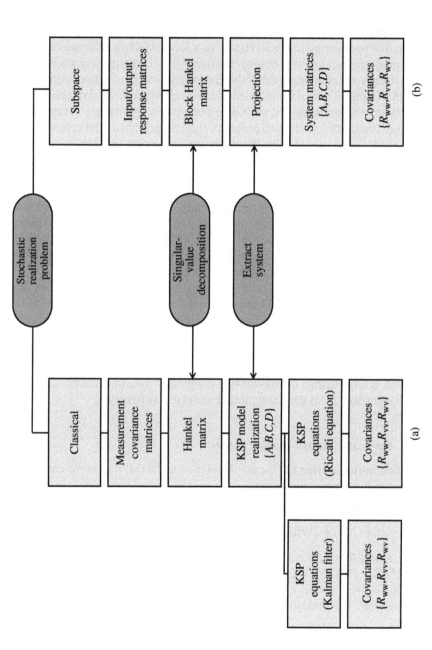

Figure 7.1 Stochastic realization problem: (a) classical and (b) Subspace approaches.

7.2 Stochastic Realization Problem

In this section, we develop stochastic realization theory from a deterministic perspective by showing first the relation to a set of Markov parameters evolving from the measurement or output covariance matrix sequence of a stationary stochastic system characterized by the Gauss–Markov model discussed previously in Chapter 3. Once we establish this representation, we show how a set of algebraic equations, that is, the KSP equations evolve that are necessary to solve the "stochastic" realization problem. Here we present a brief review of the major results necessary to solve the stochastic realization problem.

Analogous to the deterministic realization problem, there are essentially two approaches to consider: (i) the classical and the (ii) subspace. The classical approach for the assumed stationary processes can be developed either in the frequency domain by performing a multichannel spectral factorization or in the time domain from the measurement covariance sequence. Direct factorization of the power spectral density (PSD) matrix is inefficient and may not be very accurate, while realization from the covariance sequence is facilitated by efficient realization techniques and solutions of the KSP relations. Classically, we focus on the minimal stochastic realization from the measurement sequence of a LTI system driven by white noise. The solution to this problem is well known [9, 10] and outlined in Figure 7.1. The measurement covariance sequence from this system is processed by the realization algorithm to obtain a deterministic model followed by the solution of the KSP equations to obtain the remaining parameters that establish a stochastic realization. This approach is complicated by the fact that the matrix covariance sequence must first be estimated from the noisy multichannel measurements.

7.2.1 Correlated Gauss–Markov Model

We define the underlying LTI *Gauss–Markov* model (GM) with *correlated noise sources* for stationary processes as

$$x(t) = Ax(t-1) + Bu(t-1) + w(t-1)$$
$$y(t) = Cx(t) + D(t)u(t) + v(t) \tag{7.3}$$

where $x, w \in \mathcal{R}^{N_x \times 1}$, $u \in \mathcal{R}^{N_u \times 1}$, $y, v \in \mathcal{R}^{N_y \times 1}$ with $A \in \mathcal{R}^{N_x \times N_x}$, $B \in \mathcal{R}^{N_x \times N_u}$, $C \in \mathcal{R}^{N_y \times N_x}$, $D \in \mathcal{R}^{N_y \times N_u}$, $w \sim \mathcal{N}(0, R_{ww})$, $v \sim \mathcal{N}(0, R_{vv})$ and the cross-covariance given by $cov(w, v) = R_{wv}$.

Recursively, the underlying mean and variance statistics follow easily (as before):

$$m_x(t) = E\{x(t)\} = Am_x(t-1) + Bu(t-1)$$
$$P_{xx}(t) = var(\bar{x}(t)) = E\{\bar{x}(t)\bar{x}'(t)\} = AP_{\bar{x}\bar{x}}(t-1)A' + R_{ww}(t-1) \tag{7.4}$$

for $\bar{x}(t) := x(t) - m_x(t)$ and therefore

$$\bar{x}(t) = A\bar{x}(t-1) + w(t-1)$$

The corresponding recursive forms of the measurement statistics are given by

$$m_y(t) = E\{y(t)\} = Cm_x(t-1)$$
$$\Lambda_{yy}(t) := \text{cov}(y(t)) = E\{y(t)y'(t)\} = C\Pi(t)C' + R_{vv}(t) \qquad (7.5)$$

where the *state variance* is defined further by $\Pi(t) := P_{xx}(t) = P_{xx}(t,t)$. Also, since w is white Gaussian, the $\text{cov}(x(t)w(t)) = 0\ \forall t$, then we have the interesting relation that

$$P_{xx}(t,k) = \begin{cases} A^{t-k}P_{xx}(k,k) = A^{t-k}\Pi(k) & t \geq k \\ P_{xx}(k,k)(A')^{t-k} = \Pi(k)(A')^{t-k} & t < k \end{cases} \qquad (7.6)$$

7.2.2 Gauss–Markov Power Spectrum

For simplicity, in the "stationary" case of interest without loss of generality, we ignore the deterministic inputs, since they are incorporated into the process means and removed. We also assume zero initial conditions. Therefore, performing the $\mathcal{Z}$-transform of the GM model (ignoring convergence issues) and solving for the state $X(z)$ and measurement $Y(z)$, we have

$$X(z) = (zI - A)^{-1}W(z)$$
$$Y(z) = CX(z) + V(z) \qquad (7.7)$$

The corresponding *state power spectrum*[1] is therefore

$$S_{xx}(z) = E\{X(z)X^\dagger(z)\} = (zI - A)^{-1}S_{ww}(z)(z^{-1}I - A')^{-1} \qquad (7.8)$$

along with the *measurement power spectrum* for correlated noise sources given by

$$S_{yy}(z) = E\{Y(z))Y^\dagger(z)\} = E\{(CX(z) + V(z))(CX(z) + V(z))^\dagger\}$$
$$= CS_{xx}(z)C' + (zI - A)^{-1}S_{wv}(z) + S_{vw}(z)(z^{-1}I - A')^{-1} + S_{vv}(z) \qquad (7.9)$$

or inserting the expression for the state power spectrum above, we have

$$S_{yy}(z) = C(zI - A)^{-1}S_{ww}(z)(z^{-1}I - A')^{-1}C' + (zI - A)^{-1}S_{wv}(z)$$
$$+ S_{vw}(z)(z^{-1}I - A')^{-1} + S_{vv}(z) \qquad (7.10)$$

Substituting the white noise spectra for each process gives

$$S_{ww}(z) = R_{ww}; \quad S_{vv}(z) = R_{vv}; \quad S_{wv}(z) = R_{wv}$$

---

1 † is the conjugate transpose operation.

enabling a convenient vector–matrix representation with positivity properties as

$$
S_{yy}(z) = \underbrace{[C(zI - A)^{-1} \mid I_{N_y}]}_{T_y(z)} \underbrace{\begin{bmatrix} R_{ww} & | & R_{wv} \\ --- & --- & --- \\ R'_{wv} & | & R_{vv} \end{bmatrix}}_{\Lambda} \underbrace{\begin{bmatrix} (z^{-1}I - A')^{-1}C' \\ --- \\ I_{N_y} \end{bmatrix}}_{T'_y(z^{-1})} \geq 0
$$

(7.11)

or simply

$$
S_{yy}(z) = T_y(z) \times \Lambda \times T'_y(z^{-1}) \geq 0
$$

(7.12)

7.2.3 Gauss–Markov Measurement Covariance

In general, the *measurement covariance* sequence for the correlated GM model can be obtained from the inverse $\mathcal{Z}$-transform of the PSD or directly from the relations developed previously and the whiteness of the noise sources, that is,

$$
\Lambda_{yy}(t, \tau) = E\{y(t)y'(\tau)\} = E\{(Cx(t) + v(t))(Cx(\tau) + v(\tau))'\}
$$

(7.13)

Using the expression for the LTI solution to the state of the GM model

$$
x(t) = A^{t-\tau}x(\tau) + \sum_{n=\tau}^{t-1} A^{t-n-1}w(n) \quad \text{for} \quad t > \tau
$$

(7.14)

and substituting, the measurement covariance becomes

$$
\Lambda_{yy}(t, \tau) = E\left\{ \left(CA^{t-\tau}x(\tau) + C\sum_{n=\tau}^{t-1} A^{t-n-1}w(n) + v(t) \right)(Cx(\tau) + v(\tau))' \right\}
$$

(7.15)

Expanding this expression and noting that the state and noise sources are uncorrelated ($E\{xv'\} = E\{vx'\} = E\{wx'\} = 0$), therefore, we obtain

$$
\Lambda_{yy}(\ell) = CA^{t-\tau}\underbrace{E\{x(\tau)x'(\tau)\}}_{\Pi(\tau)}C' + C\sum_{n=\tau}^{t-1} A^{t-n-1}\underbrace{E\{w(n)v'(\tau)\}}_{R_{wv}(n,\tau)}
$$

$$
+ \underbrace{E\{v(t)v'(\tau)\}}_{R_{vv}(t,\tau)}
$$

(7.16)

Applying the whiteness property of the GM model noise sources, that is,

$$R_{wv}(t, \tau) = R_{wv}(\tau)\delta(t - \tau) \quad \text{and} \quad R_{vv}(t, \tau) = R_{vv}(\tau)\delta(t - \tau) \tag{7.17}$$

and incorporating the shifting property of the Kronecker delta function, we obtain

$$\Lambda_{yy}(t, \tau) = \begin{cases} CA^{t-\tau}\Pi(\tau)C' + CA^{t-\tau-1}R_{wv}(\tau) & t > \tau \\ C\Pi(\tau)C' + R_{vv}(\tau) & t = \tau \end{cases}$$

For the stationary case with lag ℓ, we have $t \to \tau + \ell$ and $\Lambda_{yy}(t, \tau) \to \Lambda_{yy}(\ell)$, so that for $t > \tau$,

$$\Lambda_{yy}(\ell) = CA^{\ell}\Pi(\ell)C' + CA^{\ell-1}R_{wv}(\ell) = CA^{\ell-1}(A\Pi(\ell)C' + R_{wv}(\ell))$$

and therefore for constant state and noise covariances ($\Pi(\ell) \to \Pi$, $R_{wv}(\ell) \to R_{wv}$ and $R_{vv}(\ell) \to R_{vv}$, the *stationary measurement covariance* at lag ℓ is given by

$$\Lambda_\ell := \Lambda_{yy}(\ell) = \begin{cases} CA^{\ell-1}(A\Pi C' + R_{wv}) & \text{for } \ell > 0 \\ C\Pi C' + R_{vv} & \text{for } \ell = 0 \\ \Lambda_{yy}(-\ell) & \text{for } \ell < 0 \end{cases} \tag{7.18}$$

enabling us to define the underlying *stochastic realization problem* as

GIVEN a set of noisy data, $\{y(t)\}$, $t = 1, \ldots, K$; with the measurement covariance sequence Λ_ℓ of Eq. (7.18) or equivalently the power spectrum $S_{yy}(z)$ of Eq. (7.11), **FIND** the underlying Gauss–Markov model parameter set,

$$\Sigma = \{A, B, C, D, \Pi, R_{ww}, R_{wv}, R_{vv}\}$$

such that $\Pi \geq 0$.

7.2.4 Stochastic Realization Theory

The major constraint in obtaining a solution to the stochastic realization problem resides in the fact that the realization must provide an estimate not only of the set of $\{A, B, C, D, \Pi\}$ matrices, but also these matrices guarantee that the subsequent realization of the set $\{R_{ww}, R_{wv}, R_{vv}\}$ satisfies the positive semidefinite property of a covariance matrix, that is,

$$\Lambda := \text{cov}\left(\begin{bmatrix} w_m \\ v_m \end{bmatrix} [w_n' \mid v_n']\right) = \begin{bmatrix} R_{ww} & \mid & R_{wv} \\ --- & & --- \\ R_{wv}' & \mid & R_{vv} \end{bmatrix}\delta_{mn} \geq 0 \tag{7.19}$$

That is, attempts to construct all realizations with identical covariance sequences by the choice of all possible positive definite matrices Π will generally not succeed because all Πs do *not* correspond to the required set $\{R_{ww}, R_{wv}, R_{vv}\}$ satisfying the positive semidefinite property of Eq. (7.19).

However, it has been shown that if the measurement PSD is positive semidefinite on the unit circle and if the underlying GM system is completely observable, then the $\Lambda \geq 0$ admits a factorization, that is,

$$\Lambda = \begin{bmatrix} \mathcal{W} \\ -- \\ \mathcal{V} \end{bmatrix} [\mathcal{W}' \mid \mathcal{V}'] \geq 0 \quad \text{for} \quad S_{yy}(e^{j\Omega}) \geq 0; \quad (A, C) \quad \text{(Observable)}$$

(7.20)

Thus, the stochastic realization problem *always* has a solution when there is a $\Pi \geq 0$ such that $\Lambda \geq 0$ and is guaranteed from the fact that the power spectrum is positive semidefinite ($S_{yy}(z)|_{z=e^{j\Omega}} \geq 0$) on the unit circle (see [28, 29] for proof). This result is available in the generalized KSP lemma [15] ensuring that at least one solution to the stochastic realization problem exists, that is, there exists spectral factorization such that

$$S_{yy}(z) = \left([C(zI - A)^{-1} \mid I_{N_y}] \begin{bmatrix} \mathcal{W} \\ -- \\ \mathcal{V} \end{bmatrix} \right) \left([\mathcal{W}' \mid \mathcal{V}'] \begin{bmatrix} (z^{-1}I - A')^{-1}C' \\ --- \\ I_{N_y} \end{bmatrix} \right) \geq 0$$

(7.21)

which leads to

$$S_{yy}(z) = \underbrace{[C(zI - A)^{-1}\mathcal{W} + \mathcal{V}]}_{H_y(z)} \times \underbrace{[\mathcal{W}'(z^{-1}I - A')^{-1}C' + \mathcal{V}']}_{H_y'(z^{-1})}$$

(7.22)

or simply

$$S_{yy}(z) = H_y(z) \times H_y'(z^{-1}) \qquad \text{(Spectral Factors)} \qquad (7.23)$$

With this in mind, and the fact that the Markov parameters correspond to the covariance sequence, then theoretically, if we create a Hankel matrix from the measurement covariance sequence $\{\Lambda_\ell\}$; $\ell = 0, 1, \ldots, \infty$ of Eq. (7.18) and perform a deterministic realization as in the previous chapter, we obtain the so-called KSP *model* defined by $\Sigma_{KSP} := \{A, B, C, D\}$ with corresponding transfer function $T_{KSP}(z) = C(zI - A)^{-1}B + D$ such that the measurement power spectrum can be expressed as a *sum decomposition*:

$$S_{yy}(z) = S^+(z) + S^-(z^{-1}) - \Lambda_0 = C(zI - A)^{-1}B + D + D'$$
$$+ B'(Z^{-1}I - A')^{-1}C'$$

(7.24)

where $S^+(z) = \sum_{j=0}^{\infty} \Lambda_j z^{-j}$. Therefore, from Eq. (7.18), we have a solution (but difficult) to the stochastic realization problem using *spectral factorization* techniques by

$$S_{yy}(z) = H_y(z) \times H_y'(z^{-1}) = S^+(z) + S^-(z^{-1}) - \Lambda_0 \qquad (7.25)$$

performing the sum decomposition and then the factorization.

The relationship between this realization of the covariance sequence and the stochastic realization is defined by the so-called KSP equations obtained directly from deterministic realization theory or by equating the spectral factorization of Eq. (7.21) to the sum decomposition of Eq. (7.24) (see [1, 4, 15] for proof), since

$$\Lambda_\ell = \begin{cases} \underbrace{C}_{\hat{C}} \; \underbrace{A^{\ell-1}}_{\hat{A}} \; \underbrace{(A\Pi C' + R_{wv})}_{\hat{B}} & \text{for } \ell > 0 \\[2em] \underbrace{C\Pi C' + R_{vv}}_{\hat{D}+\hat{D}'} & \text{for } \ell = 0 \end{cases} \qquad (7.26)$$

where $\hat{\Sigma}_{KSP} = \{\hat{A}, \hat{B}, \hat{C}, \hat{D}\}$ is the underlying KSP model.

This relationship can be stated formally as follows:

GIVEN a minimal realization $\hat{\Sigma}_{KSP}$ of $\{\Lambda_j\}$, THEN the model set $\{\hat{A}, \hat{B}, \hat{C}, \hat{D}, \Pi, R_{ww}, R_{wv}, R_{vv}\}$ is a *stochastic realization* of the measurement sequence, if there exists a $\Pi \geq 0$ such that the following *KSP equations* are satisfied (see [15, 30] for proof):

$$\Pi - \hat{A} \, \Pi \, \hat{A}' = R_{ww}$$
$$\hat{D} + \hat{D}' - \hat{C} \, \Pi \, \hat{C}' = R_{vv}$$
$$\hat{B} - \hat{A} \, \Pi \, \hat{C}' = R_{wv} \qquad (7.27)$$

The *key*, therefore, is to investigate just how to determine such a Π !

This completes the section on classical stochastic realization theory. Next we investigate a variety of ways that can be applied to extract a realization from noisy measurement data.

7.3 Classical Stochastic Realization via the Riccati Equation

The classical approach to solve the stochastic realization problem evolves directly from the basic theory required to ensure that such a realization satisfies the positivity constraints imposed by the multichannel power spectrum of Eq. (7.21) from Section 7.2. Thus, satisfying these constraints leads from the deterministic realization of the KSP model, Σ_{KSP}, to a Riccati equation via the KSP equations. This set of relations then basically defines a solution to the classical stochastic realization problem. The algorithms follow.

We begin by constructing the Hankel matrix from the covariance sequence $\{\Lambda_\ell\}; \ell = 0, \ldots, K$ for some finite $K > 2N_x$. We define the $(KN_y \times KN_y) \times (KN_y \times KN_y)$ block *covariance (Hankel) matrix* by

$$\Lambda_{K,K} := \begin{bmatrix} \Lambda_1 & \Lambda_2 & \cdots & \Lambda_K \\ \Lambda_2 & \Lambda_3 & \cdots & \Lambda_{K+1} \\ \vdots & \vdots & \cdots & \vdots \\ \Lambda_K & \Lambda_{K+1} & \cdots & \Lambda_{2K-1} \end{bmatrix} \tag{7.28}$$

where the covariance sequence is characterized by the Markov parameters: $\Lambda_\ell = CA^{\ell-1}\bar{B}$ with $\bar{B} := (A\Pi C' + R_{wv})$ and $\Pi \geq 0$.

Any deterministic realization algorithm applied to this sequence such that the Hankel factorization yields a KSP realization is reasonable. For instance, the classical Ho–Kalman–SVD approach yields

$$\Lambda_{K,K} = \begin{bmatrix} CA\bar{B} & CA^2\bar{B} & \cdots & CA^{K-1}\bar{B} \\ CA^2\bar{B} & CA^3\bar{B} & \cdots & CA^{K-2}\bar{B} \\ \vdots & \vdots & \cdots & \vdots \\ CA^{K-1}\bar{B} & CA^K\bar{B} & \cdots & CA^{2K-2}\bar{B} \end{bmatrix}$$

$$= \underbrace{\begin{bmatrix} C \\ CA \\ \vdots \\ CA^{K-2} \end{bmatrix}}_{\mathcal{O}_K} \underbrace{\begin{bmatrix} A\bar{B} \mid A^2\bar{B} \mid \cdots \mid A^{K-1}\bar{B} \end{bmatrix}}_{\bar{C}_K} \tag{7.29}$$

or simply

$$\Lambda_{K,K} = \mathcal{O}_K \times \bar{C}_K \tag{7.30}$$

and therefore, the "estimated" KSP model $\hat{\Sigma}_{KSP}$ is given by

$$\hat{A} \longrightarrow A$$
$$\hat{B} \longrightarrow A\Pi C' + R_{wv}$$
$$\hat{C} \longrightarrow C$$
$$(\hat{D} + \hat{D}') \longrightarrow C\Pi C' + R_{vv} \tag{7.31}$$

It has been shown that [4, 5, 31] any solution, say $\Pi^* > 0$, that corresponds to a spectral factorization of Eq. (7.21) for a full-rank $\mathcal{W} \in \mathcal{R}^{N_x \times N_y}$ and for a symmetric $\mathcal{V} \in \mathcal{R}^{N_y \times N_y}$ such that $\mathcal{V} \geq 0$ leads to the following unique solution of the KSP equations:

$$\Pi^* - A \, \Pi^* \, A' = \mathcal{W} \, (\mathcal{V} \, \mathcal{V}') \mathcal{W}' \qquad \text{(KSP Equations)}$$

$$D + D' - C \, \Pi^* \, C' = (\mathcal{V} \, \mathcal{V}')$$

$$B - A \, \Pi^* \, C' = \mathcal{W} \, (\mathcal{V} \, \mathcal{V}') \tag{7.32}$$

Solving for $\mathcal{W}$ in the last KSP equation of Eq. (7.32) and substituting the second equation for the $(\mathcal{V}\mathcal{V}')$ term yields

$$\mathcal{W} = (B - A \, \Pi^* \, C') \times (D + D' - C \, \Pi^* \, C')^{-1} \tag{7.33}$$

Substituting these expressions into the first equation of Eq. (7.32) shows that Π^* satisfies a discrete *Riccati equation*, that is,

$$\Pi^* = A\Pi^*A' + (B - A\Pi^*C')(D + D' - C\Pi^*C')^{-1}(B - A\Pi^*C')'$$

$$\text{(Riccati Equation)} \tag{7.34}$$

A realization obtained in this manner is *guaranteed* to be a stochastic realization, but with the computational expense of solving the discrete Riccati equation. Note that solutions of the Riccati equation are well known and have demonstrated that a unique Π^* exists yielding a stable, minimum-phase spectral factor (e.g. see [1, 4, 5, 16] for details). An iterative approach to solving the Riccati equation using the KSP equations directly is summarized in Table 7.1.

Once the $\hat{\Sigma}_{\text{KSP}}$ model is estimated from the covariance sequence using a classical realization algorithm, then the KSP equations can be developed and used to solve the Riccati equation providing $\Pi^* \geq 0$ as its unique, positive definite solution.

$$\Pi^* = \hat{A}\Pi^*\hat{A}' + (\hat{B} - \hat{A}\Pi^*\hat{C}')(\hat{D} + \hat{D}' - \hat{C}\Pi^*\hat{C}')^{-1}(\hat{B} - \hat{A}\Pi^*\hat{C}')' \tag{7.35}$$

Table 7.1 Iterative KSP algorithm (Riccati equation).

Covariance
$$(\mathcal{V}\mathcal{V}')(i) = D + D' - C\hat{\Pi}^*(i)C'$$

Gain
$$\mathcal{W}(i) = (B - A\hat{\Pi}^*(i)C')(\mathcal{V}\mathcal{V}')^{-1}(i)$$

Estimated covariance
$$\hat{\Pi}^*(i + 1) = A\hat{\Pi}^*(i)A' + \mathcal{W}(i)(\mathcal{V}\mathcal{V}')(i)\mathcal{W}'(i)$$

Initial conditions
$$\hat{\Pi}^*(0)$$

Stopping rule
$$|I - \hat{\Pi}^*(i) \times (\hat{\Pi}^*)^{-1}(i + 1)| \leq \epsilon \quad \text{for} \quad \epsilon \ll 1$$

With the Π^* available, we can then use $\hat{\Sigma}_{KSP}$ to obtain

$$(\mathcal{V} \, \mathcal{V}') = \hat{D} + \hat{D}' - \hat{C} \, \Pi^* \, \hat{C}'$$

$$\hat{\mathcal{W}} = (\hat{B} - \hat{A} \, \Pi^* \, \hat{C}') \times (\mathcal{V}\mathcal{V}')^{-1} \tag{7.36}$$

and therefore the source covariances are given by

$$R_{ww} = \mathcal{W} \, (\mathcal{V} \, \mathcal{V}') \, \mathcal{W}'$$

$$R_{vv} = (\mathcal{V} \, \mathcal{V}')$$

$$R_{wv} = \mathcal{W} \, (\mathcal{V} \, \mathcal{V}') \tag{7.37}$$

We summarize this approach as follows:

- *Perform* a realization (deterministic) to obtain $\hat{\Sigma}_{KSP}$.
- *Solve* the Riccati equation of Eq. (7.35) to obtain Π^*.
- *Calculate* $\mathcal{V}\mathcal{V}'$ and $\mathcal{W}$ from the KSP equations of Eq. (7.36).
- *Determine* the required set of covariances $\{R_{ww}, R_{vv}, R_{wv}\}$ from Eq. (7.37).

Table 7.2 Stochastic realization algorithm: Riccati equation approach.

Deterministic realization (Ho–Kalman (SVD) approach)

$\Lambda_{K,K} = \mathcal{O}_K \times \overline{\mathcal{C}}_K = (U_{N_x} \Sigma_{N_x}^{1/2}) \times ((\Sigma_{N_x}')^{1/2} V_{N_x}')$ (Observability/ Controllability Matrices)

KSP system matrix

$A = \mathcal{O}_{N_x}^{\#} \times \mathcal{O}_{N_x}^{\uparrow} = (\mathcal{O}_{N_x}' \mathcal{O}_{N_x})^{-1} \mathcal{O}_{N_x}' \times \mathcal{O}_{N_x}^{\uparrow} = \Sigma_{N_x}^{-1/2} U_{N_x}' \times \mathcal{O}_{N_x}^{\uparrow}$ (System/Process Matrix)

KSP measurement (output) matrix

$\hat{C} = \mathcal{O}(1 : N_y, 1 : N_x)$ (Output Matrix)

KSP input transmission matrix

$\hat{B} = \overline{C}(1 : N_x, 1 : N_u)$ (Input Matrix)

KSP input/output transmission matrix

$\hat{D} + \hat{D}' = \Lambda_0$ (Input–Matrix)

State covariance equation

$\Pi^* = \hat{A}\Pi^*\hat{A}' + (\hat{B} - \hat{A}\Pi^*\hat{C}')(\hat{D} + \hat{D}' - \hat{C}\Pi^*\hat{C}')^{-1}(\hat{B} - \hat{A}\Pi^*\hat{C}')'$ (Riccati Equation)

KSP equations

$(\mathcal{V} \, \mathcal{V}') = \hat{D} + \hat{D}' - \hat{C} \, \Pi^* \, \hat{C}'$

$\mathcal{W} = (\hat{B} - \hat{A} \, \Pi^* \, \hat{C}') \times (\mathcal{V} \, \mathcal{V}')^{-1}$ (KSP Equations)

Covariance-equations

$R_{ww} = \mathcal{W} \, (\mathcal{V} \, \mathcal{V}') \, \mathcal{W}'$

$R_{vv} = (\mathcal{V} \, \mathcal{V}')$ (Covariances)

$R_{wv} = \mathcal{W} \, (\mathcal{V} \, \mathcal{V}')$

So we see that performing a deterministic realization from the Hankel matrix populated by the covariance sequence $\{\Lambda_\ell\}$ yields the model $\hat{\Sigma}_{KSP}$ that is used to estimate Π^* from the corresponding Riccati equation enabling the determination of the set of covariances $\{R_{ww}, R_{vv}, R_{wv}\}$ from the KSP relations. We summarize the Riccati stochastic realization algorithm in Table 7.2.

This completes the section on the classical approach to solve the stochastic realization problem. However, before we leave this discussion, let us consider an alternative, yet equivalent, way to solve the problem using the well-known Kalman filter based on the underlying INV model developed in Chapter 4.

7.4 Classical Stochastic Realization via Kalman Filter

In this section, we present a special case of the Riccati equation solution to solve the stochastic realization problem. This approach evolves from the factorization theorem of Chapter 4 that led to the Wiener/Kalman filter equivalence resulting in the unique, steady-state solution of the Riccati equation. It is well known [5] that this steady-state solution uniquely specifies the optimal Kalman gain. The significance of obtaining a stochastic realization via the Kalman filter is twofold. First, since the Kalman gain is unique (modulo similarity transformations), so is the corresponding stochastic realization. Second, knowledge of this gain specifies a stable, minimum-phase (zeros within the unit circle), multichannel spectral factor [1, 16]. The importance of this approach compared to the previous one is that once the gain is determined, a stochastic realization is guaranteed immediately. However, the price paid for a so elegant solution is again the computational burden of solving a Riccati equation.

7.4.1 Innovations Model

The basic state estimation problem is to find the minimum variance estimate of the state vector of the GM model in terms of the currently available measurement sequence $y(t)$. We start with the innovations (INV) model rather than the Gauss–Markov model as before, since it is, in fact, equivalent to the steady-state Kalman filter for stationary processes. That is, the INV representation of the Kalman filter in "prediction form" is given by (see [32] for details)

$$\hat{x}(t+1) = A\hat{x}(t) + K_p(t)e(t) \qquad \text{(State Estimate)}$$
$$y(t) = C\hat{x}(t) + e(t) \qquad \text{(Measurement)}$$
$$\hat{y}(t) = C\hat{x}(t) \qquad \text{(Measurement Estimate)}$$
$$e(t) = y(t) - \hat{y}(t) \qquad \text{(Innovation)} \qquad (7.38)$$

where $e(t)$ is the innovations sequence and $K_p(t)$ is the *predicted Kalman gain* for correlated noise sources $\text{cov}(w, v)$ with state error covariance $\tilde{P}(t)$ given by

$$R_{ee}(t) = C\tilde{P}(t)C' + R_{vv}(t) \qquad \text{(Innovations Covariance)}$$
$$K_p(t) = (A\tilde{P}(t)C' + R_{wv}(t))R_{ee}^{-1}(t) \qquad \text{(Kalman Gain)}$$
$$\tilde{P}(t+1) = A\tilde{P}(t)A' - K_p(t)R_{ee}(t)K_p'(t) + R_{ww}(t) \qquad \text{(Error Covariance)}$$

$$(7.39)$$

and we have used the simplified notation: $\hat{x}(t+1|t) \to \hat{x}(t+1)$ and $\tilde{P}(t+1|t) \to \tilde{P}(t+1)$.

7.4.2 Innovations Power Spectrum

The "transfer function" of the INV model is defined as

$$T(z) := \frac{Y(z)}{E(Z)} = C(zI - A)^{-1}K_p \tag{7.40}$$

Calculating the measurement covariance corresponding to Σ_{INV} when

$$y(t) = e(t) + \hat{y}(t)$$

gives

$$R_{yy}(\ell) = \text{cov}[y(t+\ell)y(t)] = R_{\hat{y}\hat{y}}(\ell) + R_{\hat{y}e}(\ell) + R_{e\hat{y}}(\ell) + R_{ee}(\ell) \tag{7.41}$$

Taking the $\mathcal{Z}$-transform of this relation, we obtain the measurement PSD in terms of the INV model as

$$S_{yy}(z) = S_{\hat{y}\hat{y}}(z) + S_{\hat{y}e}(z) + S_{e\hat{y}}(z) + S_{ee}(z) \tag{7.42}$$

Using the linear system relations of Chapter 2 with

$$y(t) = Cx(t) + v(t)$$

we see that

$$S_{yy}(z) = CS_{\hat{x}\hat{x}}(z)C' = T(z)S_{ee}(z)T'(z^{-1}); \quad S_{ee}(z) = R_{ee}$$
$$S_{\hat{y}e}(z) = CS_{\hat{x}e}(z) = T(z)S_{ee}(z) \quad \text{and} \quad S_{e\hat{y}}(z) = S_{ee}(z)T'(z^{-1}) \tag{7.43}$$

Thus, the measurement PSD of the INV model is given by

$$S_{yy}(z) = T(z)S_{ee}(z)T'(z^{-1}) + T(z)S_{ye}(z) + S_{ey}(z)T'(z^{-1}) + S_{ee}(z) \tag{7.44}$$

Substituting for $T(z)$ and replacing $S_{ee}(z) = R_{ee}$, the measurement power spectrum can be expressed as

$$S_{yy}(z) = [C(zI - A)^{-1} \mid I_{N_y}] \begin{bmatrix} K_p R_{ee} K_p' & \mid & K_p R_{ee} \\ -- & & -- \\ (K_p R_{ee})' & \mid & R_{ee} \end{bmatrix} \begin{bmatrix} (z^{-1}I - A')^{-1}C' \\ --- \\ I_{N_y} \end{bmatrix}$$

$$(7.45)$$

Since $R_{ee} \geq 0$, then the following square-root factorization (Cholesky decomposition) always exists as

$$R_{ee} = R_{ee}^{1/2} (R_{ee}^{1/2})'$$ (7.46)

and therefore Eq. (7.45) admits the following unique spectral factorization (Wiener solution):

$$S_{yy}(z) = [C(zI - A)^{-1} K_p \mid I_{N_y}] \begin{bmatrix} R_{ee}^{1/2} \\ -- \\ R_{ee}^{1/2} \end{bmatrix} [(R_{ee}^{1/2})' \mid (R_{ee}^{1/2})'] \begin{bmatrix} K_p'(z^{-1}I - A')^{-1} C' \\ --- \\ I_{N_y} \end{bmatrix}$$

$$= \underbrace{[C(zI - A)^{-1} K_p R_{ee}^{1/2} + R_{ee}^{1/2}]}_{T_e(z)} \times \underbrace{[(R_{ee}^{1/2})' K_p'(z^{-1}I - A')^{-1} C' + (R_{ee}^{1/2})']}_{T_e'(z^{-1})}$$

(7.47)

Thus, using Eq. (7.46) and then Eq. (7.44), we have

$$S_{yy}(z) = T_e(z) \times T_e'(z^{-1})$$ (Spectral Factors)

(7.48)

which shows that the INV model indeed admits a proper spectral factorization. To demonstrate that $T_e(z)$ is the unique, stable, minimum-phase spectral factor, it is necessary to show that $|T_e(z)|$ has all its poles within the unit circle (stable). It has been shown (e.g. [1, 4, 33]) that $T_e(z)$ does satisfy these constraints and therefore is in fact the *Wiener solution* in the stationary case. Next, we must show how the INV model (Kalman filter) relates to the Markov parameters in order to specify a stochastic realization. That is, we must show the relationship of Σ_{INV} to the corresponding set of KSP parameters.

7.4.3 Innovations Measurement Covariance

In our usual model development, we start with the Gauss–Markov representation given by the model set $\Sigma_{GM} := \{A, B, C, D, R_{ww}, R_{vv}, R_{wv}, x(0), \tilde{P}(0)\}$, and a variety of state-space problems are defined within this framework (e.g. state estimation). The same problems can be defined in terms of this model. Therefore, the model set for the INV model is defined as $\Sigma_{INV} := \{A, B, C, D, K_p, R_{ee}, x(0), \tilde{P}(0)\}$. The only problem here is that we are usually "not" given Σ_{INV}, but Σ_{GM}. From the covariance equivalence of these models, that is, both satisfy the measurement covariance relations, $\Lambda_{GM} \equiv \Lambda_{INV}$, it is easy to show that if we are given Σ_{GM} and we want to obtain Σ_{INV}, then we must develop the equivalence relations between the parameters of the model sets.

This solution is given by the set of KSP equations in terms of the INV model. The KSP equations can be solved iteratively to obtain (K_p, R_{ee}) by directly implementing the INV model for the *time-invariant* case [32]. Of course, another approach is simply to execute the steady-state Kalman filter algorithm in prediction form that is equivalent to the KSP equations with Σ_{KSP} known.

Following the same development as in Section 7.2.4 with the *measurement covariance* given by the following using the INV model rather than the GM model:

$$
\Lambda_{yy}(t, \tau) = E \left\{ \left(CA^{t-\tau}\hat{x}(\tau) + C \sum_{n=\tau}^{t-1} A^{t-n-1}K_p(n)e(n) + e(t) \right) \right.
$$
$$
\left. \times (C\hat{x}(\tau) + e(\tau))' \right\} \tag{7.49}
$$

Expanding this expression and noting that the state and INV are uncorrelated $(E\{xe'\} = E\{ex'\} = 0)$, therefore, we obtain

$$
\Lambda_{yy}(\ell) = CA^{t-\tau}\underbrace{E\{\hat{x}(\tau)\hat{x}'(\tau)\}}_{\tilde{\Pi}(\tau)}C' + C \sum_{n=\tau}^{t-1} A^{t-n-1}K_p(n)\underbrace{E\{e(n)e'(\tau)\}}_{R_{ee}(n,\tau)}
$$
$$
+ \underbrace{E\{e(t)e'(\tau)\}}_{R_{ee}(t,\tau)} \tag{7.50}
$$

Applying the whiteness property of the INV, that is,

$$
R_{ee}(t, \tau) = R_{ee}(\tau)\delta(t - \tau)
$$

and incorporating the shifting property of the Kronecker delta function, we obtain

$$
\Lambda_{yy}(t, \tau) = \begin{cases} CA^{t-\tau}\tilde{\Pi}(\tau)C' + CA^{t-\tau-1}K_p(\tau)R_{ee}(\tau) & t > \tau \\ C\tilde{\Pi}(\tau)C' + R_{ee}(\tau) & t = \tau \end{cases}
$$

For the stationary case with lag ℓ, we have $t \to \tau + \ell$ and $\Lambda_{yy}(t, \tau) \to \Lambda_{yy}(\ell)$, as before, so that for $t > \tau$,

$$
\Lambda_{yy}(\ell) = CA^\ell \tilde{\Pi}(\ell)C' + CA^{\ell-1}K_p R_{ee}(\ell)
$$
$$
= CA^{\ell-1}(A\tilde{\Pi}(\ell)C' + K_p(\ell)R_{ee}(\ell))
$$

and therefore for constant state and INV covariance ($\tilde{\Pi}(\ell) \to \tilde{\Pi}$ and $R_{ee}(\ell) \to R_{ee}$), the *stationary measurement covariance* at lag ℓ is given by

$$
\Lambda_\ell := \Lambda_{yy}(\ell) = \begin{cases} \underbrace{C}_{\tilde{C}} \underbrace{A^{\ell-1}}_{\tilde{A}} \underbrace{(A\tilde{\Pi}C' + K_p R_{ee})}_{\tilde{B}} & \text{for } \ell > 0 \\ \underbrace{C\tilde{\Pi}C' + R_{ee}}_{\tilde{D}+\tilde{D}} & \text{for } \ell = 0 \\ \Lambda_{yy}(-\ell) & \text{for } \ell < 0 \end{cases} \tag{7.51}
$$

where $\tilde{\Sigma}_{\mathrm{KSP}} = \{\tilde{A}, \tilde{B}, \tilde{C}, \tilde{D}\}$ is the underlying KSP model for this case.

This relationship between this realization of the covariance sequence and the stochastic realization is defined by the KSP equations obtained *directly* from deterministic realization theory or can also be derived by equating the spectral factorization to the sum decomposition [4]. The solution can be stated formally as

GIVEN a minimal realization $\tilde{\Sigma}_{KSP}$ of $\{\Lambda_\ell\}$, THEN the model set $\{\tilde{A}, \tilde{B}, \tilde{C}, \tilde{D}, \tilde{\Pi}, R^e_{ww}, R^e_{wv}, R^e_{vv}\}$ is a *stochastic realization* of the measurement sequence, if there exists a $\tilde{\Pi} \geq 0$ such that the following *KSP equations* are satisfied:

$$\tilde{\Pi} - \tilde{A}\,\tilde{\Pi}\,\tilde{A}' = K_p\,R_{ee}\,K_p'$$
$$\tilde{B} - \tilde{A}\,\tilde{\Pi}\tilde{C}' = K_p\,R_{ee}$$
$$\tilde{D} + \tilde{D}' - \tilde{C}\,\tilde{\Pi}\,\tilde{C}' = R_{ee} \tag{7.52}$$

Therefore, we can define a *unique* set of noise source covariance matrices in terms of the Kalman gain and INV covariance as follows:

$$R^e_{ww} := K_p\,R_{ee}\,K_p'$$
$$R^e_{wv} := K_p\,R_{ee}$$
$$R^e_{vv} := R_{ee} \tag{7.53}$$

Solving for K_p in the second KSP equation of Eq. (7.52) and substituting the last equation for the INV covariance R_{ee} yields

$$K_p = (\tilde{B} - \tilde{A}\,\tilde{\Pi}\,\tilde{C}') \times (\tilde{D} + \tilde{D}' - \tilde{C}\,\tilde{\Pi}\,\tilde{C}')^{-1} \tag{7.54}$$

Substituting these expressions into the first equation of Eq. (7.52) shows that the state error covariance $\tilde{\Pi}$ satisfies a discrete *Riccati equation*, that is,

$$\tilde{\Pi} = \tilde{A}\tilde{\Pi}\tilde{A}' + (\tilde{B} - \tilde{A}\tilde{\Pi}\tilde{C}')(\tilde{D} + \tilde{D}' - \tilde{C}\tilde{\Pi}\tilde{C}')^{-1}(\tilde{B} - \tilde{A}\tilde{\Pi}\tilde{C}')'$$
$$\text{(Riccati Equation)} \tag{7.55}$$

guaranteeing a proper stochastic realization.

As before the KSP equations can be solved iteratively: $R_{ee}(i) \to R_{ee}, K(i) \to K$ and $\tilde{\Pi}(i) \to \tilde{\Pi}$ yielding the KSP algorithm using the INV model summarized in Table 7.4 to follow.

It should also be noted in passing that from canonical realization theory the gain and INV covariance represent a canonical realization with the minimal number of parameters: $(N_x N_y)$ to characterize K_p and $(\frac{1}{2}N_y(N_y + 1))$ to characterize R_{ee} (see [6] for details).

7.4.4 Stochastic Realization: Innovations Model

As stated in Section 7.3, any deterministic realization algorithm applied to $\{\Lambda_\ell\}$ such that the *Hankel factorization* yields a reasonable approach to extract the KSP realization evolving from the INV model. Assuming the data was generated by the correlated GM model of the previous section with the measurement covariance represented in terms of its parameters, the classical Ho–Kalman–SVD approach can be applied with $\overline{B}_e := A\tilde{\Pi}C' + K_pR_{ee}$

$$\Lambda_{K,K} = \mathcal{O}_K \times \overline{C}_K$$

$$= \underbrace{\begin{bmatrix} C \\ CA \\ \vdots \\ CA^{K-2} \end{bmatrix}}_{\mathcal{O}_K} \underbrace{\left[A\overline{B}_e \mid A^2\overline{B}_e \mid \cdots \mid A^{K-1}\overline{B}_e \right]}_{\overline{C}_K} \tag{7.56}$$

and therefore, due to *covariance equivalence* of Σ_{GM} and Σ_{INV}, the "estimated" KSP model, $\tilde{\Sigma}_{KSP}$, is provided by Ho–Kalman algorithm as

$$\tilde{A} \longrightarrow A$$
$$\tilde{B} \longrightarrow A\Pi C' + R_{wv}$$
$$\tilde{C} \longrightarrow C$$
$$(\tilde{D} + \tilde{D}') \longrightarrow C\Pi C' + R_{vv} \tag{7.57}$$

Once the $\tilde{\Sigma}_{KSP}$ model is estimated from the covariance sequence using a deterministic realization algorithm, then the KSP equations can be solved to complete the stochastic realization, that is,

$$R_{ee} = \tilde{D} + \tilde{D}' - \tilde{C} \, \tilde{\Pi} \, \tilde{C}'$$
$$K_p = (\tilde{B} - \tilde{A} \, \tilde{\Pi} \, \tilde{C}') \times R_{ee}^{-1} \tag{7.58}$$

and therefore the noise source covariances become

$$R_{ww}^e = K_pR_{ee}K_p'$$
$$R_{vv}^e = R_{ee}$$
$$R_{wv}^e = K_pR_{ee} \tag{7.59}$$

We summarize this approach as follows:

- *Perform* a realization (deterministic) to obtain $\tilde{\Sigma}_{KSP}$.
- *Solve* the Riccati equation of Eq. (7.55) to obtain $\tilde{\Pi}$.
- *Calculate* R_{ee} and K from the KSP equations of Eq. (7.58).
- *Determine* the required set of covariances $\{R_{ww}^e, R_{vv}^e, R_{wv}^e\}$ from R_{ee} and K of Eq. (7.59).

Again we see that performing a deterministic realization from the Hankel matrix populated by the covariance sequence $\{\Lambda_\ell\}$ yields the model $\tilde{\Sigma}_{KSP}$ that is used to estimate $\tilde{\Pi}$ from the corresponding Riccati equation, Kalman gain, and INV covariance enabling the determination of the set of noise

Table 7.3 Stochastic realization algorithm: Kalman filter approach.

Deterministic realization (Ho–Kalman (SVD) approach)

$\Lambda_{K,K} = \mathcal{O}_K \times \overline{C}_K = (U_{N_x} \Sigma_{N_x}^{1/2}) \times ((\Sigma'_{N_x})^{1/2} V'_{N_x})$ (Observability/ Controllability Matrices)

KSP system matrix

$\tilde{A} = \mathcal{O}^{\#}_{N_x} \times \mathcal{O}^{\uparrow}_{N_x} = (\mathcal{O}'_{N_x} \mathcal{O}_{N_x})^{-1} \mathcal{O}'_{N_x} \times \mathcal{O}^{\uparrow}_{N_x} = \Sigma^{-1/2}_{N_x} U'_{N_x} \times \mathcal{O}^{\uparrow}_{N_x}$ (System/Process Matrix)

KSP measurement (output) matrix

$\tilde{C} = \mathcal{O}(1 : N_y, 1 : N_x)$ (Output Matrix)

KSP input transmission matrix

$\tilde{B} = \overline{C}(1 : N_x, 1 : N_u)$ (Input Matrix)

KSP input/output transmission matrix

$\tilde{D} = \Lambda_0$ (Input/Output Matrix)

State covariance equation

$\tilde{\Pi} = \tilde{A}\tilde{\Pi}\tilde{A}' + (\tilde{B} - \tilde{A}\tilde{\Pi}\tilde{C}')(\tilde{D} + \tilde{D}' - \tilde{C}\tilde{\Pi}\tilde{C}')^{-1}(\tilde{B} - \tilde{A}\tilde{\Pi}\tilde{C}')'$ (Riccati Equation)

KSP equations

$R_{ee} = \tilde{D} + \tilde{D}' - \tilde{C}\,\tilde{\Pi}\,\tilde{C}'$

$K_p = (\tilde{B} - \tilde{A}\,\tilde{\Pi}\,\tilde{C}') \times R_{ee}^{-1}$ (KSP equations)

Covariance-equations

$R^e_{ww} = K_p R_{ee} K'_p$

$R^e_{vv} = R_{ee}$ (Covariances)

$R^e_{wv} = K_p R_{ee}$

---

source covariances $\{R^e_{ww}, R^e_{vv}, R^e_{wv}\}$ from the KSP relations. We summarize the Kalman-filter-based stochastic realization algorithm in Table 7.3.

Consider the following example to illustrate this approach.

Example 7.1 Suppose we acquired noisy MIMO data from a 2-input/ 2-output unknown (internally) system and we estimate its output covariance sequence as follows:

$$\Lambda = \{\Lambda_0; \ \Lambda_1; \ \Lambda_2; \ \Lambda_3; \ \Lambda_4; \ \Lambda_5; \ \Lambda_6\}$$

or in terms of the measured data, we have

$$\Lambda = \left\{ \begin{bmatrix} 13.73 & 3 \\ 3 & 7 \end{bmatrix}; \begin{bmatrix} -8.73 & 1 \\ -2.64 & -0.64 \end{bmatrix}; \begin{bmatrix} 6.09 & -1.64 \\ 2.31 & -0.32 \end{bmatrix}; \begin{bmatrix} -3.77 & 1.32 \\ -1.61 & 4.64 \end{bmatrix} \right.$$

$$\left. \begin{bmatrix} 2.16 & -0.86 \\ 0.98 & -3.6 \end{bmatrix}; \begin{bmatrix} -1.18 & 0.50 \\ -0.56 & 0.22 \end{bmatrix}; \begin{bmatrix} 0.63 & -0.28 \\ 0.30 & -0.13 \end{bmatrix} \right\}$$

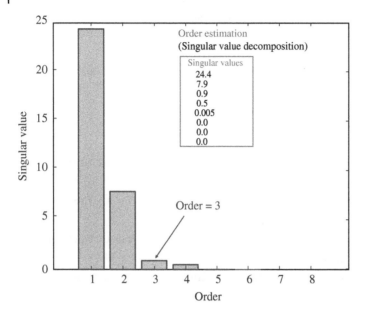

Figure 7.2 SVD order estimation of an unknown system from the Hankel covariance matrix: order selected as 3.

We first estimate the *order* of the unknown system from the Hankel matrix created from the output covariance matrices, Λ_ℓ above, and perform an SVD obtaining the singular values illustrated in Figure 7.2:

$$\Sigma = \{24.4,\ 7.9,\ 0.9^*,\ 0.5,\ 0.005,\ 0.0,\ 0.0,\ 0.0\}$$

and choose a 3rd-order system (asterisk) to represent the data.

Next we use the canonical realization algorithm of Section 6.3 on the Hankel array and extract the following KSP model (Σ_{KSP}) with the resulting observability invariants: $v_1 = 1$ and $v_2 = 2$ to give

$$A_R = \begin{bmatrix} -1 & | & 1 & 0 \\ 0 & | & 0 & 1 \\ 0.252 & | & -0.29 & -0.083 \end{bmatrix}; \quad B_R = \begin{bmatrix} -8.73 & 1 \\ -2.64 & -0.63 \\ 2.32 & -0.32 \end{bmatrix}$$

$$C_R = \begin{bmatrix} 1 & | & 0 & 0 \\ 0 & | & 1 & 0 \end{bmatrix}; \quad D_R = \begin{bmatrix} 6.86 & 1.5 \\ 1.5 & 3.5 \end{bmatrix}$$

We iteratively solved the Kalman-filter-based KSP equations of Table 7.4 completing 10-iterations for convergence at an acceptance error of $\varepsilon = 10^{-25}$

Table 7.4 Iterative KSP algorithm (Kalman filter).

Innovations covariance

$R_{ee}(i) = \tilde{D} + \tilde{D}' - \tilde{C}\tilde{\Pi}(i)\tilde{C}'$

Gain

$K_p(i) = (\tilde{B} - \tilde{A}\tilde{\Pi}(i)\tilde{C}')R_{ee}^{-1}(i)$

Estimated covariance

$\tilde{\Pi}(i+1) = \tilde{A}\tilde{\Pi}(i)\tilde{A}' + K_p(i)R_{ee}(i)K_p'(i)$

Initial conditions

$\tilde{\Pi}(0)$

Stopping rule

$|I - \tilde{\Pi}(i) \times \tilde{\Pi}^{-1}(i+1)| \leq \epsilon$ for $\epsilon \ll 1$

---

yielding the following error and INV covariances as well as the Kalman gain as follows:

$$\tilde{\pi} = \begin{bmatrix} 7.17 & 1.80 & -1.93 \\ 1.80 & 0.60 & -0.48 \\ -1.93 & -0.48 & 0.52 \end{bmatrix}; \quad R_{ee} = \begin{bmatrix} 6.56 & 1.19 \\ 1.19 & 6.4 \end{bmatrix};$$

$$K_p = \begin{bmatrix} -0.6 & 0.46 \\ -0.11 & -0.005 \\ 0.16 & -0.13 \end{bmatrix}$$

We can complete the stochastic realization of the INV model as in Table 7.3 to obtain the following: $\Sigma_{INV} = \{A, B, C, D, R_{ww}^e, R_{wv}^e, R_{vv}^e\}$

$$R_{ww}^e = \begin{bmatrix} 3.00 & 0.35 & -0.82 \\ 0.35 & 0.08 & -0.09 \\ -0.82 & -0.09 & 0.23 \end{bmatrix}; \quad R_{vv}^e = \begin{bmatrix} 6.56 & 1.2 \\ 1.2 & 6.4 \end{bmatrix};$$

$$R_{wv}^e = \begin{bmatrix} -3.36 & 2.20 \\ -0.71 & -0.16 \\ 0.89 & -0.64 \end{bmatrix}$$

completing an example of the classical stochastic realization technique using the Kalman-filter-based solution of the corresponding Riccati equation. □

7.5 Stochastic Subspace Realization: Orthogonal Projections

Subspace identification (SID) algorithms are based on the INV representation of the measurement data "combined" with the underlying deterministic model Σ_{ABCD} given by

$$\hat{x}(t+1) = A\hat{x}(t) + Bu(t) + K_p e(t) \qquad \text{(State Estimate)}$$
$$y(t) = C\hat{x}(t) + Du(t) + e(t) \qquad \text{(Measurement)}$$

$$(7.60)$$

where $e(t)$ is the innovations sequence and $K_p(t)$ is the *predicted Kalman gain* as before.

As in Section 6.4, using the expression for the LTI solution to the state of the INV model

$$\hat{x}(t) = A^t \hat{x}(0) + \sum_{n=0}^{t-1} A^{t-n-1} Bu(n) + \sum_{n=0}^{t-1} A^{t-n-1} K_p e(n) \quad \text{for} \quad t > 0 \quad (7.61)$$

and substituting for the state into the measurement equation gives

$$y(t) = CA^t \hat{x}(0) + \sum_{n=0}^{t-1} CA^{t-n-1} Bu(n) + \sum_{n=0}^{t-1} CA^{t-n-1} K_p e(n) + Du(t) + e(t)$$

$$(7.62)$$

Expanding this relation further over k-samples and collecting the terms, we obtain

$$
\begin{bmatrix} y(0) \\ y(1) \\ \vdots \\ y(k-1) \end{bmatrix} = \underbrace{\begin{bmatrix} C \\ CA \\ \vdots \\ CA^{t-k-1} \end{bmatrix}}_{\mathcal{O}_k} \hat{x}(0) + \underbrace{\begin{bmatrix} D & & \cdots & 0 \\ CB & D & \cdots & 0 \\ \vdots & \vdots & \vdots & \vdots \\ CA^{k-1}B & \cdots & CB & D \end{bmatrix}}_{\mathcal{T}_k} \begin{bmatrix} u(0) \\ u(1) \\ \vdots \\ u(k-1) \end{bmatrix}
$$

$$
+ \underbrace{\begin{bmatrix} I & & \cdots & 0 \\ CK_p & I & \cdots & 0 \\ \vdots & \vdots & \vdots & \vdots \\ CA^{k-1}K_p & \cdots & CK_p & I \end{bmatrix}}_{\Xi_k} \begin{bmatrix} e(0) \\ e(1) \\ \vdots \\ e(k-1) \end{bmatrix} \qquad (7.63)
$$

where $\mathcal{O}_k \in \mathcal{R}^{kN_y \times N_x}$ is the *observability matrix*, $\mathcal{T}_k \in \mathcal{R}^{kN_y \times kN_u}$ impulse response matrix, and $\Xi_k \in \mathcal{R}^{kN_y \times kN_y}$ is a *noise (INV) impulse response matrix* – both Toeplitz matrices [32].

Shifting these relations in time $(0 \to t)$ yields

$$
\begin{bmatrix} y(t) \\ y(t+1) \\ \vdots \\ y(t+k-1) \end{bmatrix} = \mathcal{O}_k \hat{x}(t) + \mathcal{T}_k \begin{bmatrix} u(t) \\ u(t+1) \\ \vdots \\ u(t+k-1) \end{bmatrix} + \Xi_k \begin{bmatrix} e(t) \\ e(t+1) \\ \vdots \\ e(t+k-1) \end{bmatrix} \quad (7.64)
$$

leading to the *vector* input/output relation

$$
\mathbf{y}_k(t) = \mathcal{O}_k \hat{\mathbf{x}}(t) + \mathcal{T}_k \mathbf{u}_k(t) + \Xi_k \mathbf{e}_k(t) \quad (7.65)
$$

where $\mathbf{y} \in \mathcal{R}^{kN_y \times 1}$, $\hat{\mathbf{x}} \in \mathcal{R}^{N_x \times 1}$, $\mathbf{u} \in \mathcal{R}^{kN_u \times 1}$, and $\mathbf{e} \in \mathcal{R}^{kN_y \times 1}$.

We begin by defining the data matrices for samples ranging from k-to-$(2k-1)$ – the future data as [25]. First, stack these k-vectors to create batch data (*block Hankel*) matrices to obtain the subsequent "stochastic data equation." That is, defining the block output data matrix ($\mathcal{Y}_{k|2k-1}$), block input data matrix ($\mathcal{U}_{k|2k-1}$), and block INV data matrix ($\mathcal{E}_{k|2k-1}$)[2] over K-samples is

$$
\mathcal{Y}_{k|2k-1} := \begin{bmatrix} \mathbf{y}(k) & \mathbf{y}(k+1) & \cdots & \mathbf{y}(K+k-1) \\ \mathbf{y}(k+1) & \mathbf{y}(k+2) & \cdots & \mathbf{y}(K+k) \\ \vdots & \vdots & \vdots & \vdots \\ \mathbf{y}(2k-1) & \mathbf{y}(2k) & \cdots & \mathbf{y}(K+2k-2) \end{bmatrix} \quad (7.66)
$$

with

$$
\mathcal{U}_{k|2k-1} := \begin{bmatrix} \mathbf{u}(k) & \mathbf{u}(k+1) & \cdots & \mathbf{u}(K+k-1) \\ \mathbf{u}(k+1) & \mathbf{u}(k+2) & \cdots & \mathbf{u}(K+k) \\ \vdots & \vdots & \vdots & \vdots \\ \mathbf{u}(2k-1) & \mathbf{u}(2k) & \cdots & \mathbf{u}(K+2k-2) \end{bmatrix} \quad (7.67)
$$

and

$$
\mathcal{E}_{k|2k-1} := \begin{bmatrix} \mathbf{e}(k) & \mathbf{e}(k+1) & \cdots & \mathbf{e}(K+k-1) \\ \mathbf{e}(k+1) & \mathbf{e}(k+2) & \cdots & \mathbf{e}(K+k) \\ \vdots & \vdots & \vdots & \vdots \\ \mathbf{e}(2k-1) & \mathbf{e}(2k) & \cdots & \mathbf{e}(K+2k-2) \end{bmatrix} \quad (7.68)
$$

leading to the corresponding *stochastic data equation* given by

$$
\mathcal{Y}_{k|2k-1} = \mathcal{O}_k \mathcal{X}_k + \mathcal{T}_k \mathcal{U}_{k|2k-1} + \Xi_k \mathcal{E}_{k|2k-1} \quad (7.69)
$$

where the block (Hankel) matrices are defined as

$$
\mathcal{Y}_{k|2k-1} = [\mathbf{y}_k(k)\ \mathbf{y}_k(k+1)\ \cdots\ \mathbf{y}_k(K+k-1)\,] \in \mathcal{R}^{kN_y \times K}
$$

$$
\mathcal{U}_{k|2k-1} = [\mathbf{u}_k(k)\ \mathbf{u}_k(k+1)\ \cdots\ \mathbf{u}_k(K+k-1)\,] \in \mathcal{R}^{kN_u \times K}
$$

---

2 The notation "$k|2k-1$" defines the first and last row elements of the K-column data matrices.

$$\mathcal{E}_{k|2k-1} = [\mathbf{e}_k(k) \; \mathbf{e}_k(k+1) \; \cdots \; \mathbf{e}_k(K+k-1)\,] \in \mathcal{R}^{kN_y \times K}$$

$$\mathcal{X}_k = [\mathbf{x}(k) \; \mathbf{x}(k+1) \; \cdots \; \mathbf{x}(K+k-1)\,] \in \mathcal{R}^{N_x \times K}$$

Here $\mathcal{U}_{0|k-1}$, $\mathcal{Y}_{0|k-1}$, $\mathcal{E}_{0|k-1}$ are defined as the *past* inputs and outputs (U_p, Y_p, E_p), while $\mathcal{U}_{k|2k-1}$, $\mathcal{Y}_{k|2k-1}$, $\mathcal{E}_{k|2k-1}$, and $\mathcal{X}_k$ are the *future* inputs/outputs and states $(\mathcal{U}_f, \mathcal{Y}_f, \mathcal{E}_f, \mathcal{X}_f)$, which are all *block Hankel* matrices.

$$\mathcal{U}_p := \mathcal{U}_{0|k-1}; \quad \mathcal{Y}_p := \mathcal{Y}_{0|k-1}; \quad \mathcal{E}_p := \mathcal{E}_{0|k-1}; \quad \mathcal{X}_p := \mathcal{X}_0$$

$$\mathcal{U}_f := \mathcal{U}_{k|2k-1}; \quad \mathcal{Y}_f := \mathcal{Y}_{k|2k-1}; \quad \mathcal{E}_f := \mathcal{E}_{k|2k-1}; \quad \mathcal{X}_f := \mathcal{X}_k$$

with the matrix input/output relations given by

$$\mathcal{Y}_p = \mathcal{O}_k \mathcal{X}_p + \mathcal{T}_k \mathcal{U}_p + \Xi_k \mathcal{E}_p \qquad \text{(Past)}$$

$$\mathcal{Y}_f = \mathcal{O}_k \mathcal{X}_f + \mathcal{T}_k \mathcal{U}_f + \Xi_k \mathcal{E}_f \qquad \text{(Future)} \qquad (7.70)$$

and data matrices, past and future, defined by:

$$\mathcal{D}_p := \begin{bmatrix} \mathcal{U}_p \\ --- \\ \mathcal{Y}_p \end{bmatrix} = \begin{bmatrix} \mathcal{U}_{0|k-1} \\ --- \\ \mathcal{Y}_{0|k-1} \end{bmatrix}; \quad \mathcal{D}_f := \begin{bmatrix} \mathcal{U}_f \\ --- \\ \mathcal{Y}_f \end{bmatrix} = \begin{bmatrix} \mathcal{U}_{k|2k-1} \\ --- \\ \mathcal{Y}_{k|2k-1} \end{bmatrix} \qquad (7.71)$$

The overall objective (as before) is to extract the observability matrix and therefore the system matrices Σ_{INV} from the noisy (stochastic) input/output data. That is, we seek a solution that possesses the column space (range) equivalent to $\mathcal{O}_K$ (e.g. range($\mathcal{Y}_{k|2k-1}$) $\equiv$ range($\mathcal{O}_K$)). With this property satisfied, the extraction of Σ_{INV}, after performing an SVD to obtain the equivalent column space, is enabled, but unfortunately $\mathcal{T}_K$ and Ξ_K are *unknown*. Thus, we must somehow estimate both of the response matrices, first, in order to remove them from the data equation. This leads us directly to the projection theory and orthogonality properties of the INV sequence.

Following the least-squares approach of Section 6.4, the projection operator $\mathcal{P}^3$ is orthogonal to the input space $\mathcal{U}$ such that $\langle \mathcal{U}, \mathcal{P}^\perp \rangle = \mathbf{0}$ is given by [17, 19, 34]

$$\mathcal{P}^\perp_{\mathcal{U}_{k|2k-1}} = \mathbf{I} - \mathcal{P}_{\mathcal{U}_{k|2k-1}} = \mathbf{I} - \mathcal{U}_{k|2k-1}(\mathcal{U}_{k|2k-1}\mathcal{U}_{k|2k-1})^{-1}\mathcal{U}_{k|2k-1} \qquad (7.72)$$

which is the orthogonal projection of the null (column) space of $\mathcal{U}_{k|2k-1}$ onto the null space of $\mathcal{U}$ such that

$$\langle \mathcal{U}_{k|2k-1}, \mathcal{P}^\perp_{\mathcal{U}_{k|2k-1}} \rangle \Rightarrow \mathcal{U}_{k|2k-1} \times \mathcal{P}^\perp_{\mathcal{U}_{k|2k-1}} = \mathbf{0} \qquad (7.73)$$

where $\mathcal{U}_{k|2k-1}$ is assumed to be persistently exciting or equivalently full rank, $\rho(\mathcal{U}) = kN_u$.

---

3 This projection operator is over a random vector space (Hilbert) [25] incorporating the expectation operator such that $\langle a, b \rangle = E\{a'b\}$ (see Appendix B for details).

Applying the projection operator to the stochastic data equation of Eq. (7.69) nullifies the input term, that is,

$$
\begin{aligned}
\mathcal{P}^{\perp}_{\mathcal{Y}_{k|2k-1}|\mathcal{U}_{k|2k-1}} &:= \mathcal{Y}_{k|2k-1} \times \mathcal{P}^{\perp}_{\mathcal{U}_{k|2k-1}} \\
&= \mathcal{O}_k \hat{\mathcal{X}}_k \mathcal{P}^{\perp}_{\mathcal{U}_{k|2k-1}} + \underbrace{\mathcal{T}_k \mathcal{U}_{k|2k-1}\, \mathcal{P}^{\perp}_{\mathcal{U}_{k|2k-1}}}_{0} + \Xi_k \mathcal{E}_{k|2k-1} \mathcal{P}^{\perp}_{\mathcal{U}_{k|2k-1}} \quad (7.74)
\end{aligned}
$$

The INV sequence is unaffected by the projection operator $\mathcal{P}^{\perp}_{\mathcal{U}_{k|2k-1}}$ because it is orthogonal to the inputs, past and future, therefore

$$
\mathcal{E}_{k|2k-1} \mathcal{P}^{\perp}_{\mathcal{U}_{k|2k-1}} = \mathcal{E}_{k|2k-1} \times (\mathbf{I} - \mathcal{U}_{k|2k-1}(\mathcal{U}_{k|2k-1}\mathcal{U}_{k|2k-1})^{-1}\mathcal{U}_{k|2k-1}) = \mathcal{E}_{k|2k-1}
$$

That is, from the well-known properties of the INV, we have that the future INV sequence is uncorrelated with the *past inputs* [32]

$$
\mathcal{E}_{k|2k-1} \mathcal{U}_{0|k-1} = 0
$$

Therefore, postmultiplying Eq. (7.74) by $\mathcal{U}_{0|k-1}$ and using this property, we can extract the observability matrix

$$
\mathcal{Y}_{k|2k-1} \mathcal{P}^{\perp}_{\mathcal{U}_{k|2k-1}} \mathcal{U}_{0|k-1} = \mathcal{O}_k \hat{\mathcal{X}}_k \mathcal{P}^{\perp}_{\mathcal{U}_{k|2k-1}} \mathcal{U}_{0|k-1} + \underbrace{\Xi_k \mathcal{E}_{k|2k-1} \mathcal{U}_{0|k-1}}_{0}
$$

$$
= \mathcal{O}_k \hat{\mathcal{X}}_k \mathcal{P}^{\perp}_{\mathcal{U}_{k|2k-1}} \mathcal{U}_{0|k-1} \quad (7.75)
$$

It has been shown that the corresponding *rank condition* for the stochastic realization problem is satisfied as (see [19] for details)

$$
\rho(\hat{\mathcal{X}}_k \mathcal{P}^{\perp}_{\mathcal{U}_{k|2k-1}} \mathcal{U}_{k|2k-1}) = N_x \quad (7.76)
$$

If the underlying stochastic representation incorporates process noise such as the INV and Gauss–Markov models, then the SID technique must include both "past" inputs and outputs leading to either the PO-MOESP or the N4SID approach [17, 19, 27]. These methods provide robust solutions to the stochastic realization problem in this case.

The "past" input PI-MOESP method of the previous chapter has been developed from an instrumental variable (IV) perspective in which the IV is defined as $\mathcal{I}_p := \mathcal{U}_{0|k-1}$; however, in contrast, both PO-MOESP and N4SID techniques are based on an augmented IV that incorporates the past outputs as well, that is,

$$
\mathcal{I}_{0|k-1} := \begin{bmatrix} \mathcal{U}_{0|k-1} \\ --- \\ \mathcal{Y}_{0|k-1} \end{bmatrix} \in \mathcal{R}^{k(N_u+N_y)\times K} \quad (7.77)
$$

Recall that instrumental variables are selected to remove the effects of the noise term, leaving the critical information intact. IVs must satisfy the following conditions:

- $E\{\mathcal{E}_{0|k-1}\mathcal{I}'_{0|k-1}\} = 0$ (Uncorrelated with Noise)
- $\rho(E\{\mathcal{X}_0\mathcal{I}'_{0|k-1}\}) = N_x$ (Rank Condition)

that is, the IV is to be selected such that it is uncorrelated with the noise (INV) and preserves the rank in order to extract the extended observability matrix $\mathcal{O}_k$ or the extended states $\mathcal{X}_f$.

It also follows from the orthogonality property of the INV that it is orthogonal to past outputs as well.

$$E\{\mathcal{E}_{k|2k-1}\mathcal{Y}'_{0|k-1}\} = 0$$

After applying the projection operator $\mathcal{P}^\perp_{\mathcal{U}_{0|k-1}}$ to the future measurements removing the effect of the past inputs as in Eq. (7.74), we replace the "instrument" with the augmented instrumental variable $\mathcal{I}_{0|k-1}$ such that [19]

$$\left(\mathcal{Y}_{k|2k-1}\mathcal{P}^\perp_{\mathcal{U}_{k|2k-1}}\right)\mathcal{I}'_{0|k-1} = \left(\mathcal{O}_k\hat{\mathcal{X}}_k\mathcal{P}^\perp_{\mathcal{U}_{k|2k-1}}\right)\mathcal{I}'_{0|k-1} + \underbrace{\left(\Xi_k\mathcal{E}_{k|2k-1}\mathcal{P}^\perp_{\mathcal{U}_{k|2k-1}}\right)\mathcal{I}'_{0|k-1}}_{0}$$

$$(7.78)$$

It has been shown that the rank condition is preserved under these operations [19], that is,

$$\rho\left(E\left\{\mathcal{Y}_{k|2k-1}\mathcal{P}^\perp_{\mathcal{U}_{k|2k-1}}\mathcal{I}'_{0|k-1}\right\}\right) = N_x$$

Therefore, with this choice of IV, we have that the observability matrix can be recovered from

$$\mathcal{Y}_{k|2k-1}\mathcal{P}^\perp_{\mathcal{U}_{k|2k-1}}\mathcal{I}'_{0|k-1} = \mathcal{O}_k\hat{\mathcal{X}}_k\mathcal{P}^\perp_{\mathcal{U}_{k|2k-1}}\mathcal{I}'_{0|k-1} \qquad (7.79)$$

leading to a robust implementation of the PO-MOESP algorithm to follow.

7.5.1 Multivariable Output Error State-SPace (MOESP) Algorithm

The MOESP[4] or more appropriately the PO-MOESP technique is based on performing an orthogonal LQ-decomposition on the data matrix (see Appendix C) [17–23, 25]. That is, $\mathcal{D} = L \times Q$ with L a lower block triangular matrix, $L \in \mathcal{R}^{2k(N_u+N_y)\times 2k(N_u+N_y)}$ and Q an orthogonal matrix, $Q \in \mathcal{R}^{2k(N_u+N_y)\times K}$ such that

$$\mathcal{D} := \begin{bmatrix} \mathcal{U}_{k|2k-1} \\ --- \\ \mathcal{I}_{0|k-1} \\ --- \\ \mathcal{Y}_{k|2k-1} \end{bmatrix}$$

---

4 This method is called the past outputs (PO-MOESP) technique with instrumental variable $\mathcal{I}_{0|k-1}$.

$$
= \begin{bmatrix} L_{11} & | & 0_{kN_u \times kN_y} & | & 0_{kN_u \times kN_y} \\ \hline L_{21} & | & L_{22} & | & 0_{kN_u \times kN_y} \\ \hline L_{31} & | & L_{32} & | & L_{33} \end{bmatrix} \underbrace{\begin{bmatrix} Q'_1 \\ \hline Q'_2 \\ \hline Q'_3 \end{bmatrix}}_{Q} \in \mathcal{R}^{2k(N_u + N_y) \times K} \qquad (7.80)
$$

$$
\underbrace{\phantom{\begin{bmatrix} L_{11} \end{bmatrix}}}_{L}
$$

with $L_{11} \in \mathcal{R}^{kN_u \times kN_u}$, $L_{21} \in \mathcal{R}^{k(N_u+N_y) \times kN_u}$, $L_{22} \in \mathcal{R}^{k(N_u+N_y) \times k(N_u+N_y)}$, $L_{31} \in \mathcal{R}^{kN_y \times kN_u}$, $L_{32} \in \mathcal{R}^{kN_y \times k(N_u+N_y)}$, $L_{33} \in \mathcal{R}^{kN_y \times kN_y}$; and $Q'_1 \in \mathcal{R}^{kN_u \times K}$, $Q'_2 \in \mathcal{R}^{k(N_u+N_y) \times K}$, $Q'_3 \in \mathcal{R}^{kN_y \times K}$.

Expanding each of these block rows, we obtain

$$
\begin{aligned}
\mathcal{U}_{k|2k-1} &= L_{11} Q'_1 \\
\mathcal{I}_{0|k-1} &= L_{21} Q'_1 + L_{22} Q'_2 \\
\mathcal{Y}_{k|2k-1} &= L_{31} Q'_1 + L_{32} Q'_2 + L_{33} Q'_3
\end{aligned} \qquad (7.81)
$$

Postmultiplying the stochastic data relations of Eq. (7.69) by Q_i; $i = 1, 2$ and applying the LQ-decomposition properties $Q'_i Q_i = I$ and $Q'_i Q_j = 0$ gives

$$
\begin{aligned}
\mathcal{Y}_{k|2k-1} \times Q_1 &= \mathcal{O}_k \mathcal{X}_k Q_1 + \mathcal{T}_k \mathcal{U}_{k|2k-1} Q_1 + \Xi_k \xi_{k|2k-1} Q_1 \\
&= L_{31} \underbrace{Q'_1 Q_1}_{I} + L_{32} \underbrace{Q'_2 Q_1}_{0} + L_{33} \underbrace{Q'_3 Q_1}_{0}
\end{aligned} \qquad (7.82)
$$

Similarly postmultiplying by Q_2 gives

$$
\begin{aligned}
\mathcal{Y}_{k|2k-1} \times Q_2 &= \mathcal{O}_k \mathcal{X}_k Q_2 + \mathcal{T}_k \mathcal{U}_{k|2k-1} Q_2 + \Xi_k \xi_{k|2k-1} Q_2 \\
&= L_{31} \underbrace{Q'_1 Q_2}_{0} + L_{32} \underbrace{Q'_2 Q_2}_{I} + L_{33} \underbrace{Q'_3 Q_2}_{0}
\end{aligned} \qquad (7.83)
$$

leading to the following relations:

$$
\begin{aligned}
\mathcal{U}_{k|2k-1} \times Q_1 &= L_{11}; \quad \mathcal{Y}_{k|2k-1} \times Q_1 = L_{31} \\
\mathcal{U}_{k|2k-1} \times Q_2 &= 0; \quad \mathcal{Y}_{k|2k-1} \times Q_2 = L_{32}
\end{aligned} \qquad (7.84)
$$

Substituting these relations into Eqs. (7.82) and (7.83) gives

$$
\begin{aligned}
\mathcal{Y}_{k|2k-1} \times Q_1 &= L_{31} = \mathcal{O}_k \mathcal{X}_k Q_1 + \mathcal{T}_k L_{11} + \Xi_k \xi_{k|2k-1} Q_1 \\
\mathcal{Y}_{k|2k-1} \times Q_2 &= L_{32} = \mathcal{O}_k \mathcal{X}_k Q_2 + \Xi_k \xi_{k|2k-1} Q_2
\end{aligned} \qquad (7.85)
$$

The last terms in each of these expressions are null based on the INV orthogonality properties with past inputs/outputs and future inputs, that is, we have that

$$
E\{\mathcal{E}_{k|2k-1}[\mathcal{U}'_{k|2k-1} \mid \mathcal{I}'_{0|k-1}]\} = 0 \qquad (7.86)
$$

and from the upper block rows of the LQ-decomposition (see Section 6.4), we have

$$
\begin{bmatrix} \mathcal{U}_{0|k-1} \\ \hline \mathcal{I}_{0|k-1} \end{bmatrix} = \begin{bmatrix} L_{11} & | & 0 \\ \hline L_{21} & | & L_{22} \end{bmatrix} \begin{bmatrix} Q_1' \\ \hline Q_2' \end{bmatrix} \tag{7.87}
$$

it follows that

$$
E\{\mathcal{E}_{k|2k-1}[Q_1 \mid Q_2]\} = 0 \tag{7.88}
$$

implying that the orthogonal matrices Q_i'; $i = 1, 2$ are *uncorrelated* with the future noise $\mathcal{E}_{k|2k-1}$ [19].

Therefore, nullifying the innovation terms, we have that Eq. (7.85) can be simplified as

$$
\mathcal{Y}_{k|2k-1} \times Q_1 = \mathcal{O}_k \mathcal{X}_k Q_1 + \mathcal{T}_k L_{11} + \underbrace{\Xi_k \xi_{k|2k-1} Q_1}_{0}
$$

$$
\mathcal{Y}_{k|2k-1} \times Q_2 = \mathcal{O}_k \mathcal{X}_k Q_2 + \underbrace{\Xi_k \xi_{k|2k-1} Q_2}_{0}
$$

resulting in the final set of relations enabling the extraction of the observability matrix and the KSP model, that is,

$$
\mathcal{Y}_{k|2k-1} Q_1 = L_{31} = \mathcal{O}_k \mathcal{X}_k Q_1 + \mathcal{T}_k L_{11} \tag{7.89}
$$

$$
\mathcal{Y}_{k|2k-1} Q_2 = L_{32} = \mathcal{O}_k \mathcal{X}_k Q_2 \tag{7.90}
$$

Performing the SVD of L_{32}, that is,

$$
L_{32} = [U_S \mid U_N] \begin{bmatrix} \Sigma_S & | & 0 \\ \hline 0 & | & \Sigma_N \end{bmatrix} \begin{bmatrix} V_S' \\ \hline V_N' \end{bmatrix} = \underbrace{U_S \Sigma_S V_S'}_{\text{Signal Subspace}} + \underbrace{U_N \Sigma_N V_N'}_{\text{Noise Subspace}} \tag{7.91}
$$

assuming the noise component is negligible compared to the signal component (as before) yields

$$
L_{32} = \mathcal{O}_k \hat{\mathcal{X}}_k Q_2 \approx U_S \times \Sigma_S \times V_S' \approx \underbrace{(U_S \Sigma_S^{1/2})}_{\mathcal{O}_k} \times ((\Sigma_S')^{1/2} V_S') \tag{7.92}
$$

extracting the observability matrix with a balanced stochastic realization.

The system matrices of the KSP model, $\Sigma_{\text{KSP}} = \{A, B, C, D\}$, can now be determined as before in the deterministic case using the shift-invariant approach to give

$$
A = \mathcal{O}_{N_x-1}^{\#} \times \mathcal{O}_{N_x}^{\uparrow} = \left(\Sigma_S^{-1/2} U_S'\right) \times \mathcal{O}_{N_x}^{\uparrow}; \quad C = \mathcal{O}(1 : N_y, 1 : N_x) \tag{7.93}
$$

with B and D obtained by solving a least-squares problem directly, since premultiplying Eq. (7.89) by U'_N gives

$$U'_N L_{31} = \underbrace{U'_N \mathcal{O}_k \mathcal{X}_k Q_1}_{0} + U'_N \mathcal{T}_k L_{11} = U'_N \mathcal{T}_k L_{11}$$

after applying the orthogonality conditions $U'_N \mathcal{O}_k = 0$, we have

$$U'_N L_{31} = U'_N \mathcal{T}_k L_{11}$$

$$Z := U'_N \mathcal{T}_k = U'_N L_{31} \times L_{11}^{-1} \tag{7.94}$$

Stacking and reordering the vectors leads to an overdetermined set of linear equations and a least-squares solution with

$$\begin{bmatrix} \hat{D} \\ -- \\ \hat{B} \end{bmatrix} = \mathcal{U}^{\#} \times \tilde{Z} \tag{7.95}$$

where $\mathcal{U}$ is the block coefficient matrix (see Section 6.5 for details).

Using $\Sigma_{\mathrm{KSP}} = \{A, B, C, D\}$, the corresponding INV-based (Kalman) Riccati equation can be solved, yielding a positive definite solution $\tilde{\Pi} > 0$ such that

$$\tilde{\Pi} = A\tilde{\Pi}A' + (B - A\Pi C')(D + D' - C\tilde{\Pi}C')^{-1}(B - A\tilde{\Pi}C')' \tag{7.96}$$

The INV covariance R_{ee} and Kalman gain are then given by enabling the solution of the stochastic realization problem as follows:

$$R_{ee} = (D + D' - C\tilde{\Pi}C')$$
$$K_p = (B - A\,\tilde{\Pi}\,C') \times R_{ee}^{-1} \tag{7.97}$$

and therefore, the corresponding covariances are

$$R_{ww}^e = K_p R_{ee} K_p'$$
$$R_{vv}^e = R_{ee}$$
$$R_{wv}^e = K_p R_{ee} \tag{7.98}$$

completing the stochastic realization.

Therefore, we have the *PO-Multivariable Output Error State-SPace* (PO-MOESP) [17] algorithm given by the following steps:

- Compute the LQ-decomposition of $\mathcal{D}$ of Eq. (7.80).
- Perform the SVD of L_{32} in Eq. (7.91) to extract $\mathcal{O}_k$.
- Obtain A and C from Eq. (7.93).
- Solve the least-squares problem to obtain B and D from Eq. (7.95).
- With Σ_{KSP} available, solve the INV-based (Kalman) Riccati equation of Eq. (7.96) for the positive definite state error covariance $\tilde{\Pi}$.
- Compute the INV covariance R_{ee} and Kalman gain K_p from Eq. (7.97).
- Compute the corresponding set of noise covariances $\{R_{ww}^e, R_{vv}^e, R_{wv}^e\}$ from Eq. (7.98).

Table 7.5 Stochastic realization algorithm: PAST OUTPUT-MOESP (PO-MOESP) approach.

Orthogonal projection (LQ-decomposition)

$$\mathcal{D} = \begin{bmatrix} \mathcal{U}_f \\ \mathcal{D}_p \\ \mathcal{Y}_f \end{bmatrix} = \begin{bmatrix} L_{11} & 0 & 0 \\ L_{21} & L_{22} & 0 \\ L_{31} & L_{32} & L_{33} \end{bmatrix} = \begin{bmatrix} \tilde{Q}'_1 \\ \tilde{Q}'_2 \\ \tilde{Q}'_3 \end{bmatrix} \qquad \text{(Data Matrix)}$$

Singular value decomposition

$$L_{32} = [U_S \ U_N] \begin{bmatrix} \Sigma_S & 0 \\ 0 & \Sigma_N \end{bmatrix} \begin{bmatrix} V'_S \\ V'_N \end{bmatrix}$$

$$\approx \ U_S \times \Sigma_S \times V'_S \approx \underbrace{(U_S \Sigma_S^{1/2})}_{\mathcal{O}_k} \times ((\Sigma'_S)^{1/2} V'_S) \qquad \text{(Observability Matrix)}$$

KSP system matrix

$$A = (\mathcal{O}'_{N_x} \mathcal{O}_{N_x})^{-1} \mathcal{O}'_{N_x} \mathcal{O}^{\uparrow}_{N_x} = (\Sigma_S^{-1/2} U'_S) \mathcal{O}^{\uparrow}_{N_x} \qquad \text{(System/Process Matrix)}$$

KSP measurement (output) matrix

$$C = \mathcal{O}(1 : N_y, 1 : N_x) \qquad \text{(Output Matrix)}$$

KSP input/input/output transmission matrices

$$\begin{bmatrix} \hat{D} \\ -- \\ \hat{B} \end{bmatrix} = \mathcal{U}^{\#} \times \tilde{Z} \qquad \text{(Input Matrices)}$$

State covariance equation

$$\tilde{\Pi} = A\tilde{\Pi}A' + (B - A\tilde{\Pi}C')(D + D' - C\tilde{\Pi}C')^{-1}(B - A\tilde{\Pi}C')' \qquad \text{(Riccati Equation)}$$

KSP equations

$$R_{ee} = D + D' - C\,\tilde{\Pi}\,C'$$
$$K_p = (B - A\,\tilde{\Pi}\,C') \times R_{ee}^{-1} \qquad \text{(KSP Equations)}$$

Noise covariance equations

$$R^e_{ww} = K_p R_{ee} K'_p$$
$$R^e_{vv} = R_{ee} \qquad \text{(Noise-Covariances)}$$
$$R^e_{wv} = K_p R_{ee}$$

We summarize the PO-MOESP algorithm in Table 7.5 and consider the following example of sinusoids in noise.

Example 7.2   Consider the following sinusoidal system representative of frequency ranges that might occur from vibrating machinery aboard ships navigating throughout the ocean or aircraft flying in the sky from structures like bridges or buildings or even from seismic events ranging from 0 to 25 Hz. The signal model for these events can easily be represented by sinusoids in noise:

$$y(t_k) = \sum_{i=1}^{N_s} A_i \, \sin 2\pi f_i t_k + v(t_k); \quad k = 1, \ldots, K$$

where $v \sim \mathcal{N}(0, \sigma_{vv}^2)$.

Data were synthesized by six sinusoids at the frequencies of $f_i = \{2, 5, 10, 14, 20, 22 \text{ Hz}\}$ all at unity amplitude with an signal-to-noise ratio (SNR) of 0 dB for $K = 10\,000$ samples at a sampling interval of $\Delta t = 0.02$ seconds. The simulated data are shown in Figure 7.3, indicating the noisy sequence in (a) and its corresponding power spectrum using the correlation method in (b). We note that the sinusoidal frequencies are clearly discernible at 0 dB SNR.

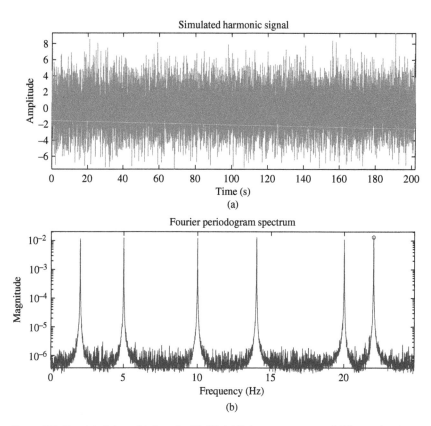

Figure 7.3 Simulated sinusoids in noise (0 dB). (a) Noisy measurement. (b) Spectral estimate (correlation method).

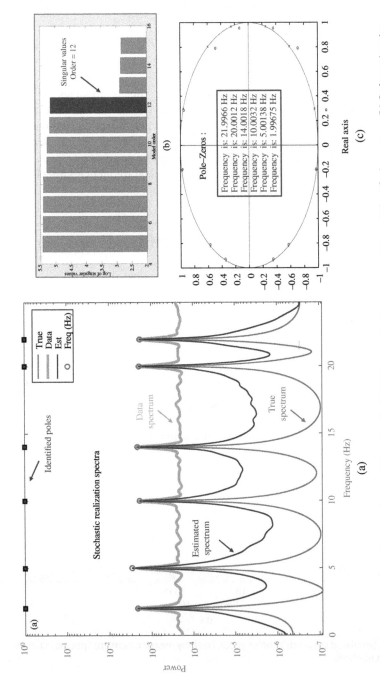

Figure 7.4 Subspace identification of sinusoids in noise (0 dB). (a) Spectral estimates: true, data and identified spectrum. (b) Order estimation using SVD: Order = 12. (c) Pole–zero plot with identified frequencies (continuous).

Both the PO-MOESP and N4SID (to follow) SID methods were applied to this data set with very similar results. The resulting spectral estimates are shown in Figure 7.4 with the estimated spectra in (a) along with the identified frequencies in (c). The SVD-method of order estimation was applied by varying the order from 5 to 15, and the correct order of 12 states or equivalently 6 sinusoids was estimated from the singular values as shown in (b). The estimated signals are shown in Figure 7.5, showing a small snapshot of both the true and estimated impulse response as well as the "fit" to the noisy data. Each indicates a reasonable estimate of both signal and measurement. The validation statistics were also quite good with a zero-mean/whiteness test result of Z-M: $0.003 < 0.022$; W-T: 3.3% out with the weighted sum-squared residual (WSSR) statistic below the threshold, again indicating a good fit to the data. The final estimated sinusoidal frequencies are shown in Table 7.6. Here we again see extremely good results at this SNR for both methods with small relative errors of ∼0.5% for N4SID and ∼1.1% for MOESP techniques – essentially identical performance. □

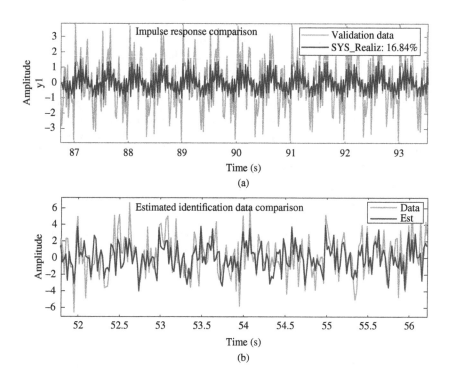

Figure 7.5 Comparison of identified sinusoids in noise (0 dB). (a) Impulse response fit. (b) Estimated data fit.

Table 7.6 Stochastic subspace identification: six (6) sinusoidal system.

		Frequency estimation/model error/validation results				
Frequency	N4SID	Relative	MOESP	Relative	Z-Mean	WSSR
Hz	Estimate	Error	Estimate	Error	3.1×10^{-2}	Below τ
2	1.997	1.50×10^{-3}	2.006	3.0×10^{-3}	1.6×10^{-8}	Yes
5	5.001	0.20×10^{-3}	4.984	3.2×10^{-3}	1.6×10^{-8}	Yes
10	10.007	0.70×10^{-3}	10.004	0.40×10^{-3}	1.6×10^{-8}	Yes
14	14.002	0.14×10^{-3}	14.009	0.06×10^{-3}	1.6×10^{-8}	Yes
20	20.001	0.05×10^{-3}	20.001	0.05×10^{-3}	1.6×10^{-8}	Yes
22	21.997	0.14×10^{-3}	22.004	0.02×10^{-3}	1.6×10^{-8}	Yes
Average	—	0.46×10^{-3}	—	1.09×10^{-3}	—	Yes

7.6 Stochastic Subspace Realization: Oblique Projections

The alternative approach to decomposing or separating the data space is to apply the oblique projection operator enabling the extraction of the stochastic state-space model while eliminating the past input/output/noise data. This technique first estimates a *state sequence* from the noisy input/output directly through the oblique projection of particular row spaces of the input/output block Hankel data matrices followed by least-squares solutions of state/residual equations to extract the underlying model Σ_{INV}. In this approach, the row space of the state sequence can be obtained using the SVD, here the *key* element of oblique SID is based on extracting:

$$\mathcal{X}_k = \underbrace{[\mathbf{x}(k)\ \mathbf{x}(k+1)\ \cdots\ \mathbf{x}(K+k-1)\,]}_{\text{State--Space Sequence}}$$

This is the underlying concept of the N4SIDapproach of the previous chapter when applied to the stochastic realization problem of this chapter.

We start with past and future input/output relations of Eq. (7.70) and include the corresponding state relation that is developed by expanding the INV state-space model, that is,

$$\hat{x}(k) = A^k \hat{x}(j) + \sum_{n=0}^{k-1} A^{k-n-1} Bu(n+j) + \sum_{n=0}^{k-1} A^{k-n-1} K_p e(n+j)$$

to determine the set of future states in terms of past states $\mathcal{X}_p$, inputs $\mathcal{U}_p$ and INV $\mathcal{E}_p$ as

$$\mathcal{X}_f = A^k \mathcal{X}_p + C_{AB}^{\leftarrow}(k)\,\mathcal{U}_p + C_{AK}^{\leftarrow}(k)\,\mathcal{E}_p \qquad (7.99)$$

where the input/innovation "reversed" controllability matrices are given by

$$C^{\leftarrow}_{AB}(k) = [A^{k-1}B \; A^{k-2}B \; \cdots \; AB \; B] \quad \text{(reversed input controllability)}$$

$$C^{\leftarrow}_{AK}(k) = [A^{k-1}K_p \; A^{k-2}K_p \; \cdots \; AK_p \; K_p]$$

$$\text{(reversed innovation controllability)} \qquad (7.100)$$

for $C^{\leftarrow}_{AB} \in \mathcal{R}^{N_x \times kN_u}$ and $C^{\leftarrow}_{AK} \in \mathcal{R}^{N_x \times kN_y}$.

The set of "stochastic" data equations for the past/future inputs/outputs/ states are summarized as follows:

$$\mathcal{Y}_p = \mathcal{O}_k \mathcal{X}_p + \mathcal{T}_k \mathcal{U}_p + \Xi_k \mathcal{E}_p \qquad \text{(Past Outputs)}$$

$$\mathcal{Y}_f = \mathcal{O}_k \mathcal{X}_f + \mathcal{T}_k \mathcal{U}_f + \Xi_k \mathcal{E}_f \qquad \text{(Future Outputs)}$$

$$\mathcal{X}_f = A^k \mathcal{X}_p + C^{\leftarrow}_{AB} \mathcal{U}_p + C^{\leftarrow}_{AK} \mathcal{E}_p \qquad \text{(Future States)} \qquad (7.101)$$

As before in the previous chapter, the orthogonal projector onto the joint past data and future input space can be constructed from the sum of two oblique projectors and the orthogonal complement as illustrated in Figure 6.7. Here the orthogonal projection is performed first enabling the projection onto the joint space $(\mathcal{D}_p \vee \mathcal{U}_f)$ that is followed by the decomposition along either $\mathcal{D}_f$ or $\mathcal{U}_p$ [25].

$$P^{\perp}_{\mathcal{Y}_f | \mathcal{D}_p \vee \mathcal{U}_f} = \underbrace{P^{\parallel}_{\mathcal{Y}_f | \mathcal{D}_p \circ \mathcal{U}_f} + P^{\parallel}_{\mathcal{Y}_f | \mathcal{U}_f \circ \mathcal{D}_p} +}_{\text{Oblique Projections}} \quad \underbrace{P^{\perp}_{\mathcal{Y}_f | \mathcal{U}_f \vee \mathcal{D}_p}}_{\text{Orthogonal Complement Projection}} \qquad (7.102)$$

Applying the oblique projection operator $P_{\bullet | \mathcal{D}_p \circ \mathcal{U}_f}$ to the "future outputs" equation above, projecting each term individually onto the past data $(\mathcal{D}_p)$ along the future inputs $(\mathcal{U}_f)$, we have

$$\mathcal{Y}_f \times P^{\parallel}_{\bullet | \mathcal{D}_p \vee \mathcal{U}_f} = (\mathcal{O}_k \mathcal{X}_f + \mathcal{T}_k \mathcal{U}_f + \Xi_k \mathcal{E}_f) \times P^{\parallel}_{\bullet | \mathcal{D}_p \circ \mathcal{U}_f}$$

$$\mathcal{Y}_f \times P^{\parallel}_{\bullet | \mathcal{D}_p \vee \mathcal{U}_f} = \mathcal{O}_k \mathcal{X}_f \times P^{\parallel}_{\bullet | \mathcal{D}_p \circ \mathcal{U}_f} + \mathcal{T}_k \mathcal{U}_f \times P^{\parallel}_{\bullet | \mathcal{D}_p \circ \mathcal{U}_f} + \Xi_k \mathcal{E}_f \times P^{\parallel}_{\bullet | \mathcal{D}_p \circ \mathcal{U}_f}$$

$$P^{\parallel}_{\mathcal{Y}_f | \mathcal{D}_p \vee \mathcal{U}_f} = \mathcal{O}_k P^{\parallel}_{\mathcal{X}_f | \mathcal{D}_p \circ \mathcal{U}_f} + \mathcal{T}_k \underbrace{P^{\parallel}_{\mathcal{U}_f | \mathcal{D}_p \circ \mathcal{U}_f}}_{0} + \Xi_k \underbrace{P^{\parallel}_{\mathcal{E}_f | \mathcal{D}_p \circ \mathcal{U}_f}}_{0} \qquad (7.103)$$

where the second from last term is null by the property of an oblique projection $(P_{\mathcal{U}_f | \mathcal{D}_p \circ \mathcal{U}_f} = 0)$, while the last term is null because the INV are orthogonal to the inputs and past outputs $(\mathcal{U}_p, \mathcal{U}_f, \mathcal{Y}_p)$ as illustrated in Figure 7.6.

The future states $\mathcal{X}_f$ only depend on a combination of past inputs, outputs, INV and do not depend on $\mathcal{U}_f$, that is, solving the "past" state equation for $\mathcal{X}_p$ using the pseudo-inverse of the observability matrix $(\mathcal{O}^{\#}_k)$, we have

$$\mathcal{X}_p = \mathcal{O}^{\#}_k \mathcal{Y}_p - \mathcal{O}^{\#}_k \mathcal{T}_k \mathcal{U}_p - \mathcal{O}^{\#}_k \Xi_k \mathcal{E}_p \qquad (7.104)$$

Substituting this relation into the future state equation, we obtain

$$\mathcal{X}_f = A^k (\mathcal{O}^{\#}_k \mathcal{Y}_p - \mathcal{O}^{\#}_k \mathcal{T}_k \mathcal{U}_p - \mathcal{O}^{\#}_k \Xi_k \mathcal{E}_p) + C^{\leftarrow}_{AB}(k) \, \mathcal{U}_p + C^{\leftarrow}_{AK}(k) \, \mathcal{E}_p \qquad (7.105)$$

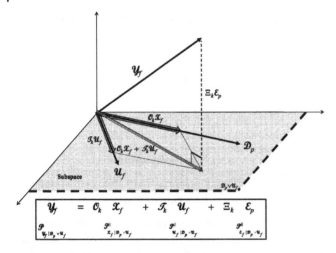

Figure 7.6 Projections: orthogonal projection of $\mathcal{Y}_f$ onto the joint subspaces $\mathcal{D}_p \vee \mathcal{U}_f$ followed by the oblique projection *onto* the past data $\mathcal{D}_p$ *along* the future inputs $\mathcal{U}_f$.

grouping terms gives

$$\mathcal{X}_f = [\underbrace{C_{AB}^{\leftarrow}(k) - A^k \mathcal{O}_k^\# \mathcal{T}_k \mid A^k \mathcal{O}_k^\#}_{\Theta_p} \mid C_{AK}^{\leftarrow}(k) - A^k \mathcal{O}_k^\# \Xi_k] \begin{bmatrix} \mathcal{U}_p \\ -- \\ \mathcal{Y}_p \\ -- \\ \mathcal{E}_p \end{bmatrix} \qquad (7.106)$$

or simply

$$\mathcal{X}_f = \Theta_p \times \mathcal{D}_p + \underbrace{(C_{AK}^{\leftarrow}(k) - A^k \mathcal{O}_k^\# \Xi_k) \mathcal{E}_p}_{0} \qquad (7.107)$$

showing that the future states *depend only* on the past input, output, and INV and not on the future inputs $\mathcal{U}_f$. However, because $\mathcal{X}_f$ lies in the joint (row) subspace of $\mathcal{D}_p \vee \mathcal{U}_f$ and the innovation is orthogonal to the joint space $\langle \mathcal{E}_p, \mathcal{D}_p \vee \mathcal{U}_f \rangle = 0$, then the last term of Eq. (7.107) is null, resulting in the final expression for the future state as follows:

$$\mathcal{X}_f = \mathcal{P}_{\mathcal{X}_f \mid \mathcal{D}_p \circ \mathcal{U}_f}^{\parallel} = \Theta_p \times \mathcal{D}_p \qquad (7.108)$$

as before in Eq. (6.106)[5] resulting in the final expression for the oblique projection of the future output on the past data along the future inputs:

$$\mathcal{P}_{\mathcal{Y}_f \mid \mathcal{D}_p \circ \mathcal{U}_f}^{\parallel} = \mathcal{O}_k \mathcal{P}_{\mathcal{X}_f \mid \mathcal{D}_p \circ \mathcal{U}_f}^{\parallel} = \mathcal{O}_k \times \mathcal{X}_f = \mathcal{O}_k \times (\Theta_p \times \mathcal{D}_p) \qquad (7.109)$$

---

5 Recall that the expression relating the orthogonal-to-oblique projection relation of Eq. (6.109) was applied in that proof.

This relationship shows that obliquely projecting the future data nullifies not only the inputs but also the INV, while enabling the extraction of the observability and estimated state directly from the given data.

Thus, the state vector $\mathcal{X}_f$ provides the basis for the *intersection* of the past and future subspaces as discussed in Section 6.4.2 and can be extracted from this projection. This is a special state vector. It is, in fact, a "state estimate" – the output of a Kalman filter providing the *best* estimate of the state based on the available data! In contrast to the recursive filter developed in Section 4.2 that requires the entire stochastic realization $\Sigma_{INV} = \{A, B, C, D, R^e_{ww}, R^e_{wv}, R^e_{vv}\}$ as well as the initial state and error covariance $(\hat{x}(0), \tilde{P}(0))$ to be applied, this realization, extracted using SID techniques is numerical, only requiring input/output data for its development and application. In fact, it is *not* a steady-state Kalman filter processor as in Section 4.2, but one that processes the data matrices in parallel, "column-by-column," in a batch manner [27]. This is a rather remarkable property of the SID approach!

The *numerical Kalman filter*[6] evolves directly from the oblique projection and can be placed in the following closed-form input/output structure as [27]

$$
\hat{x}(k) = [A^k - \Omega_k\mathcal{O}_k \mid C^{\leftarrow}_{AB}(k) - \Omega_k\mathcal{T}_k \mid \Omega_k]
\begin{bmatrix}
\hat{x}(0) \\
- - - \\
u(0) \\
\vdots \\
u(k-1) \\
- - - \\
y(0) \\
\vdots \\
y(k-1)
\end{bmatrix}
\tag{7.110}
$$

where $\Omega_k = (C^{\leftarrow}_{AK}(k) - A^k\tilde{P}(0)\mathcal{O}'_k)\,(L_k - \mathcal{O}_k\tilde{P}(0)\mathcal{O}'_k)$. Expanding this expression over K-samples (columns) with $\hat{\mathcal{X}}_f := [\hat{x}(k)\ \hat{x}(k+1)\ \cdots\ \hat{x}(k+K-1)]'$, we obtain

$$
\hat{\mathcal{X}}_f = [A^k - \Omega_k\mathcal{O}_k \mid C^{\leftarrow}_{AB}(k) - \Omega_k\mathcal{T}_k \mid \Omega_k]
\begin{bmatrix}
\hat{\mathcal{X}}(0) \\
- - - \\
\mathcal{U}_{0|k-1} \\
- - - \\
\mathcal{Y}_{0|k-1}
\end{bmatrix}
\tag{7.111}
$$

or grouping the data as before, we have

$$
\hat{\mathcal{X}}_f = [A^k - \Omega_k\mathcal{O}_k \mid (C^{\leftarrow}_{AB}(k) - \Omega_k\mathcal{T}_k|\Omega_k)]
\begin{bmatrix}
\hat{\mathcal{X}}(0) \\
- - - \\
\mathcal{D}_{0|k-1}
\end{bmatrix}
\tag{7.112}
$$

for $\hat{\mathcal{X}}(0) := [\hat{x}_0(0)\ \hat{x}_1(0)\ \cdots\ \hat{x}_{K-1}(0)]'$

---

6 It should be noted that the classical structural form of the Kalman filter of Section 4.2 was transformed for the N4SID development here (see [27] for details).

This is the fundamental structure of the numerical Kalman filter with implicit initial conditions embedded, but never explicitly computed by the algorithm since they are not required. The main point is that the estimated state sequence $\hat{X}_f$ is extracted *directly* from the input/output data and does *not* require the underlying innovations model Σ_{INV}.

The development of the numerical Kalman filter is quite complex and lengthy; therefore, it is excluded from this discussion; however, we refer the interested reader to the text [27] or paper [35] where all of the details are carefully developed.

Performing the SVD on Eq. (7.109), we obtain the partitioned signal and noise relationship as

$$
\mathcal{P}^{\parallel}_{\mathcal{Y}_f|\mathcal{D}_p \circ \mathcal{U}_f} = \mathcal{O}_k \hat{X}_f = [U_S \mid U_N]
\begin{bmatrix}
\Sigma_S & \mid & 0 \\
- & - & - \\
0 & \mid & \Sigma_N
\end{bmatrix}
\begin{bmatrix}
V'_S \\
-- \\
V'_N
\end{bmatrix}
$$

$$
= \underbrace{U_S \Sigma_S V'_S}_{\text{Signal Subspace}} + \underbrace{U_N \Sigma_N V'_N}_{\text{Noise Subspace}} \tag{7.113}
$$

Again assuming the noise is "negligible" ($\Sigma_N \ll \Sigma_S$) and selecting a *balanced stochastic realization*[7] ($\Sigma_S = \Sigma_S^{1/2} \times (\Sigma'_S)^{1/2}$), we obtain the critical N4SID subspace relations:

$$
\mathcal{P}^{\parallel}_{\mathcal{Y}_f|\mathcal{D}_p \circ \mathcal{U}_f} \approx \underbrace{U_S \Sigma_S^{1/2}}_{\mathcal{O}_k} \underbrace{(\Sigma'_S)^{1/2} V'_S}_{\hat{X}_f} \in \mathcal{R}^{kN_y \times K} \quad \text{(Projection)}
$$

$$
\mathcal{O}_k = U_S \Sigma_S^{1/2} \in \mathcal{R}^{kN_y \times N_x} \quad \text{(Observability)}
$$

$$
\hat{X}_f = \Sigma_S^{1/2} V'_S \in \mathcal{R}^{N_x \times K} \quad \text{(States)} \tag{7.114}
$$

Thus, the oblique projection operator enables the extraction of the observability matrix as well as the estimated states leading to the extraction of a stochastic realization Σ_{INV} of the process LQ-decomposition and insight discussed in Section 6.4.

7.6.1 Numerical Algorithms for Subspace State-Space System Identification (N4SID) Algorithm

For the stochastic realization problem from noisy input/output data, we have developed the *parallel* or *oblique* projection method based on both past

---

7 The "balanced" stochastic realization is more complex than its deterministic counterpart and requires that the minimum and maximum Riccati equation solutions be diagonalized (see [25] for details).

$(\mathcal{U}_{0|k-1}, \mathcal{Y}_{0|k-1})$, present $(\mathcal{U}_{k|k}, \mathcal{Y}_{k|k})$, and future $(\mathcal{U}_{k+1|2k-1}, \mathcal{Y}_{k+1|2k-1})$ data. It is based on the underlying concept that the projection of the "future outputs" depends solely on the past inputs, outputs, and INV as well as the "future" inputs that evolve directly from the available data. Since the state sequence is available from the SVD of the noisy input/output data matrices, it is used to establish a solution to the least-squares problem enabling the SID to be accomplished [17–27, 35, 36].

As in Section 6.5, the numerically efficient approach to extract the system Σ_{INV} from input/output data is to employ the LQ-decomposition of the block input/output Hankel data matrices. First, after eliminating the noise (INV) of Eq. (7.107) performing the LQ-decomposition of the (reordered) past and future data matrices, then we have that

$$
\mathcal{D}_{0|2k-1} := \underbrace{\begin{bmatrix} \mathcal{U}_{0|k-1} \\ --- \\ \mathcal{U}_{k|k} \\ --- \\ \mathcal{U}_{k+1|2k-1} \\ --- \\ \mathcal{Y}_{0|k-1} \\ --- \\ \mathcal{Y}_{k|k} \\ --- \\ \mathcal{Y}_{k+1|2k-1} \end{bmatrix}}_{D} = \underbrace{\begin{bmatrix} L_{11} & 0 & 0 & 0 & 0 & 0 \\ - & - & - & - & - & - \\ L_{21} & L_{22} & 0 & 0 & 0 & 0 \\ - & - & - & - & - & - \\ L_{31} & L_{32} & L_{33} & 0 & 0 & 0 \\ - & - & - & - & - & - \\ L_{41} & L_{42} & L_{43} & L_{44} & 0 & 0 \\ - & - & - & - & - & - \\ L_{51} & L_{52} & L_{53} & L_{54} & L_{55} & 0 \\ - & - & - & - & - & - \\ L_{61} & L_{62} & L_{63} & L_{64} & L_{65} & L_{66} \end{bmatrix}}_{L} \times \underbrace{\begin{bmatrix} Q'_1 \\ --- \\ Q'_2 \\ --- \\ Q'_3 \\ --- \\ Q'_4 \\ --- \\ Q'_5 \\ --- \\ Q'_6 \end{bmatrix}}_{Q}
$$

$$(7.115)$$

with $L_{11} \in \mathcal{R}^{kN_u \times kN_u}$, $L_{22} \in \mathcal{R}^{N_u \times N_u}$, $L_{33} \in \mathcal{R}^{(k-1)N_u \times (k-1)N_u}$, $L_{44} \in \mathcal{R}^{kN_y \times kN_y}$, $L_{55} \in \mathcal{R}^{N_y \times N_y}$, $L_{66} \in \mathcal{R}^{(k-1)N_y \times (k-1)N_y}$; $L_{ij}, i \neq j$ deduced from L_{ii} (lower tri-angular); and $Q'_1 \in \mathcal{R}^{kN_u \times K}$, $Q'_2 \in \mathcal{R}^{N_u \times K}$, $Q'_3 \in \mathcal{R}^{(k-1)N_u \times K}$, $Q'_4 \in \mathcal{R}^{kN_y \times K}$, $Q'_5 \in \mathcal{R}^{N_y \times K}$, $Q'_6 \in \mathcal{R}^{(k-1)N_y \times K}$.

Recall that for the stochastic realization using the N4SID approach both Kalman state[8] sequences $\hat{\mathcal{X}}_k$ and $\hat{\mathcal{X}}_{k+1}$ are required to extract the INV model from the input/output data using the least-squares approach [35]. Therefore, a second oblique projection based on projecting the future (shifted) output $\mathcal{Y}_{k+1|2k-1}$ onto the (shifted) data $\mathcal{D}_{0|k}$ along the (shifted) future inputs $\mathcal{U}_{k+1|2k-1}$, that is, $\mathcal{P}^{\parallel}_{\mathcal{Y}_{k+1|2k-1}|\mathcal{D}_{0|k}} \circ \mathcal{U}_{k+1|2k-1}$ is required (see Figure 6.8b).

---

8 Recall that $\hat{\mathcal{X}}_k \equiv \mathcal{X}_f$ and $\hat{\mathcal{X}}_0 \equiv \mathcal{X}_p$.

In terms of the LQ-decomposition matrices, the oblique projection operators are given by (see [35] for more details)

$$
\mathcal{P}^{\parallel}_{\mathcal{Y}_{k|2k-1}|D_{0|k-1}} \circ \mathcal{U}_{k|2k-1}
$$

$$
= \underbrace{[\mathcal{L}_{\mathcal{U}_{0|k-1}} \mid \mathcal{L}_{\mathcal{U}_{k|2k-1}} \mid \mathcal{L}_{\mathcal{Y}_{0|k-1}}]}_{\text{Unknown}} \times \begin{bmatrix} \mathcal{U}_{0|k-1} \\ --- \\ \mathcal{U}_{k|2k-1} \\ --- \\ \mathcal{Y}_{0|k-1} \end{bmatrix} = L_{5:6,1:4} \, L^{-1}_{1:4,1:4} \begin{bmatrix} \mathcal{U}_{0|2k-1} \\ --- \\ \mathcal{Y}_{0|k-1} \end{bmatrix}
$$

(7.116)

where the "notation" $L_{i:j,m:n}$ defines a block submatrix of L extracted from the ith-to-jth rows and mth-to-nth columns with $\mathcal{L}_{\mathcal{U}_{0|k-1}}$, $\mathcal{L}_{\mathcal{U}_{k|2k-1}}$ and $\mathcal{L}_{\mathcal{Y}_{0|k-1}}$ the *unknown* projection matrices to be obtained from this set of linear equations through backsubstitution due to the lower triangular structure of the L_{ii}-matrices [27].

The second required projection as in Eq. (6.121) is obtained by solving

$$
\underbrace{[\mathcal{L}_{\mathcal{U}_{0|k}} \mid \mathcal{L}_{\mathcal{U}_{k+1|2k-1}} \mid \mathcal{L}_{\mathcal{Y}_{0|k}}]}_{\text{Unknown}} \times \begin{bmatrix} \mathcal{U}_{0|k} \\ --- \\ \mathcal{U}_{k+1|2k-1} \\ --- \\ \mathcal{Y}_{0|k} \end{bmatrix} = L_{6:6,1:5} \, L^{-1}_{1:5,1:5} \begin{bmatrix} \mathcal{U}_{0|2k-1} \\ --- \\ \mathcal{Y}_{0|k} \end{bmatrix}
$$

(7.117)

where $\mathcal{L}_{\mathcal{U}_{0|k}}$, $\mathcal{L}_{\mathcal{U}_{k+1|2k-1}}$, and $\mathcal{L}_{\mathcal{Y}_{0|k}}$ are the *unknown* projection matrices to be obtained from this set of linear equations also through backsubstitution.

With the solution matrices $\{\mathcal{L}_{\mathcal{U}_{0|k-1}}, \mathcal{L}_{\mathcal{U}_{k|2k-1}}, \mathcal{L}_{\mathcal{Y}_{0|k-1}}\}$ and $\{\mathcal{L}_{\mathcal{U}_{0|k}}, \mathcal{L}_{\mathcal{U}_{k+1|2k-1}}, \mathcal{L}_{\mathcal{Y}_{0|k}}\}$ now *available*, the observability matrix can be extracted using the SVD in terms of the LQ-decomposition as

$$
\mathcal{P}^{\parallel}_{\mathcal{Y}_{k|2k-1}|D_{0|k-1}} \circ \mathcal{U}_{k|2k-1} = [\mathcal{L}_{\mathcal{U}_{0|k-1}} \mid 0 \mid \mathcal{L}_{\mathcal{Y}_{0|k-1}}] \, L_{1:4,1:4} \times Q'_{1:4} =
$$

$$
[U_S \mid U_N] \begin{bmatrix} \Sigma_S & \mid & 0 \\ - & - & - \\ 0 & \mid & \Sigma_N \end{bmatrix} Q'_{1:4} \times \begin{bmatrix} V'_S \\ -- \\ V'_N \end{bmatrix} \approx \underbrace{U_S \Sigma_S V'_S}_{\text{Signal Subspace}} = \underbrace{U_S \Sigma_S^{1/2}}_{\mathcal{O}_k} \underbrace{(\Sigma'_S)^{1/2} V'_S}_{\hat{x}_f}
$$

(7.118)

assuming the noise is "negligible" ($\Sigma_N \ll \Sigma_S$) and selecting a *balanced stochastic realization* ($\Sigma_S = \Sigma_S^{1/2} \times (\Sigma'_S)^{1/2}$), we obtain the critical N4SID subspace relations as before:

$$
\mathcal{P}^{\parallel}_{\mathcal{Y}_{k|2k-1}|D_{0|k-1}} \circ \mathcal{U}_{k|2k-1}
$$

$$
= \mathcal{O}_k \, \hat{x}_k = \underbrace{[\mathcal{L}_{\mathcal{U}_{0|k-1}} \mid 0 \mid \mathcal{L}_{\mathcal{Y}_{0|k-1}}]}_{\text{Known}} \, L_{1:4,1:4} \times Q'_{1:4} \quad \text{(Oblique Projection)}
$$

$$\mathcal{P}^{\|}_{\mathcal{Y}_{k+1|2k-1}|\mathcal{D}_{0|k}}\,^{\circ}\mathcal{U}_{k+1|2k-1}$$

$$= \mathcal{O}_{k-1}\,\hat{\mathcal{X}}_{k+1} = \underbrace{[\mathcal{L}_{\mathcal{U}_{0|k}} \mid 0 \mid \mathcal{L}_{\mathcal{Y}_{0|k}}]}_{\text{Known}} \times L_{1:5,1:5} \times Q'_{1:5} \quad \text{(Shifted Projection)}$$

$$(7.119)$$

where $\mathcal{O}_{k-1} \in \mathcal{O}_k$ is the extended observability matrix with the *last block row eliminated*.

From these projections, we can now estimate *both* estimated state sequences, $\hat{\mathcal{X}}_{k+1}$ and $\hat{\mathcal{X}}_k$ in order to solve the least-squares problem, and extract Σ_{INV}. It is clear from the oblique projection that this can easily be accomplished from the SVD, since

$$\hat{\mathcal{X}}_k = \mathcal{O}_k^\# \times \mathcal{P}^{\|}_{\mathcal{Y}_{k|2k-1}|\mathcal{D}_p}\,^{\circ}\mathcal{U}_f = (\Sigma_{S_k}^{-1/2}U_{S'_k}) \times \mathcal{P}^{\|}_{\mathcal{Y}_{k|2k-1}|\mathcal{D}_p}\,^{\circ}\mathcal{U}_f$$

$$\hat{\mathcal{X}}_{k+1} = \mathcal{O}_{k-1}^\# \times \mathcal{P}^{\|}_{\mathcal{Y}_{k+1|2k-1}|\mathcal{D}_p}\,^{\circ}\mathcal{U}_f = (\Sigma_{S_{k-1}}^{-1/2}U'_{S_{k-1}}) \times \mathcal{P}^{\|}_{\mathcal{Y}_{k+1|2k-1}|\mathcal{D}_p}\,^{\circ}\mathcal{U}_f \quad (7.120)$$

where $\Sigma_{S_{k-1}}^{-1/2} \in \Sigma_{S_k}^{-1/2}$ and $U_{S_{k-1}} \in U_{S_k}$ are the singular values and orthogonal matrices (respectively) with the last block row eliminated (as above for the observability matrix).

Therefore, in terms of the LQ-matrices, we have that the estimated states are given by

$$\hat{\mathcal{X}}_k = \Sigma_{S_k}^{-1/2}U'_{S_k} \times [\mathcal{L}_{\mathcal{U}_{0|k-1}} \mid 0 \mid \mathcal{L}_{\mathcal{Y}_{0|k-1}}] \times L_{1:4,1:4} \times Q'_{1:4}$$

$$\hat{\mathcal{X}}_{k+1} = \Sigma_{S_{k-1}}^{-1/2}U'_{S_{k-1}} \times [\mathcal{L}_{\mathcal{U}_{0|k}} \mid 0 \mid \mathcal{L}_{\mathcal{Y}_{0|k}}] \times L_{1:5,1:5} \times Q'_{1:5} \quad (7.121)$$

and

$$\hat{\mathcal{U}}_{k|k} = L_{2:2,1:4} \times Q'_{1:4}$$

$$\hat{\mathcal{Y}}_{k|k} = L_{5:5,1:5} \times Q'_{1:5} \quad (7.122)$$

With these states of the numerical Kalman filter now available from the SVD and pseudo-inversions, we can define the underlying "batch" state-space (INV) model as

$$\hat{\mathcal{X}}_{k+1} = A\,\hat{\mathcal{X}}_k + B\,\mathcal{U}_{k|k} + \xi_{\omega_k}$$

$$\hat{\mathcal{Y}}_{k|k} = C\,\hat{\mathcal{X}}_k + D\,\mathcal{U}_{k|k} + \xi_{v_k} \quad (7.123)$$

where

$$\hat{\mathcal{X}}_{k+1} := [\hat{x}(k+1) \;\cdots\; \hat{x}(k+K-1)]$$

$$\hat{\mathcal{X}}_k := [\hat{x}(k) \;\cdots\; \hat{x}(k+K-2)]$$

$$\hat{\mathcal{Y}}_{k|k} := [\hat{y}(k) \;\cdots\; \hat{y}(k+K-2)]$$

$$\mathcal{U}_{k|k} := [u(k) \;\cdots\; u(k+K-2)]$$

$$\xi_{\omega_k} := [\varepsilon_\omega(k) \;\cdots\; \varepsilon_\omega(k+K-2)]$$

$$\xi_{v_k} := [\varepsilon_v(k) \;\cdots\; \varepsilon_v(k+K-2)]$$

and the residuals or equivalently INV sequence and its covariance are defined by

$$\xi := \begin{bmatrix} \xi_{\omega_k} \\ --- \\ \xi_{v_k} \end{bmatrix} \quad \text{and} \quad R_{\xi\xi} := E\{\xi\xi'\} = E\left\{ \begin{bmatrix} \xi_{\omega_k} \\ --- \\ \xi_{v_k} \end{bmatrix} [\xi_{\omega_k} \mid \xi_{v_k}]' \right\} \quad (7.124)$$

More compactly, we now have

$$\underbrace{\begin{bmatrix} \hat{\mathcal{X}}_{k+1} \\ --- \\ \hat{\mathcal{Y}}_{k|k} \end{bmatrix}}_{\text{known}} = \begin{bmatrix} A & \mid & B \\ --- & & --- \\ C & \mid & D \end{bmatrix} \underbrace{\begin{bmatrix} \hat{\mathcal{X}}_k \\ --- \\ \mathcal{U}_{k|k} \end{bmatrix}}_{\text{known}} + \begin{bmatrix} \xi_{\omega_k} \\ --- \\ \xi_{v_k} \end{bmatrix} \quad (7.125)$$

which can be solved as a least-squares problem as

$$\begin{bmatrix} \hat{A} & \mid & \hat{B} \\ --- & & --- \\ \hat{C} & \mid & \hat{D} \end{bmatrix} = \min_{ABCD} \left\| \begin{bmatrix} \hat{\mathcal{X}}_{k+1} \\ --- \\ \hat{\mathcal{Y}}_{k|k} \end{bmatrix} - \begin{bmatrix} A & \mid & B \\ --- & & --- \\ C & \mid & D \end{bmatrix} \begin{bmatrix} \hat{\mathcal{X}}_k \\ --- \\ \mathcal{U}_{k|k} \end{bmatrix} \right\|^2 \quad (7.126)$$

where $\| \ \|$ is the Frobenius matrix norm.

Therefore, the *least-squares solution* follows as

$$\begin{bmatrix} \hat{A} & \mid & \hat{B} \\ --- & & --- \\ \hat{C} & \mid & \hat{D} \end{bmatrix} = \left(\begin{bmatrix} \hat{\mathcal{X}}_{k+1} \\ --- \\ \hat{\mathcal{Y}}_{k|k} \end{bmatrix} \begin{bmatrix} \hat{\mathcal{X}}_k \\ --- \\ \mathcal{U}_{k|k} \end{bmatrix}' \right) \left(\begin{bmatrix} \hat{\mathcal{X}}_k \\ --- \\ \mathcal{U}_{k|k} \end{bmatrix} \begin{bmatrix} \hat{\mathcal{X}}_k \\ --- \\ \mathcal{U}_{k|k} \end{bmatrix}' \right)^{-1} \quad (7.127)$$

with the corresponding "least-squares" residual (INV) covariances estimated by

$$\hat{R}_{\xi\xi} := E\{\hat{\xi}\hat{\xi}'\} = \begin{bmatrix} R^e_{\omega\omega} & \mid & R^e_{\omega v} \\ --- & & --- \\ R^e_{\omega v}{}' & \mid & R^e_{vv} \end{bmatrix} = \begin{bmatrix} KR_{ee}K' & \mid & KR^{1/2}_{ee} \\ --- & & --- \\ (KR^{1/2}_{ee})' & \mid & R_{ee} \end{bmatrix} \quad (7.128)$$

and therefore

$$\hat{R}_{ee} \approx R^e_{vv} \qquad \qquad \text{(Innovations Covariance)}$$

$$\hat{K} \approx R^e_{\omega v}\hat{R}^{-1/2}_{ee} \qquad \text{(Kalman Gain)} \qquad (7.129)$$

It is interesting to note that the N4SID method does *not* require the Riccati equation solution as in both classical and MOESP approaches. That is, the *full solution* of the stochastic realization problem given by $\Sigma_{\text{INV}} = \{\hat{A}, \hat{B}, \hat{C}, \hat{D}, R^e_{\omega\omega}, R^e_{\omega v}, R^e_{vv}\}$ is obtained through the solution of the least-squares relations [35].

This *oblique* projection algorithm is termed N4SID and can be summarized by the following steps [25, 35]:

- *Compute* the LQ-decomposition of the augmented data matrix $\mathcal{D}$ of Eq. (7.115).
- *Solve* for the unknown $\mathcal{L}$-matrices of Eqs. (7.116) and (7.117) to compute the projections.
- *Perform* the SVD of the projections: $\mathcal{P}^{\parallel}_{\mathcal{Y}_f|\mathcal{D}_p} \circ \mathcal{U}_f$ in Eq. (7.113) to extract $\mathcal{O}_k$ and $\mathcal{O}_{k-1}$.
- *Compute* $\hat{\mathcal{X}}_{k+1}$ and $\hat{\mathcal{X}}_{k+1}$ of Eq. (7.120).
- *Construct* the "batch" state-space system of Eq. (7.125).
- *Extract* A, B, C, D by solving the least-squares problem as in Eq. (7.126).
- *Estimate* $R^e_{\omega\omega}, R^e_{\omega\upsilon}, R^e_{\upsilon\upsilon}$ from the residual (INV) sequence as in Eq. (7.128).
- *Compute* the residual (INV) covariance and Kalman gain from Eq. (7.129).

completing the oblique projection solution to the stochastic realization problem.

We summarize the N4SID algorithm in Table 7.7. Next we consider the relationship between these algorithms.

7.6.2 Relationship: Oblique (N4SID) and Orthogonal (MOESP) Algorithms

The relationship of the oblique-projection based N4SID method to that of the and orthogonal projection-based PO-MOESP method is well established [36]. Essentially, both techniques use identical instrumental variables $\mathcal{I}_p$ as shown in Eq. (7.77); therefore, there exists a well-defined relationship[9] between them. This relationship can be seen by examining projections of each, that is, the relations between orthogonal and oblique projection operators.

The *oblique projection operator* ($\mathcal{P}_{\text{oblique}}$) can be expressed in terms of the common instrumental variable $\mathcal{I}_p$ of past inputs and outputs as [17, 37, 38]

$$\mathcal{P}_{\text{oblique}} = \mathcal{P}^{\perp}_{\mathcal{U}_f} \mathcal{I}'_p (\mathcal{I}_p \mathcal{P}^{\perp}_{\mathcal{U}_f} \mathcal{I}'_p)^{-1} \mathcal{I}_p \mathcal{P}^{\perp}_{\mathcal{U}_f} \tag{7.130}$$

which is applied directly to the future data $\mathcal{Y}_f$, while the *orthogonal projection operator* ($\mathcal{P}_{\text{orthogonal}}$) is given by

$$\mathcal{P}_{\text{orthogonal}} = \underbrace{\mathcal{P}^{\perp}_{\mathcal{U}_f} \mathcal{I}'_p (\mathcal{I}_p \mathcal{P}^{\perp}_{\mathcal{U}_f} \mathcal{I}'_p)^{-1} \mathcal{I}_p \mathcal{P}^{\perp}_{\mathcal{U}_f}}_{\mathcal{P}_{\text{oblique}}} \times \mathcal{P}^{\perp}_{\mathcal{U}_f} \tag{7.131}$$

Therefore, we see that

$$\mathcal{P}_{\text{orthogonal}} = \mathcal{P}_{\text{oblique}} \times \mathcal{P}^{\perp}_{\mathcal{U}_f} \tag{7.132}$$

as shown in [17, 36].

---

9 The *canonical variate analysis* (CVA) method is also included, but it is beyond the scope of our discussion [36]. Details are available [7, 25].

Table 7.7 Numerical algorithm for (stochastic) subspace state-space system identification (N4SID).

Oblique projection (LQ-decomposition)

$$
D_{0|2k-1} =
\begin{bmatrix}
U_{0|k-1} \\
U_{k|k} \\
U_{k+1|2k-1} \\
\mathcal{Y}_{0|k-1} \\
\mathcal{Y}_{k|k} \\
\mathcal{Y}_{k+1|2k-1}
\end{bmatrix}
=
\begin{bmatrix}
L_{11} & 0 & 0 & 0 & 0 & 0 \\
L_{21} & L_{22} & 0 & 0 & 0 & 0 \\
L_{31} & L_{32} & L_{33} & 0 & 0 & 0 \\
L_{41} & L_{42} & L_{43} & L_{44} & 0 & 0 \\
L_{51} & L_{52} & L_{53} & L_{54} & L_{55} & 0 \\
L_{61} & L_{62} & L_{63} & L_{64} & L_{65} & L_{66}
\end{bmatrix}
\times
\begin{bmatrix}
Q_1' \\
Q_2' \\
Q_3' \\
Q_4' \\
Q_5' \\
Q_6'
\end{bmatrix}
\quad \text{(Data Matrix)}
$$

$\mathcal{L}$-Matrix solutions

$$
\left\{ \mathcal{L}_{U_{0|k-1}}, \mathcal{L}_{U_{k|2k-1}}, \mathcal{L}_{\mathcal{Y}_{0|k-1}} \right\} ; \quad \left\{ \mathcal{L}_{U_{0|k}}, \mathcal{L}_{U_{k+1|2k-1}}, \mathcal{L}_{\mathcal{Y}_{0|k}} \right\}
\qquad (\mathcal{L}\text{-Matrices})
$$

Singular-value decomposition

$$
P^{\|}_{\mathcal{Y}_{k|2k-1} | D_{0|k-1} \, \circ U_{k|2k-1}} = [U_S \mid U_N]
\begin{bmatrix}
\Sigma_S & \mid & 0 \\
- & - & - \\
0 & \mid & \Sigma_N
\end{bmatrix}
Q'_{1:4} \times
\begin{bmatrix}
V_S' \\
-- \\
V_N'
\end{bmatrix}
\approx U_S \Sigma_S V_S' \quad \text{(Projection Matrix)}
$$

Observability/state extractions

$$
P^{\|}_{\mathcal{Y}_{k|2k-1} | D_{0|k-1} \, \circ U_{k|2k-1}} = \mathcal{O}_k \hat{\mathcal{X}}_k = [\mathcal{L}_{U_{0|k-1}} \mid 0 \mid \mathcal{L}_{\mathcal{Y}_{0|k-1}}] L_{1:4,1:4} Q'_{1:4}
$$

$$
P^{\|}_{\mathcal{Y}_{k+1|2k-1} | D_{0|k} \, \circ U_{k+1|2k-1}} = \mathcal{O}_{k-1} \hat{\mathcal{X}}_{k+1} = [\mathcal{L}_{U_{0|k}} \mid 0 \mid \mathcal{L}_{\mathcal{Y}_{0|k}}] L_{1:5,1:5} Q'_{1:5}
$$

(Oblique Projections)

$$
\mathcal{O}_k = U_S \Sigma_{S_k}^{1/2}; \quad \mathcal{O}_{k-1} = U_{S_{k-1}} \Sigma_{S_{k-1}}^{1/2};
$$

(Observability Matrices)

$$
\hat{\mathcal{X}}_k = \mathcal{O}_k^{\#} \times P^{\|}_{\mathcal{Y}_{k|2k-1} | D_p \, \circ U_f} = (\Sigma_{S_k}^{-1/2} U_{S_k}') \times P^{\|}_{\mathcal{Y}_{k|2k-1} | D_p \, \circ U_f}
$$

$$
\hat{\mathcal{X}}_{k+1} = \mathcal{O}_{k-1}^{\#} \times P^{\|}_{\mathcal{Y}_{k+1|2k-1} | D_p \, \circ U_f} = (\Sigma_{S_{k-1}}^{-1/2} U_{S_{k-1}}') \times P^{\|}_{\mathcal{Y}_{k+1|2k-1} | D_p \, \circ U_f}
$$

(State Estimates)

System matrices

$$
\begin{bmatrix}
\hat{A} & | & \hat{B} \\
- & - & - \\
\hat{C} & | & \hat{D}
\end{bmatrix}
=
\left(
\begin{bmatrix}
\hat{\mathcal{X}}_{k+1} \\
--- \\
\hat{\mathcal{Y}}_{k|k}
\end{bmatrix}
\begin{bmatrix}
\hat{\mathcal{X}}_k \\
--- \\
U_{k|k}
\end{bmatrix}'
\right)
\left(
\begin{bmatrix}
\hat{\mathcal{X}}_k \\
--- \\
U_{k|k}
\end{bmatrix}
\begin{bmatrix}
\hat{\mathcal{X}}_k \\
--- \\
U_{k|k}
\end{bmatrix}'
\right)^{-1}
$$

Covariance equations

$$
\hat{R}_{\xi\xi} = E\{\hat{\xi} \hat{\xi}'\} =
\begin{bmatrix}
R_{\omega\omega}^e & | & R_{\omega\upsilon}^e \\
- & - & - \\
R_{\omega\upsilon}^e{}' & | & R_{\upsilon\upsilon}^e
\end{bmatrix}
=
\begin{bmatrix}
K R_{ee} K' & | & K R_{ee}^{1/2} \\
- & - & - \\
(K R_{ee}^{1/2})' & | & R_{ee}
\end{bmatrix}
$$

Kalman filter equations

$$
\hat{R}_{ee} \approx R_{\upsilon\upsilon}^e; \qquad \hat{K} \approx R_{\omega\upsilon}^e \hat{R}_{ee}^{-1/2}
$$

(Innovations Covariance/ Kalman Gain)

In the literature, this relationship has also been developed in terms of the full-rank (left) weighting matrix $\mathcal{W}_L$ and (right) column matrix $\mathcal{W}_R$ such that the rank of the projection matrix $\mathcal{P}$ is preserved, that is,

$$
\rho(\mathcal{I}_p \times \mathcal{W}_R) = \rho(\mathcal{I}_p) = N_x
$$

It has been shown that these subspace methods all start with a "weighted projection" $\mathcal{W}_L \mathcal{P} \mathcal{W}_R$ and employ the singular value decomposition (SVD) to either extract the states (right singular vectors) or to extract the observability matrix (left singular vectors) in order to identify Σ_{INV} [36]. The relationship of these weighting matrices for the N4SID and PO-MOESP methods again demonstrates that the only difference between the orthogonal and oblique projection approaches lies in the *extra* orthogonal projection operation $\mathcal{P}_{\mathcal{U}_f}^{\perp}$, that is,

$$\mathcal{W}_L[N4SID] \rightarrow \mathcal{W}_L[PO - MOESP]; \quad \mathcal{W}_R \rightarrow \mathcal{W}_R \times \mathcal{P}_{\mathcal{U}_f}^{\perp} \qquad (7.133)$$

7.7 Model Order Estimation and Validation

SID is based on the underlying premise that a MIMO "black-box" state-space model is *fit* to input/output data – deterministic or random. Pragmatically, the majority of data are noisy or uncertain especially if acquired from a physical system; however, deterministic data can evolve through simulation notably of highly complex processes. In this section, we discuss state-space MOE and validation techniques for random data in contrast to the deterministic case developed in Section 6.6. In that section, the emphasis was on the SVD approach for order estimation, since the rank of the Hankel matrices evolving from systems theory enables the determination of system order directly. This approach is certainly viable in the stochastic case especially when impulse response data is replaced by covariance data. Therefore, we initially provide a brief review of these techniques for continuity. In fact, we will show how they evolved in the array signal processing array as a means of estimating the number of sources impinging on a multichannel sensor array [39] based on a statistical rather than deterministic approach to order estimation. These methods are statistical in nature following the Akaike information criterion (AIC) and minimum description length (MDL) or similarly the Bayesian information criterion (BIC) approaches or deterministic based on the Hankel singular values or the best rank approximation as discussed previously. These approaches are convenient for both classical and SID, since the SVD has become an integral part of these algorithms and the singular values are available.

As before, the SID must first provide the order to extract the innovations model Σ_{INV} and then validate its performance, that is, for the stochastic realization problem, we must do the following steps:

- *Estimate* the system order or equivalently the number of states such that $K \gg N_x$.
- *Validate* the performance of the estimated models $\hat{\Sigma}_{INV}(n); \ n = 1, \ldots, N$ over the N-ensemble of data.

We start with the order estimation problem and then proceed to validate the performance of the identified models.

7.7.1 Order Estimation: Stochastic Realization Problem

Order estimation for MIMO systems has evolved specifically from the identification problem [40] where the first step is to estimate or guess the system order based on physical principles and then attempt to validate prediction error or INV through residual testing of the estimated model. Much effort has been devoted to this problem [40–43] with the most significant statistical results evolving from the information theoretical point of view [39, 44, 45]. We consider some equivalent order estimation methods based on the SVD method applied to data, covariance, or projection matrices. The methods can be statistical in nature following the AIC and MDL (BIC) approaches or deterministic based on the Hankel singular values or the best rank approximation [39, 44, 46] as before. These approaches are convenient for both classical and SID solutions to the stochastic realization problem.

We know from "theoretical" systems (realization) theory that the key element is determining the order or number of states of the underlying system under investigation. For the stochastic problem, order estimation is again based on the *rank condition* that evolves from the Hankel covariance data matrices (classical) or the projection matrices (subspace) with their observability/controllability-like factors. However, the root of achieving a robust and reliable estimate still lies in the SVD of the covariance data/projection matrices. Classically, we have

$$\Lambda_{K,K} = \begin{bmatrix} \Lambda_1 & \cdots & \Lambda_K \\ \vdots & \vdots & \vdots \\ \Lambda_K & \cdots & \Lambda_{2K-1} \end{bmatrix} = \underbrace{[U_S \mid U_N]}_{U} \begin{bmatrix} \Sigma_S & \mid & 0 \\ - & - & - \\ 0 & \mid & \Sigma_N \end{bmatrix} \underbrace{\begin{bmatrix} V_S' \\ -- \\ V_N' \end{bmatrix}}_{V}$$

$$= \underbrace{U_S \Sigma_S V_S'}_{\text{Signal}} + \underbrace{U_N \Sigma_N V_N'}_{\text{Noise}} \qquad (7.134)$$

where the covariance sequence is characterized by the Markov parameters $\Lambda_\ell = CA^{\ell-1}(A\Pi C' + R_{wv})$ with $\Pi \geq 0$ of the unknown Gauss–Markov or INV models.

Ideally, $\Sigma_N \ll \Sigma_S$ and the number of nonzero singular values residing in Σ_S reveal the order or number of states of the underlying system; however, in contrast to the deterministic case, this is no longer true for noisy data. Although, in theory, it is a simple matter to perform the SVD of $\Lambda_{K,K}$ and count the number of nonzero singular values to obtain N_x, since $\rho(\Lambda_{K,K}) = \rho(\Sigma) = \rho(\Sigma_S) = N_x$. Therefore, the key *first step* in SID lies in the extraction of the singular values either employing the classical Hankel covariance matrix of the SVD-based stochastic realization algorithm of Section 7.4 or the SVD-based projection matrices of the subspace algorithms of Section 7.5 or 7.6.

Let us consider classical properties of the Hankel covariance matrix that enable a solution to theorder estimation problem. This set of covariance matrix singular values allows us to estimate its rank and therefore the minimal number of states of the underlying system. However, when noise or computational error occurs, then it is necessary to apply judgment when evaluating these singular values. One approach is simply to search for a meaningful change in these values, referred to as a "gap" or "jump," selecting the order corresponding to this abrupt change as illustrated in Figures 6.1 and 6.2 for a 4th-order or 6th-order system, respectively.

As before, another approach is to employ *best rank approximation* given by the ratio of the summation of singular values up to the selected order over all the values, that is, $\beta(\%)$ based on the ratio of singular values [47]:

$$\beta(\hat{N}_x) := \left(\frac{\sum_{k=1}^{\hat{N}_x} \sigma_k}{\sum_{k=1}^{K} \sigma_k} \right) \times 100 \tag{7.135}$$

The *model order estimator* (MOE) is given by [48]

$$\hat{N}_x = \max \left\{ N : \log \hat{\sigma}_N > (\log \hat{\sigma}_{\max} + \log \hat{\sigma}_{\min}) \right\} \tag{7.136}$$

and is specified as the largest integer corresponding to the singular value (σ_{N_x}) *greater than* the geometric mean of the largest and smallest singular values extracted. Here the idea is to formally search for the *gap* or *jump* in singular values.

Again with the order is selected, then the error in this choice can be specified by the $\hat{N}_{x+1}$ singular value $\sigma_{N_{x+1}}$ as a particular metric or the so-called truncation or *SVD model error* metric defined by the summation over all of the remaining singular values as [49, 50]

$$\mathcal{M}_\epsilon := 2 \sum_{k=N_x+1}^{K} \sigma_k \tag{7.137}$$

Also recall that the *Hankel norm* is a measure of the energy stored in a system and is specified by the "largest" singular value of the Hankel matrix:

$$\| \mathcal{H}_{\mathcal{K},\mathcal{K}} \| = \max\{\sigma_k\} = \sigma_1; \quad k = 1, \dots, K$$

while the $\| \mathcal{H}_{K,K} \|_2$ norm is the RMS sum of the impulse responses or covariance sequences [50]. In any case, it is clear that the Hankel data/covariance or projection matrices are critical entities in identifying a system from measured data.

Most of the statistical validation techniques to follow are primarily founded on an "error-based" metric where the error was defined in the previous chapter by the *model residual error* of Eq. (6.135). In the stochastic case, the *prediction errors* or *INV* are defined as the difference between the measured output and its prediction:

$$e(t) = y(t) - \hat{y}(t) \qquad \text{(Prediction Error)} \qquad (7.138)$$

providing the critical entity enabling the subsequent development of statistical metrics. The corresponding *mean-squared (prediction) error* is given by

$$\text{MSE} := \frac{1}{K} \sum_{t=1}^{K} e'(t)e(t) := \frac{1}{K} e'e \qquad \text{(Mean - Squared Prediction Error)}$$

$$(7.139)$$

We summarize the metrics available from the SVD of the Hankel/projection matrices as follows:

1) *Hankel order estimation* (*Best Rank*, AIC, MDL) of Eqs. (7.135), (7.155), and (7.156) $\Rightarrow \hat{N}_x$
2) SVD *model order estimation* (MOE) of Eq. (7.136)
3) *Hankel singular values*: $\sigma_{ABC} = \sqrt{\Sigma_{AC} \times \Sigma_{AB}}$
4) *Hankel norm* (energy): $\| \mathcal{H}_{K,K'} \| = \max\{\sigma_i\}; \ i = 1, \dots, N_x$
5) SVD *model error*: $\mathcal{M}_\epsilon = 2 \sum_{k=\hat{N}_x+1}^{K} \sigma_k.$
6) *Mean-squared (prediction) error*: $\text{MSE} := \frac{1}{K} e'e.$

These resulting singular values hold the same credence in the stochastic case and are evaluated and analyzed identically. Next we consider some statistical methods that have been applied to the order estimation problem.

7.7.1.1 Order Estimation: Statistical Methods

Here we consider some statistical methods of estimating the order using the singular values from a Hankel or equivalently[10] Toeplitz covariance matrix ($\Lambda_{K,K} = E\{Y_K Y_K'\}$). These metrics have evolved from the information theoretic approach using the Kullback–Leibler (KL) approach (see [32] for more details).

The *information* (self) contained in the occurrence of the event ω_i such that $Z(\omega_i) = z_i$ is

$$\mathcal{I}(z_i) = -\log_b \ \text{Pr}(Z(\omega_i) = z_i) = -\log_b \text{Pr}_Z(z_i) = -\log_b \ \text{Pr}(z_i) \qquad (7.140)$$

where b is the base of the logarithm, which results in different units for information measures (e.g. base 2 → bits, while base e → implies nats). The entropy or *average* information is defined by

$$\mathcal{H}(z_i) := E_Z\{\mathcal{I}(z_i)\} = \sum_{i=1}^{K} \mathcal{I}(z_i) \text{Pr}(z_i) = -\sum_{i=1}^{K} \text{Pr}(z_i)\log_b \ \text{Pr}(z_i) \qquad (7.141)$$

after substitution for $\mathcal{I}$.

---

10 Recall that these matrices are easily transformed to each other using the block antidiagonal identity matrix [25, 51].

Mutual information is defined in terms of the information available in the occurrence of the event $Y(\omega_j) = y_j$ about the event $Z(\omega_i) = z_i$ or

$$\mathcal{I}(z_i; y_j) = \log_b \frac{\Pr(z_i|y_j)}{\Pr(z_i)} = \log_b \Pr(z_i|y_j) - \log_b \Pr(z_i) \qquad (7.142)$$

Using these concepts, we take the information theoretic approach to MOE following [44, 45]. Since many models are expressed in terms of their probability distributions, quality or "goodness" can be evaluated by their *similarity* to the true underlying probability distribution generating the measured data. This is the fundamental concept embedded in this approach.

Suppose $\Pr(z_i)$ is the *true* discrete probability distribution and $\Pr(m_i)$ is that of the model, then the *Kullback–Leibler information* (KLI) quantity of the true distribution relative to the model is defined by using

$$\mathcal{I}_{KL}(\Pr(z_i); \Pr(m_i)) := E_Z \left\{ \ln \frac{\Pr(z_i)}{\Pr(m_i)} \right\} = \sum_{i=1}^{K} \Pr(z_i) \ln \frac{\Pr(z_i)}{\Pr(m_i)}$$

$$= \sum_{i=1}^{K} \Pr(z_i) \ln \Pr(z_i) - \sum_{i=1}^{K} \Pr(z_i) \ln \Pr(m_i) \qquad (7.143)$$

here we chose $\log_b \equiv \log_e \equiv \ln$. The KLI possesses some very interesting properties that we state without proof (see [45] for details) such as

1) $\mathcal{I}_{KL}(\Pr(z_i); \Pr(m_i)) \geq 0$
2) $\mathcal{I}_{KL}(\Pr(z_i); \Pr(m_i)) = 0 \Leftrightarrow \Pr(z_i) = \Pr(m_i)$ $\forall i$
3) The negative of the KLI is the *entropy*, $\mathcal{H}_{KL}(\Pr(z_i); \Pr(m_i))$

The second property implies that as the *model* approaches the *true* distribution, then the value of the KLI approaches *zero* (minimum). Thus, investigating Eq. (7.143), we see that the first term is a constant specified by the true distribution; therefore, we only need to estimate the *average* value of the model relative to the true distribution, that is,

$$\mathcal{L}(m_i) := E_Z\{\ln \Pr(m_i)\} = \sum_{i=1}^{K} \Pr(z_i) \ln \Pr(m_i) \qquad (7.144)$$

where $\mathcal{L}(m_i)$ is defined as the *average log-likelihood* of the random variable of value $\ln \Pr(m_i)$. Clearly, the *larger* the average log-likelihood, the *smaller* the KLI implying a *better* model.

The third property, *entropy*, is approximately equal to $\frac{1}{K}$ times the probability that the relative frequency distribution of K measurements obtained from the assumed model equals the true distribution.

From the *law of large numbers*, it follows that [45]

$$\mathcal{L}(m) := E_Y\{\ln \Pr(m)\} \to \frac{1}{K} \sum_{k=1}^{K} \ln \Pr(m(y_k)) \quad \text{for} \quad K \to \infty \qquad (7.145)$$

That is, if the event ω_i occurs at the kth-measurement, y_k, then $\Pr(m(y_k)) = \Pr(m_i)$ converges to the average log-likelihood. Further, if k_i is the number of times $\ln \Pr(m)$ takes on the value $\ln \Pr(m_i)$, then Eq. (7.145) becomes the log-likelihood of the model, that is,

$$\mathcal{L}(m_i) = \frac{1}{K} \sum_{i=1}^{K} k_i \ln \Pr(m_i) \tag{7.146}$$

and Eqs. (7.145) and (7.146) are identical in the limit. Thus, the conclusion from this discussion is that the *larger* $\mathcal{L}(m_i)$, the *better* the model. With this background information defined, let us develop the AIC.

The basic concept of the AIC is that the average log-likelihood of the model, $\mathcal{L}(\cdot)$, where the average is with respect to the measurement data, is used for the metric of goodness-of-fit as developed earlier in the general case. A more natural estimate seems to be the maximum log-likelihood; however, it is a *biased* estimator for this problem, which tends to overestimate the statistic [45]. This tendency has been observed especially when the number of "free parameters" (or states) in the model, $\Theta = \{\theta_1, \ldots, \theta_{N_\Theta}\}$, is large. Akaike [44] observed that the bias is closely related to the number of free parameters and developed an asymptotically unbiased estimator of $\mathcal{L}(\cdot)$ defined verbally by

$$\text{AIC} = -2 \times (\text{maximum log-likelihood of model})$$
$$+ 2 \times (\text{number of free parameters})$$

where historical reasons account for the 2 [44, 45]. A model that minimizes the AIC is defined as the minimum AIC estimate or MAICE. This definition implies that if there are several models that have approximately the same AIC, then the one with the "fewest" free parameters (or states) should be selected. This choice is called *parsimony* in the identification literature [40].

From the model-based perspective, we restrict the distributions of truth and model to the same inherent structure (e.g. AR, MA, ARMA, state-space). We assume that the *true* model parameters (or states) are represented by Θ_{true}, while those of the estimated model are Θ. In terms of these parameters, the corresponding distributions are, respectively, $\Pr(y(t)|\Theta_{true})$ and $\Pr(y(t)|\Theta)$ with $y(t)$ the discrete measurement data generated by the true model distribution. Therefore, if we define the AIC in terms of log-likelihood and the N_Θ-parameter set, Θ, then

$$\text{AIC}(N_\Theta) = -2 \times \mathcal{L}(\Theta) + 2 \times N_\Theta \tag{7.147}$$

where $\mathcal{L}(\Theta)$ is the *average log-likelihood*.

Next, let us investigate the set of prediction errors or INV. If the INV are given by $\{e(t)\}$, then the log-likelihood function is specified by the following relations, since the $\{e(t)\}$ are independent. From Eq. (7.145), we have the average log-likelihood

$$\mathcal{L}(K, \Theta) = E_e\{\ln \Pr(e(t)|\Theta)\} = \frac{1}{K} \sum_{t=1}^{K} \ln \Pr(e(t)|\Theta) \qquad (7.148)$$

which in the limit approaches the ensemble *mean log-likelihood* function

$$\lim_{K \to \infty} \mathcal{L}(K, \Theta) \to E_e\{\ln \Pr(e(t)|\Theta)\} \qquad (7.149)$$

For the prediction error (INV) models, the AIC is given by

$$\text{AIC}(N_\Theta) = -2 \times \mathcal{L}(\Pr(e(t)|\Theta)) + 2 \times N_\Theta \qquad (7.150)$$

but from Eq. (7.145), we have that

$$E_e\{\ln \Pr(e(t)|\Theta)\} \approx \frac{1}{K} \sum_{t=1}^{K} \ln \Pr(e(t)|\Theta) \quad \text{for large} \quad K \qquad (7.151)$$

Now into vector notation for ease of development, $\mathbf{e} = [e(1) \cdots e(K)]'$ and following [52], then for AR(N_a) model, the criterion with $N_\Theta = N_a$ becomes

$$\text{AIC}(N_\Theta) = -2\mathcal{L}(\Pr(\mathbf{e}|\Theta)) + 2K_\Theta = -\ln\left(\frac{1}{K}\mathbf{e}'\mathbf{e}\right) + 2\frac{N_\Theta}{K} \qquad (7.152)$$

where N_Θ is the number of free parameters, K the number of data samples, and $\left(\frac{1}{K}\mathbf{e}'\mathbf{e}\right)$ is the (estimated) prediction error variance. Model order is determined by selecting the order corresponding to the minimum value of the criteria (MAICE) as before.

In the estimation problem, several families of $\Pr(e(t)|\Theta)$ with different forms of $\Pr(e(t)|\Theta)$ are given and we are required to select the best "fit." The AIC still provides a useful estimate of the entropy function in this case. Here the AIC is calculated for various model orders and the value with the minimum AIC is selected, that is, the MAICE. The AIC is shown in Table 7.8.

Similar to the AIC is the MDL order estimator [53, 54]. For the AR(N_Θ) model, we have

$$\text{MDL}(N_\theta) = -\ln\left(\frac{1}{K}\mathbf{e}'\mathbf{e}\right) + \frac{N_\Theta}{2}\ln K \qquad (7.153)$$

The difference between the AIC and MDL is in the final free parameter penalty term for overestimating the order. The AIC tends to overestimate the order, while the MDL parameter penalty term increases with the data length K and order N_Θ. It has also been shown that the MDL is a more consistent order estimator, since it converges to the true order as the number of measurements increase [54].

Table 7.8 Stochastic realization: statistical tests.

Test	Criterion	Remarks
Best rank	$\beta(\hat{N}_x) := \left(\frac{\sum_{k=1}^{\hat{N}_x} \sigma_k}{\sum_{k=1}^{K} \sigma_k} \right) \times 100$	Choose $\hat{N}_x$ value
Model order estimate	$\hat{N}_x = \max\{N : \log \hat{\sigma}_N > (\log \hat{\sigma}_{max} + \log \hat{\sigma}_{min})\}$	Select $\hat{N}_x$ value (Gap)
Model error fit	$\epsilon_i = \sqrt{\frac{\sum_{t=1}^{K} \epsilon_i^2(t)}{\sum_{t=1}^{K} y_i^2(t)}}; i = 1, \ldots, N_y$	Locate knee of curve
Model error	$\mathcal{M}_\epsilon = 2 \sum_{k=\hat{N}_x+1}^{K} \sigma_k$	Calculate error value/compare
Mean-squared (prediction) Error:	$\text{MSE} = \frac{1}{K} \mathbf{e}'\mathbf{e}$	Calculate MSE value/compare
Zero-mean	$\hat{\mu}_\epsilon(i) \overset{> \text{Reject}}{\underset{< \text{Accept}}{}} 1.96 \sqrt{\frac{\hat{R}_{\epsilon\epsilon}(i)}{K}}$ for $i = 1, \ldots, N_y$	Check against threshold
Whiteness	$I_{\rho_{\epsilon\epsilon}} = \hat{\rho}_{\epsilon\epsilon}(i,k) \pm \frac{1.96}{\sqrt{K}}$ for $K > 30$	Check 95% of covariances lie within bounds
Weighted sum-squared residual	$\rho_\epsilon(\ell) \overset{> \, \mathcal{H}_1}{\underset{< \, \mathcal{H}_0}{}} N_\epsilon K + 1.96 \sqrt{2 N_\epsilon K}$	Check threshold (below for white)
Signal error test	$\hat{y}(t) = C\hat{x}(t) + Du(t)$	Visual check. Overlay (y vs $\hat{y}$) for fit
Final prediction error	$\text{FPE}(N_\Theta) = \left(\frac{K+1+N_\Theta}{K-1-N_\Theta} \right) \left(\frac{1}{K} \mathbf{e}'\mathbf{e} \right)$	Penalties for overparameterization reduced for large K
Akaike information criteria	$\text{AIC}(N_\Theta) = -\ln \left(\frac{1}{K} \mathbf{e}'\mathbf{e} \right) + 2\frac{N_\Theta}{K}$	Penalties for overparameterization biased – vary N_Θ. Knee (choose minimum)
Minimum description length	$\text{MDL}(N_\Theta) = -\ln \left(\frac{1}{K} \mathbf{e}'\mathbf{e} \right) + \frac{N_\Theta}{2} \ln K$	Penalties for overparameterization consistent – vary N_Θ. Knee (choose minimum)

There are quite a wide variety of order estimators. For instance, it can also be shown that the final prediction error (FPE) criterion given by [44]

$$\text{FPE}(N_\Theta) = \left(\frac{K + 1 + N_\Theta}{K - 1 - N_\Theta}\right)\left(\frac{1}{K}e'e\right) \tag{7.154}$$

is asymptotically equivalent to the AIC(N_Θ) for large K (data samples).

Other tests are also utilized to aid in the selection process. The signal *error test* of Table 7.8 consists of identifying models for a sequence of increasing orders, exciting each model with the exogenous input to produce an estimated response, $\hat{y}(t)$, and overlaying the estimated and the actual response ($y(t)$ vs $\hat{y}(t)$) to select a "reasonable" fit to the data. That is, these tests are performed by producing the estimated (filtered or smoothed) response

$$\hat{y}(t) = C\hat{x}(t) + Du(t) \qquad \text{(signal error test)}$$

Before we close this discussion, it is of interest to illustrate the application of the information criteria to a problem in source localization that developed the idea of determining the number of states or sources, in this case impinging on a sensor array [39]. In this case, the time series data are captured from the sensors and a multichannel covariance matrix estimated, which is then evaluated (rank) using the SVD approach. The AIC based on the eigenvalues obtained from the SVD is given by

$$\text{AIC}(\hat{N}_x) = -K \ln\left[\frac{\prod_{i=\hat{N}_x+1}^{N_y} \lambda_i}{\left(\frac{1}{N_y-\hat{N}_x}\sum_{i=\hat{N}_x+1}^{N_y} \lambda_i\right)^{N_y-\hat{N}_x}}\right] + 2\hat{N}_x(2N_y - \hat{N}_x) \tag{7.155}$$

where $\hat{N}_x$ is the selected order estimate, N_y is the number of measurement channels, K is the number of data samples, and λ_i is the ith eigenvalue obtained from the SVD.

The MDL criterion is similar to the AIC except for the final term that penalizes for overestimation of the order (parsimony) and is given by

$$\text{MDL}(\hat{N}_x) = -K \ln\left[\frac{\prod_{i=\hat{N}_x+1}^{N_y} \lambda_i}{\left(\frac{1}{N_y-\hat{N}_x}\sum_{i=\hat{N}_x+1}^{N_y} \lambda_i\right)^{N_y-\hat{N}_x}}\right] + \frac{1}{2}\hat{N}_x(2N_y - \hat{N}_x)\ln K \tag{7.156}$$

Both of these metrics can be applied to the order estimation problem with the caveat that an input or excitation sequence is assumed to be sinusoidal (plane wave).

We list some of the more popular order tests in Table 7.8 and refer the interested reader to [55] for more details. We will illustrate the MAICE procedure when we develop an application. Next we consider validating the identified deterministic model.

7.7.2 Model Validation

Once the order is selected, the SID completed and the "black-box" model Σ_{ABCD} identified, the question of its validity must be answered to decide just "how good" the model fits the data. For prediction error methods, a wealth of statistical techniques exists as mentioned for the stochastic SID case. We again rely on the identified model response compared to the system response data to calculate the fitting error.

The response of the identified model $\Sigma_M := \{A_M, B_M, C_M, D_M\}$ is easily calculated using the available inputs that were used to extract it from the data. We define the *model response,* $\mathbf{y}_M(t)$ as before in state-space form as

$$x(t) = A_M x(t-1) + B_M u(t-1) \qquad \text{(State Response)}$$

$$y_M(t) = C_M x(t) + D_M u(t) \qquad \text{(Measurement Response)}$$
$$(7.157)$$

or equivalently as a *convolution response*

$$y_M(t) = C_M A_M^t x(0) + \sum_{k=0}^{t-1} C_M A_M^{t-k-1} B_M u(k-1) + D_M \delta(t)$$

$$\text{(Convolution Response)} \qquad (7.158)$$

or directly from the *impulse response*

$$y_M(t) = H_t = C_M A_M^{t-1} B_M + D_M \delta(t);$$

$$t = 1, \ldots, K \qquad \text{(Impulse Response)} \qquad (7.159)$$

We can now use any of these representations to estimate the *model (residual) error* or simulation error defined by [40] as

$$\epsilon(t) := \mathbf{y}(t) - \mathbf{y}_M(t) \qquad \text{(Model Residual Error)}$$
$$(7.160)$$

and define the *model error metric* for the ith-measurement channel as

$$\varepsilon_i = \sqrt{\frac{\sum_{t=1}^{K} \epsilon_i^2(t)}{\sum_{t=1}^{K} \mathbf{y}_i^2(t)}}; \quad i = 1, \ldots, N_y \qquad \text{(Model Error Metric)}$$

$$(7.161)$$

to evaluate the performance of the model response $\{y_M(t)\}$ relative to the response data $\{y(t)\}$ [19, 27].

This metric can also be calculated over an N-ensemble of data for *each* measurement channel as well to provide statistics enabling a reasonable estimate of model quality. Here we have generated a data ensemble $\{y_n(t)\}; n = 1, \ldots, N$ with the corresponding *ensemble model error* for the nth realization defined by

$$\epsilon_n(t) := \mathbf{y}_n(t) - \mathbf{y}_M^n(t) \qquad \text{(ensemble model error)} \qquad (7.162)$$

Ensemble statistics and confidence intervals can be estimated from the sample mean and variance estimates *across* the realizations $\{\epsilon_n(t)\}; \quad n = 1, \ldots, N$ as

$$\mu_\epsilon(t) := \hat{E}\{\epsilon_n(t)\} = \frac{1}{N}\sum_{n=1}^{N} \epsilon_n(t)$$

$$\sigma_\epsilon(t) := \hat{E}\{\epsilon_n(t) - \mu_\epsilon(t)\} = \frac{1}{N}\sum_{n=1}^{N} \epsilon_n^2(t) - \mu_\epsilon^2(t) \qquad (7.163)$$

enabling a confidence interval estimate

$$I_\epsilon(t) = \mu_\epsilon(t) \pm 1.96\sigma_\epsilon(t) \quad \text{for} \quad N > 30$$

for *each* measurement channel realization providing another means of validating the identified model.

Another statistic that can be applied to validate the performance of the identified model is the *median absolute deviation* (MAD) statistic because of its inherent robustness property relative to the usual ensemble mean/standard deviation approach [56]

$$\text{MAD}(t) := \gamma \mathcal{M}_t(|\epsilon_n(t) - \mathcal{M}_i(\epsilon_i)|) \qquad (7.164)$$

where $\epsilon_n(t) := \{\epsilon_n(0) \cdots \epsilon_n(t)\}$ is the nth realization of the set of model errors up to time t, $\mathcal{M}_t$ is the median of the data, and $\gamma = 1.4826$ is a constant based on the normalized (assumed Gaussian) data. A confidence limit that can be used for outlier detection can also be calculated as

$$\mathcal{M}_t - \beta \times \text{MAD}(t) < \epsilon_n(t) < \mathcal{M}_t + \beta \times \text{MAD}(t) \qquad (7.165)$$

for β is a threshold equivalent to a confidence limit (e.g. $\beta = 1, 2$).

If it is not possible to create a deterministic data ensemble through simulation or experiment, then the original measurement can be partitioned to create an ensemble, or bootstrap data resampling techniques may be applied as well [57].

7.7.2.1 Residual Testing

A very common methodology for model validation consists of testing the *model error residual sequence*, which is specified in Eq. (6.135). The theoretical aspects of these tests were developed in Section 4.2, and we summarize them succinctly with the $\varepsilon(t) \rightarrow e(t)$:

- *Zero-mean testing* of the residual (INV) of Section 4.2.4. The component-wise *zero-mean test* on each component model residual error ϵ_i is given by

$$\hat{\mu}_\epsilon(i) \underset{\underset{\text{Accept}}{<}}{\overset{\overset{\text{Reject}}{>}}{}} 1.96\sqrt{\frac{\hat{R}_{\epsilon\epsilon}(i)}{N}} \quad \text{for} \quad i = 1, \ldots, N_y$$

at the 5% significance level, where $\hat{\mu}_\epsilon(i)$ is the sample mean and $\hat{R}_{\epsilon\epsilon}(i)$ is the sample variance (assuming ergodicity).

- *Whiteness testing* of the model residual error of Section 4.2.4. The whiteness confidence interval estimate is

$$I_{\rho_{\epsilon\epsilon}} = \hat{\rho}_{\epsilon\epsilon}(i,k) \pm \frac{1.96}{\sqrt{N}} \quad \text{for} \quad N > 30$$

under the null hypothesis, 95% of the normalized covariance $\hat{\rho}_{\epsilon\epsilon}(i,k)$ values must lie within this confidence interval for each *component* residual sequence to be considered statistically uncorrelated or equivalently white.

- *WSSR testing* of the model residual (INV) of Section 4.2.4. The weighted statistic aggregates all of the model residual error *vector* information over some finite window of length M and is given by

$$\rho_\epsilon(\ell) \underset{\underset{\mathcal{H}_0}{<}}{\overset{\overset{\mathcal{H}_1}{>}}{}} N_\epsilon M + 1.96\sqrt{2N_\epsilon M} \tag{7.166}$$

We summarize the steps in order estimation and model validation:

- *Generate* ensemble data: through experiments or simulation, if not possible, use partitioning or bootstrap techniques.
- *Create* Hankel covariance or input/output data matrices.
- *Perform* an SVD of the Hankel matrix extracting the observability matrix (classical) or extended observability matrix (subspace).
- *Estimate* the order from the extracted singular values and extract the corresponding statistics (Hankel order, Hankel singular values, Hankel norm, Hankel SVD model error).
- *Identify* the model Σ_{ABCD} using classical realization or SID techniques.
- *Estimate* model (simulation) error(s) and its statistics over each realization of the ensemble.
- *Validate* model performance using ensemble statistics and error metrics.

We now have the tools in place to perform system identification and evaluate the identified model performance. Consider the following example.

Example 7.3 Reconsider the 3-mass (spring–damper) mechanical system illustrated previously in Figure 6.2 from [50] where $m_i = 1; k_i = 3, k_4 = 0$; and $d_i = 0.01k_i, d_4 = 0; i = 1, \ldots, 3$. These parameters along with the input/

output transmission matrices led to the state-space representation $(N_d = 3, N_x = 6)$ developed in Example 6.3.

$$\dot{\mathbf{x}}(t) = A\mathbf{x}(t) + B\mathbf{u}(t)$$

$$\mathbf{y}(t) = C\mathbf{x}(t) + D\mathbf{u}(t)$$

The MCK system is excited "randomly" with a Gaussian white noise sequence at an SNR of -10 dB creating a persistent excitation while mimicking a typical shaker-table experiment [58]. Exciting the system with this excitation yields the noisy 3-output response shown in Figure 7.7 along with its accompanying Fourier spectra. The N4SID stochastic realization technique was performed as before. Discretizing the response with a sampling interval of $\Delta t = 0.01$ second, creating the input/output Hankel matrices and decomposing with the SVD as in Eq. (7.113), reveals a 6th-order system as before. The plot of the singular values is shown in Figure 7.8b. The corresponding metrics are as follows:

$$
\begin{aligned}
\text{Hankel order estimation}: \quad & \hat{N}_x = 6 \\
\text{Hankel singular values (log)}: \quad & \sigma_i = \{5.8,\ 5.5,\ 4.2,\ 4.1,\ 3.0,\ 2.8 - 31\} \\
\text{Hankel norm (energy)}: \quad & \| \mathcal{H}_{K,K'} \| = 164 \\
\text{SVD model error}: \quad & \mathcal{M}_\epsilon = 3.4 \times 10^{-14} \\
\text{Mean} - \text{squared error}: \quad & \text{MSE} = 2.1 \times 10^{-4} \\
\text{Final prediction error}: \quad & \text{FPE} = 3.8 \times 10^{-23} \\
\text{Akaike information criterion}: \quad & \text{AIC} = -51.0 \\
\text{Minimum description length}: \quad & \text{MDL} = -7.3 \times 10^{5}
\end{aligned}
$$

The output power spectra of each channel and the average power spectrum of the identified model are compared to the true average spectrum in Figure 7.8a, indicating a reasonable realization of the data. Again the "identified" modal frequencies of the system estimated from the "discrete" identified system matrix and transformation from the discrete ($\mathcal{Z}$-domain) to the continuous ($\mathcal{S}$-domain) given by ($s = \frac{1}{\Delta t} \ln z$) yields $f_i = \{0.497 \text{ Hz}, 0.343 \text{ Hz}, 0.123 \text{ Hz}\}$ and are shown in Figure 7.8c. To confirm that the realization *matches* the modal frequency estimates of the system closely, the signal error test of Table 7.8 was performed. The estimated channel impulse responses are overlaid onto the true impulse response data in Figure 7.9 – visually indicating a reasonable realization, even with such a low SNR.

Next we calculate the various validation metrics to observe how well they predict/validate a realization of this data. The results are shown in Table 7.9. From the table, we examine the average results of each of the statistics. Over an ensemble of 100-realizations and stochastic SIDs, the average residual error is 8.9×10^{-10}, a bit surprising at the low -10 dB SNR. The calculated *model errors* are reasonable for the corresponding average, standard deviation, and MAD statistics on the order of 10^{-3}, corresponding to less than 1% error.

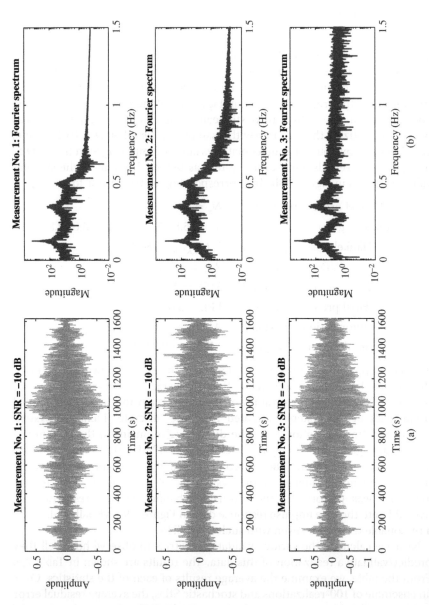

Figure 7.7 Response of 3-output mechanical system to white noise excitation (−10 dB). (a) Channel white noise responses. (b) Fourier transforms of channel responses.

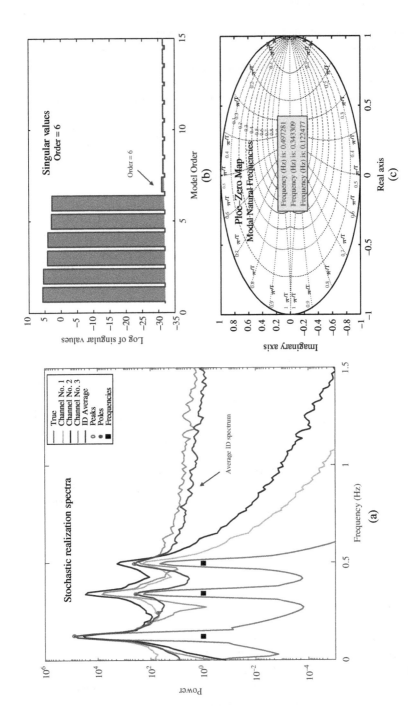

Figure 7.8 N4SIDsubspace identification of 3-output mechanical system excited by white noise (−10 dB). (a) Power spectra: outputs/average spectra indicated with poles (boxes). (b) Singular values of the projection matrix. (c) Frequencies with poles/zeros.

Table 7.9 Stochastic subspace identification: 3-mass mechanical system.

			Model error/validation results				
Channel	Residual	Average	Deviation	MAD	Z-mean	White	WSSR
No.	Error	Model error	Model error	Model error	2.2×10^{-2}	%-Out	Below τ
1	13.6×10^{-10}	5.1×10^{-3}	5.9×10^{-3}	5.0×10^{-3}	1.6×10^{-7}	0.38%	Yes
2	2.8×10^{-10}	4.7×10^{-3}	5.3×10^{-3}	4.5×10^{-3}	1.2×10^{-5}	58.0%	Yes
3	9.3×10^{-10}	5.1×10^{-3}	5.9×10^{-3}	4.9×10^{-3}	1.5×10^{-4}	4.97%	Yes
Average	8.9×10^{-10}	4.9×10^{-3}	5.7×10^{-3}	4.8×10^{-3}	—	—	Yes

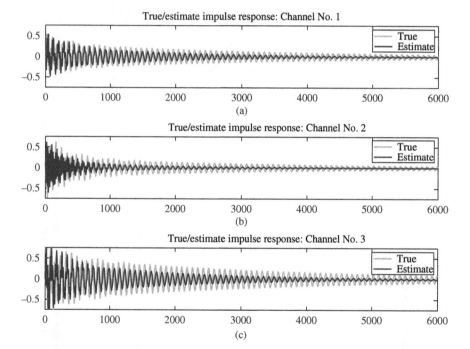

Figure 7.9 N4SID-subspace validation of 3-output mechanical system: signal estimate tests (impulse responses).

We also performed a zero-mean/whiteness *residual tests* depicted in Figure 7.10 with the reasonable results in Table 7.9: (Z-M test: $1.6 \times 10^{-7} <$ 2.2×10^{-2}; $1.2 \times 10^{-5} < 2.2 \times 10^{-2}$; $1.5 \times 10^{-4} < 2.2 \times 10^{-2}$) and (W-test: 0.38% out; 58% out; 4.9% out). Note the W-test results for channels 1 and 3 were very good (< 5%), validating a good fit; however, the result for channel 2

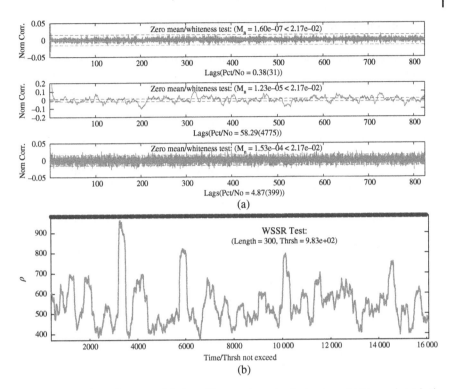

Figure 7.10 N4SID-subspace validation of 3-output mechanical system. (a) Channel residual zero-mean/whiteness tests. (b) WSSR test.

was poor, and this is confirmed by the signal error test of Figure 7.9. The WSSR test in Figure 7.10b confirms a white residual sequence validating the model fit to the data and acceptable identification substantiating the "matches" shown in the figures.

This completes the example. ☐

This completes the section on order estimation and stochastic model validation techniques. Next we consider a case study that applies these methods to a synthesized data set of a more complex mechanical system – a vibrating structure.

7.8 Case Study: Vibration Response of a Cylinder: Identification and Tracking

In this section, we study the application of the stochastic SID of a vibrating cylindrical structure represented by a LTI, MIMO. The structure is excited by a random input, specifically an impact device. The investigation of the vibrational response of this cylindrical object in a noisy environment (a pipe in air) to impact excitations provided by a cam-driven, hammer-like surface, randomly excited by an electric motor is of interest in order develop a real-time processing system capable of on-board performance and monitoring. We model this system as a bar and measure its response by a triaxial accelerometer to capture a set of dominant resonant (modal) frequencies. These modal frequencies are of critical interest when monitoring the underlying system for any anomalies during operation. This object was selected for investigation because it is well known (theoretically) and can be used to represent a large number of simple structures (e.g. fight vehicles, missiles) along with its well-defined modal response. From a SID perspective, we have an N_x-order state-space system characterizing the usual mass-damper-spring (MCK) mechanical structure that can be represented as an MIMO, LTI system. Here we focus on the dominant modes and the corresponding modal (resonant) frequencies representing the first six (modes) that capture the majority of the response signal of interest. The input is tacitly assumed to be white random noise.

Theoretically, we model the pipe as a beam in the free–free configuration (not pinned) leading to the following closed-form formulation for the fundamental modal frequency given by the expression [59]

$$f_0 = \frac{1}{2\pi} \left[\frac{22.373}{L^2} \right] \sqrt{\frac{E \times I}{\rho}}$$
$$f_1 = 2.757 \times f_0$$
$$f_2 = 5.404 \times f_0$$

For the cylindrical object (1.125 in. OD copper pipe), we have: E is the modulus of elasticity (1.6×10^7 N m^{-2}), I is the area moment of inertia (0.0244 in.^{4}), L is the length (60 in.), and ρ is the density (1.41×10^{-4} lb s^2). Therefore, we have the following "theoretical" modal frequencies:

$$f_i = \{51.7, 143.5, 281.3, 465.0, 694.7, 970.2, 1291.7, 1435.3, 2813.1 \text{ Hz}\}$$

that we will use to compare our results.

A simple free–free (not pinned) experiment for this case study is shown in Figure 7.11 where we observe the 60 in./1.125 in. OD copper pipe along with the excitation system (cam-driven hammer) signal conditioner (10 kHz filter), accelerometer, and digitizing oscilloscope. Data were sampled at 10 kHz and decimated to 5 kHz, yielding a 2.5 kHz Nyquist frequency for spectral analysis. This system was designed to excite the cylinder and measure its response acquiring approximately 1-million samples to be buffer windowed (sectioned) synthesizing a real-time acquisition system.

The basic approach for a *single-channel* accelerometer is to excite the pipe with the repetitive hammer blows at a high rate and measure its response.

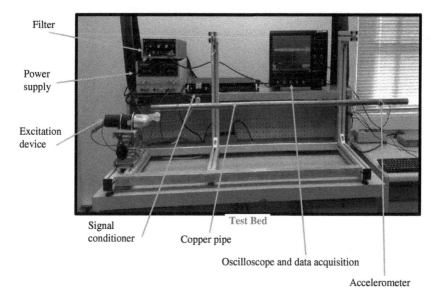

Figure 7.11 Cylindrical object (copper pipe) vibrational response experiment: pipe, cam-driven hammer excitation, triaxial accelerometer, signal conditioner (filter), data acquisition system (digitizing oscilloscope).

Once the noisy data is digitized, it is preprocessed by bandpass filtering (8th-order Butterworth) between 40 and 1.25 khz followed by a 3rd-order equalization (whitening) filter to enhance the low-band modal frequencies. Its spectrum is estimated along with the corresponding spectral peaks to extract the modal frequencies. Since we are primarily interested in a real-time application, we have investigated a variety of spectral estimators ranging from the simple discrete Fourier transform (DFT) to the sophisticated MUltiple SIgnal Classification (MUSIC) approach requiring an SVD of a Toeplitz covariance matrix [32]. Once the spectral peaks are extracted *locating* the corresponding modal frequencies, they are either provided as input to a frequency tracker for enhancement and/or simply stored for eventual probability density function (PDF) estimation using a histogram or smoothed kernel density estimator for monitoring and eventual detection. This approach is illustrated in Figure 7.12.

After analyzing this data and demonstrating the capability aimed at an eventual real-time system, the MIMO experiment was performed under the same conditions and the results were compared to validate both the single and multiple channel results. A full analysis of the multiple channel system was performed, demonstrating the capability of extracting the modal frequencies of high interest. Typical preprocessed response data is shown in Figure 7.13 with the acquired signals (XYZ-channels) in (a) and the corresponding power spectra in (b). Here we observe the modal frequencies shown at the major spectral

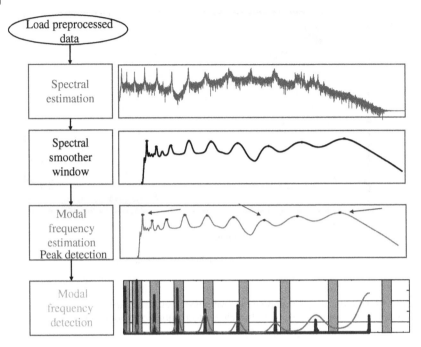

Figure 7.12 Analysis of cylindrical object (pipe) single-channel data: Fourier spectrum, frequency smoothing, modal frequency estimation (peaks), and histogram.

peaks capturing the dominant resonances validating the experimental goals. Again we are primarily focused on the first six (6) dominant modal frequencies to characterize the cylinder.

The MIMO stochastic subspace identification algorithm (N4SID) was applied to the response data and the results are summarized in Figure 7.14. In (a), we see the estimated impulse response (average) and its corresponding power spectrum in (b). We observe the "average" identified power spectrum in (b) as well as the extracted modal (poles) frequencies (squares) from the identified model system matrix $(\hat{A})$. The SVD-order test is shown in (c). This test does not reveal a well-defined order; however, based on our prior knowledge of the cylinder and its modes, we decided to select a 28th-order (14-mode) representation as adequate. The estimated modal frequencies obtained from the discrete-to-continuous transform $(\mathcal{Z} \rightarrow \mathcal{S})$ are also listed in the figure. We note that they are quite reasonable and are given in Table 7.9 with their corresponding relative (to the theoretical frequencies) errors. Finally, we validate the realization by comparing the raw, preprocessed, and identified spectra along with the corresponding (identified) poles shown in Figure 7.15. Here it is clear

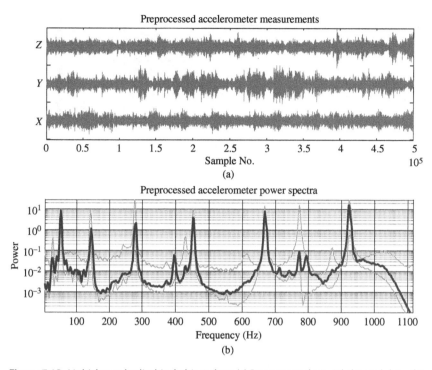

Figure 7.13 Multichannel cylindrical object data: (a) Preprocessed triaxial channel data. (b) Triaxial power spectra.

from the figure and the table that the subspace identifier has extracted the primary modal frequencies from the noisy MIMO vibrational data.

In order to investigate the potential real-time performance of the subspace approach, the preprocessed data is buffered into $20K$-sample windows with the N4SID method applied using a 28th-order model to fit the MIMO data buffers. Using the identified 14-mode model, the power spectrum is estimated along with the poles ("+") obtained from the transformed eigenvalues of the A-matrix as shown in Figure 7.16a. At each step (window buffer), the estimated poles are displayed in Figure 7.16b as circles with embedded "+" symbols along with the theoretical poles shown as thick bars in the figure. Tracking is achieved (visually) for a given modal frequency, if the circles lie within or near the thick bars. Even with the limited buffer data, the N4SID approach appears to reliably "track" the poles mimicking the expected performance of this approach for a real-time system.

The ensemble statistics for the 14-mode identification are shown in Figure 7.17 where we see the corresponding scatter plot in (a) along with the

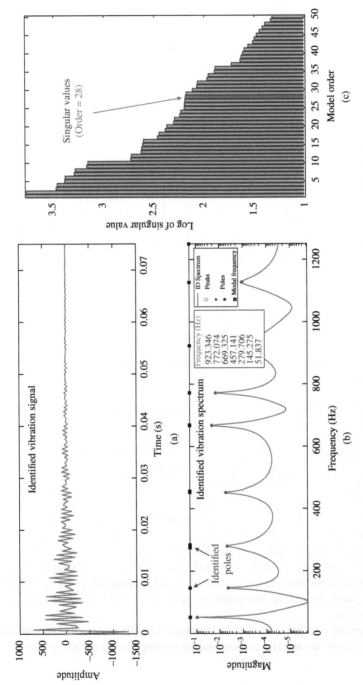

Figure 7.14 Cylindrical object multichannel identification results: (a) Identified average impulse response. (b) Identified average power spectrum and modal frequencies. (c) Order estimation (singular values) with $N_x = 28$ corresponding to 14-modal frequencies.

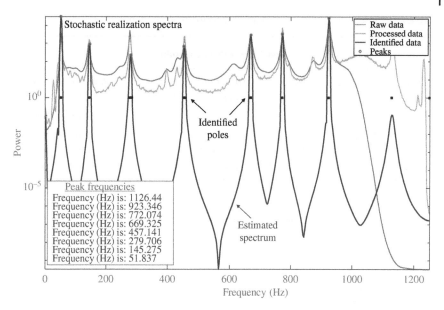

Figure 7.15 Stochastic realization of cylindrical object (pipe) power spectral: raw, processed, and identified spectra with poles and modal frequencies.

corresponding 99.9% confidence interval about the mean modal frequency. Clearly, the estimates are reasonably precise as shown in (a) and in Table 7.9 as well. In Figure 7.17b, we observe the corresponding modal frequency histogram with most of the identified frequency bins heavily populated indicating very high probabilities of the "theoretical" modal frequencies estimated by the subspace tracker.

Postprocessing of the tracking frequencies are shown in Figure 7.18 for the 14-mode tracks. It is clear from the figures that by postprocessing (off-line), the modal frequency tracks are improved from those provided on-line primarily because of the availability of the "batch" statistics. That is, the *batch* postprocessor performing both outlier correction and tracker smoothing is superior to the on-line tracks of Figure 7.18 primarily because the ensemble statistics of 25 data windows are now available to improve the estimates as shown in the figure. The light diamonds in the figure are the theoretical or considered the "true" values of the poles – those circles on or close to these represent excellent estimates of the modal frequencies. The darker diamonds are the batch median estimates giving a visual precision estimate of the modal frequencies.

Next we calculate the various validation metrics to observe how well they predict/validate a reasonable realization of this data. The results are shown in Table 7.10. From the table, we examine the average results of each of the statistics. Over an ensemble of 25-realizations and stochastic SIDs, the average

Table 7.10 Stochastic subspace identification: six (6) modal frequencies.

		Model error/validation results				
Channel	Residual	Average	Deviation	MAD	Z-mean	WSSR
No.	Error	Model error	Model error	Model error	3.1×10^{-2}	Below τ
1	1.3×10^{-11}	3.3×10^{-3}	4.2×10^{-3}	3.3×10^{-3}	1.6×10^{-8}	Yes
2	1.3×10^{-11}	3.3×10^{-3}	4.1×10^{-3}	3.3×10^{-3}	4.5×10^{-9}	Yes
3	1.3×10^{-11}	3.1×10^{-3}	3.8×10^{-3}	3.1×10^{-3}	1.1×10^{-8}	Yes
AVG	1.3×10^{-11}	3.2×10^{-3}	4.0×10^{-3}	3.2×10^{-3}	1.2×10^{-8}	Yes

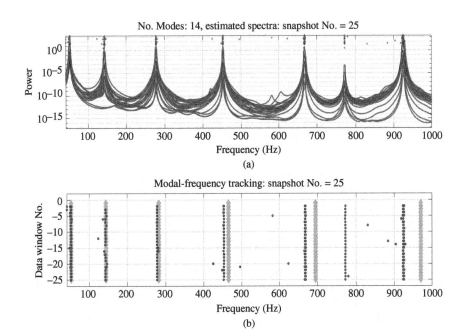

Figure 7.16 Subspace identification of cylindrical object, 14-mode model: (a) Estimated transfer function spectra of identified modal models and raw modal frequency estimates (+). (b) Modal frequency tracking (raw) results (circles with embedded +) and "truth" bands (thick bars).

residual error is 1.3×10^{-11} as expected since the synthesized signal is essentially deterministic (100 dB SNR). The calculated *model errors* are reasonable for the corresponding average, standard deviation, and MAD statistics on the order of 10^{-3} corresponding to less than 1% error. We also performed a zero-mean test with the results quite good along with the WSSR test, and again

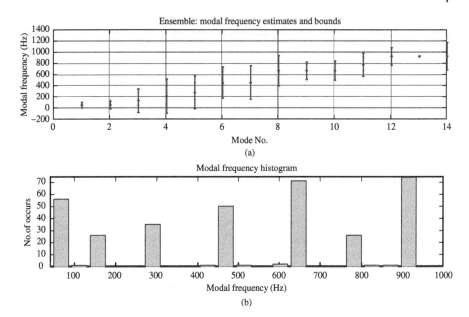

Figure 7.17 Postprocessing of modal frequency tracking (ensemble) statistics: (a) Modal frequency scatter plot with 99.9% confidence limits about the mean. (b) Modal frequency histogram indicating high probabilities in identified modal frequency bins.

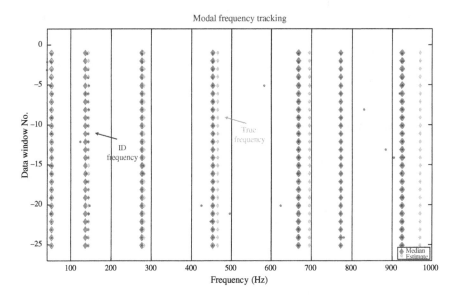

Figure 7.18 Postprocessed modal frequency tracking of cylindrical object data for an identified 14-mode model: Diamonds represent theoretical (light) and median (dark) values.

the results were quite reasonable substantiating the "matches" shown in the figures. This completes the case study of a vibrating cylindrical (pipe) structure.

7.9 Summary

In this chapter, we evolved from the deterministic to the stochastic realization problem essentially based on the extended (correlated noise) Gauss–Markov representation of a MIMO stochastic process. After introducing the correlated Gauss–Markov model and all of the underlying properties in terms of its power spectrum and covariance sequences, we theoretically defined the stochastic realization problem. Next we introduced a set of equivalent solutions to this problem leading to both the generic Riccati equation solution and the INV-based solution in terms of the steady-state Kalman filter. Both of these approaches perform a Hankel-matrix-based deterministic solution replacing the impulse response matrices with output covariance matrices leading to the so-called KSP model and solution to the corresponding KSP equations. The INV model and its underlying properties were also examined in terms of power spectrum and covariance sequences and their relationships to the KSP model and equations. Stochastic subspace realization methods were then introduced by examining the underlying orthogonal and oblique projection techniques leading to the respective MOESP methods and the N4SID approach [60–63]. We showed the theoretical relationship between these techniques or essentially between orthogonal and oblique projections. After developing these methods and applying them to both synthesized (sinusoids in noise) and actual measurement data from a vibrating cylindrical object (copper pipe), we evaluated their performance and illustrated their overall effectiveness leading to potential real-time solutions.

MATLAB NOTES

MATLAB offers a suite of system identification, performance analysis, and statistical evaluation methods for this problem. The System Identification toolbox is the premier tool enabling the solution to a variety of identification problems ranging from LTI systems as in these realization problems as well as gray-box to nonlinear systems. From our perspective in this text, the N4SID method (**n4sid**) is the workhorse of the subspace techniques, which include the MOESP and CVA techniques employing the equivalence of these algorithms through the choice of weighting matrices (**n4weight**) – a simple flag that is set to select the preferred choice. Residual analysis SVD-order estimation is embedded into all of these MIMO state-space methods. Model error and model fitting analysis is also available to select, identify, and validate model performance. Such techniques are residual analysis check for whiteness (**resid**) as well as comparing the identified model to the data (**compare**) or to a reference signal (**goodnessoffit**) and using the prediction error analysis (**pe**). Validation commands available to analyze/compare fits are **fpe**, **aic**, **selstruc** (MDL), and **predict**. Parameter estimates as well as their uncertainties (error bounds) are also available using the (**present**) command. Other implementations of subspace algorithms exist and can be extracted from the texts of [25, 27] (disk).

References

1 Faurre, P. (1976). Stochastic realization algorithms. In: *System Identification: Advances and Case Studies* (ed. R. Mehra and D. Lainiotis), 1–23. New York: Academic Press.
2 Akaike, H. (1974). Stochastic theory of minimal realization. *IEEE Trans. Autom. Control* 19: 667–674.
3 Akaike, H. (1975). Markovian representation of stochastic processes by canonical variables. *SIAM J. Control* 13 (1): 162–173.
4 Denham, M. (1974). On the factorization of discrete-time rational spectral density matrices. *IEEE Trans. Autom. Control* AC-19: 535–537.
5 Tse, E. and Wiennert, H. (1975). Structure determination and parameter identification for Multivariable stochastic linear systems. *IEEE Trans. Autom. Control* 20: 603–613.
6 Candy, J., Bullock, T., and Warren, M. (1979). Invariant system description of the stochastic realization. *Automatica* 15: 493–495.
7 Larimore, W.E. (1990). Canonical variate analysis in identification, filtering and adaptive control. In: *Proceedings of the 29th Conference on Decision and Control*, Hawaii, USA, 596–604.
8 Aoki, M. (1990). *State Space Modeling of Time Series*, 2e. London: Springer.
9 Mehra, R. (1970). On the identification of variances and adaptive Kalman filtering. *IEEE Trans. Autom. Control* AC-15: 175–184.
10 Mehra, R. (1971). On-line identification of linear dynamic systems with applications to Kalman filtering. *IEEE Trans. Autom. Control* AC-16: 12–20.
11 Ho, B. and Kalman, R. (1966). Effective reconstruction of linear state variable models from input/output data. *Regelungstechnik* 14: 545–548.
12 Candy, J., Warren, M., and Bullock, T. (1977). Realization of an invariant system description from Markov sequences. *IEEE Trans. Autom. Control* AC-23 (12): 93–96.
13 Chen, C. (1984). *Linear System Theory and Design*. New York: Holt-Rinehart & Winston.
14 Candy, J. (1976). Realization of invariant system descriptions from Markov sequences. PhD dissertation. Gainesville, FL: University Florida Press.

15 Popov, V. (1973). *Hyperstability of Control Systems*. New York: Springer.
16 Willems, J. (1971). Least squares stationary optimal control and the algebraic Riccati equation. *IEEE Trans. Autom. Control* AC-16: 621–634.
17 Viberg, M. (1995). Subspace-based methods for the identification of linear time-invariant systems. *Automatica* 31 (12): 1835–1851.
18 Soderstrom, T. and Stoica, P. (1989). *System Identification*. Englewood Cliffs, NJ: Prentice-Hall.
19 Verhaegen, M. and Verdult, V. (2007). *Filtering and System Identification: A Least-Squares Approach*. Cambridge: Cambridge University Press.
20 Verhaegen, M. (1994). Identification of the deterministic part of MIMO state space models given in innovations form from input-output data. *Automatica* 30 (1): 61–74.
21 Verhaegen, M. and Dewilde, P. (1992). Subspace model identification: Part I: the output-error state-space model identification model identification class of algorithms. *Int. J. Control* 56 (5): 1187–1210.
22 Verhaegen, M. and Dewilde, P. (1992). Subspace model identification: Part II: analysis of the elementary output-error state space model identification algorithm. *Int. J. Control* 56 (5): 1211–1241.
23 Verhaegen, M. and Dewilde, P. (1993). Subspace model identification: Part III: analysis of ordinary output-error state space model identification algorithm. *Int. J. Control* 58: 555–586.
24 Van Der Veen, A., Deprettere, E., and Swindlehurst, A. (1993). Subspace-based methods for the identification of linear time-invariant systems. *Proc. IEEE* 81 (9): 1277–1308.
25 Katayama, T. (2005). *Subspace Methods for System Identification*. London: Springer.
26 Katayama, T. (2010). Subspace identification of combined deterministic-stochastic systems by LQ decomposition. In: *Proceedings of the 2010 American Control Conference*, Volume 3, 2941–2946.
27 van Overschee, P. and De Moor, B. (1996). *Subspace Identification for Linear Systems: Theory, Implementation, Applications*. Boston, MA: Kluwer Academic Publishers.
28 Gokberg, I. and Krein, M. (1960). Systems of integral equations on a half line with kernels depending on the difference of arguments. *Uspekh: Mat. Nau.* 13: 217–287.
29 Youla, D. and Tissi, P. (1966). N-port synthesis via reactance extractation. *IEEE Int. Conv. Rec.* 14: 183–205.
30 Glover, K. (1984). All optimal Hankel-norm approximations of linear multivariable systems and their L^{∞}-error bounds. *Int. J. Control* 39: 1115–1193.
31 Glover, K. and Willems, J.C. (1974). Parameterizations of Linear Dynamical systems: canonical forms and identifiability. *IEEE Trans. Autom. Control* 19: 640–646.

32 Candy, J. (2006). *Model-Based Signal Processing.* Hoboken, NJ: John Wiley/IEEE Press.

33 Sage, A. and Melsa, J. (1971). *Estimation Theory with Applications to Communcations and Control.* New York: McGraw-Hill.

34 Sharf, L. (1991). *Statistical Signal Processing: Detection, Estimation and Time Series Analysis.* Reading, MA: Addison-Wesley.

35 van Overschee, P. and De Moor, B. (1994). N4SID: numerical algorithms for state space subspace system identification. *Automatica* 30: 75–93.

36 van Overschee, P. and De Moor, B. (1995). A unifying theorem for three subspace system identification algorithms. *Automatica* 31 (12): 1853–1864.

37 Behrens, R. and Scharf, L. (1994). Signal processing applications of oblique projection operators. *IEEE Trans. Signal Process.* 42 (6): 1413–1424.

38 Kayalar, S. and Weinert, H. (1989). Oblique projections: formulas, algorithms and error bounds. *Math. Control Signals Syst.* 2 (1): 33–45.

39 Wax, M. and Kailath, T. (1985). Detection of signals by information theoretic criteria. *IEEE Trans. Acoust. Speech Signal Process.* 33 (2): 387–392.

40 Ljung, L. (1999). *System Identification: Theory for the User*, 2e. Englewood Cliffs, NJ: Prentice-Hall.

41 Ljung, L. and Soderstrom, T. (1983). *Theory and Practice of Recursive Identification.* Cambridge: MIT Press.

42 Soderstrom, T. and Stoica, P. (1989). *System Identification.* New York: Academic Press.

43 Norton, J. (1986). *An Introduction to Identification.* New York: Academic Press.

44 Akaike, H. (1974). A new look at the statistical model identification. *IEEE Trans. Autom. Control* 19 (6): 716–723.

45 Sakamoto, Y., Ishiguro, M., and Kitagawa, G. (1986). *Akaike Information Criterion Statistics.* Boston, MA: Kluwer Academic.

46 Rissanen, J. (1978). Modeling by the shortest data description. *Automatica* 14: 465–471.

47 Golub, G. and van Loan, C. (1996). *Matrix Computations*, 3e. Baltimore, MD: Johns Hopkins University Press.

48 Bauer, D. (2001). Order estimation for subspace methods. *Automatica* 37: 1561–11573.

49 Moore, B. (1981). Principal component analysis in linear systems: controllability, observability, and model reduction. *IEEE Trans. Autom. Control* AC-26 (1): 17–32.

50 Gawronski, W. (2004). *Advanced Structural Dynamics and Active Control of Structures.* New York: Springer.

51 Juang, J. (1994). *Applied System Identification.* Upper Saddle River, NJ: Prentice-Hall PTR.

52 Kay, S. (1988). *Modern Spectral Estimation.* Englewood Cliffs, NJ: Prentice-Hall.

53 Hayes, M. (1996). *Statistical Digital Signal Processing and Modeling.* New York: Wiley.

54 Haykin, S. (1993). *Adaptive Filter Theory.* Englewood Cliffs, NJ: Prentice-Hall.

55 Jategaonkar, R., Raol, J., and Balakrishna, S. (1982). Determination of model order for dynamical systems. *IEEE Trans. Syst. Man Cybern.* 12 (1): 56–62.

56 Pham Gia, T. and Hung, T. (2001). The mean and median absolute deviations. *Math. Comput. Model.* 34: 921–936.

57 Zoubir, A. and Iskander, D. (2004). *Bootstrap Techniques for Signal Processing.* Cambridge: Cambridge University Press.

58 Reynders, E. (2012). System identification methods for (operational) modal analysis: review and comparison. *Arch. Comput. Methods Eng.* 19 (1): 51–124.

59 Irvine, T. (1999). *Application of the Newton-Raphson Method to Vibration Problems.* Vibrationdata Publications.

60 Swevers, J., Adams, M., and De Moor, B. (1987). A new direct time domain identification method for linear systems using singular value decomposition. In: *Proceedings of the 12th International Seminar on Modal Analysis, I2-3,* 1–20.

61 deMoor, B., Vandewalle, J., Vandenberghe, L., and Mieghem, P. (1988). A geometrical strategy for the identification of state space models of linear multivariable systems with singular value decomposition. In: *IFAC Proceedings 88,* 700–704.

62 Mohanty, N. (1986). *Random Signals, Estimation, and Identification.* New York: Van Nostrand.

63 Candy, J. (2016). *Bayesian Signal Processing: Classical, Modern and Particle Filtering Methods,* 2e. Hoboken, NJ: Wiley/IEEE Press.

Problems

7.1 From the correlated Gauss–Markov model, derive the state covariance equations of Eq. (7.6).

7.2 Starting with the nonstationary measurement covariance $\Lambda_{yy}(t, s)$:
a) Derive the expression for $t \geq s$ as well as $t < s$.
b) Suppose the underlying process is "stationary," derive the corresponding measurement covariance $\Lambda_{yy}(t + \ell, s)$ for lag ℓ.

7.3 For the "correlated" Gauss–Markov model given by

$$x(t) = Ax(t-1) + w(t-1)$$
$$y(t) = Cx(t) + v(t)$$

a) Derive the expression for the state power spectrum $S_{xx}(z)$.
b) Derive the expression for the measurement power spectrum $S_{yy}(z)$.

7.4 Derive the Riccati equation starting with the KSP equation of Eq. (7.27) without solving for the $\mathcal{K}$-term directly and backsubstituting as before. (*Hint*: $(RR')^{-1}(RR') = I$.)

7.5 Suppose we are given a realization of the KSP model Σ_{KSP}, starting with the predictor–corrector Kalman filter of Table 4.1 develop the steady-state relations and show that

a) The error covariance equations satisfy a Riccati equation.
b) From these relations, develop the corresponding KSP equations for the Kalman filter. (*Hint*: Use the innovations model.)

7.6 Derive the Kalman filter expressions incorporating "correlated" noise sources ($E\{wv\}$). (*Hint*: Use prediction form.)

7.7 Suppose we are given the scalar covariance sequence:

$$\Lambda_{6,6} = \{6.5, 2.5, 0.833, 0.277, 0.0926, 0.0386\}$$

a) Perform a deterministic realization of this sequence. What is your identified model, Σ_{ABCD}?
b) Verify that a particular realization is $\Sigma_{ABCD} = \{1/3, 5/4, 2, 13/4\}$.
c) Develop the KSP equations and find $\tilde{\Pi}$.
d) Calculate the innovations covariance and gain: R_{ee} and K_p.
e) With Σ_{INV} now available, validate that this is a viable stochastic realization. (*Hint*: Calculate the output covariance using this model.)

7.8 In a simple radar system, a large-amplitude, narrow-width pulsed sinusoid is transmitted by an antenna. The pulse propagates through space at light speed ($c = 3 \times 10^8$ m s^{-1}) until it hits the object being tracked. The object reflects a portion of the energy. The radar antenna receives a portion of the reflected energy. By locating the time difference between the leading edge of the transmitted and received pulse Δt, the distance or range from the radar r can be determined; that is,

$$r = \frac{c\Delta t}{2}.$$

The pulse is transmitted periodically; that is,

$$r(t) = \frac{c\Delta(t)}{2}$$

Assume that the object is traveling at a constant velocity ρ with random disturbance $w \sim N(0,100 \text{ m s}^{-1})$. Therefore, the range of the object at $t + 1$ is

$$r(t + 1) = r(t) + T\rho(t), \quad T = 1 \text{ m s}^{-1}$$

where T is the sampling interval between received range values. The measured range is contaminated by noise

$$y(t) = r(t) + n(t) \quad \text{for} \quad n \sim N(0, 10)$$

a) Develop a model to simulate this radar system.
b) Design a range estimator using the MBP and assuming the initial target range is $r(0) \approx 20 \pm 2$ km.
c) Identify a stochastic realization from this synthesized data set using the classical realization approach (covariance matrices) to obtain Σ_{INV}.
d) Using the same synthesized data set apply a subspace approach to extract a stochastic realization Σ_{INV}.
e) How do the realizations compare?

7.9 We are given a deterministic impulse response sequence from an unknown (black-box) SISO-system simulation with

$$\mathcal{H}_t = \{0, 1, -1, 0, 1\}; \quad t = 1, \dots, 5$$

The physical system is measured experimentally with the output given by

$$\mathcal{Y}_t = \{0.141, 1.142, -0.933, -0.121, 1.072\}$$

with a known measurement uncertainty of $\sigma = 0.1$.

a) Perform a deterministic identification on the impulse response data, $\mathcal{H}_t$.
b) Perform a deterministic identification on the experimental data, $\mathcal{Y}_t$.
c) Calculate the impulse responses from each of the identified models. How do they compare? Are they close?
d) Identify stochastic realizations from the data sets using a subspace approach to obtain Σ_{INV}.
e) How do all of these realizations compare? (*Hint*: Calculate the impulse responses from each of the identified models.)

7.10 Develop the AIC for an autoregressive model of order N_a given by

$$y(t) + a_1 y(t-1) + \cdots + a_{N-a} y(t - N_a) = e(t)$$

where y is the output and e is the prediction error for a data set of K-samples.

7.11 Suppose we are given the following deterministic system and asked to investigate its performance:

$$x(t) = Ax(t-1) + Bu(t-1)$$
$$y(t) = Cx(t) + Du(t)$$

with

$$A = \begin{bmatrix} 0 & 1 \\ -0.7 & 1.5 \end{bmatrix}; \quad B = \begin{bmatrix} 0.25 \\ 0.625 \end{bmatrix}; \quad C = [0.5\ 0.5]; \quad D = [1\ 0]$$

a) Generate impulse response data $\mathcal{H}_t$ with $\Delta t = 0.1$ second and perform a deterministic identification.
b) Add zero-mean random Gaussian noise with $\sigma_v = 0.001$ and let $u(t)$ be a unit pseudo-random noise sequence to synthesize noisy experimental data, $\mathcal{Y}_t$ and perform a deterministic identification.
c) Calculate the impulse responses from each of the identified models. How do they compare to the set synthesized in (a)? Are they close?
d) Identify stochastic realizations from the data sets using a subspace approach to obtain Σ_{INV}.
e) How do all of these realizations compare? (*Hint*: Compare the impulse responses from each of the identified models.)

7.12 Consider the following example problem from [27] with

$$x(t) = Ax(t-1) + Bu(t-1)$$
$$y(t) = Cx(t) + Du(t)$$

with

$$A = \begin{bmatrix} 0.603 & 0.603 & 0 & 0 \\ -0.603 & 0.603 & 0 & 0 \\ 0 & 0 & -0.603 & -0.603 \\ 0 & 0 & 0.603 & -0.603 \end{bmatrix}; \quad B = \begin{bmatrix} 0.924 \\ 2.758 \\ 4.317 \\ -2.644 \end{bmatrix}$$

and

$$C = [-0.575\ 1.075 - 0.523\ 0.183]; \quad D = -0.714$$

a) Generate impulse response data $\mathcal{H}_t$ with $\Delta t = 1$ second and perform a deterministic identification.

b) Add zero-mean random Gaussian noise with $\sigma_v = 0.05$ to the synthetic response data to create noisy experimental data, $\mathcal{Y}_t$ and perform a deterministic identification with $u(t)$ a unit pseudo-random noise sequence.

c) Calculate the impulse responses from each of the identified models. How do they compare to the set synthesized in (a)? Are they close?

d) Identify stochastic realizations from the data sets using a subspace approach to obtain Σ_{INV}.

e) How do all of these realizations compare? (*Hint*: Compare the impulse responses from each of the identified models.)

7.13 Consider the design of an identifier and estimator for a series RLC circuit (second-order system) excited by a pulse train. Using Kirchhoff's voltage law, we can obtain the circuit equations with $i = C(de/dt)$:

$$\frac{d^2 e}{dt^2} + \frac{R}{L}\frac{de}{dt} + \frac{1}{LC}e = \frac{1}{LC}e_{in}$$

where e_{in} is a unit pulse train. The dynamic equations can be placed in state-space form by choosing $x := [e \mid de/dt]'$ and $u = e_{in}$:

$$\frac{dx}{dt} = \begin{bmatrix} 0 & 1 \\ -\frac{1}{LC} & -\frac{R}{L} \end{bmatrix} x + \begin{bmatrix} 0 \\ -\frac{1}{LC} \end{bmatrix} u + \begin{bmatrix} 0 \\ -\frac{1}{LC} \end{bmatrix} w$$

where $w \sim N(0, R_{ww})$ is used to model component inaccuracies. A high-impedance voltmeter is placed in the circuit to measure the capacitor voltage e. We assume that it is a digital (sampled) device contaminated with noise of variance R_{vv}; that is,

$$y(t) = e(t) + v(t)$$

where $v \sim N(0, R_{vv})$.

For this problem, we have the following parameters: $R = 5\ \text{k}\Omega$, $L = 2.5\ \text{H}$, $C = 0.1\ \mu\text{F}$, and $\Delta t = 0.1\ \text{ms}$. We assume that the component inaccuracies can be modeled using $R_{ww} = 0.01$, characterizing a deviation of ± 0.1 V uncertainty in the circuit representation. Finally, we assume that the precision of the voltmeter measurements are ($e \pm 0.2$ V), the two standard deviation value, so that $R_{vv} = 0.01\ (\text{V})^2$. Summarizing the circuit model, we have the continuous-time representation

$$\frac{dx}{dt} = \begin{bmatrix} 0 & 1 \\ -4 & -2 \end{bmatrix} x + \begin{bmatrix} 0 \\ -4 \end{bmatrix} u + \begin{bmatrix} 0 \\ -4 \end{bmatrix} w$$

and the discrete-time measurements

$$y(t) = [1 \mid 0]\, x(t) + v(t)$$

where

$$R_{ww} = \frac{(0.1)^2}{\Delta t} = 0.1 \ (\text{V})^2 \quad \text{and} \quad R_{vv} = 0.01 \ (\text{V})^2$$

We first convert the system into a sampled-data (discrete) representation yielding the following discrete-time Gauss–Markov model:

$$x(t) = \begin{bmatrix} 0.98 & 0.09 \\ -0.36 & 0.801 \end{bmatrix} x(t-1) + \begin{bmatrix} -0.019 \\ -0.36 \end{bmatrix} u(t-1) + \begin{bmatrix} -0.019 \\ -0.36 \end{bmatrix} w(t-1)$$

$$y(t) = [1 \mid 0]x(t) + v(t)$$

where

$$R_{ww} = 0.1(\text{V})^2 \quad \text{and} \quad R_{vv} = 0.01(\text{V})^2$$

a) Using the discrete model, simulate the RLC circuit for a unit pulse train and generate the measured output voltage, $y(t)$ (measurement model).

b) Using this synthesized data, identify a "black-box" state-space model from this data using a subspace method. Extract a stochastic realization Σ_{INV}.

c) Repeat this identification using a pseudo-random sequence as the deterministic input. Compare these identifications.

d) Design a Kalman filter from the "known" discrete state-space model and apply it to the data generated in (a).

e) Design a Kalman filter from the identified "black-box" state-space model and apply it to the data generated in (a).

f) Compare the Kalman filter designs. Do they perform equivalently? How does their design criteria compare? (zero-mean, whiteness).

7.14 Consider the design of a processor for a storage tank containing plutonium-nitrate [32]. We would like to design Kalman filter capable of predicting the amount (mass) of plutonium present in the tank at any instant in order to estimate the total mass at the end of each day. The level of solution in the tank at any time is attributed to the dynamics of evaporation of the water and nitric acid as well as radiolysis effects. The process model relating the solution *mass* $m(t)$ to the *plutonium concentration* $P(t)$, that is,

$$\frac{d}{dt}m(t) = -K_H - K_r m(t)P(t)$$

where m is the solution mass (kg), K_H is the evaporation rate of the water and acid solution, (kg d^{-1}), K_r is the radiolysis constant, (kg (water)/day per kg (Pu)), P is the plutonium concentration kg (Pu)/kg (solution).

Since the change in plutonium concentration is very slow due to the radiolysis effects, we assume that it is constant, $P(t) = P \quad \forall t$.

A typical measurement system for a tank solution is the pneumatic bubbler. The bubbler measures density, ρ, and the level h of liquid based on differential pressures.

$$p_1 = g\rho(t)(h(t) - H) + p_3 \quad \text{and} \quad p_2 = g\rho(t)h(t) + p_3$$

where $p_1, p_2, p_3, H, h(t)$ are the respective pressures and heights shown in the figure and $\rho(t)$ and g are the corresponding density and gravitational constant. It follows, therefore, that the *pressure differentials* are

$$\Delta P_A = p_2 - p_1 = g\rho(t)H \quad \text{and} \quad \Delta P_B = p_2 - p_3 = g\rho(t)h(t)$$

Thus, the bubbler measures (indirectly) the density and height from these pressure measurements. The gauges measure ΔP_A and ΔP_B to give

$$\rho(t) = \frac{\Delta P_A}{gH} \quad \text{and} \quad h(t) = \frac{\Delta P_B}{\rho(t)g}$$

We are interested in the solution mass at a given time, so the model is still not complete. This mass is related to the density and level by

$$m(t) = a\rho(t) + b\frac{\Delta P_B}{g} \quad \text{or} \quad \Delta P_B = \frac{g}{b}m(t) - \frac{a}{b}g\rho(t)$$

We now have our measurement model expressed in terms of the differential pressure readings ΔP_A and ΔP_B.

We define the state vector as $x' = [m \ \rho]$. Since we do not have a meaningful model to use for the relationship of density to the solution mass, we model it simply as a random walk

$$\frac{d}{dt}\rho(t) = w(t)$$

where w is white with variance $R_{w_c w_c}$.

Summarizing the process and measurement models in state-space form, we have

$$\frac{d}{dt}\begin{bmatrix} m(t) \\ \rho(t) \end{bmatrix} = \begin{bmatrix} -K_r P & 0 \\ 0 & 0 \end{bmatrix}\begin{bmatrix} m(t) \\ \rho(t) \end{bmatrix} + \begin{bmatrix} 1 \\ 0 \end{bmatrix}u(t) + \begin{bmatrix} w_1(t) \\ w_2(t) \end{bmatrix}$$

where u is a step function of amplitude $-K_H$. The corresponding measurement model is

$$\begin{bmatrix} \Delta P_A \\ \Delta P_B \end{bmatrix} = \begin{bmatrix} g/b & -(a/b)g \\ 0 & gH \end{bmatrix}\begin{bmatrix} m(t) \\ \rho(t) \end{bmatrix} + \begin{bmatrix} v_1(t) \\ v_2(t) \end{bmatrix}$$

Finally, we must develop models for the process and measurement uncertainties. The process uncertainty can be adjusted using the simulation, but as a starting value, we select $R_{w_c w_c} = I$, based on judgments of the density and accuracy of the model parameters. The process uncertainty can be adjusted using the simulation, but as a starting value we select $R_{w_c w_c} = I$, based on judgments of the density and accuracy of the model parameters. The measurements can be estimated from manufacturer's specifications of the measurement instrumentation given by $K_r = -8.1 \times 10^{-5}$ kg Da^{-1}, $K_H = 0.5$ kg Da^{-1}, $H = 2.5$ m, $a = 0.021$ m^3, $b = 0.328$ m^2, $g = 9.8$ m s^{-2}, $P = 0.173$, $m_p = 1448$ kg m^{-3}, $m_m = 983$ kg. The uncertainty in the measurements is approximated from typical specification. Using these model parameter values, we estimate the nominal ΔP_A and ΔP_B to be 3.6×10^4 and 3.0×10^4 N m^{-2}. From the set of model parameters, the process and measurement models are given by

$$\frac{d}{dt}x(t) = \begin{bmatrix} -1.43 \times 10^{-5} & 0 \\ 0 & 0 \end{bmatrix} x(t) + \begin{bmatrix} 1 \\ 0 \end{bmatrix} u(t) + w(t)$$

and the measurement model as

$$y(t) = \begin{bmatrix} 29.8 & -0.623 \\ 0 & 24.9 \end{bmatrix} x(t) + v(t)$$

For our gauges, we have that the measurement covariance matrix is

$$R_{vv} = \text{diag} [5.06 \times 10^4 \ 1.4 \times 10^5] \quad (\text{N}^2 \text{ m}^{-4})$$

We implement a discrete system using the matrix exponential series approximation for the discrete-time matrices, resulting in the discrete model at a sampling interval of 0.1 *day* as

$$x(t) = \begin{bmatrix} 0.999 & 0 \\ 0 & 1 \end{bmatrix} x(t-1) + \begin{bmatrix} 1 \\ 0 \end{bmatrix} u(t-1) + w(t-1)$$

$$y(t) = \begin{bmatrix} 29.8 & -0.623 \\ 0 & 24.9 \end{bmatrix} x(t) + v(t)$$

$R_{ww} = \text{diag} [10 \ 10]$,

$R_{vv} = \text{diag} [5.06 \times 10^4 \ 1.4 \times 10^5]$ with initial conditions

$\hat{x}(0|0) = [988 \ 1455 \ ]'$ and $\tilde{P}(0|0) = \text{diag} [0.01 \ 0.01]$.

a) Using the discrete model, simulate the storage tank for a 25-day period and generate the measured outputs.

b) Using this synthesized data, identify a "black-box" state-space model from this data by a subspace method. Extract a stochastic realization Σ_{INV}.

c) Repeat this identification using a pseudo-random sequence as the deterministic input. Compare these identifications.

d) Design a Kalman filter from the "known" discrete state-space model and apply it to the data generated in (a).

e) Design a Kalman filter from the identified "black-box" state-space model and apply it to the data generated in (a).

f) Compare the Kalman filter designs. Do they perform equivalently? How do their design criteria compare (zero-mean, whiteness)?

8

Subspace Processors for Physics-Based Application

8.1 Subspace Identification of a Structural Device

Mechanical systems operating in noisy environments lead to low signal-to-noise ratios (SNRs) creating a challenging signal processing problem to monitor a structural device in real time. To detect/classify a particular type of device from noisy vibration data, it is necessary to identify signatures that make it unique. Resonant (modal) frequencies emitted offer a signature characterizing its operation. The monitoring of structural modes to determine the condition of a device under investigation is essential, especially if it is a critical entity of an operational system. The development of a model-based scheme capable of the on-line tracking of structural modal frequencies by applying both system identification methods to extract a modal model and state estimation methods to track their evolution is discussed. An application of this approach to an unknown structural device is discussed, illustrating the approach and evaluating its performance.

One approach is to recognize that unique modal frequencies (e.g. sinusoidal lines) appear in the estimated power spectrum that are solely characteristic of the device under investigation [1]. Therefore, this study is based on constructing a "black-box" model of the device that captures these physical features that can be exploited to "diagnose" whether the particular device subsystem is operating normally from noisy vibrational data. Measures of anomalies can deteriorate significantly if noise is present – a common situation in an operational environment. Multiple sensors (such as accelerometers for vibrations, microphones for acoustics, strain gauges for stress, and thermocouples for temperature) in a structure provide additional information about the system for condition and performance [2]. This implies that the application of a multi-channel (multi-input, multi-output) system representation, which is most easily handled in state-space form without restrictions to single-channel spectral representations, is required.

Model-Based Processing: An Applied Subspace Identification Approach, First Edition. James V. Candy.
© 2019 John Wiley & Sons, Inc. Published 2019 by John Wiley & Sons, Inc.

Here, the vibrational signature of the structurally unknown device is measured (directly or remotely) and provided as a noisy input to a modal subspace identifier that is used to track modal frequencies that can be used to decide whether it is vibrating normally. If so, it provides an input to the device classifier to decide on the particular class or subsystem anomaly. In this application, we limit our discussion to the modal frequency identification and tracking problem (see [3] for detection details).

8.1.1 State-Space Vibrational Systems

Most structural vibrational systems are multiple input/multiple output (MIMO) systems that are easily captured within a state-space framework. For instance, a linear, time-invariant mechanical system can be expressed as a second-order vector–matrix, differential equation given by

$$M\ddot{\mathbf{d}}(\tau) + C_d\dot{\mathbf{d}}(\tau) + K\mathbf{d}(\tau) = B_p\mathbf{p}(\tau) \tag{8.1}$$

where $\mathbf{d}$ is the $N_d \times 1$ displacement vector, $\mathbf{p}$ is the $N_p \times 1$ excitation force, and M, C_d, K, are the $N_d \times N_d$ lumped mass, damping, and spring constant matrices characterizing the vibrational process model, respectively.

Defining the $2N_d$-state vector in terms of the displacement and its derivative as $\mathbf{x}(\tau) := [\mathbf{d}(\tau) \mid \dot{\mathbf{d}}(\tau)]$, then the continuous-time state-space representation of this process can be expressed as

$$\dot{\mathbf{x}}(\tau) = \underbrace{\begin{bmatrix} 0 & | & I \\ --- & | & --- \\ -M^{-1}K & | & -M^{-1}C_d \end{bmatrix}}_{A_c} \mathbf{x}(\tau) + \underbrace{\begin{bmatrix} 0 \\ M^{-1}B_p \end{bmatrix}}_{B_c} \mathbf{p}(t)$$

The corresponding measurement or output vector relation can be characterized by

$$\mathbf{y}(\tau) = C_c\mathbf{x}(\tau) + D_c\mathbf{u}(\tau) \tag{8.2}$$

where the output or measurement vector is $\mathbf{y} \in \mathcal{R}^{N_y \times 1}$ completing the MIMO vibrational model.

One of the most expository representations of a mechanical system is its modal representation, where the modes and mode shapes expose the internal structure and its response to various excitations [4–8]. This characterization can easily be found from state-space systems by transforming the coordinates to modal state-space, which is accomplished through an eigen-decomposition in the form of a *similarity transformation*. Here, the system matrices $\Sigma := \{A, B, C, D\}$ are transformed to modal coordinates by the transformation matrix T_M constructed of the eigenvectors of the underlying system that yields an "equivalent" system from an input/output perspective; that is, the transfer functions and impulse responses are identical [6–8].

For complex modal case, which is quite common in structural dynamics, the eigenvalues are complex, but still distinct [4–8]. The system matrix can be decomposed using an eigen-decomposition, which now yields complex eigen-pairs along with the corresponding complex eigenvectors to specify the modal similarity transformation matrix T_M. Applying this transformation to the system matrix A_c that leads to the modal state-transition matrix for the complex eigen-system as

$$\Phi_M(\tau, \tau_0) = e^{\Lambda(\tau - \tau_0)} = \exp\left\{ \underbrace{\begin{bmatrix} \Lambda_1 & & 0 \\ & \ddots & \\ 0 & & \Lambda_{N_d} \end{bmatrix}}_{\Lambda} (\tau - \tau_0) \right\}$$

$$\text{for} \quad \Lambda_i = \begin{bmatrix} \sigma_i & \omega_i \\ -\omega_i & \sigma_i \end{bmatrix} \tag{8.3}$$

Thus, the continuous-time, complex modal state-space system is given by

$$\dot{\mathbf{x}}(\tau) = A_M \mathbf{x}(\tau) + B_M \mathbf{u}(\tau)$$
$$\mathbf{y}(\tau) = C_M \mathbf{x}(\tau) + D_M \mathbf{u}(\tau) \tag{8.4}$$

where $A_M \in \mathcal{R}^{N_x \times N_x}$, $B_M \in \mathcal{R}^{N_x \times N_u}$, $C_M \in \mathcal{R}^{N_y \times N_x}$, $D_M \in \mathcal{R}^{N_y \times N_u}$ and $N_x = 2 \times N_d$.

Sampling this system leads to a discrete-time state-space representation that can be transformed back to the continuous-time domain for our application.

The generic linear, time-invariant *state-space* model is defined by its system matrix A, input transmission matrix B, output or measurement matrix C, and direct input feed-through matrix D for *discrete-time* systems as

$$x(t + 1) = Ax(t) + Bu(t) \qquad \text{(State)}$$
$$y(t) = Cx(t) + Du(t) \qquad \text{(Output)} \tag{8.5}$$

for the state $x \in R^{N_x \times 1}$, input $u \in R^{N_u \times 1}$, and output $y \in R^{N_y \times 1}$.

Corresponding to this representation is the *impulse response* termed the Markov parameters [4–8].

$$H_t = CA^{t-1}B + D\delta(t) \tag{8.6}$$

for δ the Kronecker delta function.

As in Chapter 3, expanding the state equations, we have

$$x(t) = A^t x(0) + \sum_{k=0}^{t-1} A^{t-k-1} Bu(k-1); \quad t = 0, 1, \ldots, K \tag{8.7}$$

with the output is given by

$$y(t) = CA^t x(0) + \sum_{k=0}^{t-1} CA^{t-k-1} Bu(k-1) + D\delta(t) \tag{8.8}$$

Expanding this relation further and collecting terms, we obtain

$$\begin{bmatrix} y(0) \\ y(1) \\ \vdots \\ y(K-1) \end{bmatrix} = \underbrace{\begin{bmatrix} C \\ CA \\ \vdots \\ CA^{t-K-1} \end{bmatrix}}_{\mathcal{O}} x(0) + \underbrace{\begin{bmatrix} D & & \cdots & 0 \\ CB & D & \cdots & 0 \\ \vdots & \vdots & \ddots & \vdots \\ CA^{K-1}B & \cdots & CB & D \end{bmatrix}}_{\mathcal{T}} \begin{bmatrix} u(0) \\ u(1) \\ \vdots \\ u(K-1) \end{bmatrix}$$

$$\tag{8.9}$$

where $\mathcal{O}$ is the *observability matrix* of linear systems theory of Section 3.4 [6–8]. Shifting these relations in time leads to the vector input/output relation:

$$y_K(\ell) = \mathcal{O}_K x(\ell) + \mathcal{T}_K u_K(\ell) \tag{8.10}$$

Combining these m-vectors of Eq. (8.10) to create a batch data (*block Hankel*) matrix over the K-samples, we obtain input/state/output block matrices as

$$\mathcal{Y}_{\ell,m;K} := [\mathbf{y}_m(\ell) \cdots \mathbf{y}_m(\ell + K - 1)] \tag{8.11}$$

$$\mathcal{U}_{\ell,m;K} := [\mathbf{u}_m(\ell) \cdots \mathbf{u}_m(\ell + K - 1)] \tag{8.12}$$

$$\mathcal{X}_{\ell,m;K} := [\mathbf{x}_m(\ell) \cdots \mathbf{x}_m(\ell + K - 1)] = [\mathbf{x}(\ell) \, A\mathbf{x}(\ell) \cdots A^{K-1}\mathbf{x}(\ell)] \tag{8.13}$$

and, therefore, we have the *data equation* that relates the system model to the data (input and output matrices):

$$\mathcal{Y}_{\ell,m;K} = \mathcal{O}_K \mathcal{X}_{\ell,m;K} + \mathcal{T}_K \mathcal{U}_{\ell,m;K} \tag{8.14}$$

This expression represents the fundamental relationship for the input–state–output of a linear, time-invariant (LTI) state-space system.

8.1.1.1 State-Space Realization

In this subsection, we briefly review the deterministic state-space realization approach based on the system theoretic properties of the underlying Hankel array discussed in Section 6.2. The core of this suite of techniques is founded on the Hankel matrix defined in terms of the impulse response of a LTI system given by

$$H_t = \begin{cases} C A^{t-1} B & \text{for} \quad t > 0 \\ D & \text{for} \quad t = 0 \end{cases} \tag{8.15}$$

leading to the $(N_x \times 2N_x - 1)$-*Hankel* matrix.

Recall that $\mathcal{H}$ can be factored in terms of the observability $(\mathcal{O})$ and controllability $(\mathcal{C})$ matrices, that is, [9]

$$\mathcal{H}_{N_x,N_x} = \mathcal{O}_{N_x} \times \mathcal{C}_{N_x} = \underbrace{\begin{bmatrix} C \\ CA \\ \vdots \\ CA^{N_x-1} \end{bmatrix}}_{\mathcal{O}} \times \underbrace{[B \quad AB \quad \cdots \quad A^{N_x-1}B]}_{\mathcal{C}} \qquad (8.16)$$

Some of the most important system theoretic properties of Hankel matrices are discussed in Section 6.2. The most significant relative to this application are as follows:

- For a minimal realization, Σ the rank of the Hankel matrix satisfies:

  $\rho[\mathcal{H}_{m,K}] \leq \min\{\rho[\mathcal{O}_m], \rho[\mathcal{C}_K]\} \leq N_x$ \quad (Rank Property)

- $\rho[\mathcal{H}_{m,K}] = \rho[\mathcal{H}_{m+1,K}] = \rho[\mathcal{H}_{m,K+1}] = N_x$ \quad (Shift Property)
- Such a minimal realization is *stable* if: the eigenvalues of the system matrix A lie within the unit circle: $\lambda[A] \leq 1$ \quad (Stability)
- The corresponding observability and controllability Gramians defined below satisfy the positivity constraints:

  $\Sigma_{AC} = A'\Sigma_{AC}A + CC' \geq 0$ \quad (Observability Gramian)

  $\Sigma_{AB} = A\Sigma_{AB}A' + BB' \geq 0$ \quad (Controllability Gramian)

- A *balanced realization* for a stable system A matrix is defined for $\Sigma_{AB} = \Sigma_{AC} = \Sigma_K$ where Σ is a diagonal matrix such that

  $$\Sigma_K = \text{diag}[\sigma_1, \sigma_2, \ldots, \sigma_K] \quad \text{for} \quad \sigma_1 \geq \sigma_2 \geq \cdots \sigma_K$$

and Σ_K is given by the singular value decomposition (SVD) of the Hankel matrix:

$$\mathcal{H}_{K,K} = \mathcal{U}_K \Sigma_K \mathcal{V}_K$$

Recall that from the shift-invariant property of the Hankel matrix that we have

$$\mathcal{H}_{K,K}^{\uparrow} := \mathcal{O}_K^{\uparrow} \mathcal{C}_K = (\mathcal{O}_K A) \times \mathcal{C}_K \quad \text{and} \quad \mathcal{H}_{K,K}^{\leftarrow} := \mathcal{O}_K \mathcal{C}_K^{\leftarrow} := \mathcal{O}_K \times (A\mathcal{C}_K)$$

$$(8.17)$$

since

$$\mathcal{O}_K^{\uparrow} = \mathcal{O}_K \times A = \underbrace{\begin{bmatrix} CA \\ CA^2 \\ \vdots \\ CA^K \end{bmatrix}}_{\mathcal{O}_K^{\uparrow}} \quad \text{and}$$

$$\mathcal{C}_K^{\leftarrow} = A \times \mathcal{C}_K = \underbrace{[AB \quad A^2B \quad \cdots \quad A^KB]}_{\mathcal{C}_K^{\leftarrow}}$$

Thus, the *realization problem* is

GIVEN a set of MIMO transfer function matrices $\{H_{ij}(z)\}$ or impulse response matrices $\{H_k(t)\}$ with corresponding Markov parameters, $H_k = CA^t B + D\delta(t);\ t = 0, 1, \ldots, K - 1$, FIND the underlying state-space system, $\Sigma := \{A, B, C, D\}$.

8.1.2 Deterministic State-Space Realizations

A solution to the *deterministic realization problem* is obtained by performing the SVD of the Hankel matrix, $\mathcal{H}_{m,K}$, where $m \geq N_x$ and $K \geq N_x$ with N_x the dimension of the minimum realization (system) or equivalently the underlying *true* number of states; therefore,

$$\mathcal{H}_{m,K} = [\mathcal{U}_{N_x} \mid \mathcal{U}_N]\begin{bmatrix} \Sigma_{N_x} & \mid & 0 \\ - & - - & - \\ 0 & \mid & 0 \end{bmatrix}\begin{bmatrix} \mathcal{V}'_{N_x} \\ -- \\ \mathcal{V}'_N \end{bmatrix} = \mathcal{U}_{N_x}\Sigma_{N_x}\mathcal{V}'_{N_x} \tag{8.18}$$

for $\Sigma_{N_x} = \text{diag}[\sigma_1,\ \sigma_2,\ \ldots,\ \sigma_{N_x}]$

From the factorization of the Hankel matrix and its SVD, we have that

$$\mathcal{H}_{m,K} = \mathcal{O}_m \times \mathcal{C}_K = \underbrace{\left(\mathcal{U}_{N_x}\Sigma_{N_x}^{1/2}\right)}_{\mathcal{O}_{N_x}}\underbrace{\left((\Sigma_{N_x}^{1/2})'\mathcal{V}'_{N_x}\right)}_{\mathcal{C}_{N_x}} \tag{8.19}$$

where Σ_{N_x} is the matrix square root obtained by a Cholesky decomposition [9].

From shifting properties (above) of the observability (or controllability) matrix, we can obtain the system matrices as [10–15]

$$A = \mathcal{O}_{N_x}^{\#} \times \mathcal{O}_{N_x}^{\uparrow};\quad B = \mathcal{C}(1 : N_x, 1 : N_u);\quad \text{and}\quad C = \mathcal{O}(1 : N_y, 1 : N_x) \tag{8.20}$$

where N_x is the number of states, N_u is the dimension of the input vector $\mathbf{u}$, and N_y that of the output or measurement vector $\mathbf{y}$ with the *notation* $1 : N \rightarrow 1, 2, \ldots, N$ or equivalently *select* the corresponding rows (columns) 1-to-N of the matrix.

This completes the fundamental background information required to develop the subsequent techniques developed.

8.1.2.1 Subspace Approach

The fundamental subspace identification approach is to extract the state-space representation from input/output data extending the realization from impulse response data – still assumed *deterministic*. Input/output data can be handled in a manner similar to the impulse response data discussed previously. In this case, we must return to the data matrices.

We are given input/output data corresponding to a LTI system with vector inputs $\mathbf{u} \in \mathcal{R}^{N_u \times 1}$ and vector outputs $\mathbf{y} \in \mathcal{R}^{N_y \times 1}$ along with discrete-time samples, $t = 0, 1, \ldots, K$.

Suppose we have k-data samples such that $k > N_x$, then the corresponding block Hankel matrices can be created directly from Eq. (8.10) with the shift k to give both *vector* input/output (state) relations.

$$\mathbf{y}_k(t) = \mathcal{O}_k \mathbf{x}(t) + \mathcal{T}_k \mathbf{u}(t) \tag{8.21}$$

and the corresponding *matrix* input/output (state) equation as

$$\mathcal{Y}_{k|2k-1} = \mathcal{O}_k \mathcal{X}_k + \mathcal{T}_k \mathcal{U}_{k|2k-1} \tag{8.22}$$

where the matrices are defined in Eq. (8.10).

The *initial states* are given by

$$\mathcal{Y}_{0|k-1} = \mathcal{O}_k \mathcal{X}_0 + \mathcal{T}_k \mathcal{U}_{0|k-1} \tag{8.23}$$

Here $\mathcal{U}_{0|k-1}$, $\mathcal{Y}_{0|k-1}$ are the *past* inputs and outputs, while $\mathcal{U}_{k|2k-1}$, $\mathcal{Y}_{k|2k-1}$ are the *future* inputs and outputs, which are all *block Hankel* matrices [10–15].

The *augmented*, input/output, *data matrix* D along with its corresponding LQ-decomposition is constructed as

$$\mathcal{D}_{0|k-1} = \begin{bmatrix} \mathcal{U}_{0|k-1} \\ ---- \\ \mathcal{Y}_{0|k-1} \end{bmatrix} = \begin{bmatrix} I & | & 0 \\ --- & \\ \mathcal{T}_k & | & \mathcal{O}_k \end{bmatrix} \begin{bmatrix} \mathcal{U}_{0|k-1} \\ --- \\ \mathcal{X}_0 \end{bmatrix} = \begin{bmatrix} L_{11} & | & 0 \\ --- & \\ L_{21} & | & L_{22} \end{bmatrix} \begin{bmatrix} Q'_1 \\ --- \\ Q'_2 \end{bmatrix} \tag{8.24}$$

These expressions enable the orthogonal decomposition of $\mathcal{Y}_{0|k-1}$; therefore, it follows from Eq. (8.24) that (see Section 6.4 for details)

$$\mathcal{Y}_{0|k-1} Q_2 = \mathcal{O}_k \times \mathcal{X}_0 \times Q_2 = L_{22} \tag{8.25}$$

Therefore, performing the SVD of L_{22}, that is,

$$L_{22} = [U_1 \mid U_2] \begin{bmatrix} \Sigma_1 & | & 0 \\ - & - & - \\ 0 & | & 0 \end{bmatrix} \begin{bmatrix} V'_1 \\ -- \\ V'_2 \end{bmatrix} \tag{8.26}$$

yields

$$\mathcal{O}_k \mathcal{X}_0 Q_2 = U_1 \times \Sigma_1 \times V'_1 = \underbrace{(U_1 \Sigma_1^{1/2})}_{\mathcal{O}_k} \times \underbrace{(\Sigma_1^{1/2} V'_1)}_{C_k} \tag{8.27}$$

and as before the system matrices are A, B, C, D can be extracted by

$$A = \mathcal{O}_{N_x}^{\#} \times \mathcal{O}_{N_x}^{\uparrow}; \quad C = \mathcal{O}(1 : N_y, 1 : N_x) \tag{8.28}$$

The input transmission an direct feed-through matrices B and D can be obtained by solving a set of overdetermined linear equations (least-squares) from the following relations:

$$\mathcal{L} \times \begin{bmatrix} \hat{D} \\ -- \\ \hat{B} \end{bmatrix} = U_2' L_{21} L_{11}^{-1} \tag{8.29}$$

where $\mathcal{L}$ is a block coefficient matrix (see Section 6.4 for the details).

Therefore, we have the *Multivariable Output Error State-SPace* (MOESP) algorithm given by as follows:

- Compute the LQ-decomposition of $\mathcal{D}$ of Eq. (8.24).
- Perform the SVD of L_{22} in Eq. (8.26) to extract $\mathcal{O}_k$.
- Obtain A and C from Eq. (8.28).
- Solve the least-squares problem to obtain B and D from Eq. (8.29).

8.1.3 Vibrational System Processing

The overall approach to modal-frequency estimation/tracking is based on the development of robust subspace identification techniques that can be applied to solve this problem as rapidly as possible for potential real-time applications. The main idea is to preprocess a section or window of digitized data and perform a system (vibrational) identification followed by an extraction of the underlying modes from the identified model producing raw estimates of the corresponding modal frequencies. Once the "raw" modal frequencies in each window are extracted, a sequential tracking algorithm (Kalman filter) is applied to "smooth" the estimates that are eventually to be input to an anomaly detection algorithm for monitoring performance.

Typical processing of the system under investigation entails the following steps to:

- *Preprocess* (outliers, whitening, bandpass filtering, normalization) data relative to the targeted vibrational system information.
- *Identify* (subspace) underlying vibrational model from the data sets in state-space form $\Sigma = \{A, B, C, D\}$.
- *Transform* identified model to the modal state-space representation $\Sigma_M = \{A_M, B_M, C_M, D_M\}$.
- *Extract* modal frequencies and mode shapes from the modal state-space model.
- *Track* (Kalman filter) the raw frequencies sequentially in real time.
- *Postprocess* (outlier/tracker) the on-line tracking data.
- *Estimate* the modal frequency statistics to evaluate performance.

The modal frequency estimator/tracker is based on the following Gauss–Markov representation that evolves directly from a finite difference representation of the *instantaneous frequency* changes:

$$\dot{f}(t) \approx \frac{f(t_{k+1}) - f(t_k)}{\Delta t_k}$$

rewriting this expression gives

$$f(t_{k+1}) = f(t_k) + \Delta t_k \dot{f}(t_k) \tag{8.30}$$

Assuming that the frequency change is constant over the sampling interval ($\dot{f}(t_{k+1}) \approx \dot{f}(t_k)$) and the model uncertainty is characterized by Gaussian process noise leads to the following set of discrete-time, Gauss–Markov stochastic equations:

$$f(t_{k+1}) = f(t_k) + \Delta t_k \dot{f}(t_k) + w_1(t_k) \qquad \text{(Frequency)}$$
$$\dot{f}(t_{k+1}) = \dot{f}(t_k) + w_2(t_k) \qquad \text{(Rate)} \tag{8.31}$$

where w is zero-mean, Gaussian with $\mathbf{w} \sim \mathcal{N}(0, R_{ww})$. The corresponding measurement is also contaminated with instrumentation noise represented by zero-mean, Gaussian uncertainties as

$$y(t_k) = f(t_k) + v(t_k) \qquad \text{(Measurement)}$$

such that $v \sim \mathcal{N}(0, R_{vv})$. A combination of both process and measurement systems can be placed in a discrete-time ($t_k \rightarrow t$), state-space framework by defining the state vector $x(t) := [f(t) \mid \dot{f}(t)]'$.

With this frequency model established, we know that the optimal solution to the state estimation or frequency tracking problem is provided by the Kalman filter developed in Section 4.1. We restrict the processor to reach steady state; that is, the Kalman gain is a constant that can be obtained directly from the discrete Riccati equation solution to give the frequency tracker relations [9, 16].

$$\hat{f}(t+1) = \hat{f}(t) + K_1 e(t+1)$$
$$\hat{\dot{f}}(t+1) = \hat{\dot{f}}(t) + K_2 e(t+1)$$
$$\hat{y}(t+1) = \hat{f}(t+1) \tag{8.32}$$

where K is now a *precalculated* constant. It was found after a large number of runs that this approach leads to a very robust tracking solution that is desirable for on-line operations.

After all of the operations have been performed, the subspace approach has essentially evolved from a set of noisy MIMO data, to identifying a state-space model at *each* data window and extracting its modal frequencies and shapes. Each of the modal models is available for archiving (if desired) and the resulting set of modal frequencies is available for postprocessing. The data that consists of the identified set of modal frequencies and shapes for each of the N_f-frequencies are available for postprocessing consisting of *outlier* detection/correction and further *smoothing* using the frequency tracker (as above).

This completes the discussion of the subspace approach, and next we apply it to our "unknown" device and evaluate its performance.

8.1.4   Application: Vibrating Structural Device

In this section, we discuss the application of the subspace approach to a structurally "unknown" device, that is, a complex, stationary structure (black-box) with no rotating parts that is subjected to random excitation with accelerometer sensors placed on its surface and around its periphery. We do have some prior information about its modal response from historical tables and use this information as targeted modes (frequencies and shapes) to evaluate the validity and performance of these results as well as guiding any preprocessing of the acquired data.

The device under test was subjected to random excitations by placing a stinger or motor-driven rod perpendicular to the base of the structure. A suite of 19 triaxial accelerometers were positioned strategically about the device surface as well as a single sensor allocated to measure the excitation time series. In total, an array of 57-accelerometer channels acquired a set of 10-minute-duration data at a 6.4 kHz sampling frequency. The data were subsequently downsampled to 0.9 kHz in order to focus on the targeted modal frequencies (<400 Hz). From the state-space perspective, we have a targeted system of up to a maximum of 14-modes or 28-states with an array of 57 channels of time series measurements and 1-channel of an excitation measurement.

The raw data (downsampled) represents the expected data windows acquired from the real-time acquisition system. The windowed time series were preprocessed by performing outlier detection/correction, whitening filter (optional), bandpass filtering, normalization prior to performing the system identification. Once preprocessed, the input–output data were provided to the subspace algorithm that enabled the identification of a discrete-time state-space model, $\Sigma = \{A, B, C, D\}$, which was then transformed to the modal state-space, $\Sigma_M = \{A_M, B_M, C_M, D_M\}$, providing both modal frequencies and mode shape information, that is, modal eigenvalues and eigenvectors. Outliers were detected/corrected from these raw modal frequency estimates and provided as input to the frequency tracker (steady-state Kalman filter [9]) enabling a "smoothed" sequential estimate in real time. After the data set is processed, modal frequencies and mode shapes extracted, the data are then postprocessed to improve the estimates even further by applying a "batch" tracker (Kalman filter) with the improved ensemble statistics, since the entire

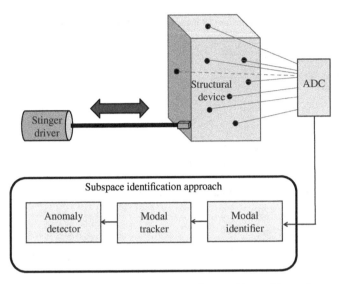

Figure 8.1 Structural device experimental setup: Motor-driven stinger random vibrations, MIMO ADC measurements, subspace identification: identifier, tracker, detector system.

track of data is now available. These results are then analyzed and provided for eventual anomaly detection (see Figure 8.1).

With all of this information available, we first performed a suite of subspace identifications by specifying the number of modes ranging from 9 to 16, and 20, 25, for our targeted number was 13/14 modal frequencies from the historical tables. For each number of these modes, we generated the corresponding frequency tracks one for each modal frequency as shown in Figure 8.2 for a 13-mode (26-state) identification. Here we observe the estimated power spectrum of the identified model for each data window in (a) along with the identified modal frequencies (+) in order to illustrate any clustering. In Figure 8.2b, we see the resulting *on-line* outlier corrected/tracks for each of the identified modal frequencies. Therefore, an ensemble of frequency estimates resulted for each track available for postprocessing statistical analysis. Comparisons of the ensemble averages of the identified modal frequencies are shown in Table 8.1. When the number of modes selected was less than or equal to 14, the tracks were reasonably stable, but only 10–11 of the target frequencies were essentially captured (within reasonable bounds) by the subspace identifier. Increasing the order greater than 14 enabled another of the targeted modes to be identified (11 → 12), but created an excessive number of "extraneous" modes with wild frequency tracks. It is clear from the table that orders less than 12 are not capable of reasonably estimating 10 or more modes and that those orders of 12 and above can capture at least 10 modes.

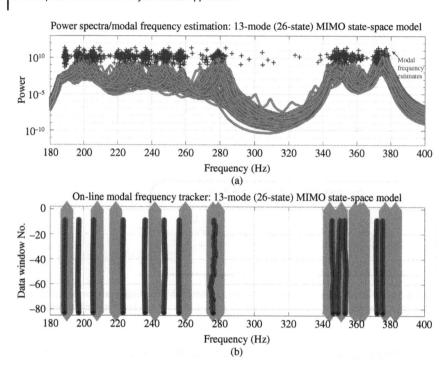

Figure 8.2 Subspace identification of 13-mode model: (a) Estimated transfer functions of identified modal models and raw modal frequency estimates (+). (b) Sequential (on-line) modal frequency tracker (thick line) results with outlier corrections (10-sample initialization).

These frequencies along with their accompanying statistics are used to determine which of the model orders selected enable a "reasonable" estimate of the targeted frequencies as shown in Table 8.1. Next we are able to select the model order based on the calculated 1σ standard deviations along with the corresponding percentage relative error[1] statistics based on the targeted modal frequencies. The average results, standard deviations and error percentages, are also shown in the table. Even though the 15 and 16 modal identifications provided somewhat superior statistical estimates of the modal frequencies and the fact that an additional targeted mode was identified, the erratic behavior of their extraneous modes caused concern for eventual on-line failure detection; therefore, they were not considered viable candidates. Based on these concerns, we selected the 13-mode (26-state) model for our on-line, subspace algorithm providing not only the most reasonable trade-off of deviation/error (± 1.8 Hz, 0.83%) but also a practical window time (3300-samples) of

---

1 Percentage relative error ($\%\epsilon$) is given by $\frac{\text{True}-\text{Estimate}}{\text{True}} \times 100$.

Table 8.1 Modal identification ensemble statistics.

Modal frequency estimates (relative error(ε))					
Frequency (Hz)	$12 \pm \sigma(\%\epsilon)$	$13 \pm \sigma(\%\epsilon)$	$14 \pm \sigma(\%\epsilon)$	$15 \pm \sigma(\%\epsilon)$	$16 \pm \sigma(\%\epsilon)$
190	189±0.2(0.5)	189±0.2(0.5)	189±0.1(0.5)	188±0.2(1.1)	188±0.1(1.1)
208	207±0.4(0.5)	206±0.5(1.0)	205±0.4(1.4)	204±0.5(1.9)	203±0.4(2.4)
219	229±2.3(4.6)	223±0.7(0.5)	221±1.2(0.9)	216±0.8(1.4)	213±0.5(2.7)
242	238±0.8(1.7)	236±0.8(2.1)	241±1.7(0.4)	247±1.0(2.1)	243±0.7(0.4)
260	260±2.1(0.0)	256±0.8(0.4)	258±3.5(0.8)	254±0.9(2.3)	259±1.1(0.4)
276	—	276±2.6(0.0)	—	268±3.1(2.9)	276±2.1(0.0)
279	281±6.3(0.7)	—	279±5.1(0.0)	284±3.8(1.8)	—
344	347±0.7(0.9)	344±10.8(0.0)	346±7.6(0.6)	348±1.7(1.2)	349±1.4(1.5)
351	353±0.6(0.6)	350±0.6(0.3)	352±0.6(0.6)	353±0.6(0.6)	354±0.8(0.3)
359	—	354±1.4(1.4)	359±1.9(0.0)	359±1.9(0.0)	359±1.1(0.0)
362	—	—	—	—	362±0.9(0.0)
364	371±0.3(0.0)	372±0.2(2.4)	372±0.4(2.2)	373±0.1(2.5)	373±0.1(2.5)
377	376±0.3(0.3)	375±0.3(0.5)	377±9.2(0.0)	376±0.3(0.3)	378±0.4(0.3)
383	—	—	—	—	—
Average $\sigma(\%\epsilon)$	±1.4 (0.98%)	±1.8 (0.83%)	±3.00 (0.67%)	±1.24 (1.55%)	±0.8 (0.97%)

3.6–7.1 s/identification, which is based on the required number of samples for the identification algorithm (ID time × sampling interval). This time is quite reasonable for a 900 Hz sampling frequency and vibrational monitoring of the system. Note that the higher the order, the number of samples for subspace identification increases.

Postprocessing of the tracking frequencies were shown in Figure 8.3 for the 13-mode comparison and Figure 8.4 for the 12 – –14-mode tracks. It is clear from the figures that by postprocessing (off-line), the modal frequency tracks are improved from those provided on-line primarily because of the availability of the "batch" statistics. That is, the *batch* postprocessor performing both outlier correction and tracker smoothing is superior to the on-line tracks primarily because the ensemble statistics of 83 data windows are now available to improve the estimates.

The ensemble statistics for the 13-mode identification are shown in Figure 8.3, where we see the scatter plot along with the corresponding 99.9% confidence interval about the mean modal frequency. Clearly, the estimates are reasonably precise as shown in Figure 8.3a and in Table 8.1 as well. In

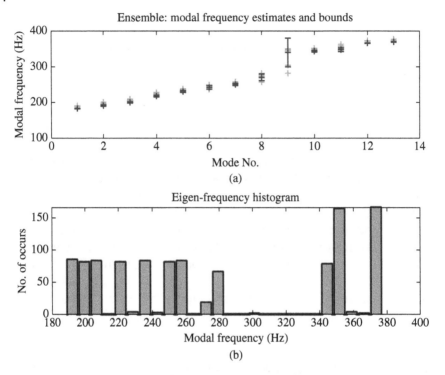

Figure 8.3 Postprocessing modal frequency tracking ensemble statistics: (a) Modal frequency scatter plot with 99.9% confidence limits about the mean. (b) Modal frequency histogram indicating high probabilities in identified modal frequency bins.

Figure 8.3b, we observe the corresponding modal frequency histogram with most of the identified frequency bins heavily populated, indicating very high probabilities of the modal frequencies estimated by the subspace tracker.

8.1.5 Summary

This study has illustrated the development of a model-based modal tracking scheme capable of on-line processing of structural vibrational responses. By applying both system identification methods to extract a modal model and state estimation techniques to track the modal frequencies, a sequential anomaly detector was also developed subsequently and applied successfully to noisy anomalous data (see Ref. [3] for details).

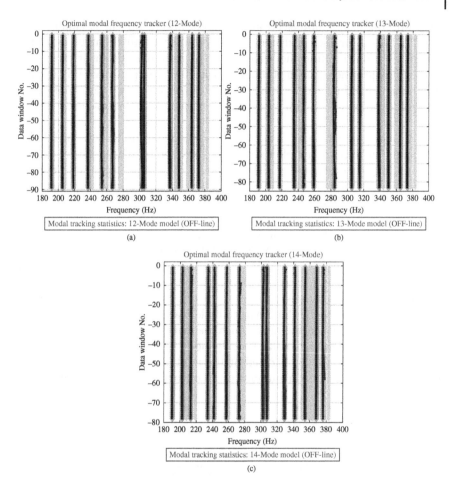

Figure 8.4 Postprocessed modal frequency tracking of (a) 12-mode, (b) 13-mode, (c) 14-mode identified state-space model results.

8.2 MBID for Scintillator System Characterization

Scintillator/photomultiplier systems are prevalent in many instruments to measure a variety of nuclear events. When γ energy from a nuclear event irradiates the scintillator, event radiation interacts with the scintillator material generating photons, which are detected by the photomultiplier tube (PMT)

[17, 18]. The PMT photoelectrons or photocathode current is amplified in the PMT by a number of dynode stages producing a total charge out for a given radiation flux into the scintillator. The resulting current is converted into a voltage when passed through a typical load impedance. This voltage is attenuated and provided as input to a digitizer through a variety of signal conditioners (filters, amplifiers, receivers) before analog-to-digital (A/D) conversion.

In this section, we concentrate on the development of analytic models for the scintillator/photomultiplier system and develop both a (i) statistical simulator and (ii) a Kalman filter (KF) to extract the desired signal information while rejecting the noise and uncertainty. We first develop the model analytically, transform it to the model-based framework (state-space [8, 9]), and then use experimental data to perform both a subspace identification and a MBID to fit a model and apply it to our processing problem.

A diagram of a scintillation/photomultiplier system is shown in Figure 8.5. As an incident α particle interacts with the scintillator crystal, typically sodium iodide (NaI), the kinetic energy of the particle is converted into detectable light through the "prompt fluorescence" property of the particular crystal. The light (photons) then strike the thin photocathode material of the photomultiplier, causing it to emit photoelectrons, which are focused (electrostatically) onto a series of electron multiplier stages or dynodes, which amplify the converted energy that is eventually collected at the multiplier anode. These dynodes are electrodes that emit a number of secondary electrons in response to

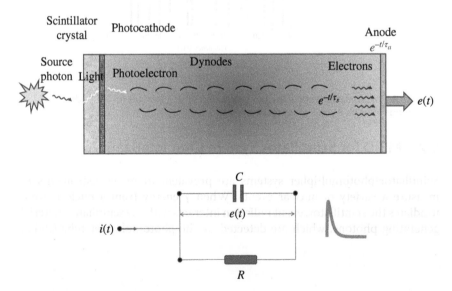

Figure 8.5 Scintillator/photomultiplier system with equivalent circuit model.

the absorption of a single photoelectron, thereby amplifying the original photoelectron from stage to stage with a typical gain factor of 10^7 [18]. These electrons are then collected at the anode of the PMT producing the output voltage pulse. It is this pulse that is of most interest, since it is proportional to the original incident photon energy emitted from the unknown source.

The shape of the voltage pulse at the PMT output is governed by the prompt fluorescence intensity of the scintillator crystal at a time t following the initial excitation with an exponential decay time τ_s such that the intensity is given (simply) by

$$i(t) = i_0 \, e^{-t/\tau_s} \tag{8.33}$$

which is usually a few nanoseconds. A full description of the output pulse must also take the rise time at the anode into account, which is usually around 3–4 times larger (faster) than its fall τ_a. Then, the overall voltage output pulse of the scintillator system is given by [17]

$$i(t) = i_0(e^{-t/\tau_a} - e^{-t/\tau_s}) \tag{8.34}$$

Using the parameters for a plastic scintillator from [17] ($\tau_s = 1.7$ ns, $\tau_a = 0.2$ ns, $i_0 = 1$), the resulting pulse is shown in Figure 8.6. Next, we develop the model of Eq. (8.34) from an equivalent electrical circuit representation.

8.2.1 Scintillation Pulse Shape Model

Following Knoll [17], the shape of the voltage pulse at the anode of the PMT following an event depends on the time constant of the anode circuit and the decay time of the scintillator. For this application, we would like a very fast output pulse, and, therefore, the design calls for the anode circuit time constant to be much smaller than that of the scintillator decay time. The equivalent anode circuit (ideal) can be realized as a parallel resistor/capacitor connection with the input current modeling the fluorescence as a simple exponential as in Eq. (8.33). Here the lumped capacitance (C) represents that of the anode and cable connections as well as the circuit input capacitance, while the resistance (R) is a physical resistor or the input impedance of the circuit load. The input electron current arriving at the PMT anode is given by Eq. (8.33) with the initial current as a function of the total charge collected and given by

$$i(t) = \frac{Q}{\tau_s} \tag{8.35}$$

Thus, writing the node equation using Kirchhoff's current law, we have that total input current is the sum of the currents through the resistance and capacitance as

$$i(t) = i_R(t) + i_C(t) = C\frac{de(t)}{dt} + \frac{e(t)}{R} \tag{8.36}$$

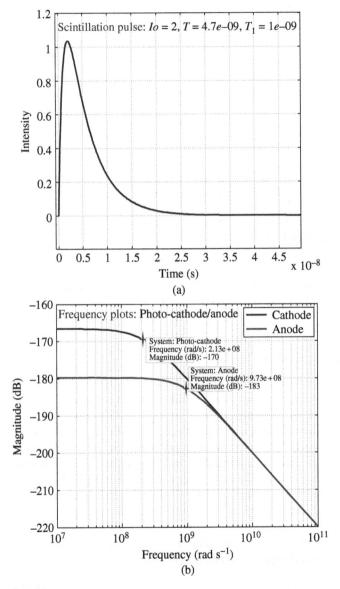

Figure 8.6 Typical scintillator/photomultiplier response: (a) Pulse. (b) Bode plot.

The solution of this differential equation follows using Laplace transforms to give the output voltage signature for the scintillator as

$$e(t) = K_{as}[e^{-t/\tau_a} - e^{-t/\tau_s}] \quad \text{for} \quad K_{as} = \frac{Q}{C}\left(\frac{\tau_a}{\tau_a - \tau_s}\right) \tag{8.37}$$

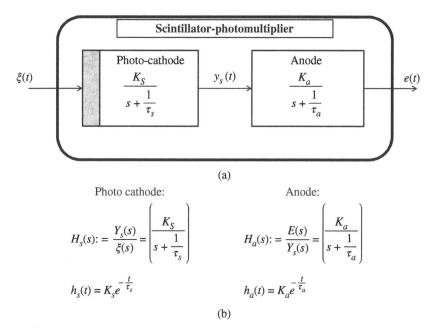

Figure 8.7 Cascaded scintillator/photomultiplier system model: (a) Diagram. (b) Transfer function/impulse response.

Thus, the output voltage pulse $e(t)$ has the identical form of the physical model of Eq. (8.34).

8.2.2 Scintillator State-Space Model

In order to develop the simulation for the scintillator, we choose to use the state-space representation by developing the underlying differential equation governing our scintillator model of Eq. (8.36), which is easily obtained from the cascade of subsystems shown in Figure 8.7.

The photocathode system is governed by the differential equation that evolves from the transfer function

$$\frac{di(t)}{dt} = -\frac{1}{\tau_s}i(t) + K_s\xi(t) \qquad \text{(Photoelectron Current)}$$

$$y_s(t) = i(t) \qquad \text{(Output Current)} \qquad (8.38)$$

where ξ is the input excitation (assumed to be an impulse-like signal).

The anode of the photomultiplier is modeled by the RC-circuit and is governed by the Kirchhoff current relation of Eq. (8.36)

$$\frac{de(t)}{dt} = \frac{1}{\tau_a}e(t) + K_i i(t) \quad \text{for } \tau_a = RC \qquad \text{(Anode Voltage)}$$

$$y_a(t) = e(t) \qquad \text{(Output Voltage)} \qquad (8.39)$$

If we define the state vector $\mathbf{x} := [i(t) \mid e(t)]'$, then we obtain the vector–matrix equation

$$\dot{\mathbf{x}}(t) = \begin{bmatrix} -\frac{1}{\tau_s} & \mid & 0 \\ - & - & - \\ 0 & \mid & -\frac{1}{\tau_a} \end{bmatrix} \mathbf{x}(t) + \begin{bmatrix} K_s \\ - \\ \frac{K_a}{C} \end{bmatrix} u(t) \qquad \text{(States)}$$

$$y(t) = [0 \mid 1]\mathbf{x}(t) \qquad\qquad \text{(Measurement)} \qquad (8.40)$$

or simply

$$\frac{d\mathbf{x}(t)}{dt} = A_c \mathbf{x}(t) + B_c u(t)$$
$$y(t) = C_c \mathbf{x}(t) \qquad\qquad (8.41)$$

the usual state-space representation of a LTI system with A_c, B_c, C_c the respective $N_x \times N_x, N_x \times N_u, N_y \times N_x$ continuous-time system, input and output matrices [9] with the corresponding impulse response, $H_t = C_c e^{A_c(t-\tau)}B_c$ and transfer function given by $H(s) = C_c(sI - A_c)^{-1}B_c$ completing the state-space representation of the scintillator–photomultiplier system.

8.2.3 Scintillator Sampled-Data State-Space Model

Sampled-data A/D conversion implies that $t \to t_k$ and therefore the state-transition matrix is $\Phi(t_k, t_0) = e^{A_c(t-t_0)}$ with state solution [19, 20]

$$\mathbf{x}(t_k) = \Phi(t_k, t_{k-1})\mathbf{x}(t_{k-1}) + \int_{t_{k-1}}^{t_k} \Phi(t_k, \tau)B_c \mathbf{u}(\tau)d\tau \qquad (8.42)$$

Define the sampled-data system with the input assumed piecewise-constant, that is, $u(t_{k-1}) \le u(t) < u(t_k)$ and $B := \int_{t_{k-1}}^{t_k} \Phi(t_k, \tau)B_c \, d\tau$ leading to the discrete sampled-data system given by

$$\mathbf{x}(t_k) = A\mathbf{x}(t_{k-1}) + B\mathbf{u}(t_{k-1}) \qquad (8.43)$$

where $A := e^{A_c(t_k - t_{k-1})} = e^{A\Delta t_k}$ for $\Delta t_k := t_k - t_{k-1}$ with $B := \int_{t_{k-1}}^{t_k} e^{A(t_k-\tau)} \, d\tau \times B_c$.

The corresponding sampled measurement system given by

$$y(t_k) = C\mathbf{x}(t_k) \qquad (8.44)$$

We developed an impulse response simulation of this system, and the results are shown in Figure 8.8, where we see the states, the current i and the voltage v and the output y.

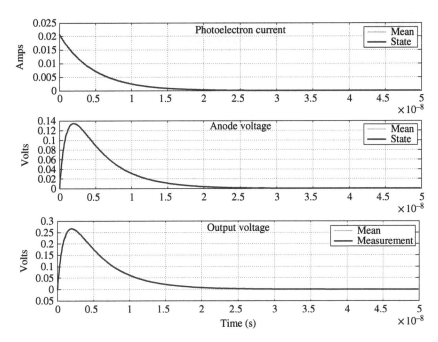

Figure 8.8 Deterministic state-space model output for cascaded scintillator/photomultiplier system simulation.

8.2.4 Gauss–Markov State-Space Model

Since we are using digitized data, then a sampled-data representation of this continuous-time system will be a better approach to characterizing this system. Therefore, using the same approach of the deterministic system given above, we obtain

$$\mathbf{x}(t_k) = A\mathbf{x}(t_{k-1}) + B\mathbf{u}(t_{k-1}) + \mathbf{w}(t_{k-1})$$
$$\mathbf{y}(t_k) = C\mathbf{x}(t_k) + v(t_k) \tag{8.45}$$

where both additive process and measurement noises are zero-mean, Gaussian processes with respective covariance matrices $R_{ww}(t_{k-1})$, $R_{vv}(t_k)$. The corresponding statistics are given by (see [9] for more details)

$$\mathbf{m}_x(t_k) = A\mathbf{m}_x(t_{k-1}) + B\mathbf{u}(t_{k-1})$$
$$\mathbf{P}_{xx}(t_k) = A\mathbf{P}_{xx}(t_{k-1})A' + R_{ww}(t_{k-1})$$
$$\mathbf{m}_y(t_k) = C\mathbf{m}_x(t_k)$$
$$\mathbf{R}_{yy}(t_k) = C\mathbf{P}_{xx}(t_k)C' + R_{vv}(t_k) \tag{8.46}$$

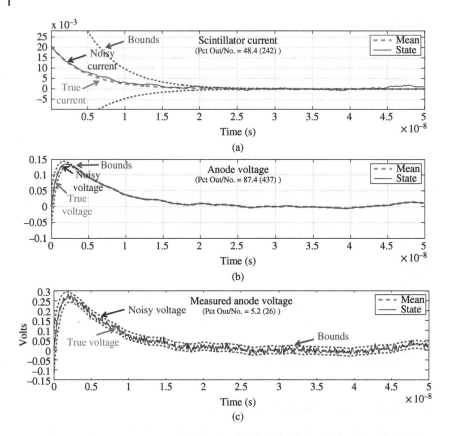

Figure 8.9 Stochastic state-space model output for cascaded scintillator/photomultiplier system simulation.

We performed a Gauss–Markov simulation of the scintillation/photomultiplier system of Section 8.2.4 using the set of parameters estimated (see next subsection) from our average PMT data set: $K_{ab} = 1.98$, $\tau_a = 1.03 \times 10^{-9}$ seconds, $\tau_s = 4.7 \times 10^{-9}$ seconds. We used the impulse excitation and added zero-mean, Gaussian noise with respective covariances. The results for the current, voltage, and measurement output, respectively, are shown in Figure 8.9.

8.2.5 Identification of the Scintillator Pulse Shape Model

We applied a Numerical algorithms for (4) state-space Subspace IDentification subspace identification (N4SID) technique to measured scintillator response data. The Hankel singular values indicated a second-order system. The raw and identified responses are shown in Figure 8.10a along with the

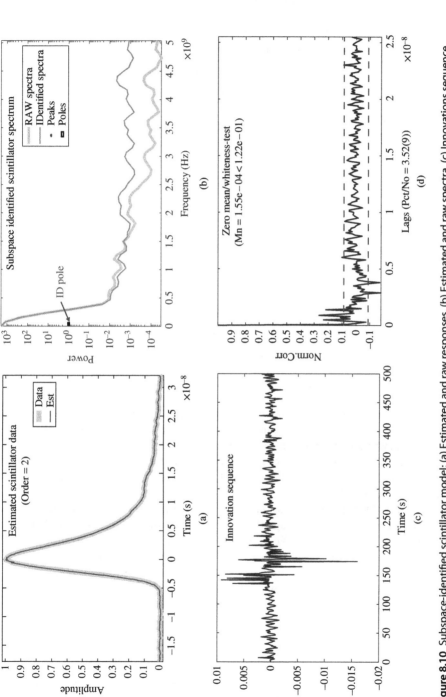

Figure 8.10 Subspace-identified scintillator model: (a) Estimated and raw responses. (b) Estimated and raw spectra. (c) Innovations sequence from N4SID algorithm. (d) Zero-mean/whiteness test results: (0.000 16 < 0.122/3.5% out).

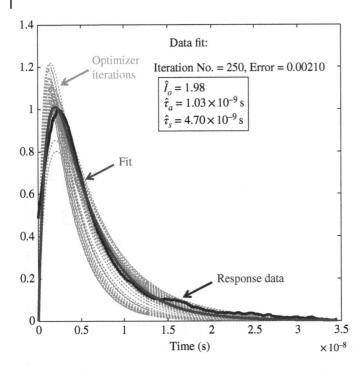

Figure 8.11 Parameter estimation for scintillator voltage response.

corresponding power spectra in (b). The resulting innovations sequence from the identification is shown in Figure 8.10c with a "good fit" indicated from the zero-mean/whiteness test results of (0.000 16 < 0.122/3.5% out) shown in (d). Therefore, the subspace identifier is able to capture the model quite well, and the Kalman filter can be constructed directly using this "black-box" model.

Another approach is to use a MBID technique with the physics-based model embedded requiring estimates of the parameters (K_{ab}, τ_a, τ_s) from the measured response (averaged) of our scintillator and photomultiplier system shown in Figure 8.7. We applied a nonlinear optimization technique using the well-known Nelder–Mead algorithm. The results of the fit are shown in Figure 8.11 illustrating a reasonable estimate of the response. The final parameter estimates are quite good extracting the truth parameters as $K_{ab} = 1.98, \tau_a = 1.03 \times 10^{-9}$ seconds, $\tau_s = 4.7 \times 10^{-9}$ seconds and are used in the simulation model for each of our photomultiplier channels.

8.2.6 Kalman Filter Design: Scintillation/Photomultiplier System

Now that we have developed a stochastic representation of the scintillator/photomultiplier system, which is commonplace in practice with the underlying uncertainties (noise, parameters, etc.), we can develop a processor capable of using these models and extracting the desired signals from the noisy, uncertain measurement data. Under the Gauss–Markov assumptions, there exists an optimal model-based solution – the Kalman filter of Chapter 4 and Table 4.1 [9].

Using the synthesized noisy scintillator/photomultiplier data shown in Figure 8.9, we applied the KF of Table 4.1 with the results shown in Figure 8.12. The performance of the processor is optimum, when the innovations sequence is deemed statistically white. The results of this run are white, indicating the desired performance. In Figure 8.12, we observe the results of the estimates: the current and voltage outputs. Here we see the smoothed estimates with the predicted uncertainties. The current is reasonable, but the predicted error bounds are clearly underestimated while the voltages lie well within the predicted

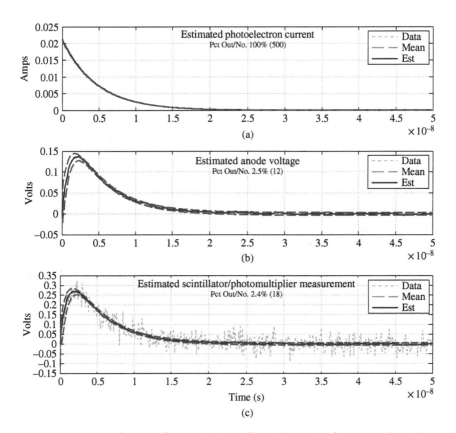

Figure 8.12 Estimated outputs for cascaded scintillator/photomultiplier system data: (a) Photoelectron current (100% out). (b) Anode voltage (2.5% out). (c) Measured output voltage (2.4% out).

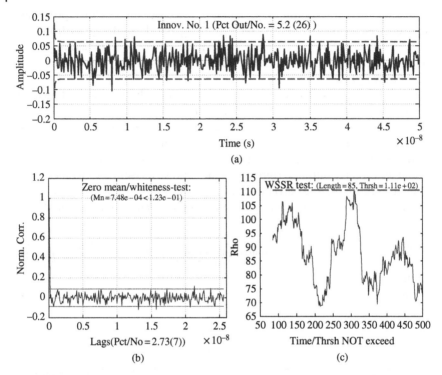

Figure 8.13 Performance metrics for Kalman filter: (a) Innovation (5.2% out). (b) Whiteness test (0.0008 < 0.1200 and 2.7% out). (c) WSSR test (below threshold).

uncertainty bounds. In Figure 8.13, we see the underlying performance metrics for this realization of the KF with the innovations sequence lying within the predicted bounds and the optimality test (zero-mean/bounded covariance) well with the prescribed bounds (0.000 08 < 0.120 00/2.7% out) with the weighted-sum squared residual (WSSR) statistic below the threshold both indicating a "tuned" optimal processor for this data realization (see Section 4.2.4 for details).

8.2.6.1 Kalman Filter Design: Scintillation/Photomultiplier Data

In this section, we applied the identified model of Section 8.2.6 to the noisy uncertain measurement data of Figure 8.9. Here the parameters of the KF are adjusted to "track" the raw impulse data and the results are shown in Figures 8.14 and 8.15. In Figure 8.14, we observe the estimation of the photoelectron current, anode voltage and output voltage of the PMT unit. The estimates are quite reasonable, and the predicted statistics (after retuning the KF) also track the estimated signals. Of course, these results are not considered valid unless the performance metrics (zero-mean/white

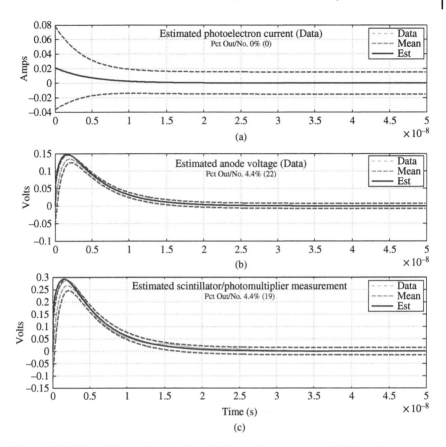

Figure 8.14 Kalman filter estimation for scintillator voltage response: (a) Photoelectron current (0.0% out). (b) Anode voltage (4.4% out). (c) Scintillator/photomultiplier output voltage (4.4% out).

innovations) are met. This performance is shown in Figure 8.15, where we see that the criteria are met satisfactorily with the innovations sequence lying within the predicted bounds, the zero-mean/whiteness test being satisfied (0.0012 < 0.123/4.9% out), and the WSSR statistic lying beneath its predicted threshold. Thus, we have developed a Kalman filter capable of extracting the noisy scintillation/photomultiplier data and predicting the underlying statistics.

8.2.7 Summary

We have discussed the development of a processor for a scintillator/photomultiplier device through simulation, MBID, and application to

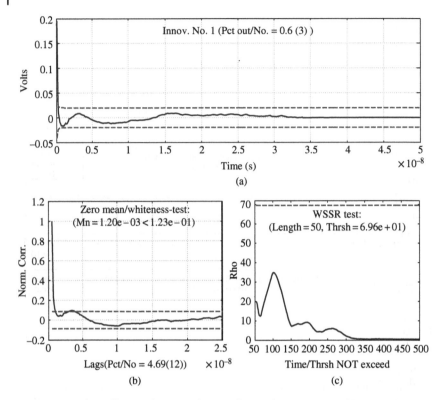

Figure 8.15 Kalman filter performance for scintillator voltage response: (a) Innovations sequence (0.6% out). (b) Zero-mean/whiteness test (0.0012 < 0.123/4.7% out). (c) WSSR statistic (below threshold).

noisy measurement data. Here, we observe how the subspace/MBID enables the design and application of a Kalman filter (see [21] for more details).

8.3 Parametrically Adaptive Detection of Fission Processes

The detection of special nuclear material (SNM) is becoming more and more of a concern as terrorist activities throughout the world have increased dramatically. Sequential detection techniques have been proven to provide a highly reliable detection process with high confidence minimizing false alarms capable of making decisions in a timely manner [22, 23]. We discuss a sequential Bayesian methodology to cast the basic decision problem into this framework. Here we exploit the fissioning nature of the threat to obtain unique signatures that can be utilized for SNM detection.

Previous work in the SNM detection area applying Bayesian techniques has been applied to the radiation detection problem [17, 24] where a sequential algorithm coupled to a parameter estimation scheme enables the detection of threat materials based on the γ-rays emitted by the targeted radionuclide [25–29]. Here we concentrate on neutron emissions measured directly by a neutron multiplicity counter that counts the arrivals.

As will be shown, the implementation of a sequential Bayesian detector for this problem requires a priori information about the unknown source(s) that is being targeted for detection much the same as a sequential radiation detection scheme [25–29]. Therefore, an estimation scheme must be *embedded* in the detector to provide updated parameter estimates of the source in order to be successful. The parameter estimator also incorporates its own independent Bayesian framework.

The development of a Bayesian particle filter (PF) or equivalently sequential Markov chain Monte Carlo (MCMC) processor for a fissioning process is quite a challenging problem due to its underlying stochastic nature. This effort is based on the fundamental theory and modeling [30–32]. Even though the underlying theory is complex, it can still be captured by stochastic modeling. Here a Bayesian approach to the problem is developed based on the underlying time-interval probability distribution [30].

8.3.1 Fission-Based Processing Model

A neutron arriving at a detector at energy level α and arrival time t_m can be characterized as a single impulse $\alpha_m \delta(t - t_m)$. A train of neutrons is defined as a set of arrivals that do *not* overlap in time. *Interarrival times* are defined as $\tau_m := t_m - t_{m-1}$ for $m = 0, 1, \ldots, M$ and the complete set of interarrivals by $T_M := \{\tau_0, \tau_1, \ldots, \tau_M\}$. These arrivals are measured by a neutron multiplicity counter that is basically a sophisticated neutron detector that evaluates the interarrival time probability distribution of neutrons emitted spontaneously by fissionable materials. Recall that the radioactive decay of each unstable nucleus produces multiple neutrons that can interact with other nuclei exciting them to energy levels enabling them to split into smaller unequal fission fragments, which are also unstable and decay even further (emitting neutrons) toward stable nuclei. The detection of these neutrons, which can pass through heavy shielding, provides a methodology to detect SNM. The key issue with threat materials is that the number of neutrons released are produced by a single decay defining its multiplication. These "correlated" neutrons offer a unique SNM signature that indicates both its multiplication and the mass of spontaneous fission isotopes. Thus, the *neutron multiplicity counter* or *neutron detector* is a stochastic measurement system creating an estimated time-interval probability distribution that is used in the neutron detection process to alarm on

SNM. Thus, the essential ingredient of these measurements is its underlying time-interval probability distribution.

A simplistic model of the multiplication process consists of a fission chain generated by a spontaneous fission under the assumption that the source is characterized by an inhomogeneous Poisson process with a varying fission rate [33, 34]. During a single fission chain v-neutrons are emitted with probability P_v. These emissions are slowed in a moderator and diffuse exponentially in time as thermal neutrons. Then the probability that n of the v neutrons are absorbed in the detector and converted into electrical pulses is

$$\Pr[v; T_M] = P_v \begin{pmatrix} v \\ n \end{pmatrix} \epsilon^n (1 - \epsilon)^{v-n} \tag{8.47}$$

where ϵ is the probability of neutron detection (detection efficiency). Mathematically, we can represent the N_v-neutron burst sequence emitted by

$$\eta(\tau_m; \Theta_S) = \sum_{i=1}^{N_v} \alpha_i \delta(\tau - \tau_m(\Theta_S)); \quad m = 0, \ldots, M; \tau_m < T_M \tag{8.48}$$

for α the ith energy of the neutron at the mth interarrival in the Mth time interval. Typically, we ignore the neutron energy and concentrate on the interarrival, since the source information is contained in $\tau_m(\Theta_S)$.

For spontaneous fissions, the quantity, k_{eff}, is the average number of neutrons from one fission that initiates another [24]. Any remaining neutrons are absorbed or escape. The value of k_{eff} specifies how a chain will proceed. For instance, the $k_{\text{eff}} = 1$ (critical mass) leads to a fission level that is constant and is typical to power plant operation, while $k_{\text{eff}} > 1$ (super-critical mass) for an event implies that there may be k_{eff}-events to follow, which is typical in weapons applications. For the latter, the number of fission reactions *increase* exponentially. We will use Eq. (8.48) in developing the subsequent fission detection schemes to follow.

8.3.2 Interarrival Distribution

Theoretically, the conditional distribution of interarrival times τ conditioned on a set of source parameters Θ (following [30, 31]) is given by

$$\Pr[\tau|\Theta] = \underbrace{R_1 r_0 n_0}_{\text{Time between chain initiations}} + \underbrace{\frac{F_S}{R_1} \sum_{n=2}^{\infty} e_n(\epsilon) \left(\sum_{k=1}^{n-1} k \, e^{-k\lambda\tau} \right) \lambda b_0(\tau)}_{\text{Time between neutrons in same chain}}$$

$$\tag{8.49}$$

where R_1 is the count rate; r_0 is the probability that no neutrons are detected within the time interval τ; n_0 is the probability of zero counts in the time interval

τ; τ is the time interval or interarrival time; F_S is the fission rate; $e_n(\epsilon)$ is the probability of detecting n neutrons from the same fission chain; λ is the inverse of the diffusion timescale; and $b_0(\tau)$ is the probability of no counts in the time interval τ.

Embedded in Eq. (8.49) is a set of various relations that capture the time-interval probability:

$$F_S = \frac{N_A}{A} \frac{\ln 2 \, t_{1/2}}{t_{1/2} \, t_{1/2}^{SF}} m_S$$

$$R_1 = \epsilon \, q \, \mathcal{M} \bar{v}_S \, F_S$$

for m_S mass of the source; ϵ is the detection efficiency; p is the probability that a neutron induces a fission; q is the escape probability ($q = 1 - p$); $\mathcal{M}$ is the system multiplication; $\bar{v}_S$ is the average neutron count from a spontaneous fission; $\bar{v}_I$ is the average neutron count from an induced fission; N_A is Avogadro's number; A is the atomic weight; and $t_{1/2}$ is the half-life with *multiplication* is given by

$$\mathcal{M} = \frac{1}{1 - p\bar{v}_I} = \frac{1}{1 - k_{\text{eff}}}$$

for k_{eff} the *effective multiplication* and *detection efficiency* approximated by

$$\epsilon = \frac{a}{b\mathcal{M}_e^2 + c\mathcal{M}_e}; \quad 0 \leq \epsilon \leq 0.04; \quad a, b, c \quad \text{fit parameters}$$

where $\mathcal{M}_e = q \times \mathcal{M}$. The *probability* of detecting n-neutrons of the v emitted with probability P_v is given by

$$e_n(\epsilon) = \sum_{v=n}^{\infty} P_v \binom{v}{n} \epsilon^n (1 - \epsilon)^{v-n}$$

The following probabilities complete the distribution:

$$r_0 = \frac{F_S}{R_1} \sum_{n=1}^{\infty} e_n(\epsilon) \left(\sum_{k=0}^{n-1} e^{-k\lambda\tau} \right)$$

and

$$b_0(\tau) = \exp\left[-F_S \int_0^\tau \frac{1}{1 - e^{-\lambda t}} \left(1 - \sum_{v=0}^{\infty} P_v (1 - \epsilon(1 - e^{-\lambda t}))^v \right) dt \right]$$

$$n_0(\tau) = r_0(\tau) \times b_0(\tau)$$

Since the objective is to "decide" whether or not a fissioning source is present, we require a priori knowledge of the source parameters: mass, multiplication, detection efficiency, and diffusion timescale parameters. Notationally, we define the source parameters as $m_S, k_{\text{eff}}, \epsilon$, and λ, respectively, and note their intimate relations in the overall probability distribution function.

8.3.3 Sequential Detection

In order to develop a sequential processor, we must test the binary hypothesis that the measured interarrival times have evolved from a fissioning SNM threat. The basic decision problem is simply stated as

GIVEN a set of uncertain neutron multiplicity detector interarrival measurements $\{\tau_m\}$; $m = 0, 1, \ldots, M$ from an unknown source, DECIDE whether or not the source is a threat (SNM). If so, "extract" its characteristic parameters, Θ to "classify" its type.

We are to test the hypothesis that the set of measured neutron interarrivals T_M have evolved from a threat or nonthreat source. Therefore, we specify the hypothesis test by

$$\mathcal{H}_0 : \quad T_M = \mathcal{T}_B(\underline{m}; \Theta_b) + \mathcal{T}_V(\underline{m}) \qquad \text{(Nonthreat)}$$

$$\mathcal{H}_1 : \quad T_M = \mathcal{T}_S(\underline{m}; \Theta_s) + \mathcal{T}_B(\underline{m}; \Theta_b) + \mathcal{T}_V(\underline{m}) \qquad \text{(Threat)} \qquad (8.50)$$

where $\mathcal{T}_S$ is the unknown source interarrivals with parameters Θ_s, $\mathcal{T}_B$ is the background interarrivals (cosmic rays, etc.) with parameters Θ_b, $\mathcal{T}_V$ is the zero-mean, Gaussian measurement (instrumentation) interarrival noise, $T_M := \{\tau_0, \tau_1, \ldots, \tau_M\}$ and $\underline{m} := 0, 1, \ldots, M$.

The fundamental approach of classical detection theory to solving this binary decision problem is to apply the Neyman–Pearson criterion of maximizing the detection probability for a specified false-alarm rate [35] with the parameters Θ *known*. The result leads to a *likelihood ratio* decision function defined by [35, 36]

$$\mathcal{L}(T_M; \Theta) := \frac{\Pr[T_M | \Theta; \mathcal{H}_1]}{\Pr[T_M | \Theta; \mathcal{H}_0]} \underset{\underset{\mathcal{H}_0}{<}}{\overset{\overset{\mathcal{H}_1}{>}}{\mathcal{T}}} \qquad (8.51)$$

with threshold $\mathcal{T}$. This expression implies a "batch" decision; that is, we gather the M interarrivals T_M, calculate the likelihood (Eq. (8.51)) over the entire batch of data, and compare it to the threshold $\mathcal{T}$ to make the decision.

8.3.4 Sequential Processor

An alternative to the batch approach is the sequential method that can be developed by expanding the likelihood ratio for each interarrival to obtain the recursion or equivalently *sequential likelihood ratio* for the mth interarrival as follows:

$$\mathcal{L}(T_m; \Theta) = \mathcal{L}(T_{m-1}; \Theta) \times \frac{\Pr[\tau_m | T_{m-1}, \Theta; \mathcal{H}_1]}{\Pr[\tau_m | T_{m-1}, \Theta; \mathcal{H}_0]}; \quad m = 0, \ldots, M \qquad (8.52)$$

with $\Pr[\tau_0 | T_{-1}, \Theta; \mathcal{H}_\ell] = \Pr[\tau_0 | \Theta; \mathcal{H}_\ell]$, the *prior* under each hypothesis.

The Wald *sequential probability-ratio test* (SPRT) is [22, 23]

$$\begin{aligned}
\mathcal{L}(T_m; \Theta) & > \mathcal{T}_1(m) \quad \text{Accept } \mathcal{H}_1 \\
\mathcal{T}_0(m) \le \mathcal{L}(T_m; \Theta) & \le \mathcal{T}_1(m) \quad \text{Continue} \\
\mathcal{L}(T_m; \Theta) & < \mathcal{T}_0(m) \quad \text{Accept } \mathcal{H}_0
\end{aligned} \tag{8.53}$$

where the thresholds are specified in terms of the false alarm (P_{FA}) and miss (P_M) probabilities as

$$\mathcal{T}_0(m) = \frac{P_M(m)}{1 - P_{FA}(m)}, \quad \mathcal{T}_1(m) = \frac{1 - P_M(m)}{P_{FA}(m)} \tag{8.54}$$

These thresholds are determined from a receiver operating characteristic (ROC) curve (detection vs false-alarm probabilities) obtained by simulation or a controlled experiment to calculate the decision function [37]. That is, an operating point is selected from the ROC corresponding to specific detection (or equivalently miss) and false-alarm probabilities specifying the required thresholds, which are calculated according to Eq. (8.54) for each parameter update.

A reasonable approach to this problem of making a reliable decision with high confidence in a timely manner is to develop a sequential detection processor (see Figure 8.16). At each neutron arrival (τ_m), we *sequentially update* the decision function and compare it to the thresholds to perform the detection – "neutron-by-neutron." Here as each neutron is monitored producing the interarrival sequence, the processor takes each interarrival measurement and attempts to "decide" whether or not it evolves from a threat or nonthreat. For each interarrival, the decision function is "sequentially" updated and compared to the detection thresholds obtained from the ROC curve operating point enabling a rapid decision. Once the threshold is crossed, the decision

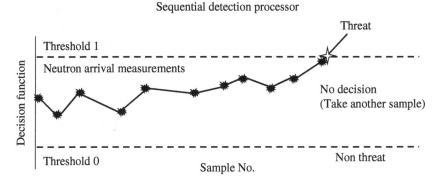

Figure 8.16 As each individual interarrival is extracted, it is discriminated, estimated, the decision function calculated, and compared to thresholds to DECIDE if the targeted threat is detected. Quantitative performance and sequential thresholds are determined from the estimated ROC curve and the selected operating point (detection/false-alarm probability).

(threat or nonthreat) is made and the arrival is processed; however, if not enough data is available to make the decision, then another measurement is obtained.

For our problem, we typically have information about the background, disturbance, and noise parameters, but we rarely have the source information. Therefore, we still can make a decision, but require estimates of the unknown parameters, that is, $\hat{\Theta} \to \Theta$. In this case, we must construct a *composite* or *generalized* likelihood ratio test (GLRT). Therefore, from the batch likelihood decision function, we can consider the two cases of the GLRT: (i) parameters are *random* or (ii) parameters are *deterministic* but *unknown*.

Here we assume that Θ is deterministic but unknown. Therefore, the approach is to estimate the unknown parameter vector $\hat{\Theta} \to \Theta$ under each hypothesis and proceed with the simple testing. A *maximum likelihood estimate* $\hat{\Theta}_{ML}$ can be used to create the GLRT such that

$$\mathcal{L}(T_M; \Theta) = \frac{\max\limits_{\Theta_1} \Pr[T_M | \Theta_1; \mathcal{H}_1]}{\max\limits_{\Theta_0} \Pr[T_M | \Theta_0; \mathcal{H}_0]} \tag{8.55}$$

This is the approach we employ *initially*. The batch solution for the GLRT can also be extended to the sequential case as before giving the solution by simply replacing $\hat{\Theta}_{ML} \to \Theta$, that is,

$$\mathcal{L}(T_m; \hat{\Theta}) = \mathcal{L}(T_{m-1}; \hat{\Theta}) \times \frac{\Pr[\tau_m | T_{m-1}, \hat{\Theta}_1; \mathcal{H}_1]}{\Pr[\tau_m | T_{m-1}, \hat{\Theta}_0; \mathcal{H}_0]}; \quad m = 0, 1, \dots, M \tag{8.56}$$

Anticipating Gaussian models (exponential family [38]) for our unknown parameters, we develop the logarithmic form of the sequential likelihood decision function. Simply taking the natural logarithm of Eq. (8.56), that is, $\Lambda(T_M; \Theta) := \ln \mathcal{L}(T_M; \Theta)$, we obtain the *log-likelihood* sequential decision function as

$$\Lambda(T_m; \hat{\Theta}) = \Lambda(T_{m-1}; \hat{\Theta}) + \ln \Pr[\tau_m | T_{m-1}, \hat{\Theta}_1; \mathcal{H}_1]$$
$$- \ln \Pr[\tau_m | T_{m-1}, \hat{\Theta}_0; \mathcal{H}_0] \tag{8.57}$$

Using these formulations, we develop the detection algorithm for our problem next. We should note that we only consider the "threat detection problem" in this section.

8.3.5 Sequential Detection for Fission Processes

Here we start with the results of Section 8.3.4 and incorporate the physics of the fission process. For fission detection, we start with the simple neutron model of Eq. (8.48) at interarrival time τ_m leading to the subsequent (sequential) hypothesis test:

$$\mathcal{H}_0 : \tau_m = \mathcal{T}_B(m; \Theta_b) + \mathcal{T}_V(m) \qquad \text{(Nonthreat)}$$

$$\mathcal{H}_1 : \tau_m = \mathcal{T}_S(m; \Theta_s) + \mathcal{T}_B(m; \Theta_b) + \mathcal{T}_V(m) \qquad \text{(Threat)} \qquad (8.58)$$

The sequential detection solution for this problem with unknown source parameters follows directly from the GLRT results of Eq. (8.56).

To implement the processor, we must first determine the required conditional probabilities in order to specify the decision function, that is,

$$\Pr[\tau_m | T_{m-1}, \hat{\Theta}; \mathcal{H}_1] = \Pr[\mathcal{T}_S(m; \Theta_S) | T_{m-1}, \hat{\Theta}_S; \mathcal{H}_1]$$

$$+ \Pr[\mathcal{T}_B(m; \Theta_b) | T_{m-1}, \Theta_b; \mathcal{H}_1] + \Pr[\mathcal{T}_V(m)] \qquad (8.59)$$

and under the null hypothesis

$$\Pr[\tau_m | T_{m-1}, \Theta; \mathcal{H}_0] = \Pr[\mathcal{T}_B(m; \Theta_b) | T_{m-1}, \Theta_b; \mathcal{H}_0] + \Pr[\mathcal{T}_V(m)] \qquad (8.60)$$

where the Gaussian interarrival noise is distributed as $\mathcal{T}_V \sim \mathcal{N}(0, \sigma_{vv}^2)$ and the known background disturbances are ignored, while the interarrival distribution is specified (instantaneously at τ_m) by Eq. (8.49) with $\tau_m \to \tau$ to give

$$\Pr[\mathcal{T}_S(m; \Theta_S) | T_{m-1}, \hat{\Theta}_S; \mathcal{H}_1] = R_1 r_0 n_0$$

$$+ \frac{F_S}{R_1} \sum_{n=2}^{\infty} e_n(\epsilon) \left(\sum_{k=1}^{n-1} k e^{-k\lambda \tau_m} \right) \lambda b_0(\tau_m) \qquad (8.61)$$

$$\Pr[\mathcal{T}_V(m)] = \frac{1}{\sqrt{2\pi\sigma_{vv}^2}} \exp\left\{ -\frac{1}{2} \frac{\mathcal{T}_V^2(m)}{\sigma_{vv}^2} \right\} \qquad (8.62)$$

Therefore, the log-likelihood ratio becomes

$$\Lambda(T_m; \Theta) = \Lambda(T_{m-1}; \Theta) + \ln\left[R_1 r_0 n_0 + \frac{F_S}{R_1} \sum_{n=2}^{\infty} e_n(\epsilon) \left(\sum_{k=1}^{n-1} k e^{-k\lambda \tau_m} \right) \lambda b_0(\tau_m) \right.$$

$$\left. + \frac{1}{\sqrt{2\pi\sigma_{vv}^2}} \exp\left\{ -\frac{1}{2} \frac{\mathcal{T}_V^2(m)}{\sigma_{vv}^2} \right\} \right] - \ln\left(\frac{1}{\sqrt{2\pi\sigma_{vv}^2}} \right) + \frac{1}{2} \frac{\mathcal{T}_V^2(m)}{\sigma_{vv}^2} \qquad (8.63)$$

We can also extend the problem to the *random* case. Suppose we assume that each of the independent source parameters is governed by a *random walk/random constant* model, that is, the parameters are assumed piecewise constant and subjected to zero-mean, Gaussian, random uncertainties, w_Θ, with covariance, $R_{w_\Theta w_\Theta}$ driving the process [9], then in this case we have that the parameters, $\Theta \sim \mathcal{N}(\Theta_0, P_0 + R_{w_\Theta w_\Theta})$, where Θ_0 is the initial mean with P_0 its corresponding covariance. Thus, the corresponding multivariate prior distribution is given by

$$\Pr[\Theta(\tau_m)] = (2\pi)^{-N_\Theta/2} |P_0 + R_{w_\Theta w_\Theta}|^{-1/2}$$

$$\times \exp\{-1/2 \, (\Theta - \Theta_0)'(P_0 + R_{w_\Theta w_\Theta})^{-1}(\Theta - \Theta_0)\} \qquad (8.64)$$

and the log-likelihood ratio is

$$
\Lambda(T_m; \Theta) = \Lambda(T_{m-1}; \Theta) + \ln \Pr[\tau_m, \Theta(\tau_m) | T_{m-1}; \mathcal{H}_1]
$$
$$
- \ln \Pr[\tau_m | T_{m-1}, \Theta; \mathcal{H}_0] \tag{8.65}
$$

where the second term can be expanded further by applying Bayes' rule to give

$$
\Pr[\tau_m, \Theta(\tau_m) | T_{m-1}; \mathcal{H}_1] = \Pr[\tau_m | T_{m-1}, \Theta(\tau_m); \mathcal{H}_1] \times \Pr[\Theta(\tau_m) | T_{m-1}; \mathcal{H}_1] \tag{8.66}
$$

Substituting this expression for the source distribution of Eq. (8.59) gives

$$
\Pr[\mathcal{T}_S(m; \Theta_S) | T_{m-1}, \hat{\Theta}_S; \mathcal{H}_1] \rightarrow \Pr[\tau_m | T_{m-1}, \Theta(\tau_m); \mathcal{H}_1]
$$
$$
\times \Pr[\Theta(\tau_m) | T_{m-1}; \mathcal{H}_1] \tag{8.67}
$$

With this in mind, the sequential log-likelihood can be calculated directly by substituting the prescribed distributions into Eq. (8.65) to give

$$
\Lambda(T_m; \Theta) = \Lambda(T_{m-1}; \Theta) + \ln \left(R_1 r_0 n_0 + \frac{F_S}{R_1} \sum_{n=2}^{\infty} e_n(\epsilon) \left(\sum_{k=1}^{n-1} k e^{-k\lambda \tau_m} \right) \lambda b_0(\tau_m) \right.
$$
$$
+ (2\pi)^{-N_\Theta/2} |P_0 + R_{w_\Theta w_\Theta}|^{-1/2}
$$
$$
\times \exp\{-1/2\, (\Theta - \Theta_0)'(P_0 + R_{w_\Theta w_\Theta})^{-1}(\Theta - \Theta_0)\}
$$
$$
\left. + \frac{1}{\sqrt{2\pi\sigma_{vv}^2}} \exp \left\{ -\frac{1}{2} \frac{\mathcal{T}_V^2(m)}{\sigma_{vv}^2} \right\} \right)
$$
$$
+ \frac{1}{2} \ln(2\pi\sigma_{vv}^2) - \frac{1}{2} \frac{\mathcal{T}_V^2(m)}{\sigma_{vv}^2} \tag{8.68}
$$

This completes the development of the sequential Bayesian detection approach for fission processes. Next we must consider the parameter estimation problem in more detail.

8.3.6 Bayesian Parameter Estimation

In order to implement the GLRT of Section 8.3.5, we must estimate the unknown parameters Θ at each arrival. We first develop the batch scheme and then its sequential version similar to the sequential Bayesian detector of the previous section. Here we develop the Bayesian parameter estimator that can be applied to the following problem:

GIVEN a set of uncertain multiplicity counter (interarrival time) measurements, T_M; FIND the "best" estimate $\hat{\Theta}$ of the unknown fission source parameters, Θ.

From a statistical perspective, we would like to estimate the posterior distribution of source parameters Θ given the entire interarrival data set T_M or

$\Pr[\Theta|T_M]$. Applying Bayes' theorem, we have that

$$\Pr[\Theta|T_M] = \frac{\Pr[T_M|\Theta] \times \Pr[\Theta]}{\Pr[T_M]} \tag{8.69}$$

Due to the sequential nature of our problem, that is, the neutron multiplicity counter measures each neutron arrival time individually – neutron-by-neutron, we require a *sequential version*.

8.3.7 Sequential Bayesian Processor

It can be shown [39] that a sequential Bayesian solution can be developed for the posterior. Starting with the first term of Eq. (8.69) and applying Bayes' rule, we have

$$\Pr[T_M|\Theta] = \Pr[\tau_M, T_{M-1}|\Theta] = \Pr[\tau_M|T_{M-1}, \Theta] \times \Pr[T_{M-1}|\Theta] \tag{8.70}$$

and for the denominator term we have

$$\Pr[T_M] = \Pr[\tau_M, T_{M-1}] = \Pr[\tau_M|T_{M-1}] \times \Pr[T_{M-1}] \tag{8.71}$$

Substituting Eqs. (8.70) and (8.71) into Eq. (8.69) and grouping terms, we obtain

$$\Pr[\Theta|T_M] = \underbrace{\left(\frac{\Pr[\tau_M|T_{M-1}, \Theta]}{\Pr[\tau_M|T_{M-1}]} \right)}_{W(\tau_m)} \times \underbrace{\left(\frac{\Pr[T_{M-1}|\Theta] \times \Pr[\Theta]}{\Pr[T_{M-1}]} \right)}_{\Pr[\Theta|T_{M-1}]} \tag{8.72}$$

or

$$\Pr[\Theta|T_M] = W(\tau_m) \times \Pr[\Theta|T_{M-1}] \tag{8.73}$$

which is the Bayesian sequential form of the *posterior distribution*. If we further assume that the interarrivals are Markovian with the current arrival depending only on the previous, that is, $(\tau_m, T_{M-1}) \to (\tau_m, \tau_{m-1})$, then we have the desired expression for sequentially propagating the posterior as

$$\Pr[\Theta|\tau_m] = W(\tau_m) \times \Pr[\Theta|\tau_{m-1}] \tag{8.74}$$

where

$$W(\tau_m) = \frac{\Pr[\tau_m|\tau_{m-1}, \Theta]}{\Pr[\tau_m|\tau_{m-1}]} \tag{8.75}$$

Here we assumed that the parameter vector is a random constant Θ with no associated dynamics to construct a sequential Bayesian processor. However, in the real-world case, it is clear that when measuring neutron interarrivals from an unknown source, then there can easily be variations or uncertainties associated with each parameter. Perhaps a more reasonable model for these parametric variations is the *random walk/constant* introduced in Section 8.3.6 [39].

That is, we know in continuous time that the walk is given by $\frac{d}{dt}\Theta(\tau) = w_\Theta(\tau)$ and by taking first differences to approximate the derivative, we can obtain a sampled-data representation [39] as

$$\Theta(\tau_m) = \Theta(\tau_{m-1}) + \Delta\tau_m \; w_\Theta(\tau_{m-1}) \tag{8.76}$$

where $\Delta\tau_m := \tau_m - \tau_{m-1}$ for the parametric uncertainty included in $\Theta(\tau_m) \sim \mathcal{N}(m_\Theta(\tau_m), R_{\Theta,\Theta}(\tau_m) + R_{w_\theta w_\theta}(\tau_m))$, since $w_\Theta \sim \mathcal{N}(0, R_{w_\theta w_\theta}(\tau_m))$. The variations of each Θ-parameter can be controlled by its initial guess $\Theta(\tau_0)$ and variance, $R_{\Theta,\Theta}(\tau_m) + R_{w_\theta w_\theta}(\tau_m)$.

Physically we have that $R_{w_\theta w_\theta} = \text{diag}[\sigma_{MM}^2 \; \sigma_{k_{\text{eff}}k_{\text{eff}}}^2 \; \sigma_{\epsilon\epsilon}^2 \; \sigma_{\lambda^{-1}\lambda^{-1}}^2]$ and the subsequent search can include a bounded uniform variate for each initial value, $\Theta_0 = \mathcal{U}[a,b]$ enabling a more pragmatic approach to modeling the parameter set with their individual accompanying uncertainties. The statistics of the *random walk/constant* model are given by the sequential *Gauss–Markov structure*

$$\Theta(\tau_m) = \Theta(\tau_{m-1}) + \Delta\tau_m \; w_\Theta(\tau_{m-1}) \; \text{(State)}$$
$$m_\Theta(\tau_m) \quad = m_\Theta(\tau_{m-1}) \quad\quad \text{(Mean)}$$
$$R_{\Theta\Theta}(\tau_m) = R_{\Theta\Theta}(\tau_{m-1}) + R_{w_\theta w_\theta}(\tau_{m-1}) \; \text{(Variance)} \tag{8.77}$$

With this model in mind, we rederive the sequential Bayesian processor as before starting with the *batch* approach. We would like to estimate $\Theta(\tau_m)$ with the complete parameter set defined by $\Theta_M := \{\Theta(\tau_0), \Theta(\tau_1), \ldots, \Theta(\tau_M)\}$. The batch posterior is given by Bayes' theorem as before:

$$\Pr[\Theta_M | T_M] = \frac{\Pr[T_M | \Theta_M] \times \Pr[\Theta_M]}{\Pr[T_M]} \tag{8.78}$$

The first term can be decomposed by applying Bayes' rule as

$$\Pr[T_M | \Theta_M] = \Pr[\tau_m, T_{M-1} | \Theta(\tau_m), \Theta_{M-1}]$$
$$= \Pr[\tau_m | T_{M-1}, \Theta(\tau_m), \Theta_{M-1}]$$
$$\times \Pr[T_{M-1} | \Theta(\tau_m), \Theta_{M-1}] \tag{8.79}$$

The interarrival time τ_m is independent Θ_{M-1} and T_{M-1}, so that the first term in Eq. (8.79) becomes $\Pr[\tau_m | \Theta(\tau_m)]$, while the second term simplifies to $\Pr[\tau_{M-1} | \Theta_{M-1}]$, since the parameter vector $\Theta(\tau_m)$ is assumed independent of the past measurement data. Therefore, we have

$$\Pr[T_M | \Theta_M] = \Pr[\tau_m | \Theta(\tau_m)] \times \Pr[T_{M-1} | \Theta_{M-1}] \tag{8.80}$$

The second term of Eq. (8.78) can be decomposed similarly as

$$\Pr[\Theta_M] = \Pr[\Theta(\tau_m), \Theta_{M-1}] = \Pr[\Theta(\tau_m) | \Theta_{M-1}] \times \Pr[\Theta_{M-1}] \tag{8.81}$$

and the decomposition of $\Pr[T_M]$ in the denominator is given in Eq. (8.70).

Substituting these relations (Eqs. (8.70), (8.80), and (8.81)) into Eq. (8.78), and assuming a Markovian process as before, we obtain

$$\Pr[\Theta_M | T_M] = (\Pr[\tau_m | \Theta(\tau_m)] \times \Pr[T_{M-1} | \Theta_{M-1}])$$
$$\times (\Pr[\Theta(\tau_m) | \Theta(\tau_{m-1})] \times \Pr[\Theta_{M-1}]) / \Pr[\tau_m | \tau_{m-1}] \times \Pr[T_{M-1}]$$

$$(8.82)$$

Now grouping the terms, the desired posterior distribution becomes

$$\Pr[\Theta_M | T_M] = \left(\frac{\Pr[\tau_m | \Theta(\tau_m)] \times \Pr[\Theta(\tau_m) | \Theta(\tau_{m-1})]}{\Pr[\tau_m | \tau_{m-1}]} \right)$$
$$\times \left(\frac{\Pr[T_{M-1} | \Theta_{M-1}] \times \Pr[\Theta_{M-1}]}{\Pr[T_{M-1}]} \right)$$

$$(8.83)$$

or simply (replacing $\tau_m \rightarrow T_M$)

$$\Pr[\Theta_m | T_m] = W(\tau_m) \times \Pr[\Theta_{m-1} | T_{m-1}] \tag{8.84}$$

where

$$W(\tau_m) := \frac{\Pr[\tau_m | \Theta(\tau_m)] \times \Pr[\Theta(\tau_m) | \Theta(\tau_{m-1})]}{\Pr[\tau_m | \tau_{m-1}]} \tag{8.85}$$

which is the sequential Bayesian solution for the dynamic parametric model (random walk/constant). A particle filter is an implementation of this recursion [39, 40].

8.3.8 Particle Filter for Fission Processes

Particle filtering is a technique that evolves from the "importance sampling" approach to statistical sampling of data. The key idea is to select particles (or parameters in our problem) from the regions of highest probabilities or equivalently regions of *highest importance*. Once the resulting importance weight is determined, the desired posterior distribution is approximated by a nonparametric probability mass function (PMF) as

$$\hat{\Pr}[\Theta_M | T_M] = \sum_i \mathcal{W}_i(\tau_m) \, \delta(\Theta(\tau_m) - \Theta_i(\tau_m)) \tag{8.86}$$

where $\mathcal{W}_i$ is the *normalized* weighting function given as the ratio of the posterior at interarrival time τ_m and the designed importance distribution q as

$$W_i(\tau_m) := \frac{\Pr[\Theta(\tau_m) | T_m]}{q[\Theta(\tau_m) | T_m]} = \frac{\Pr[\tau_m | \Theta(\tau_m)]}{q[\tau_m | \Theta(\tau_m)]}$$
$$\times \frac{\Pr[\Theta(\tau_m)]}{q[\Theta(\tau_m) | \Theta_{m-1}, \tau_m]}$$

$$(8.87)$$

The normalized weight is simply

$$\mathcal{W}_i(\tau_m) := \frac{W_i(\tau_m)}{\sum_i W_i(\tau_m)} \tag{8.88}$$

The "bootstrap" processor is the most popular technique [39]. Here the proposal is selected as the *transition* prior and the weighting function becomes simply the likelihood

$$W(\tau_m) = W(\tau_{m-1}) \times \Pr[\tau_m | \Theta(\tau_m)] \tag{8.89}$$

With these relations in mind, the sequential Bayesian processor can be developed for our problem. We start with the basic bootstrap technique to estimate the unknown source parameters that will eventually become part of the log-likelihood decision function. Initially, we assume the prior distributions are uniformly distributed with bounds selected over some pragmatic intervals $\mathcal{U}(a, b)$. The dynamic parameter updates are given by the random walk/constant model of Eq. (8.76) driven by zero-mean, Gaussian noise with covariance $R_{w_\Theta w_\Theta}$ with initial mean (constant) $\Theta(\tau_0)$ and corresponding parametric covariance $R_{\Theta\Theta}(\tau_m)$. The likelihood distribution embeds the "fission physics" of Eq. (8.49).

The measurement prediction model is based on the current parameter estimates and its analytic time-interval likelihood distribution. As a new neutron interarrival τ_m is available at the output of the multiplicity counter, the predicted parameter estimate $\hat{\Theta}(\tau_m | \tau_{m-1})$ is made and provided as input to the physics likelihood probability distribution $\mathcal{L}_\Theta(\hat{\Theta}(\tau_m)) := \Pr(\tau | \Theta)$ for the prediction.

After the predicted interarrival $\hat{\tau}_m$ is available, the update and resampling steps are performed and the next measurement is awaited. The bootstrap algorithm performs the following steps shown in Table 4.9.

8.3.9 SNM Detection and Estimation: Synthesized Data

In this section, we investigate the performance of the sequential Bayesian detector/estimator on simulated data. We assume a set of parameters to statistically synthesize a set of arrivals used to calculate the interarrivals $\{\tau_m\}$; $m = 0, 1, \dots, M$. We start with the results from the sequential Bayesian detection algorithm that incorporates the Bayesian parameter estimator as part of its inherent structure as illustrated in Figure 8.17. Note that the unknown source parameter vector to be estimated has the following physical parameters (from the interarrival PDF): $\Theta := [m_S \; k_{eff} \; \epsilon \; \lambda^{-1}]^T$ – the mass of the source, effective multiplication (k_{eff}), detector efficiency, and diffusion time, respectively.

The bootstrap PF was applied to a synthesized neutron arrival sequence shown in Figure 8.17. We note from the figure that the interarrival data is

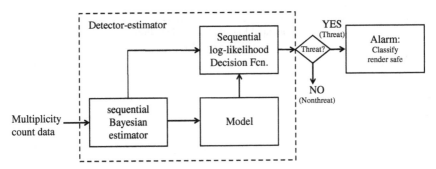

Figure 8.17 Fission detection/estimation processing: Count data, fission parameter estimation, sequential detection and alarm.

processed by the sequential Bayesian estimator to provide predicted estimates of the source parameters, $\hat{\Theta}(\tau_m|\tau_{m-1})$. These estimates are then input to the physics-based likelihood to predict the corresponding PDF which is a part of the log-likelihood decision function. These predicted parameters are also provided directly as individual inputs to the log-likelihood decision function. Once the decision function is calculated at τ_m, it is compared to the thresholds to "decide" whether a threat is present. If so, the alarm is initiated. If not, another measurement is processed on arrival (take more data).

There exist a variety of metrics that can be applied to evaluate detection performance ranging from confusion matrices to sophisticated statistical hypothesis tests [39], but perhaps the most basic and most robust method is the calculation of the ROC curve. The ROC curve is simply a graph of detection (P_{DET}) vs false-alarm (P_{FA}) probabilities parameterized by threshold, $\mathcal{T}$ with *perfect* performance occurring when $P_{DET} = 1$ and $P_{FA} = 0$. The ROC curve provides all of the fundamental information from which most other metrics are derived. Thus, there are many individual metrics that can be extracted directly from an ROC curve including sensitivity, specificity, cost/benefit analysis along with a set of specific features such as area-under-curve (AUC) and minimum probability of error (MinE) [39].

As mentioned previously, it is necessary to calculate an ROC curve to select an operating point (detection and false-alarm probabilities) to calculate the sequential thresholds. In order to generate the ROC, we synthesize an ensemble of 30-members each consisting of 100-arrivals selected directly from a Monte Carlo simulation data set using the following source (uranium) parameters: $m_S = 25$ kg, $k_{eff} = 0.9$; $\epsilon = 0.03$; $\lambda^{-1} = 0.01$. We chose to use a SNR of 6.9 dB defined by the 10 log-ratio of the signal energy to noise energy (variance). The local ROC for each member realization was estimated, and then the *average* ROC was used for the calculation. Performance metrics such as the AUC are also calculated to assess detection performance ($AUC = 0.95$). For "perfect" performance, the detection probability is unity and false-alarm probability zero

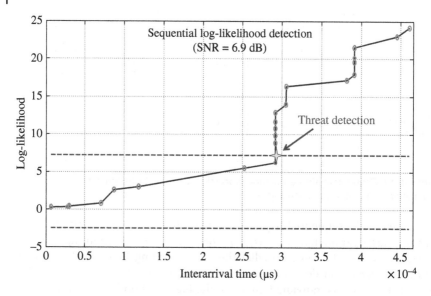

Figure 8.18 Sequential Bayesian (log-likelihood) threat detection for $(P_{FA}, P_{DET})=(0.0007, 0.916)$ and thresholds at $(\ln \tau_0, \ln \tau_1)=(-2.48, 7.23)$.

corresponding to an AUC of unity. The optimum operating point (P_{FA}, P_{DET}) is calculated by minimizing the Bayes' risk (see [39] for details) yielding a detection probability of 92% for a false-alarm probability of 0.1% at this SNR. Substituting these values into the threshold calculation of Eq. (8.54) and taking the natural logarithm give the thresholds $(\ln \tau_0, \ln \tau_1)$ as $(-2.48, 7.23)$. A typical realization of sequential Bayesian detection results for a member of the ensemble is shown in Figure 8.18 showing the decision function exceeding the upper threshold thereby indicating a threat and subsequent alarm. The ROC curve enables us to evaluate the performance of the sequential detection algorithm at various SNRs (see Figure 8.19). This performance is what we would expect to achieve at the selected thresholds. Test data is used to verify these predictions. So we see that the detection performance is quite reasonable at these SNRs indicating a feasible Bayesian detector design.

Only 100 interarrivals were investigated to observe the feasibility of this approach. We observe the uncertainty of the interarrivals caused by the randomness of the fission process. Next the physical parameters were estimated with the results shown in Figure 8.20. Again the parameter estimates are quite reasonable for this realization at 6.9 dB. For parameter estimation performance metrics, we use the average RMS-error (absolute/relative errors) of each estimator to give: mass (RMS-error = 0.03/0.28% kg); effective multiplication (RMS-error = 0.0025/0.12%); efficiency (RMS-error = 0.000 06/0.20%); and inverse diffusion time (λ) (RMS-error = 0.000 04/0.4% μs^{-1}).

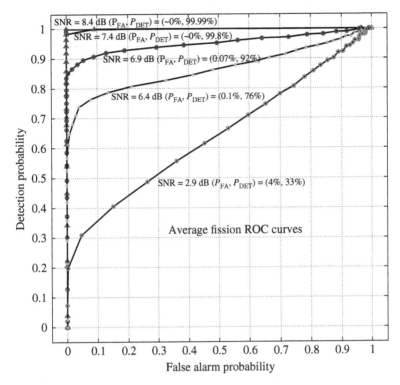

Figure 8.19 Family of sequential Bayesian detection ROC curves for various SNRs: 8.4 dB $(P_{FA}, P_{DET})=(0.0, 0.999)$, 7.4 dB $(P_{FA}, P_{DET})=(0.0, 0.9832)$, 6.9 dB $(P_{FA}, P_{DET})=(0.0007, 0.916)$, 6.4 dB $(P_{FA}, P_{DET})=(0.001, 0.76)$, 2.94 dB $(P_{FA}, P_{DET})=(0.04, 0.33)$.

Note that the true parameter value is shown as the line (dashed) and both the maximum a posterior (MAP) and conditional mean (CM) estimates are shown (arrows) on the plots. They appear to track the physical parameters quite well (small RMS errors) for this realization.

We must perform a sequence of at least 100 realizations and calculate ensemble statistics (not shown). Finally, we show a set of four snapshots (or slices) at various interarrival samples (25, 50, 75, 100) throughout the simulation. Note that each is a slice of the 3D PMFs shown for all of the physical parameters. Also note how the estimated particles coalesce (as expected) about the highest probability regions, which are annotated by the arrows in Figure 8.21.

8.3.10 Summary

We have developed a sequential Bayesian approach to the fission detection problem based on a theoretical likelihood PDF capturing the underlying physics of the fission process [30]. A sequential detection processor was

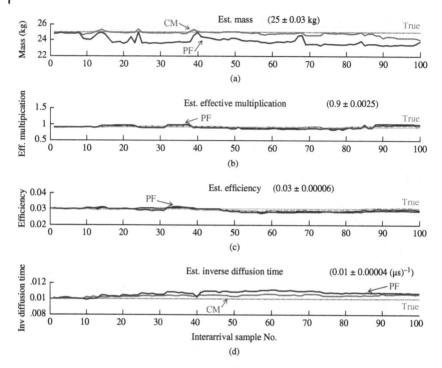

Figure 8.20 Sequential Bayesian parameter estimates (arrows) and absolute RMS error for a SNR = 6.9 dB: (a) Mass (25 ± 0.03 kg). (b) Effective multiplication (k_{eff} = 0.90 ± 0.0025). (c) Efficiency (0.03 ± 0.000 06). (d) Inverse diffusion time (λ = 0.01 ± 0.000 04 μs^{-1}).

developed based on a joint probability distribution of interarrivals and unknown source parameters using a combined random walk and random constant model of the parameter uncertainties [9]. This model was embedded along with the physics-based likelihood into both sequential detection and parameter estimation processors. The processor performed reasonably on synthesized Monte Carlo data, resulting in a successful threat (SNM) detection achieving a detection/false-alarm probability of P_{DET} = 91.7% at a P_{FA} = 0.07% in comparison to a similar theoretical approach applied to γ-ray detection of P_{DET} = 95% at a P_{FA} = 2% [25, 26]. The physics parameter estimates were also quite reasonable with relative RMS errors less than 0.4%, indicating a feasible solution to the fission detection problem. Performance metrics for both detection and parameter estimation processors were developed and assessed, implying that a combination of both sequential Bayesian γ-ray [25, 26] and neutron detection schemes have the potential of providing both a timely and accurate threat detection capability.

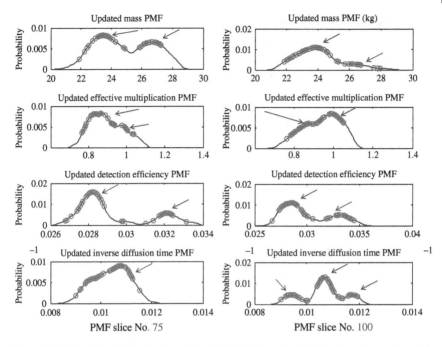

Figure 8.21 Posterior PMF of various slices (75,100 interarrivals) illustrating the multimodal (arrows) nature of the distributions as well as the coalescing of particles (circles) in highest probability regions.

8.4 Parametrically Adaptive Processing for Shallow Ocean Application

The shallow ocean is an ever-changing environment primarily due to temperature variations in its upper layers directly affecting sound propagation throughout. The need to develop processors capable of tracking these changes implies a stochastic as well as an environmentally adaptive design. Bayesian techniques have evolved to enable a class of processors capable of performing in such an uncertain, nonstationary, non-Gaussian, variable shallow ocean environment. Here we develop a sequential Bayesian processor capable of providing a joint solution to the modal function tracking and environmental adaptivity problem. The focus is on the development of both a particle filter and an UKF capable of providing reasonable performance for this problem. These processors are applied to hydrophone measurements obtained from a vertical array. The adaptivity problem is attacked by allowing the modal coefficients and/or wavenumbers to be *jointly estimated* from the noisy measurement data along with tracking of the modal functions and simultaneously enhancing the noisy pressure-field measurements.

8.4.1 State-Space Propagator

For the ocean acoustic modal function enhancement problem, we assume a horizontally stratified ocean of depth h with a *known* horizontal source range r_s and depth z_s and that the acoustic energy from a point source can be modeled as a trapped wave governed by the Helmholtz equation [41, 42]. The standard separation of variables technique and removing the time dependence leads to a set of ordinary differential equations; that is, we obtain a "depth-only" representation of the wave equation, which is an eigenvalue equation in z with

$$\frac{d^2}{dz^2}\phi_m(z) + \kappa_z^2(m)\phi_m(z) = 0, \quad m = 1, \dots, M \tag{8.90}$$

whose eigensolutions $\{\phi_m(z)\}$ are the so-called *modal functions* and κ_z is the vertical wavenumber in the z-direction. These solutions depend on the sound-speed profile $c(z)$ and the boundary conditions at the surface and bottom as well as the corresponding *dispersion* relation given by

$$\kappa^2 = \frac{\omega^2}{c^2(z)} = \kappa_r^2(m) + \kappa_z^2(m); \quad m = 1, \dots, M \tag{8.91}$$

where $\kappa_r(m)$ is the horizontal wavenumber (constant) associated with the mth mode in the r-direction and ω is the harmonic source frequency.

By assuming a known horizontal source range a priori, we obtain a range solution given by the Hankel function, $H_0(\kappa_r r_s)$, enabling the pressure-field to be represented by

$$p(r_s, z) = \sum_{m=1}^{M} \beta_m(r_s, z_s)\phi_m(z) \tag{8.92}$$

where p is the acoustic pressure; ϕ_m is the mth modal function with the modal coefficient defined by

$$\beta_m(r_s, z_s) := q\, H_0(\kappa_r r_s)\, \phi_m(z_s) \tag{8.93}$$

for (r_s, z_s) the source position and q its amplitude.

8.4.2 State-Space Model

The depth-only eigen-equation can be transformed to state-space form by defining the state vector of the mth mode as

$$\underline{\phi}_m(z) := \begin{bmatrix} \phi_m(z) \\ \frac{d}{dz}\phi_m(z) \end{bmatrix} = \begin{bmatrix} \phi_{m1}(z) \\ \phi_{m2}(z) \end{bmatrix} \tag{8.94}$$

leading to the state (vector) equation:

$$\frac{d}{dz}\underline{\phi}_m(z) = \mathbf{A}_m(z)\underline{\phi}_m(z) \tag{8.95}$$

for

$$\mathbf{A}_m(z) = \begin{bmatrix} 0 & 1 \\ -\kappa_z^2(m) & 0 \end{bmatrix} \qquad (8.96)$$

Assuming that the ocean acoustic noise can be characterized by additive uncertainties, we can extend this deterministic state equation for the M-modes, that is, $\Phi(z) := [\underline{\phi}_1(z)| \cdots |\underline{\phi}_M(z)]^T$ leading to the following $2M$-dimensional Gauss–Markov representation of the model:

$$\frac{d}{dz}\Phi(z) = \mathbf{A}(z)\Phi(z) + \mathbf{w}(z) \qquad (8.97)$$

where $\mathbf{w}(z) = [w_1 \ w_2 \ \cdots \ w_{2M}]^T$ is additive, zero-mean random noise. The system matrix $\mathbf{A}(z)$ is

$$\mathbf{A}(z) = \begin{bmatrix} \mathbf{A}_1(z) & \cdots & 0 \\ \vdots & \ddots & \vdots \\ 0 & \cdots & \mathbf{A}_M(z) \end{bmatrix} \qquad (8.98)$$

with the overall state vector given by

$$\Phi(z) = [\phi_{11} \quad \phi_{12} \ | \ \phi_{21} \quad \phi_{22} \ | \quad \cdots \quad | \ \phi_{M1} \quad \phi_{M2}]^T \qquad (8.99)$$

This representation leads to the *measurement* equations, which can be written as

$$p(r_s, z) = \mathbf{C}^T(r_s, z_s)\Phi(z) + v(z) \qquad (8.100)$$

where

$$\mathbf{C}^T(r_s, z_s) = [\beta_1(r_s, z_s) \quad 0 \ | \quad \cdots \quad | \ \beta_M(r_s, z_s) \quad 0] \qquad (8.101)$$

The random noise terms $\mathbf{w}(z)$ and $v(z)$ are assumed Gaussian and zero-mean with respective covariance matrices, $\mathbf{R}_{ww}$ and R_{vv}. The measurement noise (v) is used to represent the "lumped" effects of near-field acoustic noise field, flow noise on the hydrophone, and electronic noise. The modal noise ($\mathbf{w}$) is used to represent the "lumped" uncertainty of sound-speed errors, distant shipping noise, errors in the boundary conditions, sea state effects, and ocean inhomogeneities that propagate through the ocean acoustic system dynamics (normal-mode model).

Since the array spatially samples the pressure-field discretizing depth, we choose to discretize the differential state equations using a *central difference* approach for improved numerical stability; that is, assuming uniformly spaced hydrophones from Eq. (8.90) we have

$$\frac{d^2}{dz^2}\phi_m \approx \frac{\phi_m(z_\ell) - 2\phi_m(z_{\ell-1}) + \phi_m(z_{\ell-2})}{\Delta z_\ell^2} \qquad (8.102)$$

for $\Delta z_\ell := z_\ell - z_{\ell-1}$.

Applying this approximation to Eq. (8.90) gives

$$\phi_m(z_\ell) - 2\phi_m(z_{\ell-1}) + \phi_m(z_{\ell-2}) + \Delta z_\ell^2 \kappa_z^2(m)\phi_m(z_{\ell-1}) = 0$$

where z_ℓ is the location of the ℓth sensor. Defining the discrete modal state vector as $\phi_m(z_\ell) := [\phi_m(z_{\ell-2}) \mid \phi_m(z_{\ell-1})]'$, we obtain the following set of difference equations for the mth mode:

$$\phi_{m1}(z_\ell) = \phi_{m2}(z_{\ell-1})$$
$$\phi_{m2}(z_\ell) = -\phi_{m1}(z_{\ell-1}) + (2 - \Delta z_\ell^2 \kappa_z^2(m))\phi_{m2}(z_{\ell-1}) \tag{8.103}$$

with each of the corresponding A-submatrices now given by

$$A_m(z_\ell) = \begin{bmatrix} 0 & 1 \\ -1 & 2 - \Delta z_\ell^2 \kappa_z^2(m) \end{bmatrix}; \quad m = 1, \dots, M \tag{8.104}$$

and

$$\kappa_z^2(m) = \left(\frac{\omega^2}{c^2(z)} \right) - \kappa_r^2(m) \tag{8.105}$$

Substituting this model and combining *all* of the modes as in Eq. (8.98), the following overall *Gauss–Markov* representation of our normal-mode process and measurement is

$$\Phi(z_\ell) = A(z_\ell)\Phi(z_{\ell-1}) + w(z_\ell),$$
$$p(r_s, z_\ell) = C^T(r_s, z_s)\Phi(z_\ell) + v(z_\ell) \tag{8.106}$$

and $\Phi, w \in \mathcal{R}^{2M \times 1}, p, v \in \mathcal{R}^{1 \times 1}$ for $w \sim \mathcal{N}(0, R_{ww}), v \sim \mathcal{N}(0, R_{vv})$ with $\Phi(z_\ell) \sim \mathcal{N}(\overline{\Phi}(z_0), \overline{P}(z_0))$, $A \in \mathcal{R}^{2M \times 2M}$, $C^T \in \mathcal{R}^{1 \times 2M}$ and with $\sim$ meaning "distributed as."

This completes the normal-mode representation of the shallow ocean in state-space form, next we consider *augmenting* this model with unknown parameters to create a parametrically adaptive processor.

8.4.2.1 Augmented State-Space Models

The "parametrically adaptive" processor evolves from the normal-mode representation by defining parameter sets of interest. Variations in the ocean can be reflected, parametrically, in a number of ways. For instance, sound-speed variations are related to temperature changes especially in a shallow ocean environment directly impacting the corresponding dispersion relation of Eq. (8.91) that can be parametrically captured by the horizontal wavenumber. Besides the wavenumbers, modal variations can be reflected through the measured pressure-field relations of Eq. (8.92) that can be parametrically captured by the modal coefficients of Eq. (8.93). Therefore, we choose to use the modal coefficients as well as the horizontal wavenumbers (individually) as the parameters of interest in adapting to the changing shallow ocean environment.

Case(i): Modal Coefficients The modal coefficients of Eq. (8.93) can be used to capture modal function variations. In this case, we define the unknown *parameter vector* as

$$\theta_m(r_s, z_s) := \beta_m(r_s, z_s); \quad m = 1, \dots, M$$

and a new "augmented" state vector for the mth mode as

$$\Phi_m(z_\ell; \theta_m) := \Phi_m(z_\ell) = [\phi_{m1}(z_\ell) \; \phi_{m2}(z_\ell) \mid \theta_m(z_\ell)]^T.$$

With this choice of parameters (modal coefficients), the *augmented state* equations for the mth mode becomes

$$\phi_{m1}(z_\ell) = \phi_{m2}(z_{\ell-1}) + w_{m1}(z_{\ell-1})$$
$$\phi_{m2}(z_\ell) = -\phi_{m1}(z_{\ell-1}) + (2 - \Delta z_\ell^2 \kappa_z^2(m))\phi_{m2}(z_{\ell-1}) + w_{m2}(z_{\ell-1})$$
$$\theta_m(z_\ell) = \theta_m(z_{\ell-1}) + \Delta z_\ell w_{\theta_m}(z_{\ell-1}) \tag{8.107}$$

where we have selected a discrete *random walk* model ($\dot{\theta}_m(z) = w_{\theta_m}(z)$) based on first differences to capture the variations of the modal coefficients with additive, zero-mean, Gaussian noise of covariance $R_{w_{\theta_m} w_{\theta_m}}$.

Note that when we augment the unknown parameters into the state vector to construct the *parametrically adaptive* processor, then we assume that they are random (walks) with our precomputed initial values specified (initial conditions or means) and their corresponding covariances used to bound their uncertainty (2σ confidence bounds).

More succinctly, for the mth mode, we can write

$$\Phi_m(z_\ell) = A_m(z_{\ell-1}; \theta)\Phi_m(z_{\ell-1}) + \mathbf{w}_m(z_{\ell-1}) \tag{8.108}$$

or expanding

$$\begin{bmatrix} \phi_m(z_\ell) \\ --- \\ \theta_m(z_\ell) \end{bmatrix} = \begin{bmatrix} A_m(z_{\ell-1}) & \mid & 0 \\ - & & - - \\ 0 & \mid & 1 \end{bmatrix} \begin{bmatrix} \phi_m(z_{\ell-1}) \\ --- \\ \theta_m(z_{\ell-1}) \end{bmatrix}$$
$$+ \begin{bmatrix} W_{\phi_m}(z_{\ell-1}) \\ --- \\ W_{\theta_m}(z_{\ell-1}) \end{bmatrix} \tag{8.109}$$

where $W_{\phi_m} \sim \mathcal{N}(0, R_{W_{\phi_m} W_{\phi_m}})$, $W_{\theta_m} \sim \mathcal{N}(0, R_{W_{\theta_m} W_{\theta_m}})$, $\phi_m(0) \sim \mathcal{N}(\overline{\phi}_m(0), R_{\phi_m \phi_m})$, $\theta_m(0) \sim \mathcal{N}(\overline{\theta}_m(0), R_{\theta_m \theta_m})$.

The corresponding *nonlinear* measurement model is given by

$$p(r_s, z_\ell) = \sum_{m=1}^{M} \theta_m(z_\ell)\phi_m(z_\ell) + v(z_\ell); \quad \ell = 1, \dots, L \tag{8.110}$$

with dispersion (sound-speed)

$$c(z_\ell) = \frac{\omega}{\sqrt{\kappa_z^2(m) + \kappa_r^2(m)}}, \quad m = 1, \dots, M; \ell = 1, \dots, L \tag{8.111}$$

To complete this representation, we combine all of the modes and unknown parameters, and, therefore, the state transition is characterized by the underlying augmented state-space model as

$$\Phi(z_\ell) = A(z_{\ell-1}; \Theta)\Phi(z_{\ell-1}) + w(z_{\ell-1})$$

and the measurement is determined from the *nonlinear* pressure-field measurement model,

$$p(r_s, z_\ell) = c[\Phi(z_\ell); \Theta] + v(z_\ell) \tag{8.112}$$

Note that for this case the pressure-field is *nonlinear* in the states (modal functions) and parameters (modal coefficients), since they are multiplicands and therefore lead to non-Gaussian measurements.

Case(ii): Horizontal Wavenumbers The horizontal wavenumbers of Eq. (8.91) can be used to capture sound-speed (temperature) variations. For this case, we define the unknown *parameter vector* as

$$\theta_m(z) := \kappa_r(m); \quad m = 1, \dots, M$$

and a new "augmented" state vector as

$$\Phi_m(z_\ell; \theta_m) := \Phi_m(z_\ell) = [\phi_{m1}(z_\ell) \, \phi_{m2}(z_\ell) \mid \theta_m(z_\ell)]^T$$

With this choice of parameters (horizontal wavenumber), the augmented state equations for the mth mode become

$$\phi_{m1}(z_\ell) = \phi_{m2}(z_{\ell-1}) + w_{m1}(z_{\ell-1})$$

$$\phi_{m2}(z_\ell) = -\phi_{m1}(z_{\ell-1}) + \left(2 - \Delta z_\ell^2 \left(\frac{\omega^2}{c^2(z_\ell)} - \theta_m^2(z_{\ell-1}) \right) \right)$$

$$\times \phi_{m2}(z_{\ell-1}) + w_{m2}(z_{\ell-1})$$

$$\theta_m(z_\ell) = \theta_m(z_{\ell-1}) + \Delta z_\ell w_{\theta_m}(z_{\ell-1}) \tag{8.113}$$

where we have again selected a discrete *random walk* model ($\dot{\theta}_m(z) = w_{\theta_m}(z)$) to capture the variations of the horizontal wavenumber with additive, zero-mean, Gaussian noise of covariance $R_{w_{\theta_m} w_{\theta_m}}$. Note that even though we know that theoretically the horizontal wavenumbers are constant for each mode, we do incorporate this stochastic representation to the uncertainty inherent in the measurements and the parametric model itself.

More succinctly, for the mth mode we can write

$$\Phi_m(z_\ell) = A_m(z_{\ell-1}; \theta)\Phi_m(z_{\ell-1}) + w_m(z_{\ell-1}) \tag{8.114}$$

for

$$A_m(z_{\ell-1}; \theta) = \begin{bmatrix} 0 & 1 & \mid & 0 \\ -1 & 2 - \Delta z_\ell^2 \left(\frac{\omega^2}{c^2(z_\ell)} - \theta_m^2(z_{\ell-1}) \right) & \mid & 0 \\ - & - & & - \\ 0 & 0 & \mid & 1 \end{bmatrix}$$

The corresponding measurement model is given by

$$p(r_s, z_\ell) = \sum_{m=1}^{M} \beta_m(r_s, z_s; \theta_m(z_\ell))\phi_m(z_\ell) + \upsilon(z_\ell); \quad \ell = 1, \ldots, L \qquad (8.115)$$

with

$$\beta_m(r_s, z_s; \theta_m) := q\, H_0(\theta_m(z_\ell)r_s)\, \phi_m(z_s) \qquad (8.116)$$

and dispersion (sound-speed)

$$c(z_\ell; \theta_m) = \frac{\omega}{\sqrt{\kappa_z^2(m) + \theta_m^2(z_\ell)}}, \quad m = 1, \ldots, M; \ell = 1, \ldots, L \qquad (8.117)$$

Here the "combined" augmented model for this case leads to both a *nonlinear state and measurement space*, that is,

$$\Phi(z_\ell; \Theta) = \mathbf{a}[\Phi(z_{\ell-1}; \Theta)] + \mathbf{w}(z_{\ell-1}),$$

$$p(r_s, z_\ell) = c[\Phi(z_\ell; \Theta)] + \upsilon(z_\ell) \qquad (8.118)$$

In this case, both the propagator and the pressure-field measurements are nonlinear functions of the states (modes) and unknown parameters (wavenumbers). Note that the modal coefficients are also direct functions of the estimated wavenumbers and are adapted simultaneously. Therefore, this processor is clearly non-Gaussian similar to the previous case.

It should be noted that the initial model parameters are obtained from the prior solution of the boundary value problem (BVP) typically developed as part of the experimental design process and/or after the experiment has been executed. Here the initial "guesses" at modal coefficients and modal functions themselves are calculated based on the experimental conditions such as frequencies, current–temperature–density (CTD), archival sound-speed profiles (SSP), boundary conditions, horizontal wavenumber estimators (e.g. see [43, 44] for more details) to provide the input to the normal-mode BVP solutions (SNAP [45], KRAKEN [46], SAFARI [47]) yielding the required parameters. These parameters are then input to the state-space, measurement, and noise/uncertainty models.

8.4.3 Processors

In this section, we briefly discuss the processors for our shallow oceanic problem with details available in Section 4.3. The basic parametrically adaptive problem can be defined in terms of the mathematical models as

GIVEN, $[\{p(r_s, z_\ell)\}, \{c(z_\ell)\}]$, a set of noisy pressure-field and sound-speed measurements varying in depth along with the underlying state-space model of Eqs. (8.114), (8.115) and (8.117) with unknown parameters $\{\theta(z_\ell)\}$, FIND the "best" (minimum error variance) estimates (joint) of the modal functions and parameters, that is, $\{\hat{\phi}_m(z_\ell)\}, \{\hat{\theta}_m(z_\ell)\}; m = 1, \dots, M$ and measurements $\{\hat{p}(r_s, z_\ell)\}$.

The solution to this problem lies in the *joint state/parameter estimation* problem, that is, defining the *augmented state vector*,

$$\Phi(z_\ell; \Theta) := \begin{bmatrix} \Phi(z_\ell) \\ --- \\ \Theta \end{bmatrix}$$

and starting with the joint distribution applying Bayes' theorem, we obtain

$$\Pr[\Phi(z_\ell; \Theta)|P_\ell] = \left(\frac{\Pr[p(r_s, z_\ell)|\Phi(z_\ell; \Theta)] \times \Pr[\Phi(z_\ell; \Theta)|\Phi(z_{\ell-1}; \Theta)]}{\Pr[p(r_s, z_\ell)|P_{\ell-1}]} \right)$$
$$\times \Pr[\Phi(z_{\ell-1}; \Theta)|P_{\ell-1}] \qquad (8.119)$$

where we have assumed conditional independence and defined the set of measurements as $P_\ell := \{p(r_s, z_1), \dots, p(r_s, z_\ell)\}$.

Define the *joint weighting function* in terms of the likelihood, transition, and evidence as (see Section 2.5 for details)

$$W(z_\ell; \Theta) := \left(\frac{\Pr[p(r_s, z_\ell)|\Phi(z_\ell; \Theta)] \times \Pr[\Phi(z_\ell; \Theta)|\Phi(z_{\ell-1}; \Theta)]}{\Pr[p(r_s, z_\ell)|P_{\ell-1}]} \right)$$
$$(8.120)$$

yielding the *sequential* Bayesian posterior distribution as

$$\Pr[\Phi(z_\ell; \Theta)|P_\ell] = W(z_\ell; \Theta) \times \Pr[\Phi(z_{\ell-1}; \Theta)|P_{\ell-1}] \qquad (8.121)$$

In this study, we are interested in investigating the applicability of the particle filter (PF) of Section 4.3.6 and the UKF of Section 4.3.4 with the goal of analyzing their performance on pressure-field data obtained from the well-known Hudson Canyon experiments performed on the New Jersey shelf [43, 44]. Recall that the PF is a sequential MCMC Bayesian processor capable of providing reasonable performance for a multimodal (multiple peaked) distribution problem estimating a nonparametric representation of the posterior distribution. On the other hand, the UKF is a processor capable of representing any unimodal (single peaked) distribution using a statistical linearization technique based on sigma points that deterministically characterize the posterior.

Here we are concerned with the *joint estimation problem* consisting of setting a prior for θ and augmenting the state vector to solve the problem as defined above thereby converting the parameter estimation problem into one of optimal filtering. Thus, the particle filter estimates the weights required to specify the posterior distribution, empirically, that is,

$$\hat{\Pr}[\Phi(z_\ell; \Theta)|P_\ell] \approx \frac{1}{N_p} \sum_{i=1}^{N_p} \hat{W}_i(z_\ell; \Theta) \times \delta(\Phi(z_\ell; \Theta) - \Phi_i(z_\ell; \Theta)) \qquad (8.122)$$

The approach we chose for our problem is to estimate these weights based on the concept of importance sampling (see Section 4.3). Recall that *importance sampling* is a technique to compute statistics with respect to one distribution using random samples drawn from another. It is a method of simulating samples from a proposal or sampling (importance) distribution to be used to approximate a targeted distribution (joint posterior) by appropriate weighting. For this choice, the weighting function is defined by

$$
\mathcal{W}(z_\ell; \Theta) := \frac{\Pr[\Phi(z_\ell; \Theta)|P_\ell]}{q[\Phi(z_\ell; \Theta)|P_\ell]} \tag{8.123}
$$

where $q[\cdot]$ is the proposed sampling or *importance distribution*.

For the "sequential" case, we have the weighting function

$$
\mathcal{W}(z_\ell; \Theta) \propto \left(\frac{\Pr[p(r_s, z_\ell)|\Phi(z_\ell; \Theta)] \times \Pr[\Phi(z_\ell; \Theta)|\Phi(z_{\ell-1}; \Theta)]}{q[\Phi(z_\ell; \Theta)|\Phi(z_{\ell-1}; \Theta), P_\ell]} \right)
$$
$$
\times \mathcal{W}(z_{\ell-1}; \Theta) \tag{8.124}
$$

where $\propto$ means *proportional to* up to a normalizing constant (evidence).

There are a variety of PF algorithms available, each evolving by a particular choice of the sampling or importance distribution, but the simplest is the *bootstrap* technique that we apply to our problem. Here the importance distribution is selected as the transition prior, that is,

$$
q[\Phi(z_\ell; \Theta)|\Phi(z_{\ell-1}; \Theta), P_\ell] \to \Pr[\Phi(z_\ell; \Theta)|\Phi(z_{\ell-1}; \Theta)] \tag{8.125}
$$

and substituting into Eq. (8.124) we obtain

$$
\mathcal{W}(z_\ell; \Theta) = \Pr[p(r_s, z_\ell)|\Phi(z_\ell; \Theta)] \times \mathcal{W}(z_{\ell-1}; \Theta) \tag{8.126}
$$

Thus, we see that once the underlying posterior is available, the estimates of important statistics can be inferred directly. For instance, the MAP estimate is simply found by locating a particular particle $\hat{\phi}_i(z_\ell)$ corresponding to the maximum of the PMF, that is, $\hat{\Phi}_i(z_\ell; \Theta)_{MAP}$, while the conditional mean estimate is calculated by integrating the posterior to give $\hat{\Phi}_i(z_\ell; \Theta)_{CM}$.

For the bootstrap implementation, we need only to draw noise samples from the state and parameter distributions and use the dynamic models above (normal-mode/random walk) in Eq. (8.114) to generate the set of particles, $\{\Phi_i(z_\ell; \Theta)\} \to \{\Phi_i(z_\ell), \Theta_i(z_\ell)\}$ for $i = 1, \dots, N_p$. That is, both sets of particles are generated from the augmented models (linear/nonlinear) for each individual case (adaptive modal coefficients or adaptive wavenumbers) from

$$
\Phi_i(z_\ell; \Theta) = \begin{cases} \mathbf{A}(z_{\ell-1})\Phi_i(z_{\ell-1}) + \mathbf{w}_i(z_{\ell-1}) & \text{(Case i: modal coefficients)} \\[2ex] \mathbf{a}[\Phi_i(z_{\ell-1}; \Theta)] + \mathbf{w}_i(z_{\ell-1}) & \text{(Case ii: wavenumbers)} \end{cases}
$$
$$
\tag{8.127}
$$

while the likelihood is determined from the nonlinear pressure-field measurement model:

$$p(r_s, z_\ell) = c[\Phi_i(z_\ell; \Theta)] + v(z_\ell) \tag{8.128}$$

Assuming additive Gaussian noise, the likelihood probability is given by

$$\Pr[p(r_s, z_\ell)|\Phi_i(z_\ell)] = \frac{1}{\sqrt{2\pi R_{vv}}} \times \exp\left\{-\frac{1}{2R_{vv}}(p(r_s, z_\ell) - c[\Phi_i(z_\ell; \Theta)])^2\right\} \tag{8.129}$$

Thus, we estimate the posterior distribution using a sequential Monte Carlo approach and construct a *bootstrap particle filter* using the following steps:

- *Initialize:* $\Phi_i(0), \mathbf{w}_i \sim \mathcal{N}(0, \mathbf{R}_{ww}), W_i(0) = 1/N_p; i = 1, \dots, N_p$
- *State transition:* $\Phi_i(z_\ell; \Theta) = \begin{cases} A(z_{\ell-1})\Phi_i(z_{\ell-1}) + \mathbf{w}_i(z_{\ell-1}) & \text{(Case i)} \\ a[\Phi_i(z_{\ell-1}; \Theta)] + \mathbf{w}_i(z_{\ell-1}) & \text{(Case ii)} \end{cases}$
- *Likelihood probability:* $\Pr[p(r_s, z_\ell)|\Phi_i(z_\ell)]$ of Eq. (8.129)
- *Weights:* $W_i(z_\ell; \Theta) = W_i(z_{\ell-1}; \Theta) \times \Pr[p(r_s, z_\ell)|\Phi_i(z_\ell)];$
- *Normalize:* $\mathcal{W}_i(z_\ell; \Theta) = \frac{W_i(z_\ell; \Theta)}{\sum_{i=1}^{N_p} W_i(z_\ell; \Theta)}$
- *Resample:* $\tilde{\Phi}_i(z_\ell; \Theta) \Longleftarrow \Phi_i(z_\ell; \Theta)$
- *Posterior:* $\hat{\Pr}[\Phi(z_\ell; \Theta)|P_\ell] = \sum_{i=1}^{N_p} \mathcal{W}_i(z_\ell; \Theta) \times \delta(\Phi(z_\ell; \Theta) - \Phi_i(z_\ell; \Theta))$

- *MAP estimate:* $\hat{\Phi}_i(z_\ell; \Theta)_{MAP} = \max_i \hat{\Pr}[\Phi_i(z_\ell; \Theta)|P_\ell]$
- *MMSE (CM) estimate:* $\hat{\Phi}_i(z_\ell; \Theta)_{MMSE} = \frac{1}{N_p} \sum_{i=1}^{N_p} \mathcal{W}_i(z_\ell; \Theta) \times \Phi_i(z_\ell; \Theta).$

8.4.4 Model-Based Ocean Acoustic Processing

In this section, we discuss the development of the propagators for the Hudson Canyon experiment performed in 1988 in the Atlantic with the primary goal of investigating acoustic propagation (transmission and attenuation) using continuous-wave data [43, 44]. The Hudson Canyon is located off the coast of New Jersey in the area of the Atlantic Margin Coring project borehole 6010. The seismic and coring data are combined with sediment properties measured at that site. Excellent agreement was determined between the model and data, indicating a well-known, well-documented shallow water experiment with bottom interaction and yielding ideal data sets for investigating the applicability of an MBP to measured ocean acoustic data. The experiment was performed at low frequencies (50–600 Hz) in shallow water of 73 m depth during a period of calm sea state. A calibrated acoustic source was towed at roughly 36 m depth along the 73 m isobath radially to distances of 4–26 km. The ship speed was between 2 and 4 knots. The fixed vertical hydrophone array consisted of 24 phones spaced 2.5 m apart extending from the seafloor up to a depth of about 14 m below the surface. The CTD and SSP measurements were made at regular intervals, and the data were collected under carefully controlled conditions in the ocean environment. The normalized horizontal wavenumber spectrum for a 50 Hz temporal frequency is dominated by 5 modes occurring at wavenumbers between 0.14 and 0.21 m^{-1} with relative amplitudes increasing with increasing wavenumbers. A SNAP [45] simulation was performed, and the results agree quite closely, indicating a well-understood ocean environment.

In order to construct the state-space propagator, we require the set of parameters that were obtained from the experimental measurements and processing (wavenumber spectra). The horizontal wavenumber spectra were estimated using synthetic aperture processing [43]. Eight temporal frequencies were employed: four on the inbounds (75, 275, 575, 600 Hz) and four on the outbounds (50, 175, 375, 425 Hz). In this application, we will confine our investigation to the 50 Hz case, which is well documented, and to horizontal ranges from 0.5 to 4 km. The raw measured data was processed (sampled, corrected, filtered, etc.) and supplied for this investigation. For this investigation, we used a single snapshot of the pressure-field across the vertical array.

8.4.4.1 Adaptive PF Design: Modal Coefficients

The model-based design and development of the environmentally adaptive PF proceeds through the following steps: (i) preprocessing the raw experimental data; (ii) solving the BVP[45] to obtain initial parameter sets for each temporal frequency (e.g. modal coefficients, wavenumbers, initial conditions); (iii) state-space forward propagator simulation of synthetic data for PF analysis/design; (iv) application to measured data; and (v) PF performance analysis.

Preprocessing of the measured pressure-field data follows the usual pattern of filtering, outlier removal, and Fourier transforming to obtain the complex pressure-field as a function of depth along the array. This data along with experimental conditions (frequencies, CTD, SSP, boundary conditions, horizontal wavenumber estimators (see [43] for details) provide the input to the normal-mode BVP solutions (SNAP [45]) yielding the output parameters. These parameters are then used as input to the state-space forward propagator developed previously.

The state-space propagator is then used to develop a set of synthetic pressure-field data with higher resolutions than the original raw data (e.g. 46-element array rather than 23-element at half-wave interelement spacing). This set represents the "truth" data that can be investigated when "tuning" the PF (e.g. number of particles, covariances). Once tuned, the processors are applied directly to the measured pressure-field data (23-elements) after readjusting some of the processor parameters (covariances). Here the performance metrics are estimated and processor performance analyzed. Since each run of the PF is a random realization, that is, the process noise inputs are random, an ensemble of results are estimated with its statistics presented. In this way, we can achieve a detailed analysis of the processor performance prior to fielding and operational version. We constrain our discussion results to processing the noisy experimental pressure-field measurements.

We performed a series of "tuning" runs for both the UKF and PF. Here we primarily adjusted the process noise covariance matrix ($\mathbf{R}_{ww}$) for each of the modal functions and then executed an 100-member ensemble of realizations using these parameters. The particle filter was designed with the same parameters, and 1500-particles were used to characterize the posterior PMF at each depth. Resampling (see Section 4.3) was applied at every iteration of the PF to avoid any potential degradation.

First we investigate the enhancement capabilities of the PF in estimating the pressure-field over a 100-member ensemble shown in Figure 8.22. The resulting figures show the averaged PF estimates. We observe the raw data (DATA) as well as both MAP and conditional mean (CM) estimates. Both estimators are

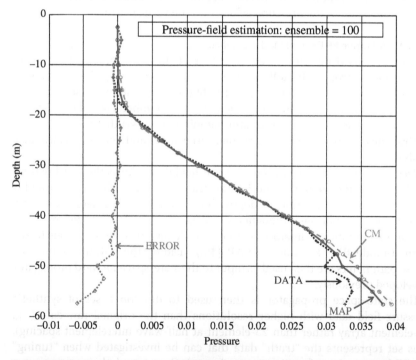

Figure 8.22 Raw/enhanced pressure-field (DATA) data from the Hudson Canyon experiment using particle filter estimators with adaptive modal coefficients: maximum a posteriori (MAP), conditional mean (CM), and the corresponding innovations (ERROR) sequence.

capable of tracking the field quite well and even filter the erratic measurements near the bottom of the channel. The innovations or residuals (ERROR) are also shown in the figure. Both estimators are capable of tracking and enhancing the pressure-field. Using classical performance (sanity tests) metrics on the innovations sequence (ERROR), the zero-mean/whiteness tests, both processors satisfy the criteria of unbiasedness (Z-M: $6.2 \times 10^{-4} < 4.9 \times 10^{-1}$) and uncorrelated innovations, that is, less than 5% exceeding the bound (6.3%). The weighted sum-squared residual (WSSR) test of Section 4.2 is also applied with satisfactory results, that is, *no* samples exceed the threshold, indicating a functionally "tuned" processor. The UKF processor also produced reasonable results for the enhanced pressure-field (not shown).

Ensemble mode tracking results are shown in Figure 8.23 for each of the modal function estimators, the PF (MAP/CM) and the UKF. In Figure 8.23 we observe that the performance of the PF appears to track the modes quite well compared to the UKF. It is interesting to note that the modal coefficient estimates are constantly being adapted (adjusted) by the processor throughout the runs attesting to the nonstationary nature of the ocean statistics as illustrated in Figure 8.24.

This completes the analysis of the Hudson Canyon experimental data for the adaptive (modal coefficient) PF processing performance.

8.4.4.2 Adaptive PF Design: Wavenumbers

As before in the modal coefficient case, we investigate the enhancement capabilities of the PF in estimating the pressure-field over a 100-member ensemble shown in Figure 8.25. Using 1500-particles, we see the raw hydrophone data (dashed line) from the experiment as well as both MAP estimates (circles) and conditional mean (CM) estimates (dotted line with circles). Both estimators appear to track the field quite well (true (mean) solution in dashes). The corresponding innovations (residual) sequence is also shown (diamonds). Classically, both estimators produced satisfactory zero-mean/statistical whiteness test as well as the WSSR tests indicating a "tuned" processor [9].

The ensemble mode tracking results are shown in Figure 8.26 for each of the modal function estimators, the PF (MAP/CM) and the UKF. In Figure 8.26, we observe that the performance of the PF appears to track the modes quite well and better than the UKF. The root-mean-squared (modal tracking) error for each mode is quite reasonable on the order of 10^{-5} again confirming their performance. It is interesting to note that the wavenumber estimates are constantly being adapted (adjusted) by the processor throughout the runs attesting to the nonstationary nature of the ocean statistics. The ensemble average wavenumber estimates are very reasonable: (0.206, 0.197, 0.181, 0.173, 0.142; (TRUE) 0.208, 0.199, 0.183, 0.175, 0.142. The PF and CM ensemble estimates are very close to the true values adapting to the changing ocean environment yet still preserving wavenumber values on average.

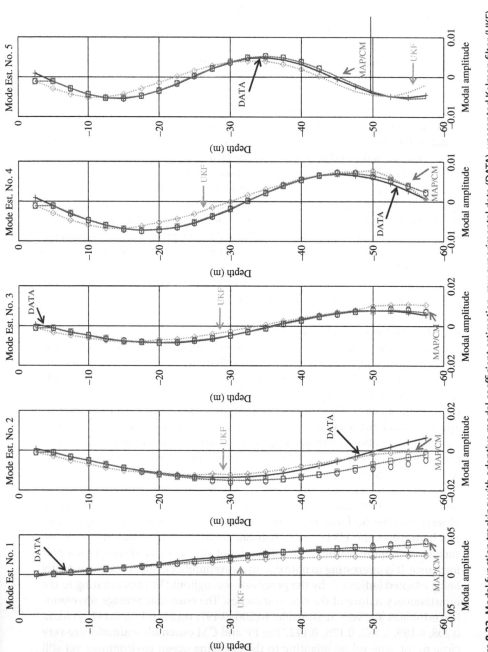

Figure 8.23 Modal function tracking with adaptive modal coefficient estimation: raw experimental data (DATA), unscented Kalman filter (UKF), maximum a posteriori (MAP) (circle), and conditional mean (CM) (square) particle filters.

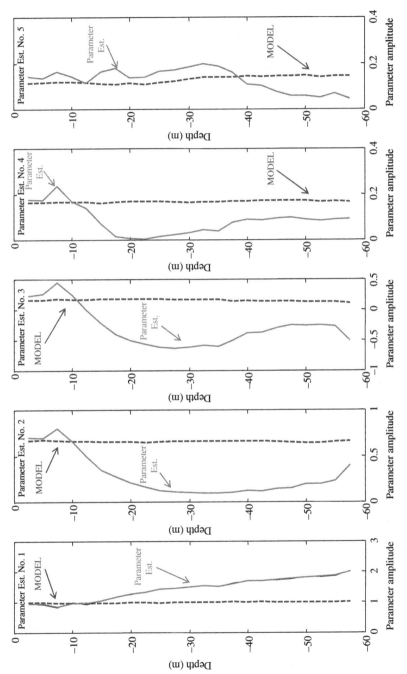

Figure 8.24 Adaptive modal coefficient parameter estimation data (MODEL) from the Hudson Canyon experiment using the MAP particle filter (Parameter Est.).

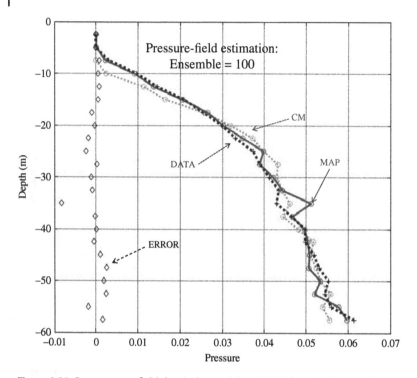

Figure 8.25 Raw pressure-field data/enhanced data (DATA) from the Hudson Canyon experiment for a 23-element hydrophone vertical array using particle filter estimators with adaptive wavenumber processing: maximum a posteriori (MAP), conditional mean (CM), and the corresponding innovations (ERROR) sequence.

We also illustrate the multimodal aspect of the oceanic data by observing the modal function posterior probability PDF estimates for mode 5 illustrated in Figure 8.27. It is clear from the plots that for each depth multiple peaks appear in the posterior estimates. The wavenumber PDF estimate corresponding to corresponding to mode 5 is shown in Figure 8.28. Again we note the multiple, well-defined peaks in the posterior distribution leading to the MAP parameter estimate.

This completes the analysis of the synthesized Hudson Canyon experiment and the PF processing performance.

8.4.5 Summary

In this study, we have discussed the development of environmentally adaptive processor s capable of tracking modes and enhancing the raw pressure-field measurements obtained from a vertical hydrophone array in shallow water. The parametric adaption was based on simultaneously estimating either the

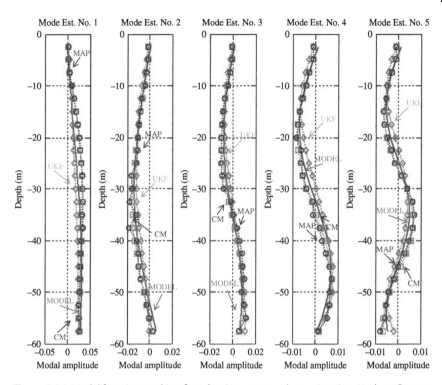

Figure 8.26 Modal function tracking for adaptive wavenumber estimation: Hudson Canyon data (MODEL) of a 23-element array, unscented Kalman filter (UKF), maximum a posteriori (MAP), and conditional mean (CM)(squares) particle filters.

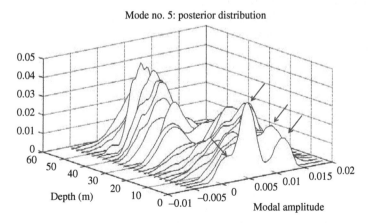

Figure 8.27 Probability mass function (PMF) posterior estimation (mode 5) surface for Hudson Canyon 23-element array data (particle vs time vs probability).

Mode no. 5: wavenumber posterior distribution

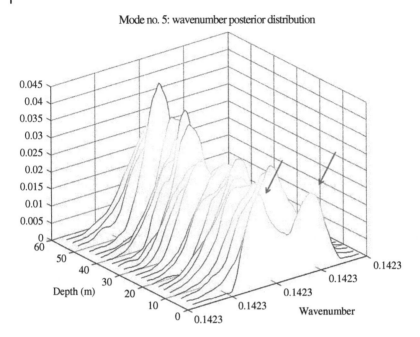

Figure 8.28 Probability mass function (PMF) posterior estimation (wavenumber 5) surface for Hudson Canyon 23-element array data (particle vs time vs probability).

modal coefficients or the horizontal wavenumbers along with the modes and pressure-field as the environmental parameters of interest. These wavenumber parameters were more challenging from a processor design perspective because of their increased sensitivity to environmental change compared to the modal coefficients. We chose a Bayesian sequential design because of the varying nature of the shallow ocean and applied a normal-mode model in state-space form to create a forward propagator. The algorithms applied were the UKF and the particle filter, both modern approaches applied to this problem (see Chapter 4). We compared their performance and found slightly better results of the PF over a 100-member ensemble. These results with more details can be found in a recent paper [48].

8.5 MBID for Chirp Signal Extraction

Chirp signals have evolved primarily from radar/sonar signal processing applications specifically attempting to estimate the location of a target in surveillance/tracking volume [49, 50] and have recently been introduced to the nondestructive evaluation (NDE) area [51]. The chirp, which is essentially a sinusoidal signal whose phase changes instantaneously at each time sample, has an interesting property in that its correlation approximates an impulse function [52, 53]. It is well known that a matched-filter detector in radar/sonar estimates the target range by cross-correlating a replicant of the transmitted chirp with the measurement data reflected from the target back to the radar/sonar receiver, yielding a maximum peak corresponding to the echo time and therefore enabling the desired range estimate.

In this application, we perform the same operation as a radar or sonar system, that is, we transmit a "chirp-like pulse" into the target medium and extract it from noisy measurements. Our problem is complicated by the presence of disturbance signals from surrounding broadcast stations as well as extraneous sources of interference in our frequency bands and, of course, the ever-present random instrumentation noise [51]. First, we model the two signals of high interest: the chirp and FSK signals and then develop a model-based identifier to extract (estimate) them from noisy data.

8.5.1 Chirp-like Signals

8.5.1.1 Linear Chirp

A *chirp signal* is a sinusoidal signal characterized by its amplitude α and phase ϕ, that is,

$$s(t) = \alpha \times \cos(\phi(t)) \tag{8.130}$$

where the instantaneous phase is defined by

$$\phi(t) = 2\pi f(t) + \phi(t_o) \tag{8.131}$$

for $f(t)$ the corresponding *instantaneous frequency*. The rate of change of the phase is defined by its derivative

$$\frac{d}{dt}\phi(t) = 2\pi f(t) =: \omega_i(t) \tag{8.132}$$

where ω_i is defined as the instantaneous *angular radian frequency*.

The frequency can be characterized by a number of different relations, one of which is the "linear" relation

$$f(t) = \beta \times t + f_o \tag{8.133}$$

where f_o is the initial (sweep) frequency at time t_o and β is the rate given by

$$\beta = \frac{f_f - f_o}{t_f} \tag{8.134}$$

where f_f is the final (sweep) frequency in the time window bounded by the final (sweep) time t_f.

For the linear chirp, the corresponding phase is found by integrating Eq. (8.132):

$$\phi(t) = \phi(t_o) + 2\pi \int_0^t f(\alpha)d\alpha = \phi(t_o) + 2\pi \left(f_o\, t + \frac{\beta\, t^2}{2} \right) \qquad (8.135)$$

The corresponding *sinusoidal chirp* signal is then given by (see spectrum in Figure 8.29a)

$$s(t) = \alpha\, \cos\left(2\pi \left(f_o \cdot t + \frac{\beta}{2} \cdot t^2 \right) \right) \qquad (8.136)$$

For our problem, we have disturbances and noise that contaminate the measurement, that is, assuming a sampled-data representation $(t \rightarrow t_k)$, we have

$$y(t_k) = s(t_k) + d(t_k) + e(t_k) + v(t_k) \qquad (8.137)$$

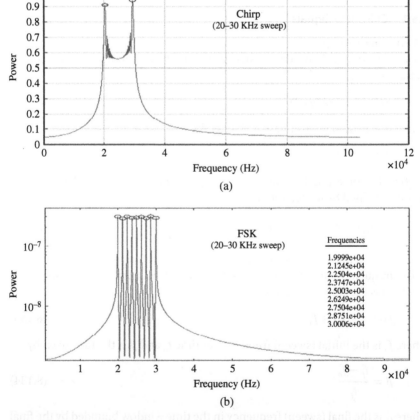

Figure 8.29 True chirp and FSK spectra (20–30 kHz sweep): (a) Chirp spectrum. (b) FSK spectrum ($\Delta f = 1.25$ kHz-steps).

where s is the transmitted chirp, d is potential broadcast disturbances, e is an extraneous disturbances, and v is a zero-mean, Gaussian random noise process with variance $R_{vv}(t_k)$. A chirp signal and spectrum with disturbances and noise is shown in Figure 8.30a for a chirp signal swept between 20 and 30 kHz with disturbances and noise illustrated in (b).

8.5.1.2 Frequency-Shift Key (FSK) Signal

An alternative for a linear frequency modulation (FM) chirp is FSK modulation deemed the simplest form of FM [54] with the spectrum illustrated in Figure 8.29b. In the *M-ary* case, *M* different frequencies are used to transmit a coded information sequence. The choice of *frequency separation* Δf is defined by the *FSK-signal*

$$s_m(t_k) = \sqrt{\frac{2\mathcal{E}_s}{\mathcal{T}}} \cos 2\pi (f_o + m\Delta f) t_k; \quad m = 0, \dots, M-1 \tag{8.138}$$

where $\mathcal{E}_s = m \times \mathcal{E}_b$ is defined as the *energy/symbol*, $\mathcal{T} = m \times \mathcal{T}_b$ is the *symbol interval* and $\Delta f = f_m - f_{m-1}$ with $\mathcal{E}_b$ is the signal *energy/bit* and $\mathcal{T}_b$ is the duration of the *bit interval*. Usually, *M-ary* FSK is used to transmit a block of $n = \log_2 M$ bits/signal. Here the *M*-FSK signals have *equal* energy $\mathcal{E}_s$.

The frequency separation Δf determines the degree of discrimination possible of the *M* transmitted signals. Each of these signals can be represented as *orthogonal* unit-vectors $\mathbf{u}_m$ scaled by $\sqrt{\mathcal{E}_s}$, that is, $\mathbf{s}_m := \sqrt{\mathcal{E}_s} \mathbf{u}_m$, where the basis functions are defined $\mathbf{b}_m(t_k) = \sqrt{\frac{2}{\mathcal{T}}} \cos 2\pi (f_o + m\Delta f) t_k$. The minimum distance between the pairs of signal vectors is given by $d = \sqrt{\mathcal{E}_s}$. We will incorporate this FSK representation to develop our model-based identifier.

In order to represent the FSK signal in the time domain, we must introduce the *gate* function $G_{t_k}(\Delta \tau)$ defined

$$G_{t_k}(\Delta \tau) := \mu(t_k - m\Delta \tau) - \mu(t_k - (m+1)\ \Delta \tau); \quad m = 0, \dots, M-1 \tag{8.139}$$

for $\mu(\cdot)$ a unit-step function and $\Delta \tau = \frac{1}{\Delta f}$ the *reciprocal* of the frequency separation or equivalently $\mathcal{T}_b$ the bit interval duration. With this operator available, we can now represent the temporal FSK signal as

$$s(t_k) = \sum_{m=0}^{M-1} s_m(t_k) \times G_{t_k}(\mathcal{T}_b)$$

$$= \sqrt{\frac{2\mathcal{E}_s}{\mathcal{T}}} \sum_{m=0}^{M-1} \cos 2\pi (f_o + m\Delta f) t_k \ [\mu(t_k - m\mathcal{T}_b) - \mu(t_k - (m+1)\mathcal{T}_b)] \tag{8.140}$$

This model will be used to extract the FSK signal.

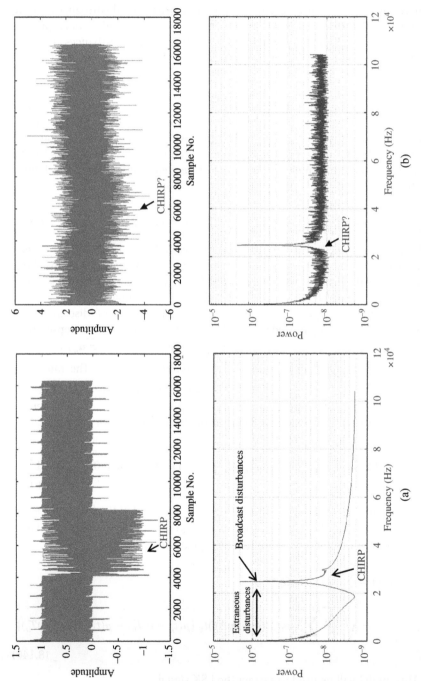

Figure 8.30 Chirp data: (a) True chirp signal and disturbance with spectrum. (b) Noisy (−40 dB) chirp signal and disturbance with spectrum.

8.5.2 Model-Based Identification: Linear Chirp Signals

Thus, the model-based linear *chirp estimation* problem can simply be stated as GIVEN a set of noisy measurement data $\{y(t_k)\}, k = 0, 1, \ldots, K$ contaminated with broadcast disturbances $\{d(t_k)\}$, extraneous disturbances $\{e(t_k)\}$ and Gaussian noise $\{v(t_k)\}$, FIND the best estimate of the chirp signal $\hat{s}(t_k)$ or equivalently the parameters $\{\hat{\alpha}, \hat{f}_o, \hat{f}_f\}$

The estimated chirp is obtained directly from its parameter estimates as

$$\hat{s}(t_k) = \hat{\alpha} \cos\left(\hat{f}_o \cdot t + 2\pi\left(\frac{\hat{\beta}}{2} \cdot t^2\right)\right) \tag{8.141}$$

for

$$\hat{\beta} = \frac{\hat{f}_f - \hat{f}_o}{t_f} \tag{8.142}$$

Thus, the solution to estimating the linear chirp in the midst of these disturbances reduces to a problem of MBID.

8.5.2.1 Gauss–Markov State-Space Model: Linear Chirp

We choose to apply the UKF of Section 4.3 as a MBID-technique to solve this nonlinear problem by modeling the parameter variations as *random walks*, that is, in state-space form we have that

$$\mathbf{f}(t_k) = \mathbf{f}(t_{k-1}) + \mathbf{w}_f(t_{k-1})$$
$$\alpha(t_k) = \alpha(t_{k-1}) + w_\alpha(t_{k-1}) \tag{8.143}$$

where $\mathbf{f} \in \mathcal{R}^{N_f \times 1}$ is the vector of frequency parameters, α is the scalar amplitude and $\mathbf{w} \in \mathcal{R}^{N_f \times 1}$ a zero-mean, multivariate Gaussian vector process with covariance $\mathbf{R}_{ww}(t_k)$. Here $\mathbf{f} = [f_o \mid f_f]'$ and $\mathbf{w}_f = [w_o \mid w_f]'$. The corresponding measurement model is given in Eq. (8.141).

Generally, the UKF algorithm provides the updated signal (parameter) estimates as[2]

$$\hat{\mathbf{f}}(t_k|t_k) = \hat{\mathbf{f}}(t_k|t_{k-1}) + \mathbf{K}_f(t_k)\epsilon(t_k|t_{k-1})$$
$$\hat{\alpha}(t_k|t_k) = \hat{\alpha}(t_k|t_{k-1}) + \mathbf{K}_\alpha(t_k)\epsilon(t_k|t_{k-1})$$
$$\epsilon(t_k|t_{k-1}) = y(t_k) - \hat{y}(t_k|t_{k-1}) - d(t_k) - e(t_k)$$
$$\hat{y}(t_k|t_{k-1}) = \hat{s}(t_k|t_k) = \hat{\alpha}(t_k|t_{k-1})$$
$$\times \cos\left(2\pi\left(\hat{f}_o(t_k|t_{k-1}) \cdot t_k + \frac{\hat{\beta}(t_k|t_{k-1})}{2} \cdot t_k^2\right)\right)$$
$$\mathbf{K}(t_k) = \mathbf{R}_{f\epsilon}(t_k|t_{k-1})\,\mathbf{R}_{\epsilon\epsilon}^{-1}(t_k|t_{k-1}) \tag{8.144}$$

---

2 The notation $\hat{\theta}(t_k|t_{k-1})$ means the estimate of θ at time t_k based on all the data up to time t_{k-1}, $Y_{k-1}, \{y(t_k)\}; k = 0, \ldots, K$.

with the instantaneous frequency (linear chirp) where the amplitude is estimated by $\hat{\alpha}(t_k|t_k)$ and

$$\hat{f}(t_k|t_{k-1}) := \hat{f}_o(t_k|t_k) \cdot t_k + \frac{\hat{\beta}(t_k|t_k)}{2} \cdot t_k^2$$

$$\hat{\beta}(t_k|t_k) = \frac{\hat{f}_f(t_k|t_k) - \hat{f}_o(t_k|t_k)}{t_f} \tag{8.145}$$

Since we are transmitting the chirp signal, we have a good estimate of the starting values for α, f_o, f_f, t_f that can be used in the signal estimation problem. An interesting property of the UKF processor is that the innovations (residual errors) should be approximately zero-mean and white (uncorrelated) indicating that the model "matches" the data.

Thus, the received data can be considered to consist of extraneous and broadcast disturbances along with random measurement noise. We developed a set of synthesized measurement data by filtering (low pass) a snip-it of raw measurements to extract the extraneous disturbance with its power spectrum shown in Figure 8.32b. Next we synthesized a unit amplitude ($\alpha = 1$) chirp and broadcast disturbances. That is, we assume a chirp signal was swept over the entire record length sweeping up from $f_o = 20$ kHz to $f_f = 30$ kHz for a period of $t_f = 19.7$ ms at a sampling interval of $\Delta t = 4.8$ μs. From the power spectrum of Figure 8.30a, we see that the extraneous disturbance is in the frequency range of 0–19 kHz (Figure 8.32) with the broadcast (sinusoid) disturbance at 2.48 kHz. For our simulation, we chose amplitudes of 0.2 and 0.5, respectively. The simulated data and spectrum (without instrumentation noise) are shown in Figure 8.32a. Note that all of the amplitude data was scaled down from actual measurements. For the simulation, we chose a measurement model of

$$y(t_k) = 1.0 \cos(2\pi(2 \times 10^4 t_k + 3 \times 10^4 t_k^2) + 0.5 \sin(2\pi\, 2.48 \times 10^4 t_k)$$

$$+0.25e(t_k) + v(t_k) \tag{8.146}$$

for $v \sim \mathcal{N}(0, 1 \times 10^{-1})$. We show the simulated noisy measurement (input to the UKF-processor) and spectrum in Figure 8.32b with instrumentation noise included obscuring the chirp signal completely.

Next we applied the UKF processor to estimate the instantaneous frequency and compared it in steady state ($\bar{f} = E\{\hat{f}(t_k|t_k)\}$) dynamically at each instant as shown in Figure 8.31. Here we see that the model can track the parameters as indicated by its estimate of the instantaneous frequency and the resulting chirp with the corresponding error. The UKF processor is "tuned," since the innovations statistics are zero-mean/white using the statistical zero-mean/whiteness test as well as the WSSR test indicating a reasonably tuned processor. The WSSR processor performs a dynamic whiteness test by sliding a finite length (parameter) window through the data and tests for whiteness as long as the statistic lies *below* the threshold (see Section 4.2 for details).

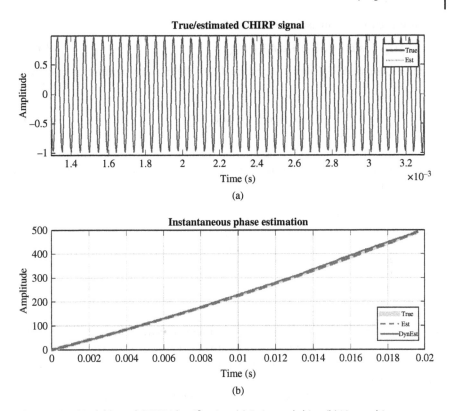

Figure 8.31 Model-based CHIRP identification: (a) Estimated chirp. (b) Linear chirp: true, steady-state, dynamic).

8.5.3 Model-Based Identification: FSK Signals

In this subsection, we develop the model-based identifier for the FSK signal discussed earlier. Recall that the FSK signal can be parameterized as

$$s_m(t_k) = \alpha \cos 2\pi (f_o + m\Delta f) t_k; \quad m = 0, \dots, M - 1 \qquad (8.147)$$

where the amplitude is $\alpha = \sqrt{\frac{2\mathcal{E}_s}{T}}$, the carrier frequency is f_o and the frequency separation is Δf. From the signal estimation perspective, we need to estimate only the amplitude and separation, since the carrier is well known as well as the number of transmitted frequencies M. Therefore, the model-based FSK *estimation* problem including the disturbances (as before) can be stated as

GIVEN a set of noisy measurement data $\{y(t_k)\}, k = 0, 1, \dots, K$ contaminated with broadcast disturbances $\{d(t_k)\}$, extraneous disturbances $\{e(t_k)\}$, and Gaussian noise $\{v(t_k)\}$, FIND the best estimate of the FSK signal $\hat{s}(t_k)$ or equivalently the parameters $\{\hat{\alpha}, \Delta\hat{f}\}$

For this case, the measurement model is

$$y(t_k) = s(t_k) + d(t_k) + e(t_k) + v(t_k) \tag{8.148}$$

and the estimated FSK signal is obtained directly from its parameter estimates as

$$\hat{s}(t_k) = \hat{\alpha} \sum_{m=0}^{M-1} \cos 2\pi (f_o + m\Delta\hat{f}) t_k \times G_{t_k}(\mathcal{T}_b) \tag{8.149}$$

Thus, the MBID solution to estimating the FSK in the midst of these disturbances again reduces to a problem of parameter estimation as before in the linear chirp case.

8.5.3.1 Gauss–Markov State-Space Model: FSK Signals

We perform a MBID applying the UKF (UKF) to this nonlinear problem by modeling the parameter variations as *random walks*, that is, in state-space form we have that

$$\Delta f(t_k) = \Delta f(t_{k-1}) + w_f(t_{k-1})$$
$$\alpha(t_k) = \alpha(t_{k-1}) + w_\alpha(t_{k-1}) \tag{8.150}$$

where Δf is the frequency separation parameter, α is the scalar amplitude, and $\mathbf{w} \in \mathcal{R}^{2\times 1}$ a zero-mean, multivariate Gaussian vector process with covariance $\mathbf{R}_{ww}(t_k)$. Here $\mathbf{w} = [w_f \mid w_\alpha]'$. The corresponding measurement model is given in Eq. (8.148).

Generally, the UKF algorithm provides the updated signal (parameter) estimates as

$$\Delta\hat{f}(t_k|t_k) = \Delta\hat{f}(t_k|t_{k-1}) + K_f(t_k)\epsilon(t_k|t_{k-1})$$
$$\hat{\alpha}(t_k|t_k) = \hat{\alpha}(t_k|t_{k-1}) + K_\alpha(t_k)\epsilon(t_k|t_{k-1})$$
$$\epsilon(t_k|t_{k-1}) = y(t_k) - \hat{y}(t_k|t_{k-1}) - d(t_k) - e(t_k)$$
$$\hat{y}(t_k|t_{k-1}) = \hat{s}(t_k|t_k) = \hat{\alpha}(t_k|t_{k-1})$$
$$\times \sum_{m=0}^{M-1} \cos 2\pi (f_o + m\Delta\hat{f}(t_k|t_{k-1})) t_k \times G_{t_k}(\mathcal{T}_b)$$
$$\mathbf{K}(t_k) = \mathbf{R}_{f\epsilon}(t_k|t_{k-1}) \, \mathbf{R}_{\epsilon\epsilon}^{-1}(t_k|t_{k-1}) \tag{8.151}$$

Since we are transmitting the FSK signal, we have a good estimate of the starting values for α, f_o, M that can be used in the signal estimation/detection problem. Recall that the processor is tuned when the innovations are zero-mean/white and not when these statistical conditions are not met. Therefore, the processor can be used as an *anomaly* detector (as before) when these conditions are not satisfied. We discuss this property subsequently.

Thus, the received data can be considered to consist of extraneous and broadcast disturbances along with random measurement noise.

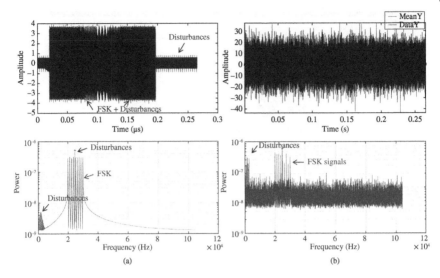

Figure 8.32 FSK data: (a) True FSK signal and disturbance with spectrum. (b) Noisy (−22 dB) FSK signal and disturbance with spectrum.

We synthesized a set of measurement data (as before for the linear chirp) by filtering (low pass) a snip-it of raw measurements to extract the extraneous disturbance with its power spectrum shown in Figure 8.32a. Next we simulated a unit amplitude ($\alpha = 1$) FSK measurement and broadcast disturbances. That is, we assume a FSK pulse signal was swept over the entire record length sweeping up from $f_o = 20$ kHz to $f_f = 30$ kHz for a period of $t_f = 0.177$ seconds at a sampling interval of $\Delta t = 4.8$ μseconds. From the power spectrum of Figure 8.32b, we see that the extraneous disturbance is in the frequency range of 0–19 kHz with the broadcast (sinusoid) disturbance at 2.48 kHz along with the random instrumentation noise signal (−22 dB SNR). For our signal/disturbance simulation, we chose amplitudes of 2.5 and 0.5, respectively. For this simulation, we have a measurement model given by

$$y(t_k) = \sum_{m=0}^{M-1} \cos 2\pi (2 \times 10^4 + 0.125 \times 10^4) t_k \, G_{t_k} (1.97 \times 10^{-3})$$
$$+0.5 \sin(2\pi \, 2.48 \times 10^4) t_k + 2.5 e(t_k) + v(t_k) \tag{8.152}$$

for $v \sim \mathcal{N}(0, 8 \times 10^1)$ for a SNR=−22 dB.

The MBID result is shown in Figure 8.33 indicating a reasonably tuned processor with the predicted measurement and estimated (average) spectrum using the parameterized signal estimates in Figure 8.33b. We note the ability of the processor to extract the desired step-frequencies with little error. This completes the section of the signal estimation problem.

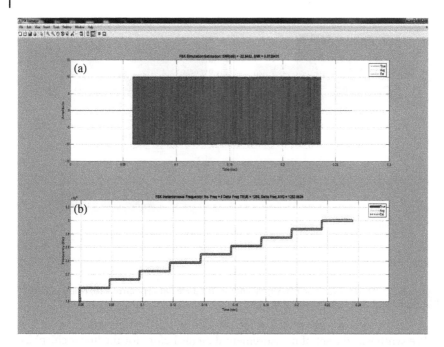

Figure 8.33 Model-based identification of FSK signal with disturbances: (a) True/estimated FSK data (−22 dB SNR). (b) Estimated FSK step-frequency (0.1252 kHz) estimates.

8.5.4 Summary

We have developed a model-based approach to processing a windowed chirp and an FSK signal in noisy data. We applied the UKF technique of Section 4.3 to perform the MBID of the critical chirp and FSK signal parameters from noisy synthesized data and demonstrated that the MBID-approach could be used to extract these signals from noisy measurements.

References

1 Pusey, H. and Pusey, S. (1991). *Focus on Mechanical Failures: Mechanisms and Detection*. Willowbrook, IL: Vibration Institute.

2 Rainieri, C. and Fabbrocino, G. (2014). *Operational Modal Analysis of Civil Engineering Structures*. New York: Springer.

3 Candy, J., Franco, S., Ruggiero, E. et al. (2017). Anomaly detection for a vibrating structure: a subspace identification/tracking approach. *J. Acoust. Soc. Am.* 142 (2): 680–696.

4 Hatch, M. (2001). *Vibration Simulation Using MATLAB and ANSYS*. Boca Raton, FL: Chapman Hall/CRC Press.

5 Gawronski, W. (2004). *Advanced Structural Dynamics and Active Control of Structures*. London: Springer.

6 Reid, J. (1983). *Linear System Fundamentals: Continuous, and Discrete, Classic and Modern*. New York: McGraw-Hill.

7 Kailath, T. (1980). *Linear Systems*. Upper Saddle River, NJ: Prentice Hall.

8 DeCarlo, R.A. (1989). *Linear Systems: A State Variable Approach with Numerical Implementation*. Engelwood Cliffs, NJ: Prentice-Hall.

9 Candy, J. (2006). *Model-Based Signal Processing*. Hoboken, NJ: Wiley/IEEE Press.

10 van Overschee, P. and De Moor, B. (1996). *Subspace Identification for Linear Systems: Theory, Implementation, Applications*. Boston, MA: Kluwer Academic Publishers.

11 Katayama, T. (2005). *Subspace Methods for System Identification*. London: Springer.

12 Verhaegen, M. and Verdult, V. (2007). *Filtering and System Identification: A Least-Squares Approach*. Cambridge: Cambridge University Press.

13 Juang, J. (1994). *Applied System Identification*. Upper Saddle River, NJ: Prentice-Hall PTR.

14 Reynders, E. (2012). System identification methods for (operational) modal analysis: review and comparison. *Arch. Comput. Methods Eng.* 19 (1): 51–124.

15 Aoki, M. (1990). *State Space Modeling of Time Series*, 2e. London: Springer.

16 Kalata, P. (1984). The tracking index: a generalized parameter for $\alpha - \beta$ and $\alpha - \beta - \gamma$ target trackers. *IEEE Trans. Aerosp. Electron. Syst.* AES-20 (2): 174–182.

17 Knoll, G. (2000). *Radiation Detection and Measurement*, 3e. Hoboken, NJ: Wiley.

18 Gilmore, G. and Hemingway, J. (2003). *Practical Gamma-Ray Spectrometry*. New York: Wiley.

19 Oppenheim, A. and Schafer, R. (1989). *Discrete-Time Signal Processing*. Englewood Cliffs, NJ: Prentice-Hall.

20 Kester, W., Sheingold, D., and Bryant, J. (2005). *Data Conversion Handbook*. Holland: Elsevier.

21 Candy, J. (2014). A Model-Based Approach to Scintillator/Photomultiplier Characterization. LLNL Report, LLNL-TR-665562.

22 Wald, A. (1945). Sequential tests of statistical hypothesis. *Ann. Math. Stat.* 16: 117–186.

23 Wald, A. (1947). *Sequential Analysis*. New York: Wiley (Reprint Dover Publications, 1973).

24 Evans, R. (1985). *The Atomic Nucleus*. New York: McGraw-Hill.

25 Candy, J., Breitfeller, E., Guidry, B. et al. (2009). Physics-based detection of radioactive contraband: a sequential Bayesian approach. *IEEE Trans. Nucl. Sci.* 56 (6): 3694–3711.

26 Candy, J., Chambers, D., Breitfeller, E. et al. (2011). Threat detection of radioactive contraband incorporating Compton scattering physics: a model-based processing approach. *IEEE Trans. Nucl. Sci.* 58 (1): 214–230.

27 Moreland, M. and Ristic, B. (2009). Radiological source detection and localization using Bayesian techniques. *IEEE Trans. Signal Process.* 57 (11): 4220–4223.

28 Jarman, K., Smith, L., and Carlson, D. (2004). Sequential probability ratio test for long-term radiation monitoring. *IEEE Trans. Nucl. Sci.* 51 (4): 1662–1666.

29 Luo, P., DeVol, T., and Sharp, J. (2009). Sequential probability ratio test using scaled time-intervals for environmental radiation monitoring. *IEEE Nucl. Sci. Symp. Conf. Rec.* N25-55: 1372–1377.

30 Prasad, M. and Snyderman, N. (2012). Statistical theory of fission chains and generalized Poisson neutron counting distributions. *Nucl. Sci. Eng.* 172: 300–326.

31 Prasad, M., Snyderman, N., Verbeke, J., and Wurtz, R. (2013). Time interval distributions and the Rossi correlation function. *Nucl. Sci. Eng.* 174: 1–29.

32 Walston, S. (2012). A Guide to the Statistical Theory of Fission Chains, 2e. Lawrence Livermore National Laboratory Report, LLNL-TR-584832.

33 Dierck, R. and Hage, W. (1983). Neutron signal multiplet analysis for the mass determination of spontaneous fission isotopes. *Nucl. Sci. Eng.* 85: 325–338.

34 Hage, W. and Cifarelli, D.M. (1985). Correlation analysis with neutron count distributions in randomly or signal triggered time intervals for assay of special fissile materials. *Nucl. Sci. Eng.* 89: 159–176.

35 Middleton, D. (1960). *An Introduction to Statistical Communication Theory*. New York: McGraw-Hill.

36 Van Trees, H. (1968). *Detection, Estimation and Modulation Theory*, Pt. 1. New York: Wiley.

37 Candy, J. and Breitfeller, E. (2013). Receiver Operating Characteristic (ROC) Curves: An Analysis Tool for Detection Performance. LLNL Report: LLNL-TR-642693.

38 Papoulis, A. and Pillai, S. (2002). *Probability, Random Variables and Stochastic Processes*, 4e. New York: McGraw-Hill.

39 Candy, J. (2016). *Bayesian Signal Processing, Classical, Modern and Particle Filtering*, 2e. Hoboken, NJ: Wiley/IEEE Press.

40 Ristic, B., Arulampalam, S., and Gordon, N. (2004). *Beyond the Kalman Filter, Particle Filters for Tracking Applications*. Boston, MA: Artech House.

41 Clay, C. (1987). Optimum time domain signal transmission and source localization in a waveguide. *J. Acoust. Soc. Am.* 81: 660–664.

42 Yardim, C., Michalopoulou, Z.-H., and Gerstoft, P. (2011). An overview of sequential Bayesian filtering in ocean acoustics. *IEEE J. Ocean. Eng.* 36 (1): 73–91.

43 Carey, W., Doutt, J., Evans, R., and Dillman, L. (1995). Shallow water transmission measurements taken on the New Jersey continental shelf. *IEEE J. Ocean. Eng.* 20 (4): 321–336.

44 Rogers, A., Yamamoto, Y., and Carey, W. (1993). Experimental investigation of sediment effect on acoustic wave propagation in shallow water. *J. Acoust. Soc. Am.* 93: 1747–1761.

45 Jensen, F. and Ferla, M. (1982). SNAP: the SACLANTCEN Normal-Mode Acoustic Propagation Model. SACLANTCEN Report, SM-121. La Spezia, Italy: SACLANT Undersea Research Centre.

46 Porter, M. (1991). The KRAKEN Normal Mode Program. Report SM-245, 183. Italy: SACLANTCEN.

47 Schmidt, H. (1987). SAFARI: Seismo-Acoustic Fast Field Algorithm for Range Independent Environments. Report SM-245, 142. Italy: SACLANTCEN.

48 Candy, J. (2015). Environmentally adaptive processing for shallow ocean applications: a sequential Bayesian approach. *J. Acoust. Soc. Am.* 138 (3): 1268–1281.

49 Klauder, J., Price, A., Darlington, S., and Alberstein, W. (1960). The theory and design of chirp radars. *Bell Syst. Tech J.* XXXIX (4): 745–808.

50 Bolger, P. (1990). *Radar Principles with Applications to Tracking Systems.* New York: Wiley.

51 Candy, J. (2016). CHIRP-Like Signals: Estimation, Detection and Processing: A Sequential Model-Based Approach. LLNL Report, LLNL-TR-12345.

52 Lancaster, D. (1965). Chirp a new radar technique. Electronics World (January 1965).

53 Ender, J. (2011). Introduction to Radar. Ruhr-Universitat Bochum Lectures, 1–179.

54 Proakis, J. (2001). *Digital Communications*, 4e. New York: McGraw-Hill.

Appendix A

Probability and Statistics Overview

A.1 Probability Theory

Defining a sample space (outcomes), Ω, a field (events), B, and a probability function (on a class of events), Pr, we can construct an *experiment* as the triple, $\{\Omega, B, \text{Pr}\}$.

Example A.1 Consider the experiment, $\{\Omega, B, \text{Pr}\}$ of tossing a fair coin, then we see that

Sample space:	Ω	$= \{H, T\}$
Events:	B	$= \{0, \{H\}, \{T\}\}$
Probability:	$\text{Pr}(H) = p$	
	$\text{Pr}(T) = 1 - p$	$\square$

With the idea of a sample space, probability function, and experiment in mind, we can now start to define the concept of a discrete random signal more precisely. We define a discrete *random variable* as a real function whose value is determined by the outcome of an experiment. It assigns a real number to each point of a sample space Ω, which consists of all the possible outcomes of the experiment. A random variable X and its realization x are written as

$$X(\omega) = x \quad \text{for} \quad \omega \epsilon \Omega \tag{A.1}$$

Consider the following example of a simple experiment.

Example A.2 We are asked to analyze the experiment of flipping a fair coin, then the sample space consists of a head or tail as possible outcomes, that is,

$$\Omega = \{0 \; H \; T\} \Rightarrow X(\omega) = x$$

$$\omega = \{H, T\}$$

Model-Based Processing: An Applied Subspace Identification Approach, First Edition. James V. Candy.
© 2019 John Wiley & Sons, Inc. Published 2019 by John Wiley & Sons, Inc.

If we assign a 1 for a head and 0 for a tail, then the random variable X performs the mapping of

$$X(\omega = H) = x(H) = 1$$
$$X(\omega = T) = x(T) = 0$$

where $x(\cdot)$ is called the sample value or realization of the random variable X. □

A *probability mass function* defined in terms of the random variable, that is,

$$P_X(x_i) = \Pr(X(\omega_i) = x_i) \tag{A.2}$$

and the *probability distribution function* is defined by

$$F_X(x_i) = \Pr(X(\omega_i) \le x_i) \tag{A.3}$$

These are related by

$$P_X(x_i) = \sum_i F_X(x_i)\delta(x - x_i) \tag{A.4}$$

$$F_X(x_i) = \sum_i P_X(x_i)\mu(x - x_i) \tag{A.5}$$

where δ and μ are the unit impulse and step functions, respectively.

It is easy to show that the distribution function is a monotonically increasing function (see Ref. [1] for details) satisfying the following properties:

$$\lim_{x_i \to -\infty} F_X(x_i) = 0$$

and

$$\lim_{x_i \to \infty} F_X(x_i) = 1$$

These properties can be used to show that the mass function satisfies

$$\sum_i P_X(x_i) = 1$$

Either the distribution or probability mass function completely describes the properties of a random variable. Given either of these functions, we can calculate probabilities that the random variable takes on values in any set of events on the real line. To complete our coin tossing example, if we define the probability of a head occurring as p, then we can calculate the distribution and mass functions as shown in the following example.

Example A.3 Consider the coin tossing experiment and calculate the corresponding mass and distribution functions. From the previous example,

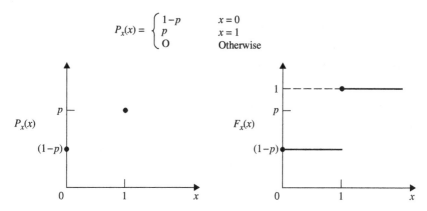

$$P_X(x) = \begin{cases} 1-p & x = 0 \\ p & x = 1 \\ 0 & \text{Otherwise} \end{cases}$$

Figure A.1 Probability mass and distribution functions for coin tossing experiment.

we have

Sample space:	Ω	$= \{H, T\}$
Events:	B	$= \{0, \{H\}, \{T\}\}$
Probability:	$P_X(x_1 = H) = p$	
	$P_X(x_0 = T) = 1 - p$	
Random variable:	$X(\omega_1 = H) = x_1 = 1$	
	$X(\omega_2 = T) = x_2 = 0$	
Distribution:	$F_X(x_i) = \begin{cases} 1 & x_i \geq 1 \\ 1-p & 0 \leq x_i \leq 1 \\ 0 & x_i < 0 \end{cases}$	

the mass and distribution functions for this example are shown in Figure A.1. Note that the sum of the mass function value must be 1 and that the maximum value of the distribution function is 1 satisfying the properties mentioned previously. □

If we extend the idea that a random variable is now a function of time as well, then we can define a stochastic process as discussed in Chapter 2. More formerly, a random or *stochastic process* is a two-dimensional function of t and ω:

$$X(t, \omega), \quad \omega \epsilon \Omega, \quad t \epsilon T \qquad (A.6)$$

where T is a set of index parameters (continuous or discrete) and Ω is the sample space.

We list some of the major theorems in probability theory and refer the reader to more detailed texts [1, 2].

Univariate: $\Pr(X) = P_X(x)$

Bivariate: $\Pr(X, Y) = P_{XY}(x, y)$

Marginal: $\Pr(X) = \sum_y P_{XY}(x, y)$

Independent: $\Pr(X, Y) = P_X(x) \times P_Y(y)$

Conditional: $\Pr(X|Y) = P_{XY}(x, y)/P_Y(y)$

Chain rule: $\Pr(X, Y, Z) = \Pr(X|Y, Z) \times \Pr(Y|Z) \times \Pr(Z)$

For a random variable, we can define basic statistics in terms of the probability mass function. The *expected value* or *mean* of a random variable X, is given by

$$m_x = E\{X\}$$

and is considered the typical or representative value of a given set of data. For this reason, the mean is called a measure of central tendency. The degree to which numerical data tend to spread about the expected value is usually measured by the *variance* or equivalently, *autocovariance* given by

$$R_{xx} = E\{(X - m_x)^2\}$$

The basic statistical measures are called *ensemble statistics* because they are measured across the ensemble ($i = 1, 2, \ldots$) of data values, that is, the expectation is always assumed over an ensemble of realizations. We summarize these statistics in terms of their mass function as

Expected value: $m_x = E\{X\}$ $= \sum_i X_i P_X(x_i)$

Nth-moment: $E\{X^n\}$ $= \sum_i X_i^n P_X(x_i)$

Nth-moment about mean: $E\{(X - m_x)^n\}$ $= \sum_i (X_i - m_x)^n P_X(x_i)$

Mean squared ($N = 2$): $E\{X^2\}$ $= \sum_i X_i^2 P_X(x_i)$

Variance: $R_{xx} = E\{(X_i - m_x)^2\} = \sum_i (X_i - m_x)^2 P_X(x_i)$

Covariance: R_{xy} $= E\{(X_i - m_x)(Y_j - m_y)\}$

Standard deviation: σ_{xx} $= \sqrt{R_{xx}}$

Conditional mean: $E\{X|Y\}$ $= \sum_i X_i P(X_i|Y)$

Joint conditional mean: $E\{X|Y, Z\}$ $= \sum_i X_i P(X_i|Y, Z)$

Conditional variance: $R_{x|y}$ $= E\{(X - E\{X|Y\})^2|Y\}$

These basic statistics possess various properties that enable them to be useful for analyzing operations on random variables; some of the more important[1] are

---

[1] Recall that independence states that the joint mass function can be factored, $\Pr(x, y) = \Pr(x) \times \Pr(y)$, which leads to these properties.

Linearity: $\quad\quad E\{ax + b\} = aE\{x\} + b = am_x + b$

Independence: $\quad E\{xy\} = E\{x\}E\{y\}$

Variance: $\quad\quad R_{xx}(ax + b) = a^2 R_{xx}$

Covariance:

Uncorrelated: $\quad E\{xy\} = E\{x\}E\{y\} \quad \{R_{xy} = 0\}$

Orthogonal: $\quad\quad E\{xy\} = 0$

Note that the expected value operation implies that for stochastic processes these basic statistics are calculated *across* the ensemble. For example, if we want to calculate the mean of a process, that is,

$$m_x(t) = E\{X(t, \omega_i) = x_i(t)\}$$

we simply take the values of $t = 0, 1, \ldots$ and calculate the mean for each value of time across ($i = 1, 2, \ldots$) the ensemble. Dealing with stochastic processes is similar to dealing with random variables except that we must account for the time indices (see Chapter 2 for more details).

Next let us define some concepts about the probabilistic information contained in a random variable. We define the (self) *information* contained in the occurrence of the random variable $X(\omega_i) = x_i$, as

$$I(x_i) = -\log_b P_X(x_i) \tag{A.7}$$

where b is the base of the logarithm, which results in different units for information measures (base $= 2 \rightarrow$ bits) and the *entropy* or *average information* of $X(\omega_i)$ as

$$H(x_i) = -E\{I(x_i)\} = \sum_i P_X(x_i)\log_b P_X(x_i) \tag{A.8}$$

Consider the case where there is more than one random variable. Then we define the *joint* mass and distribution functions of an N-dimensional random variable as

$$P_X(x_1, \ldots, x_N), \quad\quad F_X(x_1, \ldots, x_N)$$

All of the basic statistical definitions remain as before, except that we replace the scalar with the joint functions. Clearly, if we think of a stochastic process as a sequence of ordered random variables, then we are dealing with joint probability functions, that is, a collection of time-indexed random variables. Suppose we have two random variables, x_1 and x_2 and we know that the latter has already assumed a particular value, then we can define the *conditional* probability mass function of x_1 given that $X(\omega_2) = x_2$ has occurred by

$$\Pr(x_1 \mid x_2) := P_X(X(\omega_1) \mid X(\omega_2) = x_2) \tag{A.9}$$

and it can be shown from basic probabilistic axioms (see Ref. [1]) that

$$\Pr(x_1 \mid x_2) = \frac{\Pr(x_1, x_2)}{\Pr(x_2)} \tag{A.10}$$

Note also that this expression can also be written as

$$\Pr(x_1, x_2) = \Pr(x_2 \mid x_1)\Pr(x_1) \tag{A.11}$$

Substituting this equation into Eq. (A.10) gives *Bayes' rule*, that is,

$$\Pr(x_1 \mid x_2) = \Pr(x_2 \mid x_1)\frac{\Pr(x_1)}{\Pr(x_2)} \tag{A.12}$$

If we use the definition of joint mass function and substitute into the previous definitions, then we can obtain the *probabilistic chain rule* [1–3],

$$\Pr(x_1, \ldots, x_N)$$
$$= \Pr(x_1 \mid x_2, \ldots, x_N)\Pr(x_2 \mid x_3, \ldots x_N)\cdots\Pr(x_{N-1} \mid x_N)\Pr(x_N) \tag{A.13}$$

Along with these definitions follows the idea of conditional expectation, that is,

$$E\{x_i \mid x_j\} = \sum_i X_i\ \Pr(x_i \mid x_j) \tag{A.14}$$

With the conditional expectation defined, we list some of their basic properties:

1) $E_x\{X|Y\} = E\{X\}$,   if X and Y are independent
2) $E\{X\} = E_y\{E\{X|Y\}\}$
3) $E_x\{g(y)X|Y\} = g(y)E\{X|Y\}$
4) $E_{x,y}\{g(Y)X\} = E_y\{g(Y)E\{X|Y\}\}$
5) $E_x\{c|Y\} = c$
6) $E_x\{g(Y)|Y\} = g(Y)$
7) $E_{x,y}\{cX + dY|Z\} = cE\{X|Z\} + dE\{Y|Z\}$

The concepts of information and entropy can also be extended to the case of more than one random variable. We define the *mutual information* between two random variables, x_i and x_j as

$$I(x_i; x_j) = \log_b\frac{P_X(x_i \mid x_j)}{P_X(x_i)} \tag{A.15}$$

and the *average mutual information* between $X(\omega_i)$ and $X(\omega_j)$ as

$$I(X_i; X_j) = E_{x_i x_j}\{I(x_i, x_j)\} = \sum_i \sum_j P_X(x_i, x_j)I(x_i, x_j) \tag{A.16}$$

which leads to the definition of *joint entropy* as

$$H(X_i; X_j) = -\sum_i \sum_j P_X(x_i, x_j)\log_b P_X(x_i, x_j) \tag{A.17}$$

This completes the section on probability theory. Next let us consider an important multivariable distribution and its properties.

A.2   Gaussian Random Vectors

In this section, we consider the multivariable Gaussian distribution used heavily in this text to characterize Gaussian random vectors, that is, $z \sim \mathcal{N}(\mathbf{m}_z, \mathbf{R}_{zz})$ where $z \in \mathcal{R}^{N_z \times 1}$ and defined by

$$\Pr(z) = (2\pi)^{-N_z/2} |\mathbf{R}_{zz}|^{-1/2} \exp\left(-\frac{1}{2}(z - \mathbf{m}_z)' \mathbf{R}_{zz}^{-1}(z - \mathbf{m}_z)\right) \qquad (A.18)$$

where the vector mean and covariance are defined by

$$\mathbf{m}_z := E\{z\} \quad \text{and} \quad \mathbf{R}_{zz} = \text{Cov}(z) := E\{(z - \mathbf{m}_z)(z - \mathbf{m}_z)'\}$$

Certain properties of the Gaussian vectors are useful such as the following:

- *Linear transformation*: Linear transformations of Gaussian variables are Gaussian; that is, if $z \sim \mathcal{N}(\mathbf{m}_z, \mathbf{R}_{zz})$ and $y = \mathbf{A}z + \mathbf{b}$, then

$$Y \sim \mathcal{N}(\mathbf{A}\,\mathbf{m}_z + \mathbf{b}, \mathbf{A}\,\mathbf{R}_{zz}\mathbf{A}') \qquad (A.19)$$

- *Uncorrelated Gaussian vectors*: Uncorrelated Gaussian vectors are independent.
- *Sums of Gaussian variables*: Sums of independent Gaussian vectors yield Gaussian distributed vectors with mean and variance equal to the sums of the respective means and variances.
- *Conditional Gaussian vectors*: Conditional Gaussian vectors are Gaussian distributed; that is, if x and y are jointly Gaussian, with

$$\mathbf{m}_z = E\left\{\begin{matrix} x \\ y \end{matrix}\right\} = \begin{bmatrix} \mathbf{m}_x \\ \mathbf{m}_y \end{bmatrix}; \text{and} \quad \mathbf{R}_{zz} = \text{Cov}(z) = \begin{bmatrix} \mathbf{R}_{xx} & \mathbf{R}_{xy} \\ \mathbf{R}_{yx} & \mathbf{R}_{yy} \end{bmatrix}$$

then the conditional distribution for x and y is also Gaussian with conditional mean and covariance given by

$$\mathbf{m}_{x|y} = \mathbf{m}_x + \mathbf{R}_{xy}\mathbf{R}_{yy}^{-1}(y - \mathbf{m}_y)$$
$$\mathbf{R}_{x|y} = \mathbf{R}_{xx} - \mathbf{R}_{xy}\mathbf{R}_{yy}^{-1}\mathbf{R}_{yx}$$

and the vectors $x - E\{x|y\}$ and y are independent.

- *Gaussian conditional means*: Let x, y, and z be jointly distributed Gaussian random vectors and let y and z be *independent*, then

$$E\{x|y, z\} = E\{x|y\} + E\{y|z\} - \mathbf{m}_x$$

A.3   Uncorrelated Transformation: Gaussian Random Vectors

Suppose we have a Gaussian random vector, $x \sim \mathcal{N}(\mathbf{m}_x, \mathbf{R}_{xx})$ and we would like to transform it to a normalized Gaussian random vector with the mean,

$\mathbf{m}_x$, removed so that, $\mathbf{z} \sim \mathcal{N}(0, \mathbf{I})$. Assume that the mean has been removed $(\mathbf{z} \rightarrow \mathbf{z} - \mathbf{m}_x)$, then there exists a nonsingular transformation, $\mathbf{T}$, such that $\mathbf{z} = T\mathbf{x}$ and therefore

$$\mathbf{R}_{zz} = \text{Cov}(\mathbf{z}) = \text{Cov}((T\mathbf{x})(T\mathbf{x})') = T\,\mathbf{R}_{xx}T' = \mathbf{I} \tag{A.20}$$

Thus, we must find a transformation that satisfies the relation

$$\mathbf{R}_{zz} = \mathbf{I} = T\,\mathbf{R}_{xx}T' \tag{A.21}$$

Since $\mathbf{R}_{xx}$ is a positive semidefinite, symmetric matrix, it can always be factored into matrix square roots $(\mathbf{R} = \mathbf{U}\,\mathbf{U}' = \mathbf{R}_{xx}^{1/2}\mathbf{R}_{xx}^{T/2})$ using a Cholesky decomposition [4]; therefore, Eq. (A.21) implies

$$\mathbf{R}_{zz} = \mathbf{I} = (T\mathbf{U})\mathbf{U}'T' \tag{A.22}$$

or simply that

$$T = \mathbf{U}^{-1} = \mathbf{R}_{xx}^{-1/2} \qquad \text{(Inverse Matrix Square Root)} \tag{A.23}$$

and therefore

$$\mathbf{R}_{zz} = (\mathbf{R}_{xx}^{-1/2}\mathbf{U})\mathbf{U}'\mathbf{R}_{xx}^{-T/2} = \mathbf{R}_{xx}^{-1/2}\mathbf{R}_{xx}\mathbf{R}_{xx}^{-T/2} = \mathbf{I} \tag{A.24}$$

the desired result.

A.4 Toeplitz Correlation Matrices

An $N \times N$ correlation matrix, $\mathbf{R}$, is a Hermitian Toeplitz matrix $(R^H = (R^*)')$, which has the following general properties [4]:

1) The set of distinct eigenvalues, $\{\lambda_i\}$ of a Hermitian matrix are *real*.
2) The set of eigenvectors, $\{\mathbf{e}_i\}$ of a Hermitian matrix corresponding to the distinct eigenvalues, $\{\lambda_i\}$, are *orthogonal* $\rightarrow \quad \mathbf{e}_i^H\mathbf{e}_i = 0$ if $\lambda_i \neq \lambda_j$.
3) Any Hermitian matrix admits an eigen-decomposition such that

$$R = E\Lambda E^H \text{ with } E \text{ unitary } (E^{-1} = E^H).$$

4) The inverse of a Hermitian matrix can be obtained for the eigen-decomposition as

$$R^{-1} = (E\Lambda E^H)^{-1} = E^{-H}\Lambda^{-1}E^{-1} = E\Lambda^{-1}E^H = \sum_{i=1}^{N}\frac{1}{\lambda_i}\mathbf{e}_i\mathbf{e}_i^H.$$

A.5 Important Processes

We define two important processes that will be used extensively to model random signals. The first is the *uniformly* distributed process, which is specified by the mass function

$$p_X(x_i) = \frac{1}{b-a}, \qquad a < x_i < b \tag{A.25}$$

or simply

$$x \sim U(a, b)$$

where the random variable can assume any value within the specified interval. The corresponding mean and variance of the uniform random variable are given by

$$m_x = \frac{b + a}{2}, \qquad R_{xx} = \frac{(b - a)^2}{12}$$

Secondly, the *Gaussian* or *normal process* is defined by its probability mass function

$$p(X(t)) = \frac{1}{\sqrt{2\pi R_{xx}}} \exp\left\{ -\frac{1}{2} \frac{(X(t) - m_x)^2}{R_{xx}} \right\} \qquad (A.26)$$

or simply

$$x \sim N(m_x, R_{xx})$$

where m_x and R_{xx} are the respective mean and variance completely characterizing the Gaussian process, $X(t)$.

The *central limit theorem* makes Gaussian processes very important, since it states that the sum of a large number of independent random variables tends to be Gaussian, that is, for $\{x_i(t)\}$ independent then $y(t) \sim N(0, 1)$ where

$$y(t) = \sum_{t=1}^{N} \frac{(x_i(t) - m_x(t))}{\sqrt{NR_{xx}(t)}}$$

Both of these distributions will become very important when we attempt to simulate stochastic processes on the computer. Other properties of the Gaussian process that are useful are given as follows:

- 1. *Linear transformation*: Linear transformations of Gaussian variables are Gaussian; that is, if $x \sim N(m_x, R_{xx})$ and $y = ax + b$, then

$$y \sim N(am_x + b, a^2 R_{xx})$$

- 2. *Uncorrelated Gaussian variables*: Uncorrelated Gaussian variables are independent.
- 3. *Sums of Gaussian variables*: Sums of independent Gaussian variables yield a Gaussian distributed variable with mean and variance equal to the sums of the respective means and variances; that is,

$$x_i \sim N(m_x(i), R_{xx}(i)) \quad \text{and} \quad y = \sum_i k_i x(i)$$

Then

$$y \sim N\left(\sum_i k_i m_x(i), \sum_i k_i^2 R_{xx}(i) \right)$$

- 4. *Conditional Gaussian variables*: Conditional Gaussian variables are Gaussian distributed; that is, if x_i and x_j are jointly Gaussian, then

$$E\{x_i \mid x_j\} = m_{x_i} + R_{x_i x_j} R_{x_j x_j}^{-1} (x_j - m_{x_j})$$

and

$$R_{x_i \mid x_j} = R_{x_i x_i} - R_{x_i x_j} R_{x_j x_j}^{-1} R_{x_j x_i}$$

- 5. *Orthogonal errors*: The estimation error $\tilde{x}$ is orthogonal to the conditioning variable; that is, for $\tilde{x} = x - E\{x \mid y\}$, then

$$E\{y\tilde{x}\} = 0$$

- 6. *Fourth Gaussian moments*: The fourth moment of zero-mean Gaussian variables is given by

$$E\{x_1 x_2 x_3 x_4\} = E\{x_1 x_2\}E\{x_3 x_4\} + E\{x_1 x_3\}E\{x_2 x_4\} + E\{x_1 x_4\}E\{x_2 x_3\}$$

This discussion completes the introductory concepts of probability and random variables, which is extended to include stochastic processes in Chapter 2 and throughout the text.

References

1 Papoulis, A. (1965). *Probability, Random Variables and Stochastic Processes*. New York: McGraw-Hill.
2 Hogg, R. and Craig, A. (1970). *Introduction to Mathematical Statistics*. New York: MacMillan.
3 Jazwinski, A. (1970). *Stochastic Processes and Filtering Theory*. New York: Academic Press.
4 Bierman, G. (1977). *Factorization Methods for Discrete Sequential Estimation*. New York: Academic Press.

Appendix B

Projection Theory

B.1 Projections: Deterministic Spaces

Projection theory plays an important role in subspace identification primarily because subspaces are created by transforming or "projecting" a vector into a lower dimensional space – the subspace [1–3]. We are primarily interested in two projection operators: (i) orthogonal and (ii) oblique or parallel. The *orthogonal projection operator* "projects" • onto $\mathcal{Y}$ as $P_{\bullet|\mathcal{Y}}$ or its complement $P_{\bullet|\mathcal{Y}}^{\perp}$, while the *oblique projection operator* "projects" • onto $\mathcal{Y}$ "along" $\mathcal{Z}$ as $P_{\bullet|\mathcal{Y}\circ\mathcal{Z}}^{\parallel}$. For instance, orthogonally projecting the vector $\mathbf{x}$, we have $P_{\mathbf{x}|\mathcal{Y}}$ or $P_{\mathbf{x}|\mathcal{Y}}^{\perp}$, while obliquely projecting $\mathbf{x}$ along $\mathcal{Z}$ is $P_{\mathbf{x}|\mathcal{Y}\circ\mathcal{Z}}$.

Mathematically, the n-dimensional vector space $\mathcal{R}$ can be decomposed into a direct sum of subspaces such that

$$\mathcal{R} = \mathcal{R}_1 + \mathcal{R}_2 \tag{B.1}$$

then for the vector $\mathbf{r} \in \mathcal{R}^n$, we have

$$\mathbf{r} = \mathbf{r}_1 + \mathbf{r}_2 \quad \text{for} \quad \mathbf{r}_1 \in \mathcal{R}_1 \quad \text{and} \quad \mathbf{r}_2 \in \mathcal{R}_2 \tag{B.2}$$

with $\mathbf{r}_1$ defined as the projection of $\mathbf{r}$ onto $\mathcal{R}_1$, while $\mathbf{r}_2$ defined as the projection of $\mathbf{r}$ onto $\mathcal{R}_2$.

When a particular projection is termed parallel ($\parallel$) or "oblique," the corresponding *projection operator* that transforms $\mathbf{r}$ onto $\mathcal{R}_1$ along $\mathcal{R}_2$ is defined as $P_{\mathbf{r}|\mathcal{R}_1\circ\mathcal{R}_2}^{\parallel}$, then

$$\mathbf{r}_1 = P_{\mathbf{r}|\mathcal{R}_1\circ\mathcal{R}_2}^{\parallel} \quad \text{or} \quad \mathbf{r}_2 = P_{\mathbf{r}|\mathcal{R}_2\circ\mathcal{R}_1}^{\parallel} \tag{B.3}$$

with $\mathbf{r}_1$ defined as the oblique projection of $\mathbf{r}$ onto $\mathcal{R}_1$ along $\mathcal{R}_2$, while $\mathbf{r}_2$ defined as the projection of $\mathbf{r}$ onto $\mathcal{R}_2$ along $\mathcal{R}_1$.

When the intersection $\mathcal{R}_1 \cap \mathcal{R}_2 = 0$, then the direct sum follows as

$$P_{\mathbf{r}|\mathcal{R}_1\cup\mathcal{R}_1} = P_{\mathbf{r}|\mathcal{R}_1\circ\mathcal{R}_2}^{\parallel} + P_{\mathbf{r}|\mathcal{R}_2\circ\mathcal{R}_1}^{\parallel} \tag{B.4}$$

Model-Based Processing: An Applied Subspace Identification Approach, First Edition. James V. Candy.
© 2019 John Wiley & Sons, Inc. Published 2019 by John Wiley & Sons, Inc.

The *matrix projection operation* (Q) is $P \times P'$ $(Q = PP')$ a linear operator in $\mathcal{R}^n$ satisfying the properties of linearity and homogeneity. It can be expressed as an *idempotent* matrix such that $P^2 = P$ with the null space of $\mathcal{R}^n$ given by $\mathcal{N}(P) = \mathcal{R}(I_n - P)$. Here $P \in \mathcal{R}^{n \times n}$ is the projection matrix onto $\mathcal{R}(P)$ along $\mathcal{N}(P)$ iff P is idempotent. That is, the matrix P is an orthogonal projection *if and only if* it is idempotent $(P^2 = P)$ and symmetric $(P' = P)$.

Suppose that $r_1'r_2 = 0$, then $r_1 \perp r_2$ with the vectors defined as "mutually orthogonal." If $r_1'r_2 = 0 \forall r_1 \in \mathcal{R}_1$, then r_2 is orthogonal to $\mathcal{R}_1$ expressed by $r_2 \perp \mathcal{R}_1$ spanning $\mathcal{R}_2$ and called the *orthogonal complement* of $\mathcal{R}^n$. If the subspaces R_1 and R_2 are orthogonal $R_1 \perp R_2$, then their sum is called the *direct sum* $R_1 \oplus R_2$ such that

$$R^n = R_1 \oplus R_2 = R_1 \oplus R_1^\perp \quad \text{with} \quad r_1 \in R_1 \text{and } r_2 \in R_1^\perp \tag{B.5}$$

When r_1 and r_2 are orthogonal satisfying the direct sum decomposition (above), then r_1 is the *orthogonal projection* of r onto R_1.

B.2   Projections: Random Spaces

If we define the projection operator over a random space, then the random vector (w) resides in a *Hilbert space* (Ψ) with finite second-order moments, that is, [2]

$$\Psi = E\{w \mid E\{\| w \|^2\} < \infty\} \quad \text{where} \quad E\{\| w \|^2\} = \sum_i E\{w_i^2\}$$

The projection operator $\mathcal{P}$ is now defined in terms of the expectation such that the projection of $w \in \mathcal{W}$ onto $v \in \mathcal{V}$ is defined by

$$\mathcal{P}_{w|\mathcal{V}} := E\{w \mid \mathcal{V}\} = E\{w\, v'\}E\{v\, v'\}^{\#}v = \text{cov}(w, v)\text{cov}\,(v, v)^{\#}v$$

for # the pseudo-inverse operator given by $V'(VV')^{-1}$.

The corresponding projection onto the orthogonal complement $(\mathcal{V}^\perp)$ is defined by

$$\mathcal{P}_{w|\mathcal{V}}^\perp := w - E\{w \mid \mathcal{V}\} = I - E\{w\, v'\}E\{v\, v'\}^{\#}$$
$$v = I - \text{cov}(w, v)\text{cov}\,(v, v)^{\#}v$$

Therefore, the space can be decomposed into the projections as

$$\Psi = \mathcal{P}_{w|\mathcal{V}} + \mathcal{P}_{w|\mathcal{V}}^\perp$$

The oblique (parallel) projection operators follow as well for a random space, that is, the *oblique projection* operator of w onto $\mathcal{V}$ along $\mathcal{Z}$

$$\mathcal{P}_{w|\mathcal{V} \circ \mathcal{Z}}^{\parallel} := E\{\mathcal{W} \mid \mathcal{V}\}$$

and therefore, we have the operator decomposition for [2]

$$E\{\mathbf{w} \mid \mathcal{W} \cup \mathcal{V}\} = \mathbf{E}\{\mathbf{w} \mid \mathcal{W} \circ \mathcal{V}\} + \mathbf{E}\{\mathbf{w} \mid \mathcal{V} \circ \mathcal{W}\}$$

for $E\{\mathbf{w} \mid \mathcal{W} \circ \mathcal{V}\}$ is the oblique (parallel) projection of $\mathbf{w}$ onto $\mathcal{W}$ along $\mathcal{V}$ and $E\{\mathbf{w} \mid \mathcal{V} \circ \mathcal{W}\}$ is the oblique projection of $\mathbf{w}$ onto $\mathcal{V}$ along $\mathcal{W}$.

B.3 Projection: Operators

Let $A \in \mathcal{R}^{m \times n}$ then the *row space* of A is spanned by the *row vectors* $\mathbf{a}'_i \in \mathcal{R}^{1 \times n}$, while the *column space* of A is spanned by the *column vectors* $\mathbf{a}_i \in \mathcal{R}^{m \times 1}$. That is,

$$A = \begin{bmatrix} \mathbf{a}'_1 \\ \mathbf{a}'_2 \\ \vdots \\ \mathbf{a}'_m \end{bmatrix}; \mathbf{a}'_i \in \mathcal{R}^{1 \times n}; \; i = 1, \dots, m \qquad \text{(Row Vectors)}$$

and

$$A = [\mathbf{a}_1 \mid \mathbf{a}_2 \mid \cdots \mid \mathbf{a}_n]; \mathbf{a}_j \in \mathcal{R}^{m \times 1}; \; j = 1, \dots, n \qquad \text{(Column Vectors)}$$

Any vector $\mathbf{y}$ in the row space of $A \in \mathcal{R}^{m \times n}$ is defined by

$$\text{Row}(A) := \left\{ \mathbf{y} \in \mathcal{R}^{n \times 1} \mid \mathbf{y} = A'\mathbf{x} = \sum_{i=1}^{m} \mathbf{a}'_i x_i; \; \mathbf{x} \in \mathcal{R}^{m \times 1} \right\}$$

and similarly for the column space

$$\text{Column}(A) := \left\{ \mathbf{y} \in \mathcal{R}^{m \times 1} \mid \mathbf{y} = A\mathbf{x} = \sum_{j=1}^{n} \mathbf{a}_j x_j; \; \mathbf{x} \in \mathcal{R}^{n \times 1} \right\}$$

Therefore, the set of vectors $\{\mathbf{a}'_i\}$ and $\{\mathbf{a}_i\}$ provide a set of basis vectors spanning the respective row or column spaces of A and, of course, $\text{Row}(A) = \text{Column}(A')$.

B.3.1 Orthogonal (Perpendicular) Projections

Projections or, more precisely, projection operators are well known from operations on vectors (e.g. the Gram–Schmidt orthogonalization procedure). These operations, evolving from solutions to least-squares error minimization problem, can be interpreted in terms of matrix operations of their row and column spaces defined above. The orthogonal *projection operator* ($\mathcal{P}_\bullet$) is defined in terms of the row space of a matrix B by

$$\mathcal{P}_B = B'(BB')^{-1}B \qquad \text{(Orthogonal Projection Operator)}$$

$$\text{(B.6)}$$

with its corresponding orthogonal complement operator as

$$\mathcal{P}_B^\perp = (I - \mathcal{P}_B) = I - B'(BB')^{-1}B \quad \text{(Orthogonal Complement Operator)}$$
$$(\text{B.7})$$

These operators when applied to a matrix "project" the *row space* of a matrix A onto (|) the *row space* of B such that $\mathcal{P}_{A|B}$ given by

$$\mathcal{P}_{A|B} = A \times \underbrace{\mathcal{P}_B = A \times B'(BB')^{-1}B}_{\mathcal{P}_B} \quad \text{(Orthogonal Projection)} \qquad (\text{B.8})$$

and similarly onto the orthogonal complement of B such that

$$\mathcal{P}_{A|B}^\perp = A \times \underbrace{\mathcal{P}_B^\perp = A \times (I - B'(BB')^{-1}B)}_{\mathcal{P}_B^\perp} \quad \text{(Orthogonal Complement Projection)}$$

$$(\text{B.9})$$

Thus, a matrix can be decomposed in terms of its inherent orthogonal projections as a direct sum as shown in Figure B.1a, that is,

$$A = \mathcal{P}_{A|B} + \mathcal{P}_{A|B}^\perp = A \times \mathcal{P}_B + A \times \mathcal{P}_B^\perp \quad \text{(Orthogonal Decomposition)} \quad (\text{B.10})$$

Similar relations exist if the *row space* of A is projected onto the *column space* of B, then it follows that the projection operator in this case is defined by

$$\mathcal{P}_B = B(B'B)^{-1}B' \qquad \text{(Orthogonal Projection Operator)}$$
$$(\text{B.11})$$

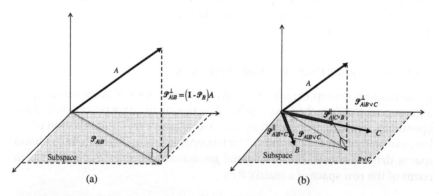

Figure B.1 Orthogonal and oblique projections. (a) Orthogonal projection of A onto B ($\mathcal{P}_{A|B}$) and its complement ($\mathcal{P}_{A|B}^\perp$). (b) Oblique projection of A onto C along B ($\mathcal{P}_{A|C \circ B}$).

along with the corresponding projection given by

$$\mathcal{P}_{A|B} = A \times \mathcal{P}_B = A \times \underbrace{B(B'B)^{-1}B'}_{\mathcal{P}_B} \qquad \text{(Orthogonal Projection)}$$

$$(B.12)$$

Numerically, these matrix operations are facilitated by the LQ-decomposition (see Appendix C) as

$$\begin{bmatrix} B \\ -- \\ A \end{bmatrix} = \begin{bmatrix} L_{11} & | & O \\ -- & | & -- \\ L_{21} & | & L_{22} \end{bmatrix} \begin{bmatrix} Q_1' \\ -- \\ Q_2' \end{bmatrix} \qquad (B.13)$$

where the *projections* are given by

$$\mathcal{P}_{A|B} = A \times \mathcal{P}_B = L_{21} \times Q_1'$$
$$\mathcal{P}_{A|B}^{\perp} = A \times \mathcal{P}_B^{\perp} = L_{22} \times Q_2' \qquad (B.14)$$

B.3.2 Oblique (Parallel) Projections

The *oblique projection* (∥) of the Row(A) onto Row(C) along (°) the Row(B) is defined by $\mathcal{P}_{A \mid C°B}$. This projection operation can be expressed as the *rows* of two nonorthogonal matrices (B and C) and their corresponding orthogonal complement. Symbolically, the projection of the rows of A onto the *joint row space* of B and C ($B \vee C$) can be expressed in terms of two oblique projections and one orthogonal complement. This projection is illustrated in Figure B.1b, where we see that the Row(A) is first orthogonally projected onto the joint row space of B and C (orthogonal complement) and then decomposed along B and C individually enabling the extraction of the oblique projection of the Row(A) onto Row(C) along (°) the Row(B).

$$\mathcal{P}_{A \mid B \vee C} = \mathcal{P}_{A \mid B°C}^{\parallel} + \mathcal{P}_{A \mid C°B}^{\parallel} + \mathcal{P}_{A \mid B \vee C}^{\perp} \qquad (B.15)$$

Pragmatically, this relation can be expressed into a more "operational" form that is applied in Section 6.5 to derive the N4SID algorithm [1, 4], that is,

$$\mathcal{P}_{A \mid C°B}^{\parallel} = [A \times \mathcal{P}_B^{\perp}][C \times \mathcal{P}_B^{\perp}]^{\#} \times C \qquad \text{(Oblique Projection)}$$

$$(B.16)$$

for # the *pseudo-inverse* operation (see Appendix C). Oblique projections also have the following properties that are useful in derivations:

$$\mathcal{P}_{B|C°B}^{\parallel} = 0$$
$$\mathcal{P}_{C|C°B}^{\parallel} = C \qquad (B.17)$$

Analogously, the orthogonal complement projection can be expressed in terms of the oblique projection operator as

$$P^{\perp}_{A|B} = P^{\parallel}_{A \mid C \circ B} \times P^{\perp}_{B} \tag{B.18}$$

Numerically, the oblique projection can be computed again using the LQ-decomposition (see Appendix C).

$$
\begin{bmatrix} B \\ -- \\ C \\ -- \\ A \end{bmatrix}
=
\begin{bmatrix}
L_{11} & | & 0 & | & 0 \\
-- & | & -- & | & -- \\
L_{21} & | & L_{22} & | & 0 \\
-- & | & -- & | & -- \\
L_{31} & | & L_{32} & | & L_{33}
\end{bmatrix}
\begin{bmatrix} Q'_1 \\ -- \\ Q'_2 \\ -- \\ Q'_3 \end{bmatrix}
\tag{B.19}
$$

where the *oblique projection* of the Row(A) onto Row(C) along the Row(B) is given by

$$P^{\parallel}_{A|C \circ B} = L_{32} \times L_{22}^{-1} \times C = L_{32} \times L_{22}^{-1} \times (L_{21}Q'_1 + L_{22}Q'_2) \tag{B.20}$$

Some handy relationships between oblique and orthogonal projection matrices are given in Table B.1 (see [1–3, 5] for more details).[1]

Table B.1 Matrix Projections.

		Projection: operators, projections, numerics	
Operation	**Operator**	**Projection**	**Numerical (LQ-decomposition)**
Orthogonal			
$P_{A|B}$	$P_{\bullet|B} = B'(BB')^{-1}B$	$A \times B'(BB')^{-1}B$	$A \times (L_{21} \times Q'_1)$
$P^{\perp}_{A|B}$	$P^{\perp}_{\bullet|B} = (I - B'(BB')^{-1}B)$	$A \times (I - B'(BB')^{-1}B)$	$A \times (L_{22} \times Q'_2)$
Oblique			
$P_{A|C \circ B}$	$P_{\bullet|C \circ B}$	$L_{32}L_{22}^{-1} \times C$	$L_{32} \times L_{22}^{-1}(L_{21}Q'_1 + L_{22}Q'_2)$
Relations			
$P_{\bullet|B}$	—	—	$Q_1Q'_1$
$P^{\perp}_{\bullet|B}$	—	—	$Q_2Q'_2$
$P_{A|C \circ B}$	$P_{\bullet|C \circ B}$	$[A \times P^{\perp}_B][C \times P^{\perp}_B]^{\#} \times C$	—
$P^{\perp}_{A|B}$	$P^{\perp}_{\bullet|B}$	$P^{\parallel}_{A \mid C \circ B} \times P^{\perp}_B$	—

---

1 Note the difference between projection operators and actual projections.

References

1 van Overschee, P. and De Moor, B. (1996). *Subspace Identification for Linear Systems: Theory, Implementation, Applications.* Boston, MA: Kluwer Academic Publishers.

2 Katayama, T. (2005). *Subspace Methods for System Identification.* London: Springer.

3 Verhaegen, M. and Verdult, V. (2007). *Filtering and System Identification: A Least-Squares Approach.* Cambridge: Cambridge University Press.

4 Behrens, R. and Scharf, L. (1994). Signal processing applications of oblique projection operators. *IEEE Trans. Signal Process.* 42 (6): 1413–1424.

5 Tangirala, A. (2015). *Principles of System Identification: Theory and Practice.* Boca Raton, FL: CRC Press.

Appendix C

Matrix Decompositions

C.1 Singular-Value Decomposition

The singular-value decomposition (SVD) is an integral part of subspace identification methods [1–4]. It is characterized by an orthogonal decomposition of a rectangular $n \times m$ matrix $Z \in \mathcal{R}^{n \times m}$

$$Z = U \times \Sigma \times V' \tag{C.1}$$

where the matrices $U \in \mathcal{R}^{n \times n}$ and $V \in \mathcal{R}^{m \times m}$ are *orthogonal* such that $U'U = I_n$, $V'V = I_m$ along with the diagonal matrix, $\Sigma = \text{diag}[\sigma_1 \ \cdots \ \sigma_N]$ with $N < \min(n, m)$. The set of ordered *singular values* are given by $\{\sigma_i\}$; $i = 1, \dots, N$ with the property that $\sigma_1 > \sigma_2 > \cdots > \sigma_N$. The singular values of Z and the respective column and row vectors of U and V, u_i and v_i are the ith *left/right singular vectors* such that [1]

$$Zv_i = \sigma_i u_i$$
$$A'u_i = \sigma_i v_i$$

which follows directly from the SVD, since $ZV = \Sigma U$ and $Z'U = \Sigma'V$.

Also premultiplying Eq. (C.1) by U' and postmultiplying by V and applying the orthogonality properties gives the relation

$$U' \times Z \times V = \Sigma = \text{diag}[\sigma_1 \ \cdots \ \sigma_N] \in \mathcal{R}^{n \times m} \quad \text{for} \quad N = \min(n, m) \tag{C.2}$$

It also follows that for a rank r matrix, the SVD can be expressed as

$$Z = U\Sigma V' = [U_r \mid U_{n-r}] \begin{bmatrix} \Sigma_r & | & 0 \\ - & - & - \\ 0 & | & 0 \end{bmatrix} \begin{bmatrix} V'_r \\ -- \\ V'_{m-r} \end{bmatrix} = U_r \Sigma_r V'_r \tag{C.3}$$

Model-Based Processing: An Applied Subspace Identification Approach, First Edition. James V. Candy.
© 2019 John Wiley & Sons, Inc. Published 2019 by John Wiley & Sons, Inc.

Numerical errors occurring in this approximation are quantified in terms of the singular values as

$$E_{abs} = \sqrt{\sum_{i=r+1}^{N} \sigma_i^2} \qquad \text{(Absolute Error)}$$

$$E_{rel} = \sqrt{\frac{\sum_{i=r+1}^{N} \sigma_i^2}{\sum_{i=1}^{N} \sigma_i^2}} \qquad \text{(Relative Error)} \qquad \text{(C.4)}$$

The SVD of a matrix provides various properties:

- The set of singular values of Z are equal to the positive square roots of the eigenvalues of $Z'Z$: $\sigma_i = +\sqrt{\lambda_i}$.
- For Z of rank r, then the singular values satisfy

$$\sigma_1 > \cdots \geq \sigma_r > \sigma_{r+1} = \cdots = \sigma_N = 0$$

- The rank condition gives

$$\rho(Z) = r$$
$$\text{null}(Z) = \text{span}\{v_{r+1}, \ldots, v_m\}$$
$$\text{range}(Z) = \text{span}\{u_1, \ldots, u_r\} \qquad \text{(C.5)}$$

- The matrix decomposition can be expressed directly in terms of the singular values as

$$Z = U_r \, \Sigma_r \, V_r' = \sum_{i=1}^{r} \sigma_i \, u_i \, v_i' \qquad \text{(C.6)}$$

- The *norms* are also expressed in terms of the singular values as

$$\| Z \|_F = \sum_{i=1}^{N} \sigma_i \quad N < \min(n, m) \qquad \text{(Frobenius Norm)}$$

$$\| Z \|_2 = \sigma_1 \qquad \text{(2-Norm)} \qquad \text{(C.7)}$$

- The *pseudo-inverse* of a matrix ($Z^{\#} = (Z'Z)^{-1}Z'$ (row space) or $Z^{\#} = Z'(ZZ')^{-1}$ (column space)) is easily determined by

$$Z^{\#} = (U \, \Sigma \, V')^{\#} = V \, \Sigma^{-1} \, U' \qquad \text{(C.8)}$$

Associated with the SVD of Eq. (C.3) is the idea of *projections*, that is, projecting a vector onto a line or plane [4]. In terms of the SVD we have the following *orthogonal projection matrices* [5]:

$$P_{Z|U} = U_1 \times U_1' = Z \times Z^{\#}$$

$$P_{Z|u}^{\perp} = U_2 \times U_2' = I - Z \times Z^{\#}$$
$$P_{Z|V} = V_1 \times V_1' = Z^{\#} \times Z$$
$$P_{Z|V}^{\perp} = V_2 \times V_2' = I - Z^{\#} \times Z \tag{C.9}$$

Finally, the SVD simultaneously diagonalizes the projection, Gramian matrix, and pseudo-inverse [2]

$$P_{Z|u} = U \times \begin{bmatrix} I & 0 \\ 0 & 0 \end{bmatrix} \times U' \qquad \text{(Projection)}$$

$$\mathcal{G}_Z := Z' \times Z = V \times \Sigma^2 \times V' \qquad \text{(Gramian)}$$

$$Z^{\#} = V \times \Sigma^{-1} \times U' \qquad \text{(Pseudo-Inverse)}$$

$$\tag{C.10}$$

C.2 QR-Decomposition

The QR-decomposition of a matrix is the matrix implementation of the well-known Gram–Schmidt vector orthogonalization procedure [4]. It is defined by the product of an orthogonal matrix $Q \in \mathcal{R}^{n \times m}$ with $m \geq n$ and an upper triangular matrix $R_q \in \mathcal{R}^{m \times m}$, that is,

$$R = Q \times R_q \tag{C.11}$$

where $Q' \times Q = I_n$ such that $Q'R = R_q$ with Q' is the orthogonal matrix transforming $R \to R_q$. It is numerically implemented using the *Householder transform*, $T = I - 2\mathbf{uu}'$ with the vector $\mathbf{u}$ defined by

$$\mathbf{u} = \pm \frac{v_1 - v_2}{\parallel v_1 - v_2 \parallel} \quad \parallel \mathbf{u} \parallel = 1 \tag{C.12}$$

C.3 LQ-Decomposition

The LQ-decomposition is the (*dual* of the QR-decomposition: $L = R'$, $Q = Q'$ where

$$Z = L \times Q \text{ with } L \text{ lower block triangular; and } Q \text{ orthogonal} \tag{C.13}$$

$$L = \begin{bmatrix} L_{11} & | & 0 \\ - & - & - \\ L_{21} & | & L_{22} \end{bmatrix} \quad \text{and} \quad Q = \begin{bmatrix} Q_1' \\ -- \\ Q_2' \end{bmatrix} \text{for} \quad Q'Q = I$$

The LQ-decomposition is a numerical (matrix) version of the *Gram–Schmidt orthogonalization* procedure [1, 4]. It contains the set of orthonormal basis vectors in Q ($Q'Q = I$) with $\mathbf{q}_i$ and L contains the coefficients of linear dependence or equivalently the relationship of the basis vectors of Z with those of Q. In fact from this viewpoint, we can think of partitioning $Q = [Q_Z \mid Q_Z^{\perp}]$ such that a rank-deficient data matrix Z can (hypothetically) be decomposed as

$$Z = LQ = [L_Z \mid 0] \begin{bmatrix} Q_Z \\ -- \\ Q_Z^{\perp} \end{bmatrix} = L_Z \times Q_Z$$

The *projection matrix* associated with the decomposition is specified by [4]

$$\mathcal{P} = Q \times Q' \qquad\qquad \text{(Projection Matrix)}$$

References

1 Golub, G. and Van Loan, C. (1989). *Matrix Computations*, 2e. Baltimore, MD: Johns Hopkins University Press.
2 Scharf, L. (1990). *Statistical Signal Processing: Detection, Estimation, and Time Series Analysis*. Reading, MA: Addison-Wesley.
3 Katayama, T. (2005). *Subspace Methods for System Identification*. London: Springer.
4 Bretscher, O. (2009). *Linear Algebra with Applications*. Upper Saddle River, NJ: Prentice-Hall International.
5 Klema, V. and Laub, A. (1980). The singular value decomposition: its computation and some applications. *IEEE Trans. Autom. Control* AC-25 (2): 164–176.

Appendix D

Output-Only Subspace Identification

In this appendix, we address the subspace identification approach for the constrained "output-only" realization problem where *no* input excitation data are available in contrast to the developments of Chapters 6 and 7. Output-only problems occur frequently in practice, especially in structural dynamics, seismic analysis, medical diagnosis to name a few prominent problem areas. For instance, in structural dynamics, large buildings, bridges, or tunnels are excited by wind or daily traffic, while earthquakes provide the excitation for seismic monitoring where medical applications are driven by unknown sources.

Subspace methods have evolved primarily from the early work of Akaike [1, 2] and Aoki [3] in stochastic realization and projection theory. The primary idea, when applied to the "output-only" problem, is to perform an orthogonal projection in a Hilbert space occupied by random vectors. That is, if $y_f(t)$ is a finite random vector of *future* outputs and $y_p(t)$ a random vector of *past* outputs, then the *orthogonal projection* of the "future output data onto the past output data" is defined by $\mathcal{P}_{y_f|y_p}$. The concept of projecting a vector onto a subspace spanned by another vector can be extended to projecting a row (column) space of a matrix onto a row (column) space of another matrix (see Appendix B). With this in mind, we define the following block Hankel matrices for future and past data as

$$\mathcal{Y}_f = \begin{bmatrix} y(k) & y(k+1) & \cdots & y(K+k-1) \\ y(k+1) & y(k+2) & \cdots & y(K+k) \\ \vdots & \vdots & \vdots & \vdots \\ y(2k-1) & y(2k) & \cdots & y(K+2k-2) \end{bmatrix} \quad \text{(Future Data)}$$

$$\mathcal{Y}_p = \begin{bmatrix} y(0) & y(1) & \cdots & y(K-1) \\ y(1) & y(2) & \cdots & y(K) \\ \vdots & \vdots & \vdots & \vdots \\ y(k-1) & y(k) & \cdots & y(K+k-2) \end{bmatrix} \quad \text{(Past Data)}$$

Model-Based Processing: An Applied Subspace Identification Approach, First Edition. James V. Candy.
© 2019 John Wiley & Sons, Inc. Published 2019 by John Wiley & Sons, Inc.

An orthogonal projection of the row space of future data $\mathcal{Y}_f$ onto the row space of past data $\mathcal{Y}_p$ is given by [1, 3]

$$P_{\mathcal{Y}_f | \mathcal{Y}_p} = \mathcal{Y}_f \times P_{\bullet | \mathcal{Y}_p} = \underbrace{E\{\mathcal{Y}_f \mathcal{Y}_p'\}}_{\mathbf{T}_k} \times \underbrace{E\{\mathcal{Y}_p \mathcal{Y}_p'\}^{-1}}_{\mathcal{T}_k} \times \mathcal{Y}_p = \mathbf{T}_k \times \mathcal{T}_k^{-1} \times \mathcal{Y}_p$$

(D.1)

where the underlying matrices are defined in terms of block Toeplitz covariance matrices, that is,

$$\mathbf{T}_k := E\{\mathcal{Y}_f \mathcal{Y}_p'\} = \begin{bmatrix} \Lambda_k & \Lambda_{k-1} & \cdots & \Lambda_1 \\ \Lambda_{k+1} & \Lambda_k & \cdots & \Lambda_2 \\ \vdots & \vdots & \ddots & \vdots \\ \Lambda_{2k-1} & \Lambda_{2k-2} & \cdots & \Lambda_k \end{bmatrix} \in \mathcal{R}^{kN_y \times kN_y}$$

(D.2)

with the data assumed to be generated by the *innovations model* of Section 7.4

$$\hat{x}(t+1) = A\hat{x}(t) + K_p(t)e(t)$$
$$y(t) = C\hat{x}(t) + e(t)$$

and measurement output covariance given by

$$\Lambda_\ell = CA^{\ell-1}\underbrace{(A\hat{\Pi}C' + K_p R_{ee})}_{\overline{B}_e}$$

Therefore, we obtain a factorization similar to that of deterministic realization theory

$$
\mathbf{T}_k = \begin{bmatrix} CA^{k-1}\overline{B}_e & A^{k-2}\overline{B}_e & \cdots & C\overline{B}_e \\ CA^k\overline{B}_e & CA^{k-1}\overline{B}_e & \cdots & CA\overline{B}_e \\ \vdots & \vdots & \ddots & \vdots \\ CA^{2k-2}\overline{B}_e & CA^{2k-3}\overline{B}_e & \cdots & CA^{k-1}\overline{B}_e \end{bmatrix} \in \mathcal{R}^{kN_y \times kN_y}
$$

$$
= \underbrace{\begin{bmatrix} C \\ CA \\ \vdots \\ CA^{k-1} \end{bmatrix}}_{\mathcal{O}_k} \underbrace{\left[A^{k-1}\overline{B}_e \mid A^{k-2}\overline{B}_e \mid \cdots \mid \overline{B}_e \right]}_{\overleftarrow{C}_{AB}(k)}
$$

(D.3)

given by the product of the *observability* matrix and the *reversed controllability* matrix

$$\mathbf{T}_k = \mathcal{O}_k \times \overleftarrow{C}_{AB}(k)$$

(D.4)

The block Toeplitz matrix of the projection $\mathcal{T}_k$ is

$$\mathcal{T}_k = E\{\mathcal{Y}_p \mathcal{Y}_p'\} = \begin{bmatrix} \Lambda_0 & \Lambda_{-1} & \cdots & \Lambda_{1-k} \\ \Lambda_1 & \Lambda_0 & \cdots & \Lambda_{2-k} \\ \vdots & \vdots & \ddots & \vdots \\ \Lambda_{k-1} & \Lambda_{k-2} & \cdots & \Lambda_0 \end{bmatrix} \in \mathcal{R}^{kN_y \times kN_y} \tag{D.5}$$

Using the innovations model, it has been shown that the augmented state vector

$$\hat{\mathcal{X}}_k := [\hat{\mathbf{x}}_k \quad \hat{\mathbf{x}}_{k+1} \quad \cdots \quad \hat{\mathbf{x}}_{K+k-1}] \in \mathcal{R}^{N_x \times KN_x}$$

satisfies the following relation [4, 5]

$$\hat{\mathcal{X}}_k = \overleftarrow{C}_{AB}(k) \times \mathcal{T}_k^{-1} \times \mathcal{Y}_p \tag{D.6}$$

and therefore, incorporating this expression into the orthogonal projection, we obtain

$$\mathcal{P}_{\mathcal{Y}_f | \mathcal{Y}_p} = \mathbf{T}_k \times \mathcal{T}_k^{-1} \times \mathcal{Y}_p = \mathcal{O}_k \underbrace{\overleftarrow{C}_{AB}(k) \times \mathcal{T}_k^{-1} \times \mathcal{Y}_p}_{\hat{\mathcal{X}}_k} = \mathcal{O}_k \times \hat{\mathcal{X}}_k \tag{D.7}$$

Thus, both the extended observability matrix and the estimated state vectors can be extracted by performing the singular value decomposition of the orthogonal projection matrix to obtain

$$\mathcal{P}_{\mathcal{Y}_f | \mathcal{Y}_p} = [U_{N_x} \mid U_{\mathcal{N}}] \begin{bmatrix} \Sigma_{N_x} & | & 0 \\ - & - & - \\ 0 & | & \Sigma_{\mathcal{N}} \end{bmatrix} \begin{bmatrix} V'_{N_x} \\ -- \\ V'_{\mathcal{N}} \end{bmatrix} = U_{N_x} \Sigma_{N_x} V'_{N_x}$$

where $\Sigma_{N_x} \gg \Sigma_{\mathcal{N}}$ and

$$\mathcal{P}_{\mathcal{Y}_f | \mathcal{Y}_p} = \mathcal{O}_k \hat{\mathcal{X}}_k = \underbrace{(U_{N_x} \Sigma_{N_x}^{1/2})}_{\mathcal{O}_{N_x}} \underbrace{((\Sigma'_{N_x})^{1/2} V'_{N_x})}_{\hat{\mathcal{X}}_k} \tag{D.8}$$

With the singular value decomposition available, we exploit the projection to extract the system and output measurement matrices as in Section 6.2

$$\hat{A} = \mathcal{O}_{N_x}^{\#} \mathcal{O}_{N_x}^{\uparrow} = (\Sigma_{N_x}^{-1/2} U'_{N_x} \times \mathcal{O}_{N_x}^{\uparrow}; \qquad \hat{C} = \mathcal{O}(1 : N_y, 1 : N_x) \tag{D.9}$$

The input transmission matrix $(\overline{B}_e)$ can be extracted by first estimating the reversed controllability matrix from Eq. (D.4)

$$\overleftarrow{C}_{AB}(k) = \mathcal{O}_{N_x}^{\#} \times \mathbf{T}_{k,k} = (\Sigma_{N_x}^{-1/2} U'_{N_x}) \times \mathbf{T}_{k,k} \tag{D.10}$$

and extracting the last k-columns to obtain

$$\hat{B} = \overline{B}_e = \overleftarrow{C}_{AB}(1 : N_x, (k-1)N_y : kN_y) \tag{D.11}$$

The input–output transmission matrix is obtained from the sample covariance estimate of $\hat{\Lambda}_0$

$$\hat{D} = \hat{\Lambda}_0 \qquad (D.12)$$

With the $\hat{\Sigma}_{KSP} = \{\hat{A}, \hat{B}, \hat{C}, \hat{D}\}$ model now available, we can solve the Riccati equation of Section 7.4 to obtain $\hat{\Pi} > 0$ and extract the remaining matrices to provide a solution to the "output-only" stochastic realization problem.

Summarizing, the *"output-only" subspace stochastic realization (OOSID)*[1] algorithm is accomplished using the following steps:

- *Create* the block Hankel matrix from the measured output sequence, $\{\mathbf{y}(t)\}$.
- *Calculate* the orthogonal projection matrix $\mathcal{P}_{\mathcal{Y}_f|\mathcal{Y}_p}$ as in Eq. (D.1).
- *Perform* the SVD of the projection matrix to obtain the extended observability and estimated state vectors in Eq. (D.8).
- *Extract* the system matrix $\hat{A}$ and output $\hat{C}$ matrices as in Eq. (D.9).
- *Calculate* the reversed controllability matrix $\overleftarrow{\mathcal{C}}_{AB}$ as in Eq. (D.10).
- *Extract* the input transmission matrix $\hat{B}$ as in Eq. (D.11).
- *Extract* the input/output matrix $\hat{D}$ from estimated output covariance $\hat{\Lambda}_0$.
- *Solve* the corresponding Riccati equation of Eq. (7.55) using $\Sigma_{KSP} = \{\hat{A}, \hat{B}, \hat{C}, \hat{D}\}$ to obtain $\hat{\Pi}$.
- *Calculate* R_{ee} and K_p from the KSP-equations of Eq. (7.52).
- *Calculate* the set of noise source covariances $\{R_{ww}^e, R_{vv}^e, R_{wv}^e\}$ using R_{ee} and K_p.

This completes the development of the output-only subspace identification algorithm.

References

1 Akaike, H. (1974). Stochastic theory of minimal realization. *IEEE Trans. Autom. Control* 19: 667–674.

2 Akaike, H. (1975). Markovian representation of stochastic processes by canonical variables. *SIAM J. Control* 13 (1): 162–173.

3 Aoki, M. (1990). *State Space Modeling of Time Series*, 2e. London: Springer.

4 van Overschee, P. and De Moor, B. (1996). *Subspace Identification for Linear Systems: Theory, Implementation, Applications*. Boston, MA: Kluwer Academic Publishers.

---

1 For structural dynamic systems only the system (A) and measurement (C) matrices are required to extract the modal frequencies and shapes [6, 7].

5 Katayama, T. (2005). *Subspace Methods for System Identification*. London: Springer.

6 Reynders, E. (2012). System identification methods for (operational) modal analysis: review and comparison. *Arch. Comput. Methods Eng.* 19 (1): 51–124.

7 Juang, J. (1994). *Applied System Identification*. Upper Saddle River, NJ: Prentice-Hall PTR.

Index

Model-Based Processing: An Applied Subspace Identification Approach, First Edition. James V. Candy.
© 2019 John Wiley & Sons, Inc. Published 2019 by John Wiley & Sons, Inc.